Multiple Stressors: A Challenge for the Future

T0145077

NATO Science for Peace and Security Series

This Series presents the results of scientific meetings supported under the NATO Programme: Science for Peace and Security (SPS).

The NATO SPS Programme supports meetings in the following Key Priority areas: (1) Defence Against Terrorism; (2) Countering other Threats to Security and (3) NATO, Partner and Mediterranean Dialogue Country Priorities. The types of meeting supported are generally "Advanced Study Institutes" and "Advanced Research Workshops". The NATO SPS Series collects together the results of these meetings. The meetings are coorganized by scientists from NATO countries and scientists from NATO's "Partner" or "Mediterranean Dialogue" countries. The observations and recommendations made at the meetings, as well as the contents of the volumes in the Series, reflect those of participants and contributors only; they should not necessarily be regarded as reflecting NATO views or policy.

Advanced Study Institutes (ASI) are high-level tutorial courses intended to convey the latest developments in a subject to an advanced-level audience

Advanced Research Workshops (ARW) are expert meetings where an intense but informal exchange of views at the frontiers of a subject aims at identifying directions for future action

Following a transformation of the programme in 2006 the Series has been re-named and re-organised. Recent volumes on topics not related to security, which result from meetings supported under the programme earlier, may be found in the NATO Science Series.

The Series is published by IOS Press, Amsterdam, and Springer, Dordrecht, in conjunction with the NATO Public Diplomacy Division.

Sub-Series

A.	Chemistry and Biology	Springer
B.	Physics and Biophysics	Springer
C.	Environmental Security	Springer
D.	Information and Communication Security	IOS Press
E.	Human and Societal Dynamics	IOS Press

http://www.nato.int/science
http://www.springer.com
http://www.iospress.nl

Series C: Environmental Security

Multiple Stressors:
A Challenge for the Future

Edited by

Carmel Mothersill

McMaster University,
Department Medical Physics & Applied Radiation Sciences,
Hamilton, Ontario, Canada

Irma Mosse

Institute of Genetics and Cytology of the National Academy of Sciences,
Minsk, Belarus

and

Colin Seymour

McMaster University,
Department Medical Physics & Applied Radiation Sciences,
Hamilton, Ontario, Canada

 Springer

Proceedings of the NATO Advanced Research Workshop on
Multipollution Exposure and Risk Assessment—A Challenge for the Future
Minsk, Belarus
1–5 October 2006

A C.I.P. Catalogue record for this book is available from the Library of Congress.

ISBN 978-1-4020-6334-3 (PB)

ISBN 978-1-4020-6335-0 (eBook)

Published by Springer,
P.O. Box 17, 3300 AA Dordrecht, The Netherlands.

www.springer.com

Printed on acid-free paper

CONTENTS

CONTENTS

PREFACE

Ecotoxiclogical risk from multiple stressors covers any situation where organisms are exposed to a combination of environmental stressors. These include physical and chemical pollutants as well as other stressors such as parasites and environmental impact (e.g., climate change or habitat loss). The combination of stressors can result in increased risk to organisms (either additive or synergistic effects) or decreased effects (protective or antagonistic effects).

The multiple stressor challenge is an international, multi-disciplinary problem requiring an international, multi-disciplinary approach. The current approach to multiple stressors is to examine one stressor at a time and assume additivity. Little work has been done on combinations of stressors such that potential interactions can be determined.

The problem is very complex. Multiple stressors pose a whole spectrum of challenges that range from basic science to regulation, policy and governance. The challenges raise fundamental questions about our understanding of the basic biological response to stressors, as well as the implications of those uncertainties in environmental risk assessment and management. In addition to the great breadth, there is also great depth in the research challenges, largely due to the complexity of the issues. From a basic science point of view, many of the mechanisms and processes under investigation are at the cutting edge of science — involving new paradigms such as genomic instability and bystander effects. The application of state-of-the-art technologies such as proteomics, transgenic organisms and biomarkers offers new opportunities for breakthroughs in scientific understanding. The problem also has a global dimension, with impacts on boundary issues, vulnerable ecosystems, vulnerable societies and developing countries. Regulatory challenges include harmonisation in risk management and regulation, being relevant for new governance mechanisms such as EU Reach and the UN/WHO Strategic Approach to International Chemical Management (SAICM). Finally, there are important socio-economic aspects connected to law (multi-causality) and stakeholder interests (both public, industry).

The NATO ARW which has led to this book is one of the very first attempts to draw relevant experts together who can address all of the above aspects. We were fortunate to be able to attract specialists with legal, ethics and economics backgrounds as well as a wide spectrum of basic and applied scientists and regulators.

The book is structured in 6 sections ranging for general introductory lectures through basic phenomenology, mechanisms, applied aspects and finally legal and ethical aspects.

We wish to acknowledge the NATO Science committee for supporting this workshop. We are very grateful for their generous support.

LIST OF CONTRIBUTORS

Aizawa Kouichi
University of Georgia, Savannah River Ecology Laboratory, Aiken, South
Carolina, USA

Ann A. Kroik
State Agrarian University, Voroshilov Street 25, Dnipropetrovsk 49600,
Ukraine

Aslan Sukru
Cumhuriyet University, Department of Environmental Engineering, 58140,
Sivas/Turkey

Audette-Stuart, Marilyne
Atomic Energy of Canada Ltd, Chalk River, ON K0J 1J0 Canada

Austin, Brian
School of Life Sciences, John Muir Building, Heriot-Watt University,
Edinburgh EH14 AS, UK

Boudagov, R.S.
Medical Radiological Research Center, RAMS, Obninsk, Russia

Bugoi, Roxana
"Horia Hulubei" National Institute of Nuclear Physics and Engineering,
P.O. Box MG-6, Bucharest 077125, Romania

Cash, Phil
Department of Medical Microbiology, University of Aberdeen, UK

Constantinescu, Bogdan
National Institute of Nuclear Physics, Horia Hulubei, P.O. Box MG-6,
077125 Bucharest, Romania

Davidchik, Valentina N.
Bach Institute of Biochemistry of the Russian Academy of Sciences,
Leninsky Prospect 33, 119071 Moscow, Russia

Demnerova, Katerina
Institute of Chemical Technology, Department of Biochemistry and
Microbiology, Technicka 3–5, 166 28 Prague 6, Czech Republic

Dikarev, Vladimir G.
Russian Institute of Agricultural Radiology and Agroecology, Kievskoe
shosse, 109 km, 249020, Obninsk, Russia

Dikareva, Nina S.
Russian Institute of Agricultural Radiology and Agroecology, Kievskoe
shosse, 109 km, 249020, Obninsk, Russia

Dinis, Maria de Lurdes
Porto University, Engineering Faculty (CIGAR), Rua Dr. Roberto Frias,
4200–465 Porto, Portugal

Dowling, Vera
Department of Biochemistry, University College Cork, Lee Maltings,
Prospect Row, Mardyke, Cork, Ireland

Dubrova, Yuri E.
University of Leicester, Department of Genetics, LE1 7RH Leicester, UK

Drozd, Tatiana
Department of Environmental and Molecular Genetics, International
Sakharov Environmental University, Dolgobrodskaya 23, 220009, Minsk,
Belarus

Dukhova, N.N.
Medical Radiological Research Center of the Russian Academy of Medical
Sciences, 249036 Obninsk, Kaluga Region, ul. Koroliova, 4, Russia

Durante, Marco
University Federico II, Department of Physics, Naples, Italy

Evseeva, Tatiana I.
Institute of Biology, Komi Scientific Center, Ural Division RAS,
Kommunisticheskaya 28, 167982, Syktyvkar, Russia

Fiúza, António
Geo-Environment and Resources Research Centre (CIGAR)
Engineering Faculty, Porto University, Rua Dr. Roberto Frias, 4200-465,
Porto, Portugal

Florko, B.V.
Joint Institute for Nuclear Research (JINR), 141980 Dubna, Moscow Region, Russia

Garcia-Bernardo, Jose
Department of Physiology, Development and Neuroscience, University of Cambridge, UK

Geras'kin, Stanislav A.
Russian Institute of Agricultural Radiology and Agroecology, Kievskoe shosse, 109 km, 24902 Obninsk, Russia

Gorova, Alla
National Mining University, K. Marks 19, Dnipropetrovs'k, 49006 Ukraine

Hinton, Thomas G.
University of Georgia, Savannah River Ecology Laboratory, Aiken, South Carolina, USA

Isayeva, V.G.
Medical Radiological Research Center of the Russian Academy of Medical Sciences, 249036 Obninsk, Kaluga Region, ul. Koroliova, 4, Russia

Kharlamova, Ganna
Kiev National Taras Shevchenko University, Kiev, Ukraine

Kharytonov, Mykola M.
State Agrarian University, Voroshilov Street 25, Dnipropetrovsk 49600, Ukraine

Kilchevsky, A.
Institute of Genetics and Cytology, Minsk, Belarus

Klimkina, Irina
National Mining University, K. Marks 19, 49600 Dnipropetrovsk, Ukraine

Kogotko, L.
Belarussian State Agricultural Academy, Gorky, Mogilev Region, Belarus

Konoplya, E.
Institute of Radiobiology, National Academy of Sciences of Belarus, Minsk, Belarus

Korogodina, V.L.
Joint Institute for Nuclear Research, 141980 Dubna, Moscow, Russia

Koroleva, Olga V.
Russian Academy of Sciences, National Center for Radiation Research and Technology, Bach Institute of Biochemistry, Leninsky prospect 33, 119071 Moscow, Russia

Korrea, Soheir
Laboratory of Mutagens and Toxigenomics, P.O. Box 29 Nasr City, Cairo, Egypt

Korrea, Soheir
National Center for Radiation Research and Technology, Egypt

Kostrova, L.N.
Institute of Genetics and Cytology, National Academy of Sciences, Minsk, Belarus

Kroik, Ann A.
State Agrarian University, Voroshilov St. 25, Dnipropetrovsk 49600, Ukraine

Kruk, A.
University of Gomel, Gomel, Belarus

Kudrjashov, V.
Institute of Radiobiology, National Academy of Sciences of Belarus, Minsk, Belarus

Kulikova, Natalia A.
Lomonsov Moscow State University, Department of Soil Science, Leninskie Gory GSP-2, 119992 Moscow, Russia

Marozik, Pavel
Institute of Genetics and Cytology, National Academy of Sciences of Belarus, Akademicheskaya 27, 220072 Minsk, Belarus

Mcdonagh, Brian
Department of Biochemistry, University College Cork, Lee Maltings, Prospect Row, Mardyke, Cork, Ireland

Melnov, Sergey
International Sakharov Environmental University, Department
of Molecular Genetics, Dolgobrodskaya 23, 220009 Minsk, Belarus

Meshcherjakova, I.S.
Research Institute of Experimental Microbiology, RAMS, Moscow, Russia

Michalik, Boguslaw
Central Mining Institute, Laboratory of Radiometry, Plac Gwarków 1,
40–166 Katowice, Poland

Mitchel, Ronald E.J.
Atomic Energy Canada, Chalk River Laboratories, K0J 1J0 Chalk River,
Canada

Molophei, V.P.
Institute of Genetics and Cytology, National Academy of Sciences, Minsk,
Belarus

Mosse, Irma
Institute of Genetics and Cytology, National Academy of Sciences
of Belarus, Akademicheskaya 27, 220072 Minsk, Belarus

Mothersill, Carmel
McMaster University, Medical Physics and Applied Radiation Sciences,
1280 Main Street W, L8S 4K1 Hamilton, Canada

Nesterenko, V.S.
Medical Radiological Research Center, Obninsk, Russia

Norkulova, K.T.
Tashkent State Technical University, University Street 2, Tashkent,
Uzbekistan

Osipov, Andreyan N.
N.I Vavilov Institute of General Genetics RAS, Gubkin Street 3, 119991
Moscow, Russia

Oudalova, Alla A.
Russian Institute of Agricultural Radiology and Agroecology, Kievskoe
shosse, 109 km, 249020, Obninsk, Russia

Oughton, Deborah
Department of Plant and Environmental Sciences, Norwegian
University of Life Sciences, P.O. Box 5003, 1432 Aas, Norway

Pärt, Peter
DG JRC, European Commission Joint Research Centre, Ispra,
Italy

Rollo, David C.
McMaster University, Department of Biology, 1280 Main Street W, L8S
4K1 Hamilton, Canada

Rosseland, Bjorn
Norwegian University of Life Sciences, Department of Plant
and Environmental Sciences, P.O. Box 5003, 1432 Aas, Norway

Saenko, Alexander S.
Medical Radiological Research Center, Korolev Street 4, 249036 Obninsk,
Russia

Salaberria, Iurgi
Institute of Environment and Sustainability, European Commission Joint
Research Centre, Ispra, Italy

Salbu, Brit
Norwegian University of Life Sciences, Department of Plant
and Environmental Sciences, P.O. Box 5003, 1432 Aas, Norway

Schofield, Paul N.
Department of Physiology, Development and Neuroscience,
University of Cambridge, UK

Segner, Helmut
Centre for Fish and Wildlife Health, Vetsuisse Faculty, University of Bern,
P.O. Box 8466, CH-3001 Bern, Switzerland

Seymour, Colin
McMaster University, Medical Physics and Applied Radiation Sciences,
1280 Main Street W, L8S 4K1 Hamilton, Canada

Sharetsky, A.N.
Medical Radiological Research Center of the Russian Academy of Medical
Sciences, 249036 Obninsk, Kaluga Region, ul. Koroliova, 4, Russia

Shchur, A.
Institute of Radiology, Mogilev Department, Mogilev, Belarus

Sheehan, David
University College Cork, Lee Maltings, Prospect Row, Mardyke Cork,
Ireland

Shpagin, D.V.
Medical Radiological Research Center of the Russian Academy of Medical
Sciences, 249036 Obninsk, Kaluga Region, ul. Koroliova, 4, Russia

Shupranova, Larisa V.
Dnipropetrovsk National University, Gagarina av.44, Dnepropetrovsk,
49600, Ukraine

Skipperud, L.
Norwegian University of Life Sciences, Department of Plant
and Environmental Sciences, P.O. Box 5003, 1432 Aas, Norway

Smith, Richard W.
McMaster University, Department of Biology, 1280 Main Street W, L8S
4K1 Hamilton, Canada

Sokolov, V.A.
Medical Radiological Research Center, RAMS, Obninsk, Russia

Stepanova, Elena V.
Bach Institute of Biochemistry of the Russian Academy of Sciences,
Leninsky prospect 33, 119071 Moscow, Russia

Surinov, B.P.
Russian Academy of Medical Sciences, Medical Radiological Research
Center, 249036 Obninsk, Kaluga Region, Russia

Tsyb, A.F.
Medical Radiological Research Center, RAMS, Obninsk, Russia

Turkman, Aysen
Dokuz Eylul University, Department of Environmental Engineering,
35160 Buca, Turkey

Tyther, Raymond
Department of Biochemistry, University College Cork, Lee Maltings,
Prospect Row, Mardyke, Cork, Ireland

Udovyk, Oleg
National Institute of Strategic Studies, Kiev, Ukraine

Yankovich, Tamara
Atomic Energy of Canada Ltd, Chalk River, ON K0J 1J0 Canada

Zamulaeva, Irina A.
Medical Radiological Research Center of RAMS, Korolev Street 4,
Obninsk, 249036, Russian Federation

Zdenek, Filip
Institute of Chemical Technology, Department of Biochemistry and
Microbiology, Technicka 3–4, 166 28 Prague 6, Czech Republic

SECTION 1

MULTIPLE STRESSORS: GENERAL OVERVIEWS

CHAPTER 1

CHALLENGES IN RADIOECOTOXICOLOGY

B. SALBU AND L. SKIPPERUD

Isotope Laboratory, Department of Plant and Environmental Sciences, Norwegian University of Life Sciences, P.O. Box 5003, N-1432 Aas, Norway

Abstract: *Radioecotoxicology* refers to responses, usually negative and detrimental responses, in living organisms exposed to radionuclides in ecosystems contaminated with artificially produced radionuclides or enriched with naturally occurring radionuclides. The key focus is put on the link between radionuclide exposures and the subsequent biological effects in flora and fauna. Radioecotoxicology is therefore an essential ingredient in impact and risk assessments associated with radioactive contaminated ecosystems. The radionuclide exposure depends on the source, release scenarios, transport, deposition and ecosystem characteristics as well as processes influencing the radionuclide speciation over time, in particular bioavailability, biological uptake, accumulation and internal distributions. Radionuclides released from a source may be present in different physico-chemical forms (e.g. low molecular mass species, colloids, pseudocolloids, particles) influencing biological uptake, accumulation, doses and biological effects in flora and fauna. Following releases from severe nuclear events, a major fraction of refractory radionuclides such as plutonium will be present as particles, representing point sources if inhaled or ingested. When organisms, and especially sensitive history lifestages, are exposed to radionuclides, free radicals are induced and subsequently, effects in several umbrella biological endpoints (e.g. reproduction and immune system failures, genetic instability and mutation, increased morbidity and mortality) may occur. However, radionuclides released into the environment rarely occur alone, but may co-occur in a mixture of other contaminants (e.g. metals, pesticides, organics, endocrine disruptors), which potentially could lead to synergisms or antagonisms. Thus, the relationships between radionuclide exposure and especially long-term effects are difficult to document and quantify, reflecting that the challenges within radioecotoxicology are multiple.

Keywords: ecotoxicology; multiple stressors; radiation

C. Mothersill et al. (eds.), Multiple Stressors: A Challenge for the Future, 3–12.
© 2007 *Springer.*

Introduction

When radionuclides are released from a source, the receiving ecosystems might be affected by *radioactive contamination*. To identify the degree of contamination, sampling and analysis are needed, and for alpha and beta emitters *radiochemistry* is needed that is the separation of radionuclides of interest from the bulk of interfering radionuclides prior to analysis. To assess the environmental impact and risks associated with the radionuclide contamination, information based on *radioecology* is needed; that is knowledge on the behaviour of radionuclides, in particular radionuclide species (Salbu, 2000; Salbu et al., 2004), in affected ecosystems. As ecosystem characteristics are essential for the behaviour of trace amounts of substances (e.g. pH in soil water, redox conditions, interacting components such as humic substances and clays) knowledge and principles from *ecology* should be implemented in radioecology. Serious consequences from radioactive contamination refer most often to negative or detrimental biological effects in exposed organisms such as man or organisms living in the affected environment. Since some organisms and some history life-stages are more susceptible to radiation exposures than others, knowledge from *radiobiologi* is essential. To evaluate biological early responses, toxicity or detrimental effects from radionuclide exposures, knowledge from *human toxicology* or *ecotoxicology* is needed. The radiation characteristics of radionuclides, their environmental mobility, bioaccumulation and doses are important determinants of the exposure and thereby the magnitude of the consequences following radionuclide releases. Thus, the phrase *radioecotoxicology* refers to the responses, usually negative or detrimental responses, in living organisms exposed to radionuclides, that is, key focus is put on the link between radionuclide exposures and the subsequent biological effects in flora and fauna.

The exposure (i.e. the radionuclide composition, their amounts and the radionuclide speciation) depends on the source, and in many cases several sources may contribute to the contamination (Fig. 1). Furthermore, the release scenarios (temperature, pressure, presence of air) may influence on the speciation of radionuclides deposited in an ecosystem. Following a severe nuclear accident a major fraction of refractory radionuclides such as plutonium is present as radioactive particles (Salbu et al., 1994, 2000; Salbu 2001; Salbu and Lind, 2005). Following an explosion under high temperature and pressure conditions, radioactive particles can be rather inert towards weathering, while during a fire the released oxidised particles are more readily dissolved Salbu et al., 2004. Thus, for radionuclides mobilised from oxidised particles, the soil to vegetation and animal transfer is rapid, while delayed for inert particles. External exposures reflect contaminated (gamma, high energy beta emitters) habitats, while the internal exposure depends on the presence

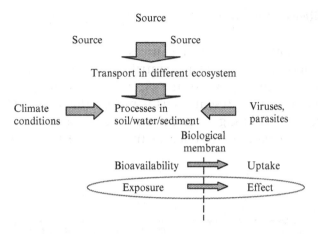

Fig. 1. Many variables influences on the exposure – biological effects relationship.

of bioavailable radionuclide species, biological uptake, accumulation and delivered doses. The radionuclide speciation depends, however, on the source term and interactions with other components during transport in air (e.g. association with soot particles), during transport in the aquatic environment (e.g. association with humic substances), and during deposition on ground (e.g. interactions with clay). These interactions may change the speciation of radionuclides released, and transformation processes influencing mobility, biological uptake, accumulation and doses take place over time. If mobile species are present, ecosystem transfer is relatively fast, whereas the ecosystem transfer is delayed if particles are present (Salbu, 2000; Salbu et al., 2000).

Relationships between exposure and especially long-term effects (responses in biological endpoints) are often difficult to document and quantify, although biological responses from molecular to ecosystem level have been identified for different organisms in contaminated areas. When organisms and especially sensitive history life-stages are exposed to radionuclides, free radicals and ROS are induced and subsequently, effects in several umbrella biological endpoints (e.g. reproduction and immune system failures, genetic instability and mutation, increased morbidity and mortality) may occur. However, radionuclides released into the environment rarely occur alone, but may co-occur in a mixture of other contaminants (e.g. metals, pesticides, organics, endocrine disruptors), which potentially could lead to effects in the same umbrella endpoints. Thus, multiple stressor exposures may induce synergetic or antagonistic effects in exposed organisms, and other factors such as climatic conditions and pathogens can also add to the stress (Salbu et al., 2005). Thus, a series of factors, including the interactions from other stressors, represents challenges within radioecotoxicology.

Sources and Release Scenarios

A significant number of nuclear sources has contributed, is still contributing, or has the potential to contribute to radioactive contamination of different ecosystem. As the Arctic ecosystems are believed to be most vulnerable, the present focus is put on this region (Fig. 2). The major sources contributing to radioactive contamination of long-lived radionuclides in Arctic ecosystems in the past are the nuclear weapon

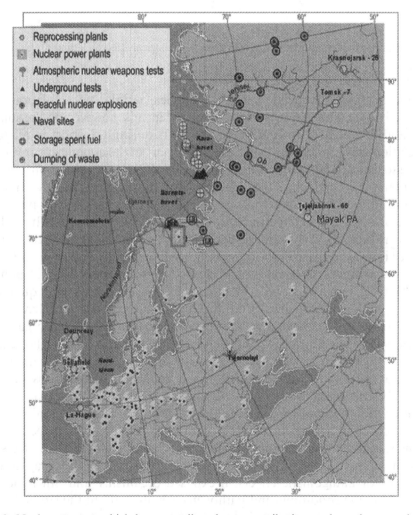

Fig. 2. Nuclear sources, which have contributed, are contributing, or have the potential to contribute to radioactive contamination of Arctic ecosystems.

tests (i.e. atmospheric tests resulting in global fallout, underground and underwater weapon tests at Novaya Zemlya resulting in local contamination), dumped liquid and solid radioactive waste in the Barents and Kara Seas and the Chernobyl accident (AMAP, 2002). Marine transport of artificially produced radionuclides from European reprocessing plants (i.e. Sellafield and Dounreay, UK, and La Hague, France, since 1950ies) and of Radium-, Lead- and Polonium-isotopes from the North Sea oil industry is ongoing (Salbu et al., 2003) together with river transport from Ob and Yenisey having large drainage areas affected from global fallout and from several nuclear installations (Mayak PA, Tomsk-7, Krasnoyarsk (Skipperud et al., 2004; Lind et al., 2006)). In addition, accidents (e.g. the US B52 bomber accident at Thule, Greenland, the COSMOS satellite accident in Canada, the Komsomolets submarine accident at Bear Island) have contributed to local contamination.

A series of sources may also potentially contribute to radionuclide contamination of the Arctic in the future; such as nuclear weapons, old land-based reactors such as the Kola NPP and Bilibino NPP, reactor fuelled submarines in operations or waiting for decommissioning, as well as spent fuel stored under non-satisfactory conditions at the Kola peninsula (AMAP, 2002).

The source and the release scenarios, including characteristics such as temperature and pressure, are important determinants of the radionuclide speciation, and thereby their mobility, biological uptake and effects, that is, consequences for affected ecosystems. Source terms are usually estimated from the inventory, for example the amount of radionuclides released; their isotopic composition; physical-chemical form of release (i.e. gas, solution, aerosols); time development of the release; release point and plume height; and the energy content of the release. Following a serious event involving nuclear fuel, however, a major fraction of released refractory radionuclides such as actinides will most probably be associated with particles (Salbu and Lind, 2005). The particle matrix, the refractory radionuclide composition and isotopic ratios will reflect the specific source (e.g. burn-up), while the release scenarios (e.g. temperature, pressures, redox conditions) will influence particle characteristics of biological significance. The composition, particle size distribution and specific activity are essential for acute respiration and skin doses, while factors influencing weathering rates such as particle size distribution, crystallographic structures, porosity, and oxidation states are essential for long-term ecosystem transfer 5]. For areas affected by particle contamination, impact and risk assessments will suffer from large uncertainties unless the impact of particles is included. Therefore, radionuclide speciation as well as processes influencing speciation, uptake, accumulation and biological effects are essential for estimating exposures to living organisms.

Ecosystem Transfer

Radionuclides released from a source may be present in different physico-chemical forms (e.g. low molecular mass species (LMM), colloids, pseudo-colloids, particles) influencing biological uptake, accumulation, doses and biological effects in flora and fauna. LMM species and colloids are believed to be mobile, while particles are easily trapped (Salbu, 2000). If mobile species are present, ecosystem transfer is relatively fast, whereas the ecosystem transfer is delayed if particles are present. Soil–water or sediment–water interactions are usually described by distribution coefficients, K_d, assumed to be constants at equilibrium. The speciation of radionuclides deposited in the environment will, however, change with time due to interactions with components in soils or sediments (Hinton et al., 2007; Skipperud et al., 2000a, b). Due to interactions with humic substances, the mobility of LMM-species is reduced. Due to weathering of particles, associated radionuclides are mobilised and the ecosystem transfer increases with time (Salbu, 2000; Skipperud et al., 2000a,b). Thus, the distribution of radionuclides between solid and solution is a time depended process and the thermodynamic constant should be replaced by a time-function.

The speciation of radionuclides is of importance for biological uptake, accumulation and biomagnification. LMM-species can cross biological membranes, directly or indirectly after interactions with ligands or carrier molecules. LMM organic ligands such as citrate may stimulate the uptake, while high molecular mass (HMM) organics (e.g. Prussian Blue) reduce uptake and are used as countermeasures for Cs-isotopes (Salbu, 2000; Salbu et al., 2004). Information on bioavailable forms is, however, still scarce. For soil-to-plant transfer, transfer coefficients; TC (m^2/kg), and for soil-plant-animal transfer aggregated transfer coefficients; T_{agg} (m^2/kg), are utilised for modelling purpose, and depend on several factors (e.g. soil types, microbial activities, plant- and animal species, dietary habits, trophic levels) and in particular on radionuclide speciation. Uptake in fish and invertebrates depends on ionic species interacting with external organs (gills, skin) or by digestive uptake. In filtering organisms, however, particles and colloids are retained and radionuclides may accumulate due to changes in bioavailability in the gut (digestion) or through phagocytosis. Bioconcentration factors (BCF) vary according to the radionuclide species in the exposure, degree of biomagnification and can be distinguished for different compartments within organisms. The link between Radionuclide speciation - K_d - T_{agg} - BCF, being time functions, represents a challenge within radioecology, with important implications for *radioecotoxicology* (Hinton et al., 2007; Salbu, 2007).

Effects from Radiation Exposure

Biological uptake, bioaccumulation and the radiation characteristics of radionuclides are important determinants of the magnitude of the environmental consequences following release. The receiving environments themselves also influence the scale of the consequences, since some organisms are more susceptible to incorporating radionuclides into exposure chains than others. Relationships between accumulation, dose and short and long-term effects (biological endpoints) are difficult to document and quantify, although biological responses from molecular to ecosystem level have been identified for different organisms in contaminated areas. When biological systems are exposed to radiation, free radicals are formed due to excitation and ionisation of water molecules in cells and Haber–Weiss and Fenton reactions (Fig. 3). Free radicals produced ($\cdot$H, $\cdot$OH) are extremely reactive, will recombine and produce various reactive compounds in cells (e.g. HO_2, H_2O_2, H_2, O_2). Free radicals and the formation of reactive oxygen species (ROS) may affect membrane integrity and damage proteins and nucleic acids (DNA, RNA). Radiation induced free radicals can be identified as ROS and as enzymatic activity of antioxidant repair enzymes for example, superoxide dismutase (SOD), catalase or glutathione cycle enzymes (glutathione reductase and peroxidase). To identify early effects, expressions of metal-responsive genes, for example, cellular antioxidants, antioxidant enzymes, heme oxygenase, metallothionein, and information of cellular integrity and protein kinetics are equally important.

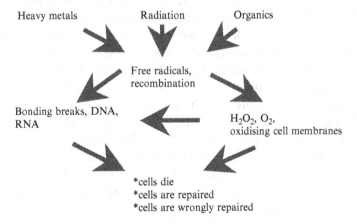

Heavy metals   Radiation   Organics

Free radicals, recombination

Bonding breaks, DNA, RNA

H_2O_2, O_2, oxidising cell membranes

*cells die
*cells are repaired
*cells are wrongly repaired

Fig. 3. Radiation induces free radicals in organisms, affecting sensitive biological endpoints: reproduction and immune system failures, genetic instability and mutation, increased morbidity and mortality. Other stressors such as metals and organics can also induce free radicals in organisms.

Dose-effect-risk relationships are based on a variety of biological endpoints ranging from molecular to ecosystem level. However, the evaluation of internal and external doses to biota using dosimetry models (e.g. equilibrium absorbed dose constant models, point source distribution models such as Loevinger's expression and Monte Carlo simulations) represents a challenge (Ulsh et al., 2003; Broion et al., 2006), in particular for uneven internal distribution of radionuclides. Recent data on Relative Biological Effectiveness (RBE) indicates that radiation with high linear energy transfers (LET) causes a greater degree of biological damage than low LET radiation for a given absorbed dose. Recent data implies also that RBEs for flora and fauna probably are different to those of humans, and calls for more research. Also at the mechanistic level there are gaps in knowledge with respect to low dose non-targeted effects of radiation such as genomic instability and bystander effects (Mothersill et al., 2006, 2007). These have been shown to occur in fish and mammals, but their impact on risk is uncertain, that is whether they are adaptive responses or they can magnify the damage.

The role of radionuclide speciation and internal distributions, exposure time associated with episodic accumulation and uneven distribution of doses (micro-dosimetry) inducing effects in sensitive biological endpoints for sensitive history life-stages for organisms is, however, still not understood, and improvement of speciation - low dose - early effect models is needed (Hinton et al., 2004). Problems also arise when benchmark concentrations are used to regulate doses and effects in the environment. The benchmark concentrations represent upper limits which, if exceeded, may result in harm to the environment. However, most benchmark values are derived from extrapolating toxicity data; from acute to chronic effects, from laboratory to field conditions, from effect concentration to no-effect concentration and from isolated test-species to complex systems. Thus, proper uncertainty estimates are essential.

However, radionuclides released into the environment rarely occur alone, but may co-occur in mixtures with other contaminants (e.g. metals, pesticides, organics, endocrine disruptors), which potentially could lead to effects in biological endpoints sensitive to radiation (Mothersill et al., 2006). One single stressor may induce multiple biological effects, if multiple interactions occur or if interactions with different biological targets take place. In mixtures with several different stressors, multiple types of interactions and interactions with multiple target sites may occur. Thus, the interactions may be concentration additive (1+1=2), synergetic (1+1=3 or 4) or antagonistic (1+1=0).

It is internationally recognised that there are severe gaps in basic knowledge with respect to biological responses from multiple stressor exposures. Identification of biological responses from low dose chronic exposure calls for early warning biomarkers, utilising modern molecular and genetic tools. Furthermore, information on dose-response relationships (on/off mechanisms), sensitivity (detection limits, thresholds), and synergetic and antagonistic effects, as well as the role of protecting agents such as antioxidants is highly needed. Development of analytical strategies, methods and biomarkers that can be utilised to increase the knowledge on biological impact from multiple stressors represents also a challenge for the future.

Conclusions

Radioecotoxicology is an essential discipline in environmental impact and risk assessments associated with radioactive contaminated ecosystems, linking radionuclide exposures to the subsequent biological effects in flora and fauna. However, a series of factors influencing the exposure must be taken into account, when doses are assessed or predicted. Similarly, a series of factors influencing the biological responses must be considered, contributing significant to the overall uncertainties in assessments. Adding the multiple stressor exposure and multiple response concept, the uncertainties increase with order of magnitude. Thus, meeting the challenges within *Radioecotoxicology* is essential to reduce the overall uncertainties in environmental impact and risk assessments for contaminated areas.

References

AMAP 2002. Arctic pollution Issues: A state of the Arctic environment report. Oslo, Norway.

AMAP 2002. AMAP Assessment 2002: Radioactivity in the Arctic. Oslo, Norway.

Brown JE, Hosseini A, Borretzen P, Thorring H. 2006. Development of a methodology for assessing the environmental impact of radioactivity in Northern Marine environments. *Mar. Pollut. Bull.* 52:1127–1137.

Hinton TG, Bedford JS, Congdon JC, Whicker FW. 2004. Effects of radiation on the environment: a need to question old paradigms and enhance collaboration among radiation biologists and radiation ecologists. *Radiat Res* 162:332–338.

Hinton T, Garten CT, Kaplan DI, Whicker FW. 2007. Major biogeochemical processes of radionuclide dispersal in terrestrial environments. *Rad. Ass.* in press.

Lind OC, Oughton DH, Salbu B, Skipperud L, Sickel M, Brown JE, Fifield LK, Tims SG. 2006. Transport of low 240Pu/239Pu atom ratio plutonium in the Ob and Yenisey Rivers to the Kara Sea. *Earth Planet. Sci. Lett.* in press.

Mothersill C, Bucking C, Smith RW, Agnihotri N, Oneill A, Kilemade M, Seymour CB. 2006. Communication of radiation-induced stress or bystander signals between fish in vivo. *Environ Sci Technol* 40:6859–6864.

Mothersill C, Salbu B, Heier LS, Teien HC, Denbeigh J, Oughton DH, Rosseland BO, Seymour CB. 2007. Multiple stressor effects of radiation and metals in salmon (Salmo salar). *J. Environ. Radioact.* April 9th epub.

Salbu B. 2000. Speciation of radionuclides in the environment. In Meyers RA, ed, *Encyclopedia of Analytical Chemistry*, John Wiley & Sons Ltd, Chishester, pp 12993–13016.

Salbu B. 2001. Actinides associated with particles. In Kudo A, ed, *Plutonium in the Environment*, First ed, Elsevier, Tokyo, pp 121–138.

Salbu B. 2007. Speciation of radionuclides – Analytical challenges within environmental impact and risk assessments. *J. Environ. Radioact.* April 30th epub.

Salbu B, Lind OC. 2005. Radioactive particles released from various nuclear sources. *Radioprotection* 40:27–32.

Salbu B, Krekling T, Oughton DH, Ostby G, Kashparov VA, Brand TL, Day JP. 1994. Hot particles in accidental releases from chernobyl and windscale nuclear installations. *Analyst* 119:125–130.

Salbu B, Janssens K, Krekling T, Simionovici A, Drakopoulos M, Raven C, Snigireva I, Snigirev A, Lind OC, Oughton DH, Adams F, Kashparov VA. 2000. X-ray absorption tomography and – XANES for characterisation of fuel particles. *ESRF Highlights* 24–25.

Salbu B, Skipperud L, Germain P, Guegueniat P, Strand P, Lind OC, Christensen G. 2003. Radionuclide speciation in effluent from La Hague reprocessing plant in France. *Health Phys.* 85:311–322.

Salbu B, Lind OC, Skipperud L. 2004. Radionuclide speciation and its relevance in environmental impact assessments. *J. Environ. Radioact.* 74:233–242.

Salbu B, Rosseland BO, Oughton DH. 2005. Multiple stressors – a challenge for the future. *J. Environ. Monit.* 7:1–2.

Skipperud L, Oughton D, Salbu B. 2000. The impact of Pu speciation on distribution coefficients in Mayak soil. *Sci. Total Environ.* 257:81–93.

Skipperud L, Oughton DH, Salbu B. 2000. The impact of plutonium speciation on the distribution coefficients in a sediment-sea water system, and radiological assessment of doses to humans. *Health Phys.* 79:147–153.

Skipperud L, Oughton DH, Fifield LK, Lind OC, Tims S, Brown J, Sickel M. 2004. Plutonium isotope ratios in the Yenisey and Ob estuaries. *Appl. Radiat. Isot.* 60:589–593.

Ulsh B, Hinton TG, Congdon JD, Dugan LC, Whicker FW, Bedford JS. 2003. Environmental biodosimetry: a biologically relevant tool for ecological risk assessment and biomonitoring. *J Environ Radioact* 66:121–139.

CHAPTER 2

THE INVOLVEMENT OF POLLUTION WITH FISH HEALTH

BRIAN AUSTIN

School of Life Sciences, John Muir Building, Heriot-Watt University, Riccarton, Edinburgh EH14 AS, Scotland, UK. e-mail: b.austin@hw.ac.uk

Abstract: It is generally accepted that pollutants, such as hydrocarbons, enter the aquatic environment by accident or deliberately, and may lead to large-scale and sudden kills of animal life, especially when the compounds are in high quantities. However, more subtle changes to the host may ensue when lesser quantities of pollutants are involved. Here, the resulting damage may include immunosuppression, physical damage to gills and epithelia, and adverse affects on metabolism. Also, there may well be increased susceptibility to various infectious diseases, including lymphocystis and ulceration. Much of the work to date has centered on laboratory studies and also surveys of polluted and clean marine sites, but it is not always possible to make firm conclusions from the data.

Keywords: fish disease; ulceration; fin rot; tail rot; pollutants; pesticides; hydrocarbons; heavy metals

Introduction

It is well established that pollutants of various kinds reach the aquatic environment either accidentally or deliberately. Large-scale releases of hydrocarbons, such as from ocean going tankers e.g. the Exxon Valdez in Alaska and the Braer in the Shetland Islands, have featured prominently on news programmes in many countries when photographs of dead sea birds amid oily beaches greet the viewer. With such examples, it is easy to associate cause, i.e. the pollution event, with effect, namely death of the animal. In addition, there are natural events when, for example, the collapse of algal blooms lead to adverse water quality and large-scale fish kills. Moreover, evidence has been obtained – and will be discussed later – for the presence of specific pollutants within the tissues of aquatic animals. However, there is a dilemma proving a relationship associating the presence of pollutants in tissues with those in the aquatic environment and the concomitant adverse effect on health. A complication is

13

C. Mothersill et al. (eds.), Multiple Stressors: A Challenge for the Future, 13–30.

that the presence of pollutants, e.g. copper, in tissues is not always correlated to the presence of the compounds in the environment. For example, the accumulation of copper to 1,936 µg/g wet weight was detected in the non-cytosolic fraction of the liver of mullet (*Mugil cephalus*) but without any evidence of environmental contamination insofar as the fish were collected from separate areas with low copper concentrations (Linde et al., 2005). Yet other fish species caught in those same areas did not reveal the presence of high copper concentrations (Linde et al., 2005). This suggests that some species have the innate ability to accumulate heavy metals regardless of whether or not the habitat is contaminated. A previous health scare concerning the presence of mercury in the liver of tuna may also be explained by the inherent ability of the species to accumulate the metal.

Pollutants in the Aquatic Environment

There is a large body of evidence demonstrating the presence of certain specified pollutants in aquatic habitats, and include:

- Heavy metals; these include arsenic, copper, zinc (Gassman et al., 1994; Han et al., 1997), cadmium, lead and mercury which occur in industrial effluents (Gassman et al., 1994; Bernier et al., 1995) and tin, tributyltin and triphenyltin that occur in anti-fouling paints used on the undersurfaces of ships to prevent bioattachment and biofouling (Horiguchi et al., 1995).

- Hydrocarbons; these may result from deliberate spillage during wartime (Evans et al., 1993; Turrell, 1994; Newton and McKenzie, 1995) or accidental discharge from tankers.

- Inorganic nitrogen as ammonia, nitrites and nitrates, which may be derived from agricultural run-off and aquaculture (Ziemann et al., 1992).

- Organic material (Grawinski and Antychowicz, 2001), including faecal debris, from drainage systems. Large numbers of bacteria may be associated with this material (Dudley et al., 1980).

- Pesticides, including dioxin (Guiney et al., 1996), 2,3,7,8-tetrachlorodibenzo-*p*-dioxin (TCDD; Grinwis et al., 2000), 1, 1, 1-trichloro-2,2-bisi (*p*-chlorophenyl) ethane (DDT) (Fitzsimons, 1995) and organochlorines (Dethlefsen et al., 1996).

- Plastics (Goldberg, 1995)

- Pulp mill effluents (Lehtinen et al., 1984; Sandstrom, 1994; Couillard and Hodson, 1996; Jeney et al., 2002)

- Toxins, such as from the collapse of dinoflagellate blooms (Noga et al., 1996).

The discharge of these pollutants may be gradual, intermittent pulses or irregular, and may be authorised or unauthorised by national or local authorities. Pollutants may be released in specified areas or dumped in an ad hoc manner. Local or more widespread adverse water quality issues may develop. It is argued that the extent and longevity of pollutants in the aquatic environment is generally unclear, and needs to be firmly clarified through effective monitoring programmes (e.g. Ibe and Kullenberg, 1995).

Evidence for the Presence of Pollutants in the Tissues of Aquatic Animals

The available data have centered largely on laboratory-based in vitro studies and investigations using animals collected from the aquatic environment. Although in vitro studies may produce interesting data, the relevance to explaining the likely outcome of pollution events to biological systems in the aquatic environment is questionable. Nevertheless, there is evidence that a wide range of pollutants, including DDT (Fitzsimons, 1995), thiocyanate (Lanno and Dixon, 1996), didecyldimethylammonium chloride (Wood et al., 1996), polychlorinated-biphenyls (PCBs; Barron et al., 2000) and creosote fractions (Sved et al., 1997) have been researched, and determined to be taken up by aquatic animals from water.

By using animals collected directly from aquatic habitats, there is good agreement between the presence of some pollutants in tissues and the levels in the environment. For example, copper and zinc have been detected in cod (*Gadus morhua*) from coastal waters of Newfoundland, Canada, in invertebrates and vertebrates from Taiwan (Han et al., 1997), and in rabbitfish (*Siganus oramin*) from the polluted waters around Hong Kong (Chan, 1995). Cadmium, lead and mercury were detected albeit in small quantities in fish from the Great Lakes (Bernier et al., 1995). It is interesting to note that the amounts of these metals in cod were below the maximum permitted levels for food (Hellou et al., 1992). Furthermore mercury, in concentrations regarded to be insufficient to cause human health problems, has been found in healthy fish and shellfish collected from the vicinity of discharges from a chlor-alkali plant in India (Joseph and Srivastava, 1993). Crustaceans, molluscs and fish from around a sewage outfall contained pesticides, i.e. chlordane, dieldrin, hexachlorobenzene and DDT (Miskiewicz and Gibbs, 1994). DDT, PCB, organochlorine and 2,3,7,8-tetrachlorobenzo-*p*-dioxin were recorded in fish from Vaike Vain Strait in Estonia (Voigt, 1994), from the North Sea (Dethlefsen et al., 1996) and the Great Lakes (Guiney et al., 1996). However, these publications did not mention any evidence that the compounds had actually harmed the aquatic animals.

Using adult walleye (*Stizostedium vitreum*) caught in polluted and relatively unpolluted sites in Wisconsin, USA during 1996 and 1997, significantly higher PCB concentrations together with more evidence of hepatic neoplasms and tumours (adenomas and carcinomas) were found in fish from the polluted areas (Barron et al., 2000). The authors considered that these deleterious effects were consistent with long-term exposure to tumour promoters, e.g. PCBs, in the environment.

Effect of Pollutants on the Health of Aquatic Animals

This is the crux of the issue – what is the effect of the pollutant on the health of the aquatic animal? It is clear that large quantities of pollutants entering the aquatic environment, such as oil from damaged tankers, over a comparatively short period of time may lead to sudden and often extensive kills. With this scenario, there is a clear link between the pollution event and mortalities among aquatic species. It is the low-level exposure to pollutants that causes the greatest discussion regarding the possible effect on the health of aquatic species. Low-level exposure may well lead to chronic damage, which may not develop for a long time (see Mayer et al., 1993) possibly long after the pollution event has ended. This complicates epidemiological investigation insofar as it is difficult to conclusively associate the pollution with ill health.

The long-term exposure to low-level pollution may lead to the accumulation of the pollutant in animals, which in turn could become weakened/immunocompromised (Vos et al., 1989; Sovenyi and Szakolczai, 1993; Sahoo et al., 2005;) and subsequently colonised by opportunist pathogens leading to clinical disease, for example, gill disease of fish (Austin and Austin, 1999). Anthropogenic factors pose recurring problems, and include:

- Chronic exposure of rainbow trout (*Oncorhynchus mykiss*) to didecyldimethylammonium chloride led to an elevation of stress factors, namely plasma cortisol, glucose and lactate, and a decrease in swimming performance (Wood et al., 1996).
- Chlorinated and aromatic hydrocarbons which accumulate in sediments, and may well lead to a decline in wild fish populations (Arkoosh et al., 1998). Such factors have been considered to be responsible in part for the decline of wild Pacific salmon populations with experimental evidence pointing to the bioaccumulation of these classes of hydrocarbons by juveniles leading to immunosuppression and increased susceptibility to disease (Arkoosh et al., 1998). Also hydrocarbons have been implicated with breakages in the double-stranded DNA, and enzyme induction in common dab (Everaarts et al., 1994).

- Damage to fins, gills, opercula and skin has followed exposure to ammonia, copper and phenol (Kirk and Lewis, 1993), cadmium (Sovenyi and Szakolczai, 1993), creosote (Sved et al., 1997), and chlorine-containing pulp mill effluents (Lindesjoo and Thulin, 1994; Lindesjoo et al., 1994; Sandstrom, 1994). Moreover, ammonia has been reported to cause the development of circular depressions and pitting in the gill epithelium of rainbow trout. Exposure to copper led to fusion of the gill lamellae and swelling of the tips of the filaments and epithelium. In comparison, phenol was observed by electron microscopy to destroy the epithelial layers as far as the cartilage (Kirk and Lewis, 1993).

- Exposure to sewage sludge has affected the growth and protein synthesis in common dab (*Limanda limanda*) (Houlihan et al., 1994) and caused liver damage in fish (Moore et al., 1996).

- Kidney damage has resulted from exposure to cadmium (Sovenyi and Szakolczai, 1993).

- Liver damage may result from contamination with cadmium (Sovenyi and Szakolczai, 1993) and thiocyanate (Lanno and Dixon, 1996).

- Sub-lethal concentrations of the pesticides atrazine and lindane (Cossarini-Dunier, 1987), heavy metals (O'Neill, 1981) including cadmium (Sovenyi and Szakolczai, 1993), sewage components (Secombes et al., 1991; 1992) and aquatic micro-organisms (Robohm et al., 1979; Evans et al., 1997) affect the immune system of fish by either stimulating or retarding anti-body production.

- Thiocyanate has been blamed as causing anaemia and interfering with thyroid function (Lanno and Dixon, 1996).

- Winter Stress Syndrome, which is characterised by reduced levels of lipids, may well result from stressors including the presence of various pollutants (Lemly, 1997).

Mill effluents have been the focus of many studies into their association with abnormalities/disease of fish (Lindesjoo and Thulin, 1994; Lindesjoo et al., 1994; Sandstrom, 1994; Jeney et al., 2002). During 1982–1987 in the Baltic Sea, the presence of abnormalities was reported, including gonad malfunction, poor embryo quality and mortalities among coastal fish species, which were considered to have been exposed to mill effluent (Sandstrom, 1994). Of relevance after 1984, the toxicity of effluent was reduced by the substitution of chlorine dioxide for chlorine. Then in 1992, the use of chlorine was eliminated altogether, which reduced the amount of organochlorines in the effluent, and led to a reduction in the level of mortalities in the fish populations (Sandstrom, 1994).

Using roach (*Rutilus rutilus*) caught from two lakes, one of which received bleached kraft mill effluent and the second of which was unpolluted, experimental infection with the digenean parasite *Rhipidocotyle fennica* led to the observation that fish from the polluted site possessed a significantly higher number of parasites after the first 2 days of infection. The explanation was that these fish from the polluted site had decreased resistance to infection. In particular, these roach had a lower leucocrit, and higher alkaline phosphatase and plasma chloride levels compared to those fish from the clean site (Jeney et al., 2002).

Dinoflagellate toxins have been blamed for major fish kills, with evidence describing erosion to the epithelium leading to ulceration (Noga et al., 1996). This observation is especially relevant insofar as ulceration is one of the most common diseases associated with pollution in the marine environment.

Pesticides, namely DDT, have been considered to be responsible for reduced hatching, blue sac disease and swim up syndrome mortalities in lake trout (*Salvelinus namaycush*) (Fitzsimons, 1995) eggs. Yet, these laboratory studies indicated that concentrations of pesticides necessary to affect the eggs were much higher than the levels found in feral fish eggs.

In Dutch coastal and estuarine habitats, flounder (*Platichthys flesus*) have a high prevalence of liver disease (pre-neoplastic) and lymphocystis, and because of this laboratory studies were conducted to determine the effect of xenobiotics, namely TCDD, oral exposure to which led to significantly enhanced immunoreactivity in the hepatocytes and endothelium of various organs and the epithelia of the digestive tract, liver and mesonephros to cytochrome P4501A but without any profound pathology (Grimwis et al., 2000). The conclusion was that flounder was relatively insensitive to the toxic effects of TCDD, but an assumption was made that exposure (to TCDD) could influence the development of tumours (Grimwis et al., 2000).

A polluted environment will inevitably contain a variety of contaminants, and it is difficult to determine which may be the trigger for a disease event. A topical example concerns pigmented salmon syndrome, which is a non-infectious haemolytic anaemia with jaundice, and occurred as an epidemic during the early 1980s in migrating Atlantic salmon (*Salmo salar*) in the River Don, Scotland. The river system received effluent from paper mills, the oil servicing industry and an airport. Ensuing experimentation reproduced the disease signs following exposure of Atlantic salmon to hydrocarbons, namely diesel and resin acids. Interestingly, the syndrome was not detected in the River Don after 1989, and most likely reflected general improvements of water quality therein (Croce et al., 1997).

Of course there is an additional problem where pollution is thought to influence the occurrence of disease but where a firm association between specific pollutants and the incidence if ill health cannot be proven. For example, there was a strong indication that the presence of proliferative kidney disease in wild brown trout and rainbow trout in southern Germany was correlated with organic pollution. However, this association was not proven (El-Matbouli and Hoffmann, 2002).

Pollution-Related Diseases

The development of measurable disease processes involves an interaction between a host, disease-causing situation, e.g. a pathogen, and inevitably a stressor (Austin and Austin, 1999). Without a stressor, the host will often not develop clinical disease (Austin and Austin, 1999). Here, the argument is that pollutants constitute the stressor. A sizeable proportion of the work correlating fish disease with pollution in the aquatic environment has involved surveys, many of which have been carried out in the North Sea (e.g. Dethlefsen and Watermann, 1980; Dethlefsen et al., 1987; 2000; McVicar et al., 1988; Vethaak and ap Rheinallt, 1992). The basic premise is that fish from polluted sites have a greater incidence of disease than specimens obtained from unpolluted areas (e.g. Dethlefsen et al., 2000). Generally, the evidence supports this premise. For example, a study of epidermal hyperplasia/papilloma, lymphocystis and skin ulcers on female dabs (*Limanda limanda*) of >3 years old from sites in and around the North Sea was carried out between 1992 and 1997. The outcome was evidence for a higher incidence of some diseases in the different sampling sites, and a suggestion by the researchers that further investigation was needed to determine the reasons for the observations and a possible link to anthropogenic factors (Dethlefsen et al., 2000). There is an obvious question concerning water flow and fish migration, i.e. currents will carry water to and from polluted and clean sites. Also, the effects of fish migration need to be considered insofar as specimens caught in polluted areas could have recently arrived from clean sites, and vice versa (Bucke et al., 1992; Vethaak et al., 1992; Jacquez et al., 1994). Notwithstanding these concerns and on the basis of probability, it is generally accepted that disease may be mediated in some way by pollution events (Vethaak and Jol, 1996).

Fish diseases, often linked to pollution, include:

- Carcinomas (Koehler et al., 2004)
- Epidermal papilloma (Dethlefsen and Waterrmann, 1980; Premdas and Metcalfe, 1994)
- Fin/tail erosion (Vethaak, 1992; Vethaak et al., 1996; Bodammer, 2000)
- Gill disease/hyperplasia (Kirk and Lewis, 1993)
- Liver disease (Malins et al., 1980; 1987; Peters et al., 1987; Vethaak et al., 1996)
- Neoplasia (Malins et al., 1980; Bucke and Feist, 1993; Depledge, 1996; Vethaak and Jol, 1996)

- Parasitic diseases (Overstreet and Howse, 1977; Pascoe and Cram, 1977; Das and Shrivastava, 1984; Khan, 1987; Möller, 1987; Khan and Thulin, 1991)
- Skin disease/ulceration (Vethaak, 1992; Vethaak and Jol, 1996) which is often associated with infection by atypical *Aeromonas salmonicida* (Austin and Austin, 1999)
- Viral diseases, principally lymphocystis (Vethaak and Jol, 1996; Vethaak et al., 1996).

The reasons for the occurrence of these conditions have been thought to include contaminated diets (Landsberg, 1995), heavy metals (Rødsaether et al., 1977), nitrogenous compounds such as ammonia (Kirk and Lewis, 1993) and nitrites (Hanson and Grizzle, 1985), pesticides (Voigt, 1994), organic material (Grawinski and Antychowicz, 2001) including sewage (Austin and Stobie, 1992) and/or unspecified pollutants (Vethaak and Jol, 1996; Bodammer, 2000). Of these, fin/tail erosion, gill disease, gill hyperplasia, and ulceration are often linked to bacterial involvement. A possible scenario is that the pollutant stresses or weakens the host allowing colonisation by micro-organisms and thus the development of clinical disease.

Viruses and carcinogens may lead to neoplasias/tumours. In this connection, certain geographical areas have been identified where the occurrence of tumours on fish and shellfish has been correlated with higher concentrations of anthropogenic compounds (Depledge, 1996). Of relevance, Koehler et al. (2004) caught juvenile and adult female flounder (*Platichthys flesus*) from a polluted river, i.e. the River Elbe, and a control site with the data revealing that the young fish had liver damage. The adult females displayed adenomas and carcinomas with a frequency of 70% in the polluted river. Exposure to carcinogens in the river was blamed on the incidence of the tumours (Koehler et al., 2004). Yet, the proof of any correlation between pollution and disease is not always documented. Surveys, such as occur with regularity in the North Sea, point to an association between pollution and disease but do not generally consider the nature or concentration of the pollutant(s). For future studies, it would be relevant to consider

- The possibility of synergism between combinations of pollutants
- The minimum exposure time to a pollutant necessary to initiate deleterious changes to the host.

CONTAMINATED FISH FOOD

Landsberg (1995) considered that toxins, i.e. from macroalgae, e.g. *Caulerpa* spp., and benthic dinoflagellates, e.g. *Gambierdiscus toxicus*, may have been responsible for large-scale mortalities in tropical reef fish in Florida, USA during 1993–1994. Examination of the fish revealed head lesions, ulcerations, fin and tail rot, and the presence of mucus on the body surface (Note: this may suggest exposure to an irritant; Austin and Austin, 1999). Amoebae, bacteria and turbellarians were found in diseased fish, although it was considered that these organisms were opportunists/secondary invaders rather than the primary cause of illness (Landsberg, 1995). In this connection, it is recognised that many diseases may develop in an already weakened host (Austin and Austin, 1999). Yet, whether or not such weakening results from pollutants is rarely if ever considered.

There is evidence that some serious pathogens may be spread through contaminated fish feed used in aquaculture. Incidences of botulism (Huss et al., 1974), mycobacteriosis (Dulin, 1979), streptococcosis (Minami, 1979) and eye disease caused by *Rhodococcus* (Claveau, 1991) have been linked to feeds containing contaminated fish products.

HEAVY METALS

There is increasing evidence that the presence of heavy metals is linked to the exacerbation of some microbial fish diseases. Copper has been singled out for attention insofar as there is evidence for its role in increasing susceptibility to *Edwardsiella tarda* (Mushiake et al., 1984) and *Vibrio anguillarum* (Rødsaether et al., 1977) infections. It has long been established that exposure to copper resulted in coagulation of the mucus layer of gills, leading to inhibition of oxygen transport and respiratory distress (Westfall, 1945) and to a reduction in the populations of lymphocytes and granulocytes in the blood leading to a reduction in phagocytosis (Mushiake et al., 1985).

Titanium dioxide has been implicated with harm to aquatic animals (Dethlefsen and Watermann, 1980; Lehtinen, 1980; Dethlefsen et al., 1987). Thus, as a result of a survey of 5,942 fish caught in Dutch coastal waters in 1986–1988, a higher incidence of epidermal hyperplasia/papillomas, lymphocystis, liver nodules and infections with *Glugea* was documented in the common dab collected from dump sites receiving titanium dioxide acids in comparison to control sites (Vethaak and van der Meer, 1991). This supported the outcome of an earlier investigation which recorded a link between the incidence of epidermal papillomas in common dabs in relation to the dumping of titanium dioxides wastes (Dethlefsen and Watermann, 1980).

HYDROCARBONS

The presence of hydrocarbons leads to impairment of mucus, defective immune systems, increased incidences of parasitism, the induction of hyperplasia and liver hypertrophy, and mortalities (Fletcher et al., 1982; Haensly et al., 1982; Khan, 1987; 1991; Lehtinen, 1980). Experimental evidence has shown that fish may develop epidermal lesions and fin erosion, or die following exposure to suspended sediments containing high molecular weight creosote fractions (Sved et al., 1997). On the contrary, low molecular weight creosote fractions led to the appearance of head lesions, specifically around the mouth, nares and opercula (Sved et al., 1997).

NITROGENOUS COMPOUNDS

Data revealed that the presence of nitrites at 6 mg/L of water increased the susceptibility of channel catfish (*Ictalurus punctatus*) to *Aeromonas hydrophila* infection (Hanson and Grizzle, 1985).

PESTICIDES

Pesticides, such as DDT and PCBs, in the aquatic environment have been linked with diseases such as "cauliflower disease", lymphocystis and ulceration (Voigt, 1994) and liver neoplasia (Moore et al., 1996). Malformations in common dab, flounder (*Platichthys flesus*), plaice (*Pleuronectes platessa*) and whiting (*Merlangus* sp.) embryos from the southern North Sea during 1984–1995 were linked to pollution with organochlorines (Dethlefsen et al., 1996). Thus as a direct result of long-term surveys, these workers considered that the malformations resulted from low water temperatures that predisposed the embryos to the effects of organochlorines.

Liver disease, including neoplasia, has been described in winter flounder (*Pleuronectes americanus*) from Boston, USA, particularly in the region of a sewage outfall (Moore et al., 1996). These workers noted that during 1987–1993, there was a reduction in the incidence of neoplasia concomitant with a decline in output of chemicals, particularly DDT and other chlorinated hydrocarbons, into the receiving waters.

ORGANIC MATERIALS/SEWAGE

The presence of some fish diseases has been linked to the presence of unknown components of sewage dumping (Siddall et al., 1994). For example as a result of a survey of 16 sites in the Dutch Wadden Sea, a higher incidence of skin ulcers and fin rot was recorded in fish caught near fresh water drainage sluices than elsewhere (Vethaak, 1992). Pollution by domestic sewage, i.e. leakage from a septic tank, was attributed to a new skin disease, which was characterised by extensive skin lesions and muscle necrosis in rainbow trout (otherwise infected with *Yersinia ruckeri* as enteric redmouth disease for which there may well have been a link with sewage sludge; Dudley et al., 1980) in Scotland during 1992 (Austin and Stobie, 1992). From diseased fish, *Serratia plymuthica* and *Pseudomonas pseudoalcaligenes* were recovered for the first time as fish pathogens. The skin lesions declined substantially after the leaking septic tank was repaired. Also, organic pollution has been blamed for a high occurrence of *S. plymuthica* infections in salmonid farms in Poland since 1996 (Grawinski and Antychowicz, 2001).

Eutrophic waters associated with faecal pollution and high levels of organic material have been attributed to the cause of disease by enteric bacteria, including *Citrobacter freundii* (Austin and Austin, 1999), *Edwardsiella tarda* (Meyer and Bullock, 1973), *Providencia rettgeri* (Bejerano et al., 1979) and *Serratia marcescens* (Baya et al., 1992). In addition, poultry faeces, which was used to fertilise fish ponds, was blamed for mass mortality in silver carp (*Hypophthalmichthys molitrix*) in Israel (Bejerano et al., 1979).

RADIATION

Despite considerable hype particularly in communities around nuclear power stations and the erratic attention of the press, there is not any hard evidence linking radiation pollution in the aquatic environment with health problems among aquatic organisms.

STRESS

Unspecified stressors have been attributed to oxygen deficiency and a statistically significant increase in the incidence of epidermal papillomas, lymphocystis and skin ulcers especially in female common dab from waters around Denmark during the summers of 1988–1993 (Mellergaard and Nielsen, 1995). Also, stress attributed to unnamed environmental factors has been associated with septicaemia in fish from Nigeria (Oladosu et al., 1994).

UNSPECIFIED CAUSES

Many articles have considered the effect of non-specific pollution on the incidence of disease. For example, Vethaak et al. (1996) discussed disease development in flounder contained in mesocosms with contaminated dredged spoil. Compared to clean systems, fish in the polluted environment displayed a higher incidence of lymphocystis and liver damage leading to neoplasia. Yet, there was no appreciable difference in the development of epidermal disease. These results indicated the health problems associated with long-term exposure to pollutants at levels comparable to those in the aquatic environment. In a parallel study, the polluted waters of the Lower

Lake of Bhopal were blamed on the incidence of tumours, i.e. fibromas, in catfish (*Heteropneustes fossilis*) (Qureshi and Prasad, 1995). Dorsal fin tissues obtained from winter flounder (*Pseudopleuronectes americanus*), which were caught from two polluted sites on the eastern seaboard of the USA, were studied by microscopy. The various pathologies observed included epithelial and mucus cell hyperplasia/hypertrophy, spongiosis and focal necrosis. A view was expressed that hypoxia could be involved in the disease process (Bodammer, 2000).

Conclusions

It is well established that pollutants enter the aquatic environment (e.g. Sved et al., 1997), and may be found in the tissues of aquatic vertebrates and invertebrates (e.g. Han et al., 1997). Moreover, some pollutants are instrumental in damaging aquatic organisms (e.g. Lanno and Dixon, 1996). However, there is only limited evidence that pollutants are actually responsible for the development of disease. Indeed, there is negative evidence that has demonstrated that the incidence of disease diminishes when pollution ceases (Sandstrom, 1994). Certainly, many surveys have reported a higher incidence of diseased animals from polluted rather than clean (=control) sites (Vethaak and ap Rheinallt, 1992). Yet, the weakness in most surveys concerns the absence of good quantitative and qualitative data about the actual pollutants. There remains serious doubt of the accuracy of information concerning the relative level of pollutants in polluted and clean sites. Moreover, it is speculative what influence water movement and fish migration patterns have on the results of the surveys. Notwithstanding, there is accumulating evidence that some pollutants immunosuppress or otherwise weaken fish (Sovenyi and Szakolczai, 1993). These weakened animals are more likely to succumb to disease. Some forms of damage, e.g. to gills and skin (Sved et al., 1997), attributable to pollutants are reminiscent of the signs associated with some infectious diseases, i.e. gill disease and ulceration (Austin and Austin, 1999), respectively. In this connection, it is relevant to mention the increasing recognition of atypical isolates of *Aeromonas salmonicida* as a cause of skin lesions/ulceration in native marine fish (e.g. Wiklund and Bylund, 1993; Nakatsugawa, 1994; Pedersen et al., 1994; Wiklund et al., 1994; Wiklund, 1995; Wiklund and Dalsgaard, 1995; Larsen and Pedersen, 1996; Austin et al., 1998). It remains to be proven whether or not this pathogen, which hitherto has been mostly associated with the disease furunculosis in salmonids (Austin and Austin, 1999), interacts mostly with marine fish that may have been already weakened by pollution.

References

Arkoosh, M.R., Casillas, E., Clemons, E., Kagley, A.N., Olson, R., Reno, P. and Stein, J.E. (1998) Effect of pollution on fish diseases: potential impacts on salmonid populations, *Journal of Aquatic Animal Health* **10**, 182–190.

Austin, B. and Stobie, M. (1992) Recovery of *Serratia plymuthica* and *Pseudomonas pseudoalcaligenes* from skin lesions in rainbow trout, *Oncorhynchus mykiss* (Walbaum), otherwise infected with enteric red mouth, *Journal of Fish Diseases* **15**, 541–543.

Austin, B., Austin, D.A., Dalsgaard, I. *et al.* (1998) Characterization of atypical *Aeromonas salmonicida* by different methods, *Systematic and Applied Microbiology* **21**, 50–64.

Austin, B. and Austin, D.A. (1999) *Bacterial Fish Pathogens: Disease of Farmed and Wild Fish.* 3rd (revised) edn, Spinger-Praxis, Godalming, England.

Barron, M.G., Anderson, M.J., Cacela, D., Lipton, J., Teh, S.J., Hinton, D.E., Zelikoff, J.T., Dikkeboom, A.L., Tillitt, D.E., Holey, M. and Denslow, N. (2000) PCBs, liver lesions, and biomarker responses in adult walleye (*Stizostedium vitreum vitreum*) collected from Green Bay, Wisconsin, *Journal of Great Lakes Research* **26**, 250–271.

Baya, A.M., Toranzo, A.E., Lupiani, B., Santos, Y. and Hetrick, F.M. (1992) *Serratia marcescens:* a potential pathogen for fish, *Journal of Fish Diseases* **15**, 15–26.

Bejerano, Y., Sarig, S., Horne, M.T. and Roberts, R.J. (1979) Mass mortalities in silver carp *Hypophthalmichthys molitrix* (Valenciennes) associated with bacterial infection following handling, *Journal of Fish Diseases* **2**, 49–56.

Bernier, J., Brousseau, P., Krzystyniak, K., Tryphonas, H. and Fournier, M. (1995) Immunotoxicity of heavy metals in relation to Great Lakes, *Environmental health Perspectives* **103** (Suppl.), 23–34.

Bucke, D. and Feist, S.W. (1993) Histopathological changes in the livers of dab *Limanda limanda* L., *Journal of Fish Diseases* **16**, 281–296.

Bucke, D., Vethaak, A.D. and Lang, T. (1992) Quantitative assessment of melanomacrophage centres (MMCs) in dab *Limanda limanda* as indicators of pollution effects on the non-specific immune system, *Marine Progress Series* **91**, 193–196.

Bodammer, J.E. (2000) Some new observations on the cytopathology of fin erosion disease in winter flounder *Pseudopleuronectes americanus, Diseases of Aquatic Organisms* **40**, 51–65.

Chan, K.M. (1995) Concentrations of copper, zinc, cadmium and lead in rabbit fish (*Siganus oramin*) collected in Victoria Harbour, Hong Kong, *Marine Pollution Bulletin* **31**, 277–280.

Claveau, R. (1991) Néphrite granulomateuse à *Rhodococcus* spp. dans un élevage de saumons de l'Atlantique (*Salmo salar*), *Médicine et Vétérinaire Québec* **21**, 160–161.

Cossarini-Dunier, M. (1987) Effects of the pesticides atrazine and lindane and of manganese ions on cellular immunity of carp, *Cyprinus carpio, Journal of Fish Biology* **31** (suppl.), 67–73.

Couillard, C.M. and Hodson, P.V. (1996) Pigmented macrophage aggregates: a toxic response in fish exposed to bleached-kraft mill effluent?, *Environmental Toxicology and Chemistry* **15**, 1844–1854.

Croce, B., Stagg, R., Everall, N., Groman, D., Mitchell, C. and Owen, R. (1997) Ecotoxicological determination of pigmented salmon syndrome – a pathological condition of Atlantic salmon associated with river pollution, *Ambio* **26**, 505–510.

Das, M.C. and Shrivastava, A.K. (1984) Fish mortality in Naini Tal Lake (India) due to pollution and parasitism, *Hydrobiology Journal* **20**, 60–64.

Depledge, M.H. (1996) Genetic ecotoxicity: an overview, *Journal of Experimental Marine Biology and Ecology* **200**, 57–66.

Dethlefsen, V. and Watermann, B. (1980), Epidermal papillomas of North Sea dab, *Limanda limanda:* histology, epidemiology and relation to dumping from TiO$_2$ industry, *ICES Special Meeting on Diseases of Commercially Important Marine Fish and Shellfish* **8**, 1–30.

Dethlefsen, V., Lang, T. and Koves, P. (2000) Regional patterns in prevalence of principal external diseases of dab *Limanda limanda* in the North Sea and adjacent areas 1992–1997, *Diseases of Aquatic Organisms* **42**, 119–132.

Dethlefsen, V., Von Westernhagen, H. and Cameron, P. (1996), Malformations in North Sea pelagic fish during the period 1984–95, *ICES Journal of Marine Sciences* **53**, 1024–1035.

Dethlefsen, V., Watermann, B. and Hoppenheit, M. (1987) Diseases of North Sea dab (*Limanda limanda* L.) in relation to biological and chemical parameters, *Archives für Fishereiwissenschaft* **37**, 101–237.

Dudley, D.J., Guentzel, M.N., Ibarra, M.J., Moore, B.E. and Sagik, B.P. (1980), Enumeration of potentially pathogenic bacteria from sewage sludges, *Applied and Environmental Microbiology* **39**, 118–126.

Dulin, M.P. (1979) A review of tuberculosis (mycobacteriosis) in fish, *Veterinary Medicine/ Small Animal Clinician May* 1979, 735–737.

El-Matbouli, M. and Hoffmann, R.W. (2002) Influence of water quality on the outbreak of proliferative kidney disease – field studies and exposure experiments, *Journal of Fish Diseases* **25**, 459–467.

Evans, M.I., Symens, P. and Pilcher, C.W.T. (1993) Short-term damage to coastal bird populations in Saudi Arabia and Kuwait following the 1991 Gulf War marine pollution, *Marine Pollution Bulletin* **27**, 157–161.

Evans, M.R., Larsen, S.-J., Riekerk, G.H.M. and Burnett, K.G. (1997) Patterns of immune response to environmental bacteria in natural populations of the red drum, *Sciaenops ocellatus* (Linnaeus), *Journal of Experimental Marine Biology and Ecology* **208**, 87–105.

Everaarts, J.M., Sleiderink, H.M., Den Besten, P.J., Halbrook, R.S. and Shugart, L.R. (1994) Molecular responses as indicators of marine pollution: DNA damage and enzyme induction of *Limanda limanda* and *Asterias rubens*, *Environmental Health Perspectives* **102**, 37–43.

Fitzsimons, J.D. (1995) A critical review of the effects of contaminants on early stage (ELS) mortality of lake trout in the Great Lakes, *Journal of Great Lakes Research* **21** (Suppl.), 267–276.

Fletcher, G.L., King, M.J., Kicenuik, J.W. and Addison, R.F. (1982) Liver hypertrophy in winter flounder following exposure to experimentally oiled sediments, *Comparative Biochemistry and Physiology* **C73**, 457–462.

Gassman, N.J., Nye, L.B. and Schmale, M.C. (1994) Distribution of abnormal biota and sediment contaminants in Biscayne Bay, Florida, *Bulletin of Marine Science* **54**, 929–943.

Goldberg, E.D. (1995) Emerging problems in the coastal zone for the twenty-first century, *Marine Pollution Bulletin* **31**, 152–158.

Grawinski, E. and Antychowicz, J. (2001) The pathogenicity of *Serratia plymuthica* for salmonid fish, *Medycyna Weterynaryjna* **57**, 187–189.

Grinwis, G.C.M., Besselink, H.T., van den Brandhof, E.J., Bulder, A.S., Engelsma, M.Y., Kuiper, R.V., Wester, P.W., Vaal, M.A., Vethaak, A.D. and Vos, J.G. (2000) Toxicity of TCDD in European flounder (*Platichthys flesus*) with emphasis on histopathology and cytochrome P450 1A induction in several organ systems, *Aquatic Toxicology* **50**, 387–401.

Guiney, P.D., Cook, P.M., Casselman, J.M., Fitzsimons, J.D., Simonin, H.A., Zabel, E.W. and Peterson, R.E. (1996) Assessment of 2,3,7,8-tetrachlorodibenzo-*p*-dioxin induced sac fry mortality in lake trout (*Salvelinus namaycush*) from different regions of the Great Lakes, *Canadian Journal of Fisheries and Aquatic Sciences* **53**, 2080–2092.

Haensly, W.E., Neff, J.M., Sharp, J.R., Morris, A.C., Bedgood, M.F. and Beom, P.D. (1982) Histopathology of *Pleuronectes platessa* L. from Aber Wrac'h and Aber Benoit, Britanny, France: long-term effects of the Amoco Cadiz crude oil spill, *Journal of Fish Diseases* **5**, 365–391.

Han, B.C., Jeng, W.L., Jeng, M.S., Kao, L.T., Meng, P.J. and Huang, Y.L. (1997) Rock-shells (*Thais clavigera*) as an indicator of As, Cu and Zn concentration on the Putai Coast of the Black-Foot disease area in Taiwan, *Archives of Environmental Contamination and Toxicology* **32**, 456–461.

Hanson, L.A. and Grizzle, J.M. (1985) Nitrite-induced predisposition of channel catfish to bacterial diseases, *Progressive Fish Culturist* **47**, 98–101.

Hellou, J., Warren, W.G., Payne, J.F., Belkhode, S. and Lobel, P. (1992) Heavy metals and other metals in three tissues of cod *Gadus morhua* from the northwest Atlantic, *Marine Pollution Bulletin* **24**, 452–458.

Horiguchi, T., Shiraishi, H., Shimizu, M., Yamazaki, S. and Morita, M. (1995) Imposex in Japanese gastropods (*Neogastropoda* and *Mesogastropoda*): effects of tributyltin and triphenyltin from anti-fouling paints, *Marine Pollution Bulletin* **31**, 402–405.

Houlihan, D.F., Costello, M.J., Secombes, C.J., Stagg, R. and Brechin, J. (1994) Effects of sewage sludge exposure on growth, feeding and protein synthesis of dab (*Limanda limanda* (L.)), *Marine Environmental Research* **37**, 331–353.

Huss, H.H., Pedersen, A. and Cann, D.C. (1974) The incidence of *Clostridium botulinum* in Danish trout farms. 1. Distribution in fish and their environment, *Journal of Food Technology* **9**, 445–450.

Ibe, A.C. and Kullenberg, G. (1995) Quality assurance-quality control (QA-QC) regime in marine pollution monitoring programmes: The GIPME perspective, *Marine Pollution Perspective* **31**, 209–213.

Jacquez, G.M., Ziskowski, J. and Rolfe, F.J. (1994) Criteria for the evaluation of alternative environmental monitoring variables: Theory and an application using winter flounder (*Pleuronectes americanus*) and Dover sole (*Microstomus pacificus*), *Environmental Monitoring and Assessment* **30**, 275–290.

Jeney, Z., Valtonen, E.T., Jeney, G. and Jokinen, E.M. (2002) Effect of pulp and paper mill effluent (BKME) on physiological parameters of roach (*Rutilus rutilus*) infected by the digenean *Rhipidocotyle fennica*, *Folia Parasitologica* **49**, 103–108.

Joseph, K.O. and Srivastava, J.P. (1993) Mercury in finfishes and shellfishes inhabiting Ennore estuary, *Fish Technology* **30**, 15–118.

Khan, R.A. (1987) Crude oil and parasites of fish, *Parasitology Today* **3**, 99–100.

Khan, R.A. (1991) Influence of concurrent exposure to crude oil and infection with *Trypanosoma murmanensis* (Protozoa: Mastigophora) on mortality in winter flounder, *Pseudopleuronectes americanus*, *Canadian Journal of Zoology* **69**, 876–880.

Khan, R.A. and Thulin, J. (1991) Influence of pollution on parasites of aquatic animals, *Advances in Parasitology* **30**, 201–238.

Kirk, R.S. and Lewis, J.W. (1993) An evaluation of pollutant induced changes in the gills of rainbow trout using scanning electron microscopy, *Environmental Technology* **14**, 577–585.

Koehler, A., Alpermann, T., Lauritzen, B. and van Noorden, C.J.F. (2004) *Clonal* xenobiotic resistance during pollution-induced toxic injury and hepatocellular carcinogenesis in liver of female flounder (*Platichthys flesus* (L.)), *Acta Histochemica* **106**, 155–170.

Landsberg, J.H. (1995) Tropical reef-fish disease outbreaks and mass mortalities in Florida, USA: what is the role of dietary biological toxins?, *Diseases of Aquatic Organisms* **22**, 83–100.

Lanno, R.P. and Dixon, D.G. (1996) Chronic toxicity of waterborne thiocyanate to rainbow trout (*Oncorhynchus mykiss*), *Canadian Journal of Fisheries and Aquatic Sciences* **53**, 2137–2146.

Larsen, J.L. and Pedersen, K. (1996) Atypical *Aeromonas salmonicida* isolated from diseased turbot (*Scophthalmus maximus*), *Acta Veterinaria Scandinavia* **37**, 139–146.

Lehtinen, K.-J. (1980) Effects on fish exposed to effluent from a titanium dioxide industry and tested with rotary-flow technique, *Ambio* **9**, 31–33.

Lehtinen, K.-J., Notini, M. and Landler, L. (1984) Tissue damage and parasite frequency in flounders, *Platichthys flesus* chronically exposed to bleached kraft pulp mill effluents, *Annales de Zoologia Fennici* **21**, 23–28.

Lemly, A.D. (1997) Role of season in aquatic hazard assessment, *Environmental Monitoring and Assessment* **45**, 89–98.

Linde, A.R., Klein, D. and Summer, K.H. (2005) Phenomenon of hepatic overload of copper in *Mugil cephalus:* role of metallothionein and patterns of copper cellular distribution, *Basic & Clinical Pharmacology & Toxicology* **97**, 230–235.

Lindesjoo, E. and Thulin, J. (1994) Histopathology of skin and gills of fish in pulp mill effluents, *Diseases of Aquatic Organisms* **18**, 81–93.

Lindesjoo, E., Thulin, J., Bengtsson, B.-B. and Tjarnlund, U. (1994) Abnormalities of a gill cover bone, the operculum, in perch *Perca fluviatilis* from a pulp mill effluent area, *Aquatic Toxicology* **28**, 189–207.

28   B. AUSTIN

Malins, D.C., McCain, B.B., Landahl, J.T. *et al.* (1980) Neoplastic and other diseases in fish in relation to toxic chemicals: an overview, *Aquatic Toxicology* 11, 43–67.

Malins, D.C., McCain, B.B., Myers, M.S. *et al.* (1987) Field and laboratory studies of the aetiology of liver neoplasms in marine fish from Puget Sound, *Environmental Health Perspectives* 71, 5–16.

Mayer, F.L., Woodward, D.L. and Adams, W.J. (1993) Chronic toxicity of pydraul 50E to lake trout, *Bulletin of Environmental Contamination and Toxicology* 51, 289–295.

McVicar, A.H., Bruno, D.W. and Fraser, C.O. (1988) Fish diseases in the North Sea in relation to sewage sludge dumping, *Marine Pollution Bulletin* 19, 169–173.

Mellergaard, S. and Nielsen, E. (1995) Impact of oxygen deficiency on the disease status of common dab *Limanda limanda, Diseases of Aquatic Organisms* 22, 101–114.

Meyer, F.P. and Bullock, G.L. (1973) *Edwardsiella tarda,* a new pathogen of channel catfish (*Ictalurus punctatus*), *Applied Microbiology* 25, 155–156.

Minami, T. (1979) *Streptococcus* sp., pathogenic to cultured yellowtail, isolated from fishes for diets, *Fish Pathology* 14, 15–19.

Miskiewicz, A.G. and Gibbs, P.J. (1994) Organochlorine pesticides and hexachlorobenzene in tissues of fish and invertebrates caught near a sewage outfall, *Environmental Pollution* 84, 269–277.

Möller, H. (1987) Pollution and parasitism in the aquatic environment, *Journal of Parasitology* 17, 353–361.

Moore, M.J., Shea, D., Hilman, R.E. and Stegeman, J.J. (1996) Trends in hepatic tumours and hydropic vacuolation, fin erosion, organic chemicals and stable isotope ratios in winter flounder from Massachusetts, USA, *Marine Pollution Bulletin* 32, 458–470.

Mushiake, K., Muroga, K. and Nakai, T. (1984) Increased susceptibility of Japanese eel *Anguilla japonica* to *Edwardsiella tarda* and *Pseudomonas anguilliseptica* following exposure to copper, *Bulletin of the Japanese Society of Scientific Fisheries* 50, 1797–1801.

Mushiake, K., Nakai, T. and Muroga, K. (1985) Lowered phagocytosis in the blood of eels exposed to copper, *Fish Pathology* 20, 49–53.

Nakatsugawa, T. (1994) Atypical *Aeromonas salmonicida* isolated from cultured shotted halibut, *Fish Pathology* 29, 193–198.

Newton, L.C. and McKenzie, J.D. (1995) Echinoderms and oil pollution: a potential stress assay using bacterial symbionts, *Marine Pollution Bulletin* 31, 453–456.

Noga, E.J., Khoo, L., Stevens, J.B., Fan, Z. and Burkholder, J.M. (1996) Novel dinoflagellate causes epidemic disease in estuarine fish, *Marine Pollution Bulletin* 32, 219–224.

Oladosu, G.A., Ayinla, O.A. and Ajiboye, M.O. (1994) Isolation and pathogenicity of a *Bacillus* sp. associated with a septicaemic condition in some tropical freshwater fish species, *Journal of Applied Ichthyology* 10, 69–72.

O'Neill, J.G. (1981) The humoral immune response of *Salmo trutta* L. and *Cyprinus carpio* L. exposed to heavy metals, *Journal of Fish Biology* 19, 197–306.

Overstreet, R.M. and Howse, H.D. (1977) Some parasites and diseases of estuarine fish in polluted habitats of Mississippi, *Annals of the New York Academy of Science* 298, 427–462.

Pascoe, D. and Cram, P. (1977) The effect of parasitism on the toxicity of cadmium to the three-spined stickleback, *Gasterosteus aculeatus* L., *Journal of Fish Biology* 10, 467–472.

Pedersen, K., Kofod, H., Dalsgaard, I. and Larsen, J.L. (1994) Isolation of oxidase-negative *Aeromonas salmonicida* from diseased turbot, *Scophthalmus maximus, Diseases of Aquatic Organisms* 18, 149–154.

Peters, N., Köhler, A. and Kranz, H. (1987) Liver pathology in fishes from the lower Elbe as a consequence of pollution, *Diseases of Aquatic Organisms* 2, 87–97.

Premdas, P.D. and Metcalfe, C.D. (1994) Regression, proliferation and development of lip papillomas in wild white suckers, *Catostomus commersoni,* held in the laboratory, *Environmental Biology of Fishes* 40, 263–269.

Qureshi, T.A. and Prasad, Y. (1995) An incidence of fibroma in a catfish, *Heteropneustes fossilis* (Bloch) caught from the Lower Lake, Bhopal, *Bionature* 15, 107–111.

Robohm, R.A., Brown, C. and Murchelano, R.A. (1979) Comparison of antibodies in marine fish from clean and polluted waters of the New York Bight: relative levels against 36 bacteria, *Applied and Environmental Microbiology* **38**, 248–257.

Rødsaether, M.C., Olafsen, J., Raa, J., Myhre, K. and Steen, J.B. (1977) Copper as an initiating factor of vibriosis (*Vibrio anguillarum*) in eel (*Anguilla anguilla*), *Journal of Fish Biology* **10**, 17–21.

Sahoo, P.K., Kumari, J. and Mishra, B.K. (2005) Non-specific immune responses in juveniles of Indian major carps, *Journal of Applied Ichthyology* **21**, 151–155.

Sandstrom, O. (1994) Incomplete recovery in a coastal fish community exposed to effluent from a modernized Swedish kraft mill, *Canadian Journal of Fisheries and Aquatic Sciences* **51**, 2195–2202.

Secombes, C.J., Fletcher, T.C., O'Flynn, J.A., Costello, M.J., Stagg, R. and Houlihan, D.F. (1991) Immunocompetence as a measure of the biological effects of sewage sludge pollution in fish, *Comparative Biochemistry and Physiology* **100C**, 133–136.

Secombes, C.J., Fletcher, T.C., White, A., Costello, M.J., Stagg, R. and Houlihan, D.F. (1992) Effects of sewage sludge on immune responses in the dab, *Limanda limanda* (L.), *Aquatic Toxicology* **23**, 217–230.

Siddall, R., Pike, A.W. and McVicar, A.H. (1994) Parasites of flatfish in relation to sewage dumping, *Journal of Fish Biology* **45**, 193–209.

Sovenyi, J. and Szakolczai, J. (1993) Studies on the toxic and immunosuppressive effects of cadmium on the common carp, *Acta Veterinaria Hungarica* **41**, 415–426.

Sved, D.W., Roberts, M.H. Jr. and Van Veld, P.A. (1997) Toxicity of sediments contaminated with fractions of creosote, *Water Research* **31**, 294–300.

Turrell, W.R. (1994) Modelling the Braer oil spill: a retrospective view, *Pollution Bulletin* **28**, 21–218.

Vethaak, A.D. (1992) Diseases of flounder (*Platichthys flesus* L.) in the Dutch Wadden Sea, and their relation to stress factors, *Netherlands Journal of Sea Research* **29**, 257–272.

Vethaak, A.D. and ap Rheinallt, T. (1992) Fish disease as a monitor for marine pollution: the case of the North Sea, *Reviews in Fish Biology and Fisheries* **2**, 1–32.

Vethaak, A.D. and Jol, J.G. (1996) Diseases of flounder *Platichthys flesus* in Dutch coastal and estuarine waters, with particular reference to environmental stress factors. 1. Epizootiology of gross lesions, *Diseases of Aquatic Organisms* **26**, 81–97.

Vethaak, A.D. and van der Meer, J. (1991) Fish disease monitoring in the Dutch part of the North Sea in relation to the dumping of waste from titanium dioxide production, *Chemical Ecology* **5**, 149–170.

Vethaak, A.D., Bucke, D., Lang, T., Wester, P., Johl, J. and Carr, M. (1992) Fish disease monitoring along a pollution transect: a case study using dab *Limanda limanda* in the German Bight, North Sea, *Marine Ecology Progress Series* **91**, 173–192.

Vethaak, A.D., Jol, J.G., Mejiboom, A. *et al.* (1996) Skin and liver diseases induced in flounder (*Platichthys flesus*) after long-term exposure to contaminated sediments in large-scale mesocosms, *Environmental Health Perspectives* **104**, 1218–1229.

Voigt, H.-R. (1994) Fish surveys in the Vaike Vain Strait between the islands of Saaremaa and Muhu, western Estonia, *Proceedings of the Estonian Academy of Science and Ecology* **4**, 128–135.

Vos, J.G., Van Loveren, H., Wester, P.W. and Vethaak, A.D. (1989) Toxic effects of environmental chemicals on the immune system, *Trends in Pharmacological Science* **10**, 289–293.

Westfall, B.A. (1945) Coagulation film anoxia in fishes, *Ecology* **26**, 283–287.

Wiklund, T. (1995) Virulence of 'atypical' *Aeromonas salmonicida* isolated from ulcerated flounder *Platichthys flesus, Diseases of Aquatic Organisms* **21**, 145–150.

Wiklund, T. and Bylund, G. (1993) Skin ulcer disease of flounder *Platichthys flesus* in the northern Baltic Sea, *Diseases of Aquatic Organisms* **17**, 165–174.

Wiklund, T. and Dalsgaard, I. (1995) Atypical *Aeromonas salmonicida* associated with ulcerated flatfish species in the Baltic Sea and in the North Sea, *Journal of Aquatic Animal Health* **7**, 218–224.

Wiklund, T., Dalsgaard, I., Erola, E. and Olivier, G. (1994) Characteristics of 'atypical', cytochrome oxidase-negative *Aeromonas salmonicida* isolated from ulcerated flounders *Platichthys flesus, Journal of Applied Bacteriology* **76**, 511–520.

Wood, A.W., Johnston, B.D., Farrell, A.P. and Kennedy, C.J. (1996) Effects of didecyldimethylammonium chloride (DDAC) on the swimming performance, gill morphology, disease resistance, and biochemistry of rainbow trout (*Oncorhynchus mykiss*), *Canadian Journal of Fisheries and Aquatic Sciences* **53**, 2424–2432.

Ziemann, D.A., Walsh, W.A., Saphore, E.G. and Fulton-Bennett, K. (1992) A survey of water quality characteristics of effluent from Hawaiian aquaculture facilities, *Journal of the World Aquaculture Society* **23**, 180–191.

CHAPTER 3

EFFECTS OF IONIZING RADIATION COMBINED WITH OTHER STRESSORS, ON NON-HUMAN BIOTA

RONALD E. J. MITCHEL*, MARILYNE
AUDETTE-STUART AND TAMARA YANKOVICH

*Atomic Energy of Canada Ltd., Chalk River, ON K0J 1J0
Canada*

Abstract: Exposure of organisms in the environment to ionizing radiation is generally considered to be harmful, regardless of the dose. This assumption derives directly from the basic assumption used for human radiation protection, that harm is directly proportional to dose, without a threshold. The consequence of combined exposures is generally unknown, but is assumed to be either additive or multiplicative. We have examined the in vivo and in vitro responses of a variety of cells and organisms. We show that exposure to one stressor can influence the outcome of a subsequent exposure to the same or another stressor. In many cases, pre-exposure to one stressor appeared to induce an adaptive response that mitigated the harm from a second stressor. These observations challenge the basic assumptions used in environmental protection strategies, suggesting that new approaches are needed.

Keywords: combined stressors; radiation; heat; chlorine; non-human biota

Introduction

In environmental toxicology, the general regulatory approach toward setting exposure limits for chemicals or ionizing radiation has been based on the approach used for human exposure, the assumption of a linear increase in risk with increasing levels of exposure, with a threshold for non-carcinogens or without a threshold for carcinogens. For example, ionizing radiation exposure limits are based on a linear-non-threshold model. At low doses, however, correlations between doses and effects are difficult to establish, and the effects of low doses are usually based on extrapolations from high doses. Recently, the general assumption of linearity, with or without a threshold, for most toxic agents has been challenged (Calabrese and Baldwin, 2000, 2003a, 2003b; Parsons, 2000, 2002, 2003; Calabrese, 2004) and it has been argued

31

C. Mothersill et al. (eds.), Multiple Stressors: A Challenge for the Future, 31–38.
© 2007 *Springer.*

that beneficial effects arise from exposure to low levels of a wide variety of agents, including ionizing radiation, that are generally considered to be detrimental at high levels. These beneficial effects arise from adaptive responses and other protective responses induced by the exposure. A recent review of the radiation-induced adaptive response documented research reports, using cells from bacteria, yeast, alga, nematodes, fish, plants, insects, amphibians, birds and mammals, including wild deer, rodents or humans, that showed nonlinear radio-adaptive processes in response to low doses of low LET radiation (Mitchel, 2006). Low doses increased cellular DNA double-strand break repair capacity, reduced the risk of cell death, reduced radiation or chemically induced chromosomal aberrations and mutations, and reduced spontaneous or radiation-induced malignant transformation in vitro.

However, predicting the actual effects of environmental radiation or other exposure to stress is complicated by the fact that organisms are often exposed to combinations of stressors. The consequence of such combined exposures is generally unknown, but conservative assumptions are usually made whereby the effects are assumed to be detrimental to the organisms. In addition, the potential negative effects are often assumed to be either additive or multiplicative. We have examined the in vivo and in vitro responses of a variety of organisms, including fungi, fish cells, amphibians and mammals, to physical (radiation, heat) or chemical (chlorine) stressors, in combination with radiation, to test the assumption of detrimental additive or multiplicative effects.

Experimental Observations and Discussion

Early evidence (Mitchel and Morrison, 1982) in a single-cell lower eukaryote indicated that exposure to one stressor (heat) could induce resistance to a subsequent exposure to both heat and ionizing radiation (Fig. 1). The authors also reported that exposure to a dose of ionizing radiation likewise induced resistance to both heat and radiation. This demonstrated that the adaptive response induced by stressors as diverse as heat and ionizing radiation was a basic protective response induced by exposure to a general stress, and was not unique to the stressor. A test (Mitchel et al., 1999) of this principle in mammals (Fig. 2) showed that, likewise, both heat and radiation could induce a protective response that protected the animals by delaying the onset of myeloid leukemia that resulted from a subsequent exposure to a high dose of ionizing radiation.

Recent experimental work conducted by Stuart and Yankovich at AECL provided other tests of this response to combined stressors. The results presented in Figs. 3–6 are yet unpublished. In Fig. 3, the data show the change in micronucleus frequency (MN, a measure of unrepaired DNA double-strand breaks) in catfish cells exposed to either heat stress alone, or

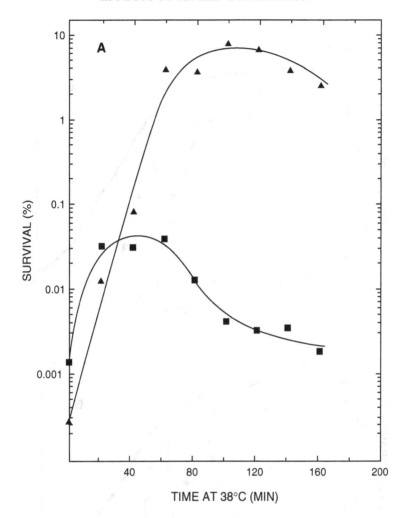

Fig. 1. Heat shock induction of resistance to the lethal effects of heat and ionizing radiation in the yeast *Saccharomyces cerevisiae*. Heat shock at 38°C. Survival after heating at 52°C for 6 min, (▲). Survival after exposure to 3 kGy, (■). (From Mitchel and Morrison, 1982).

heat stress followed by exposure to a large dose (4 Gy) of ionizing radiation. Heat alone resulted in an increase in the MN frequency that was proportional to the temperature. Subsequent exposure to a high dose of radiation resulted in an approximately additive increase in MN frequency, but only in cells exposed to temperatures up to 27°C. When the cells were exposed to 35°C and then subsequently to a high dose of radiation, there was no significant additive effect, indicating that at 35°C the cells had initiated an adaptive response that protected against the DNA damaging effects of a subsequent radiation exposure. However, it also shows that, while at lower temperatures,

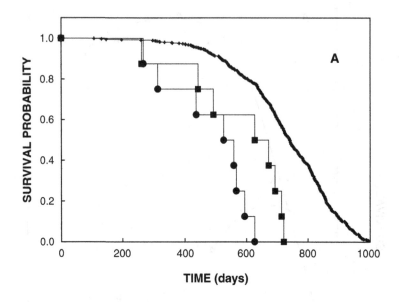

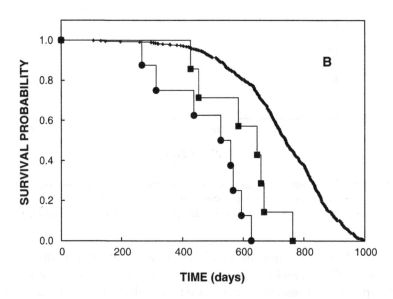

Fig. 2. Influence of adapting treatments on leukemia latency. Panel A: Survival of mice that developed myeloid leukemia after a chronic irradiation (1.0 Gy, 0.5 Gy/h) with (■) or without (•) a prior chronic 0.1 Gy (0.5 Gy/h) exposure, compared to survival of unirradiated control animals that did not develop AML (+). Panel B: Survival of mice that developed myeloid leukemia after a chronic 1.0 Gy irradiation (0.5 Gy/h) with (■) or without (•) prior whole body hyperthermia (40.5°C, 60 min) compared to unirradiated control animals that did not develop AML (+) (From Mitchel et al., 1999).

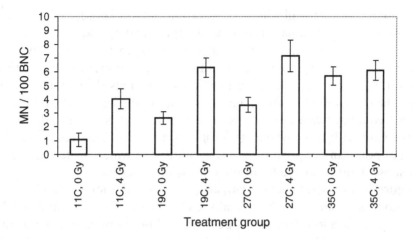

Fig. 3. Micronucleus (MN) frequencies (micronuclei/100 binucleate cells) in catfish B lymphoblasts incubated at different temperatures for 24 h prior to being exposed to a 4 Gy dose of ^{60}Co gamma radiation.

* Unpublished data obtained by M. Audette-Stuart (AECL) and T.L. Yankovich (AECL)

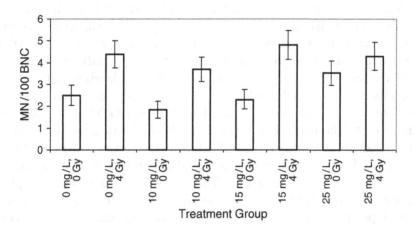

Fig. 4. Micronucleus (MN) frequencies (micronuclei/100 binucleate cells) in catfish B lymphoblasts incubated with different chlorine concentrations for 24 h prior to being exposed to a 4 Gy dose of ^{60}Co gamma radiation.

* Unpublished data obtained by M. Audette-Stuart (AECL) and T.L. Yankovich (AECL)

the effects of heat can be additive to the effects of a high radiation dose, the effects can become less than additive at higher temperatures.

This response of cells to a first stress, which protects against damage from exposure to a second radiation stress, is not restricted to heat. Figure 4 again shows MN formation in catfish cells, but in this case the cells were exposed to various concentrations of chlorine before exposure to a large dose of radiation. Exposure to chlorine concentrations up to 15 mg/L and subsequent exposure to 4 Gy of radiation produced approximately additive increases in MN. This suggests that chlorine concentrations up to 15 mg/L were not perceived as a stress by the cells, and an adaptive response was not initiated. However, when the chlorine concentration was increased to 25 mg/L, which in itself increased the MN frequency, there was no further significant increase in MN frequency when the cells were subsequently exposed to radiation. As was the case with heat, when the initial level of the first stressor exceeded a certain point an adaptive response was initiated which protected the cells from subsequent exposure to another stressor, radiation.

Figure 5 shows the effect on catfish cells when radiation was both the first and second stressor. A high dose alone (4 Gy) significantly increased MN frequency, while a low dose exposure produced no significant increase above control frequency. However, the data show that any low dose up to 250 mGy induced an adaptive response that protected the cells from a subsequent high dose (4 Gy), and the MN frequencies obtained for the combined (sequential) exposure were not different from the MN frequencies seen in control cells not exposed to radiation.

While the above experiments were conducted on laboratory animals and cells, it is important to test whether such protective adaptive responses can be detected in wild populations living in areas where they are exposed to stressors. Wild frogs were collected from three aquatic areas (Fig. 6). Two of the areas contained no significantly elevated levels of radionuclides (non-contaminated areas) and the frogs received only background radiation doses. The third area was contaminated with tritium such that the additional annual dose to the frogs was estimated at 1 mGy/year; above that received from natural background (The low LET contribution to background is also estimated at 1 mGy/year.). All frogs were large adults, indicating they had resided in the ponds for over a year. The high 4 Gy dose of radiation greatly increased MN frequency in cells from frogs taken from uncontaminated ponds. A prior low dose induced a protective response in those cells and protected them from the effects of the subsequent high dose, in a manner consistent with the other stressor data shown in Figs. 1–4. However, the cells taken from the frogs living in the radioactively contaminated pond did not respond in the same way. The large dose of radiation alone did not significantly increase MN frequency. In fact, the frequency was not different from the level in unexposed cells. A low dose prior to the high dose in these cells had no further influence. It appears that living in the pond containing the radioactivity constituted an in vivo stress to the frogs, resulting in the induction of a protective adaptive response that protected their cells from the consequences of further radiation exposure.

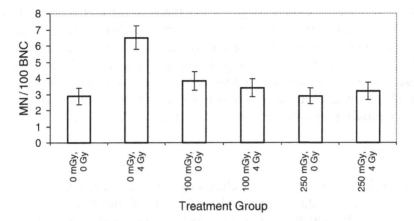

Fig. 5. Micronucleus (MN) frequencies (micronuclei/100 binucleate cells) in catfish B lymphoblasts exposed to various doses of ^{60}Co gamma radiation delivered at a dose rate of 5 mGy/min 24 h prior to exposure to a 4 Gy dose of ^{60}Co gamma radiation.

* Unpublished data obtained by M. Audette-Stuart (AECL) and T.L. Yankovich (AECL)

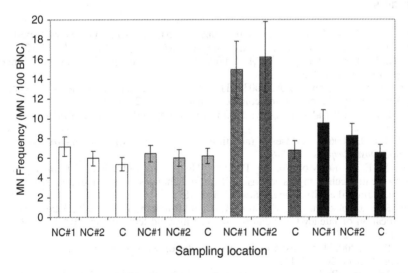

Fig. 6. Micronucleus (MN) frequencies (micronuclei/100 binucleate cells) in Leopard frog hepatocyte cultures. NC indicates cells from frogs living in a non-contaminated pond and C indicates cells from frogs living in a contaminated pond. Livers were removed from the animals and primary cell cultures were established. A fraction of each cell culture was used as controls (white bars), another was exposed to an adaptive dose only (grey bars), a third one was exposed to a high dose (doted bars) and a forth fraction was exposed to an adaptive dose followed, 3 h later, by a high dose (black bars).

* Unpublished data obtained by M. Audette-Stuart (AECL) and T.L. Yankovich (AECL)

The evidence presented suggests that the protective effects induced by exposure to a stressor are evolutionarily conserved, from lower eukaryotes up to mammals, and are therefore likely an important response necessary for life. The induced protective effects are non-specific to the stressor and exposure to one stressor will induce resistance to another stressor. These induced responses occur in cells and in whole animals, and can be detected in wild animals after a low environmental radiation exposure.

Conclusions

The data show examples of the effects of combined stressors in a variety of organisms. The data suggest that the general outcome of combined exposures to different stressors is often neither synergistic nor even additive, and that adaptive responses should be considered in assessing the risk of such exposures. It should be noted that these data are not compatible with either linear or linear non-threshold assumptions of effect versus dose.

References

Calabrese, E. J., 2004, Hormesis: from marginalization to mainstream; A case for hormesis as the default dose-response model in risk assessment. *Toxicol. Appl. Pharmacol.* **197**: 125–136.

Calabrese, E. J. and Baldwin L. A., 2000, The effects of gamma rays on longevity. *Biogerontology* **1**: 309–319.

Calabrese, E. J. and Baldwin L. A., 2003a, The hormetic dose-response model is more common than the threshold model in toxicology. *Toxicol. Sci.* **71**: 246–250.

Calabrese, E. J. and Baldwin L. A., 2003b, Toxicology rethinks its central belief. *Nature* **421**: 691–692.

Mitchel, R. E. J., 2006, Low doses of radiation are protective *in vitro* and *in vivo*: Evolutionary origins. *Dose Response* **4**: 75–90.

Mitchel, R. E. J. and Morrison D. P., 1982, Heat-shock induction of ionizing radiation resistance in saccharomyces cerevisiae. Transient changes in growth cycle distribution and recombinational ability. *Radiat. Res.* **92**: 182–187.

Mitchel, R. E. J., Jackson, J. S., McCann, R. A. and Boreham, D. R., 1999, Adaptive response modification of latency for radiation-induced myeloid leukemia in CBA/H mice. *Radiat. Res.* **152**: 273–279.

Parsons, P. A., 2000, Hormesis: an adaptive expectation with emphasis on ionizing radiation. *J. Appl. Toxicol.* **20**: 103–112.

Parsons, P. A., 2002, Radiation hormesis: challenging LNT theory via ecological and evolutionary considerations. *Health Phys.* **82**: 513–516.

Parsons, P. A., 2003, Metabolic efficiency in response to environmental agents predicts hormesis and invalidates the linear no-threshold premise: ionizing radiation as a case study. *Crit. Rev. Toxicol.* **33**: 443–449.

CHAPTER 4

ECOTOXICOLOGY – HOW TO ASSESS THE IMPACT OF TOXICANTS IN A MULTI-FACTORIAL ENVIRONMENT?

HELMUT SEGNER

Centre for Fish and Wildlife Health, Vetsuisse Faculty, University of Bern, Switzerland. email: helmut.segner@itpa.unibe.ch

Abstract: Ecotoxicology assesses the fate of contaminants in the environment and contaminant effects on constituents of the biosphere. With respect to effects assessment, current ecotoxicology uses mainly reductionistic approaches. For concluding from the reductionistic approach to the effects of toxicant exposure in a multifactorial world, ecotoxicology relies on extrapolations: (i) from suborganism and organism effect levels, as determined in laboratory tests, to ecological levels, (ii) from few laboratory test species to the broad range of species and their interactions in the ecosystem and (iii) from the analysis of the effects of single toxicants under standardized laboratory settings to the toxicant response under real world conditions, where biota are exposed to combinations of chemical, biological and physical stressors. The challenge to ecotoxicology is to identify strategies and approaches for reducing uncertainty and ignorance being inherent to such extrapolations. This chapter discusses possibilities to improve ecotoxicological risk assessment by integrating mechanistic and ecological information, and it highlights the urgent need to develop concepts and models for predicting interactions between multiple stressors.

Keywords: ecotoxicology; risk assessment; effect propagation; interspecies extrapolation; multiple stressors

Introduction

Ecotoxicology is the science of contaminants in the biosphere and their effects on constituents of the biosphere (Newman, 1998). This scientific field draws from many disciplines, for instance, from chemistry for analysing and predicting fate and transport of chemicals in the environment, from toxicology for studying mechanisms of adverse effects of chemicals in organisms, or from ecology for assessing ecological consequences of chemical pollution.

C. Mothersill et al. (eds.), Multiple Stressors: A Challenge for the Future, 39–56.
© 2007 *Springer.*

Whereas the scope of ecotoxicology is well defined on the side of environmental chemistry, some inconsistency exists on the toxicological side. Contrary to toxicology which is concerned with effects of chemicals at the level of the individual organism and its constituent parts, ecotoxicology in principal aims to assess toxic impact at the ecological rather than the individual level. In practice, however, ecological assessment is often lacking (Forbes and Forbes, 1994), instead classical toxicological studies predominate so that ecotoxicology seems to differ from human toxicology mainly by using a broader array of target species. The bias of ecotoxicology towards toxicological rather than ecological studies is not an intended one, but reflects practical, methodological as well as conceptual limitations.

Ecotoxicology is a relatively new scientific discipline (Jorgensen, 1998). During the 1950s and 1960s, public awareness was increasing that anthropogenic substances released into the environment may not just dilute and virtually disappear, but that they could accumulate in biota and man, and may lead to adverse effects. Epidemics such as, e.g. the Minamata disease in Japan – caused by food web-enrichment of organic mercury – pinpointed to the possible problems arising from environmental pollution. In 1962, Rachel Carson published the book "Silent Spring" drawing attention to the consequences of pesticide accumulation in wildlife. Ecotoxicology as a term was coined, according to Truhaut (1977), in June 1969 during a meeting of a committee of the International Council of Scientific Unions. The first textbook on ecotoxicology was then published in 1977 (Ramade, 1977), still paying much attention to human health, but subsequent textbooks increasingly focused on genuine ecotoxicological issues such as, e.g. species differences in sensitivity, ecological determinants of residues, community toxicity, or ecotoxicology as a "hierarchical science" (e.g. Moriarty, 1983; Forbes and Forbes, 1994; Walker et al., 1996; Newman, 1998). From the beginning, ecotoxicology was strongly driven by managerial and legislative needs. Thus, much emphasis was given to technological goals such as the development of standardized toxicity tests. How well regulatory testing programmes fulfilling the legislative needs do protect ecosystems from long-term, insidious decline has always been debated. It is difficult to judge upon how many pollutant-related environmental problems have been avoided due to the application of regulatory testing programmes; however, there exist a number of examples where conventional ecotoxicological approaches failed to prevent or predict the environmental problems. One such example is endocrine disruption (Sumpter and Johnson, 2006). To improve the ability of retrospective as well as predictive assessment of pollutant impact on the biosphere, ecotoxicology is confronted with a number of challenges, some of which will be addressed shortly in the present communication.

Current Approaches in Ecotoxicological Risk Assessment

A comprehensive review of ecotoxicological risk assessment is beyond the scope of this chapter, instead only a short introduction into certain principles and technologies as they are currently used will be given. Ecotoxicological risk assessment aims to estimate levels of contaminants in environmental compartments, to evaluate effects of pollutants at various levels of biological complexity, and to relate environmental exposure to environmental effects (Newman, 1998; Ahlers and Diderich, 1998; Calow and Forbes, 2003; Bradbury et al., 2004). It can be either retrospective ("from effect to pollutant") or prospective ("from pollutant to effect"; Eggen and Suter, 2007).

The retrospective approach builds on monitoring of existing chemical exposure, bioaccumulation and adverse effects in wildlife. Monitoring can start with observing exposure or bioaccumulation, and then trying to relate this to biological or ecological change, or it can start from the observation of adverse changes in wildlife, and then trying to trace this back to chemicals as cause. A number of factors obscure exposure effect-relationships in field studies, for instance, bioavailability on the exposure side, or the impact of multiple stressors on the effect side. Thus, unequivocal demonstration of cause effect-relations is difficult in field studies. Instead, usually a weight-of-evidence approach has to be taken (Rolland, 2000; Burkhardt-Holm and Scheurer, 2007). The demonstration of causative relationships is supported by the existence of temporal or spatial parallelism between exposure and effects (Downes et al., 2002) as well as by a sound epidemiological design of monitoring studies, although the latter aspect is often neglected in ecotoxicology. A number of technologies help to reveal exposure effect-relationships in retrospective studies, such as bioassay-directed fractionation and biomarkers (Brack, 2003; Segner, 2003). Bioassay-directed fractionation is a procedure combining chemical fractionation and analysis with bioassays in order to identify those chemicals within a complex environmental sample which are responsible for a measured biological activity of the sample (Brack, 2003). An example of this methodology is provided by study of Desbrow et al. (1998) on identification of the chemical nature of the estrogen-active substances in effluents of wastewater treatment plants in UK. Biomarkers are sub-organismic parameters being responsive to chemicals and thus can be used either as indicators of exposure to or effects of chemical substances (Peakall, 1994; Van der Ost, 2003). Well-known examples of biomarkers include cytochrome P4501A, which is induced by chemicals activating the arylhydrocarbon receptor, for instance dioxins, or vitellogenin, which responds to chemicals activating estrogen receptors. The concept of biomarkers has attracted much attention in ecotoxicology, and indeed biomarkers are valuable as indicators of exposure, as early warning signals of long-term or delayed toxicity, or as "signposts" for toxic modes of action, however, they are usually not predictive of adverse effects at higher levels of biological organization (Forbes et al., 2006; Hutchinson et al., 2006).

Predictive ecotoxicological risk assessment aims to estimate environmental concentrations of chemicals, to evaluate the toxic hazard and ecological effects arising from these substances, and the likelihood of adverse effects to occur in exposed biota (Newman, 1998; Calow and Forbes, 2003; Bradbury et al., 2004). Key elements in the predictive approach are the "predicted environmental concentration" (PEC) and the "predicted no effect concentration" (PNEC). A PEC can be estimated from actually measured concentrations in the environment or from mathematical modelling; a PNEC can be derived, for instance, from concentration-response determinations in single species toxicity tests in the laboratory (Ahlers and Diderich, 1998). For risk characterization, the PEC is compared to the PNEC in order to estimate the probability of adverse effects to occur. Relevant effects may range from the suborganism level over population and communities to the landscape scale, and they may vary with the broad range of potential target species in an ecosystem. The advantage of the described approach is that it is straightforward and manageable; its disadvantage is that it appears to be at least sometimes too simplistic thereby missing environmentally relevant aspects of exposure and effect, which arise from the complexity of biological and ecological systems. Shortcomings exist particularly on the effects side. PNEC values are largely based on testing of single substances in acute or (sub)chronic laboratory tests, using a few selected "model" species, and using either suborganism or organism-level endpoints. It is easy to demonstrate toxic effects on suborganism and organism-level endpoints of individuals of single species in the laboratory, but we have insufficient understanding of how to extrapolate from the analysed effect level in the laboratory test to ecological levels (effect propagation), how to extrapolate from few laboratory test species to the broad range of other species being present in the ecosystem (interspecies effect extrapolation), and how to estimate from the analysis of the effects of single toxicants under standardized laboratory settings to the toxicant response of organisms under multifactorial real world conditions, where biota are exposed to combinations of chemicals and non-chemical stressors (multiple stressor extrapolation). It is evident that such extrapolations in ecotoxicological risk assessment bear important uncertainties; however, an additional problem arises from ignorance, i.e. our inability to take unknown processes and variables into account (Hoffman-Riem and Wynne, 2002). For instance, egg shell thinning as a consequence of DDT accumulation was overlooked as long as it was unknown that this is a target of DDT action. For practical as well as for principal reasons, ecotoxicological testing will never be able and does not aim for fully reflecting environmental complexity but will always have to rely on reductionistic approaches. The problem is not the reductionism but that we need to learn more on how stressors and effects are interrelated – across and between species, across levels of biological organization, across time scales – and which processes and parameters are of key importance in determining the effects of toxicant exposure in a multifactorial environment.

Challenges in Ecotoxicological Effects Assessment

Given the limitations as discussed earlier, ecotoxicological effects assessment is confronted with a number of challenges:

- While during the early days of ecotoxicology, environmental pollution was often characterized by high levels of contaminants, acute spills, or dominance of high volume industrial chemicals, the situation has changed nowadays (Eggen et al., 2004). Enhanced regulatory practices and technical measures such as improved water treatment technologies or replacing persistent by more degradable substances successfully reduced overall environmental contamination in industrialized countries. In this situation, risks arising from low dose, chronic exposures are coming into focus. This includes questions on the importance of combined effects of chemicals at low concentrations or on combined effects of chemicals and other stressors such as altered habitat morphology or climate change. Further, although concentrations of "classical" toxicants are decreasing, at the same time, new contaminants are emerging such as pharmaceuticals and personal care products, which often show specific modes of toxic action not reliably detectable by the existing testing concepts and methodologies. The challenge is to clarify whether existing ecotoxicological concepts and tools are sufficient or how they have to be enhanced to be able to assess hazard and risks arising from the actual situation of environmental contamination.

- For many, if not the majority of existing chemicals, the available ecotoxicological information is rather limited. Even for high production volume chemicals, often not more than acute lethality data are available. At the same time, new regulations such as REACH in Europe are demanding more information on ecotoxicological properties of existing substances. It is a challenge to ecotoxicology to generate the required data but not by simply increasing the number of tests as this is confronted with many technical, economical and ethical problems, but by developing integrated testing strategies (Bradbury et al., 2004), which are taking advantage of mechanistic and ecological knowledge. Such a knowledge-based testing scheme would enable targeted testing for better chemical prioritization and hazard identification.

- Existing methodologies and concepts of ecotoxicology, despite the prefix "eco-", still reflect more toxicology than ecology. Although the emphasis on toxicology may be understandable from the historical development and pressing regulatory needs (see earlier), doubt remains that this approach might be too simplistic to inform ecological risk assessment (Calow and Forbes, 2003). The challenge is to improve this situation and to find concepts and tools that on the one hand are manageable and practical but on the other hand move ecotoxicological risk assessment closer to ecology.

The following discussion cannot provide the solution to the open questions and problems, rather it tries to point out directions to be taken in the development of new concepts and tools for ecotoxicology.

GOING MECHANISTIC

To date, ecotoxicological testing has been rather phenomenological. However, ecotoxicology has to be more than describing that effects occur, but it needs explanatory principles (Newman, 1998). Understanding of how toxic effects occur is important in extrapolation, classification, and diagnosis of effects (Eggen et al., 2004; Miracle and Ankley, 2005; Segner, 2006) and it will reduce the risk to overlook or ignore possible adverse outcome of chemical exposure (Hoffmann-Riem and Wynne, 2002), as it has been the case, for instance, with endocrine disruption (Segner, 2006; Sumpter and Johnson, 2006). Standard ecotoxicological testing relies on apical endpoints which, due to their integrative nature, lend limited insight into causative processes. Thus, to achieve more knowledge on modes of toxic action or toxic mechanisms, additional endpoints have to be considered. Often, molecular and cellular parameters are used for this purpose. However, changing the level of biological analysis does not necessarily mean to move from the description of the effect to the understanding of the underlying mechanism. Actually, the so-called mechanistic research in ecotoxicology often has been descriptive again (Moore, 2002). The same comment applies for the use of specific technologies: it is not the use of a particular technique but it is the study design and interpretation what makes the difference. For instance, during recent years much emphasis has been given to the promises of genomic methodologies in predictive, mechanism-based ecotoxicology, but as put by Miracle et al. (2003) with respect to the use of these techniques: "If a well thought-out approach is neglected during experimental design and data interpretation, then we are simply left with standard toxicology in Technicolor".

There are a number of areas of ecotoxicological effects assessment where knowledge on modes of toxic action is helpful:

- Knowledge on modes of action helps in understanding time and concentration response-relationships (questions of thresholds, hormesis, U-shaped curves of endocrine disrupting compounds, relation between acute and chronic toxicity, concentration-dependent transitions in mode of action, etc.). For instance, the information that a chemical acts through the estrogen receptor pathway explains why this substance induces irreversible (organizational) effects in developing organisms while it induces transient (activational) effects in adults (Segner et al., 2006). Mechanistic knowledge identifies the processes being affected by the chemical and from this knowledge the risk for chronic effects may be inferred, e.g. the development of cancer as a consequence of mutagenic activity of a compound (Eggen et al., 2004).

- Knowing the mode of action by which a chemical substance induces adverse effects helps in extrapolation, both across species and across biological levels. Molecular or cellular processes are often more conserved than processes at higher levels of biological organisation, what facilitates interspecies extrapolation (Segner and Braunbeck, 1998; Miracle and Ankley, 2005; Hutchinson et al., 2006). For instance, an estrogen receptor ligand in man will be also an estrogen receptor ligand in a fish, however, the physiological consequences of the receptor activation differs between man and fish. Further, if we know that a toxicant impacts a specific biological process, we may understand why certain species are more sensitive than others, and which species are at particular risk.

- Knowing modes of action assists in the assessment and prediction of mixture effects (Escher and Hermens, 2002; Eggen et al., 2004). For instance, knowing that a set of chemicals act through the same receptor pathway is important to predict that a mixture of these compounds will behave in an additive way (Silva et al., 2002). This knowledge forms also the basis of the concept of toxic equivalency factors (Safe, 1990). Vice versa, knowing molecular targets of a toxicant helps to identify biological functions at risk and helps to understand why one and the same substance may lead to multiple effects. For instance, estrogen receptors do not only function in reproductive processes, but are involved in a variety of functions (see later).

- Information on modes of action assists in prioritizing, classification and testing of chemicals. Such knowledge provides input for computational methods and structure activity relationships (Schüürmann, 1998). Knowing which processes in the organisms are affected by a substance helps in designing targeted, knowledge-based testing strategies for the substance of concern (Eggen et al., 2004; Hutchinson et al., 2006; Segner,

2006), and this would reduce uncertainty in risk assessment and guard against surprises, as they happened, for instance, in the case of endocrine disruptors (Calow and Forbes, 2003; Segner, 2006).

• Finally, an important spin-off from mechanism-oriented work is the development of diagnostic tools (Eggen and Segner, 2003). Bioassays and biomarkers have been found to be most valuable in assessing environmental contamination. Currently, only a limited set of tools is available, being indicative for rather few stressors or modes of action, but novel technologies such as genomics and proteomics may generate a broader suite of diagnostic tools.

GOING ECOLOGICAL

Ecotoxicologists succeeded in demonstrating pollutant effects at all levels of biological organisation, from molecules to ecosystems; however, they are uncertain how effects at the different levels relate to each other (Newman, 1998). A similar problem is the question of interspecies extrapolation of toxicity data. The principal dilemma of ecotoxicological effects assessment is that we have to make simplifying assumptions and we have to use reductionistic approaches but we do not know if we use the right simplifications and we do not know the rules and conditions for translating the outcome of the simplified approaches to the systems to be protected. To say it with the words of Barnthouse et al. (1987): "There is an enormous disparity between the types of data available for assessment and the type of responses of ultimate interest. The toxicological data usually have been obtained from short-term toxicity tests performed using standard protocols and test species. In contrast, the effects of concern to ecologists performing assessments are those of long-term exposures on the persistence, abundance and/or production of populations". Although this statement has been formulated almost 20 years ago, substantial progress has not been achieved since then. Since for practical reasons, it will be not possible to abandon reductionistic approaches, the question is if these concepts and methods are indeed too simplistic to inform ecological risk assessment or how we could improve them to be more effective (Calow and Forbes, 2003).

Usually ecotoxicologists follow a bottom-up approach, i.e. investigating toxic effects at the suborganism and/or organism-level and then extrapolating to the levels of populations and communities. The reason behind this approach is that, although the changes at the population/community/ecosystem levels of biological organization are the ultimate concern, they are considered to be too complex and too far removed from the causative events to be useful in diagnosis and prediction of toxic effects of chemicals. Instead, the idea is that toxic effects can be measured at the lower levels of biological organization and that these effects are prognostic for higher level consequences. In almost each review and textbook of ecotoxicology, a figure of the biological hierarchy is shown, with toxic effects propagating from the molecular through cellular and organism levels to populations or communities. What is neglected in this thinking is that there exists no linear effect propagation but at each level of the biological hierarchy, new properties emerge which are not predictable from the properties of the level below but which influence the outcome of the toxic impact (Fig. 1). For instance, a toxic cellular effect is not necessarily leading to a toxic response of the organism due to the existence of compensatory mechanisms at the supracellular level (Segner and Braunbeck, 1998). Exactly for this reason, the use of biomarkers to predict ecological effects is questionable (see earlier). Similarly, ecotoxicological test methodology puts emphasis on metrics such as survival, growth and reproduction since these changes in individual fitness are considered to be "ecologically relevant" and to influence directly the status of the population. However, translation of phenotypic variations in life history traits of individuals into demographic changes of the population depends much on the life history strategy of a particular species (Winemiller et al., 1992; Kooijman, 1998). A 50% loss of

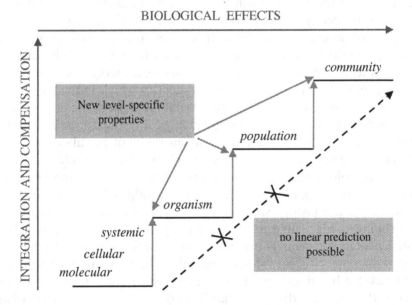

Fig. 1. The propagation of toxic effects along the levels of the biological hierarchy does not follow a linear, deterministic fashion, but the outcome of toxic exposure at a particular level of the hierarchy results on the one hand of integrating toxic and compensatory/protective mechanisms at the lower hierarchical levels and on the other hand it results on properties and processes newly emerging at the hierarchical level of concern.

fertility may have completely different demographic consequences for two species with contrasting life histories, e.g. an opportunistic and a periodic species (Gleason and Nacci, 2001). Laboratory tests often determine toxic effects for the most sensitive life stage, assuming that an adverse effect on this stage will be critical for population status. However, the most sensitive life stage might be not the most crucial factor for maintaining a viable population, since in many species, there is an overproduction of individuals at this stage what easily could compensate for the toxicant-induced losses (Newman, 2001).

This short discussion may already illustrate that inferring higher level effects from qualities of lower levels is problematic (Underwood and Peterson, 1988; Newman, 1998). What is needed to improve predictions across biological levels is to go beyond a linear thinking and to develop a better understanding of the processes and mechanisms how the different levels relate to each other. One possibility to overcome the limitations of current approaches is the use of appropriate modelling (Hutchinson et al., 2006). Models have been suggested for deriving adverse organism responses from cellular responses (Moore, 2002), but particularly also for predicting population responses from responses of individuals. For this it is important not to rely on "black box" modelling but to use physiological and ecological information. One possibility would be to combine toxicity data from laboratory tests with life history from natural populations to model population responses of the species of concern (Boxall et al., 2002). Kooijman (1998) pointed to the value of structured population modelling, in which individuals are not treated as identical copies, but accounts for the fact that individuals differ in many aspects from each other (age, size, sex, energy reserves, genetics, etc.). Modelling can be also valuable to identify which individual changes take a decisive influence on population viability (Grist et al., 2003; Gurney, 2006). Such knowledge could target toxicological testing towards the critical life history traits in order to quantify the sensitivity of population dynamics to changes in these vital traits (Calow and Forbes, 2003). While current testing in ecotoxicology relies on a fixed, pre-selected set of endpoints, a more ecological-oriented approach would select test endpoints on the basis of the analysis of the life history strategy of the species of concern. In this way, ecological knowledge could inform toxicological testing – an approach that to date has been surprisingly rarely used in ecotoxicology. Knowledge-based testing utilizing ecological information might be mutually complementary with knowledge-based testing utilizing mechanistic information (see earlier).

While the previous discussion addressed the problem of extrapolation across biological levels, another problem in informing on ecological risk from conventional single species laboratory tests is the extrapolation of toxic effects across species. When environmental contamination takes place, usually communities are exposed which comprise a large number of species differing in physiology, ecology and toxicant sensitivity. In principal, to protect an ecosystem against adverse effects of toxic substances, it would be necessary to test all the different species occurring in the ecosystem for their sensitivity towards the substance of concern. For practical reasons, however, toxicity testing is usually done only with a limited number of species, and the results from these species are then extrapolated to predict the response of other species and of the whole community. The limitations in this approach are evident. Regulatory risk assessment tries to overcome this limitations by the use of numerical "safety factors", i.e. data from standard toxicity tests are divided by fixed extrapolation factors in order to account for the uncertainties in the extrapolation from simplified laboratory tests to environmental reality and to derive a threshold value (PNEC) below which adverse ecological effects are defined to be unlikely (Ahlers and Diderich, 1998; Calow and Forbes, 2003). However, it is clear that safety factors suffer from serious limitations and represent rather a formal than a scientifically based approach. An alternative approach to extrapolate from single species toxicity tests to communities is the concept of species sensitivity distribution (SSD) (Posthuma et al., 2002). The basic assumption in the SSD concept is that the sensitivities of a set of species can be described by some kind of statistical distribution. The available ecotoxicological data on different species are seen as a subsample from this distribution and are used to calculate a concentration which is considered to be safe for most species. Thus, this concept supports prediction of toxic effects for multiple species assemblages, and it has the potential to incorporate spatial information as well as information on mixture effects (Posthuma and de Zwart, 2005).

Finally, methodologies have been developed to directly test species interactions instead of extrapolating from single species data. One such method attempting to assess ecological complexity is toxicity testing in laboratory microcosms or outdoor mesocosms (Cairns and Cherry, 1993; Newman, 1998). These systems which harbour multiple species assemblage have more ecological realism than do single species laboratory tests and still possess more tractability than do field studies. Micro- and mesocosms harbour multiple species assemblages and therefore are able to perform community impact assessments in relation to toxic exposure. The extent to which such systems represent natural systems remains a subject of debate, however (Williams et al., 2002). Another methodology for using communities in toxicity assessment is pollution-induced community tolerance (PICT) which was introduced by Blanck et al. (1988) as a tool in predictive and retrospective risk assessment. It is based on the assumption that sensitive species within a community will be replaced by more tolerant species after exposure to a toxicant, increasing the tolerance

of the whole community. Experimentally, PICT can be measured by structural
and functional parameters such as species number or photosynthetic activity
(Schmitt-Janssen and Altenburger, 2005).

GOING MULTIPLE

In their environment, organisms are exposed to multiple chemicals and multiple stressors (a stressor is defined here as any factor that extends homeostatic or protective processes beyond the limits of the normal physiological
or ecological range leading to reduced fitness: Sibly and Calow, 1989; Moore,
2002). The questions to ecotoxicologists are how toxic impact is modulated
in the presence of mixtures or in combination with other stressors, and
how the risk resulting from the interaction between chemical, physical and
biological stressors can be assessed and predicted. Exposure to one stressor
may change the response and sensitivity of the biological system to a second
stressor. Such modulations could be of particular relevance under conditions
of chronic, low-dose exposure, when chemical impact alone may provoke
only subtle changes but it may be enhanced by interaction with other stressors. However, it has to be kept in mind that biological systems have evolved
under conditions of fluctuating environments and multiple impacts. Thus,
species have developed adaptive systems which enable ecological success
in the presence of stressors, but these adaptive systems may differ in their
capacities towards specific stressors, what further complicates assessment of
combined effects. Overall, to better account for the multifactorial scenario
to which organisms, populations and communities are exposed to in their
environment, it is necessary to consider the various stressors in an integrated way instead of considering each stressor in isolation. Consequently,
van Straalen (2003) suggested that ecotoxicology should become part of a
broader scaled stress ecology.
     Empirically, stressor interactions have been shown in many studies. For
instance, season and temperature can overlay chemical induction of exposure
biomarkers (Mackay and Lazier, 1993; Behrens and Segner, 2005), the
nutritional status of exposed organisms can modify their sensitivity towards
toxicants (Lanno et al., 1989; Braunbeck and Segner, 1992), or toxicant
exposure can increase the susceptibility to degenerative and/or infectious
diseases (Bayne et al., 1985; Odum, 1985; Heugens et al., 2001; Newman, 2001;
Kiesecker, 2002). The infectious disease triad (Odum, 1985; Newman, 2001)
summarizes such interactions in that it presents the likelihood of disease
as a function of the balance between host, disease agent, and environmental factors. Pollutants, as part of the environmental milieu in which the
host and the pathogen interact, can change the balance between host and
pathogen, for instance, by altering the immunological competence of the

host and thereby rendering the host more susceptible to the pathogen, or, alternatively, by reducing survival of the infective stage of the pathogen in the environment. Vice versa, pathogens can alter response of the host to toxicants in that the diseased organism may be more sensitive to toxic impact than the healthy organism (Carlson et al., 2002; Köllner et al., 2002). Again, however, one must not neglect the capability of the host for adaptive strategies. For instance, Burki et al. (2007) showed that when rainbow trout was exposed simultaneously to a parasite and an environmental estrogen, the response of the fish was dominated by the parasite, while the estrogenic response was largely suppressed.

The multiple stressor issue has different facets in retrospective and in predictive risk assessment. In retrospective studies, a major problem is to disentangle the impact of toxicants from the influence of other factors including natural environmental change. If an adverse change is observed, this may directly result from toxicant action, it may result from factors other than toxic chemicals, or it may result from the combined action of several factors. Epidemiological and weight-of-evidence methodologies are essential to sort out the role of toxicants in complex exposure scenarios (Rolland, 2000; Burkhardt-Holm et al., 2005; Burkhardt-Holm and Scheurer, 2007). For predictive studies, principal concepts or models are needed as it is not realistic to use empirical testing for the infinite number of possible combinations. In mixtures of chemicals, the interactions may lead to amplification (synergism), reduction (antagonism) or additivity of the stand-alone effects of the individual compounds. Several models have been found useful under laboratory settings to predict the combined effects of chemicals in a mixture (e.g. Könnemann, 1981; Hermens et al., 1984; McCarthy et al., 1992; Silva et al., 2002; Altenburger et al., 2003; Monosson, 2005). How complex the assessment of the combined effect of a chemical mixture containing substances with different modes of action can be has been illustrated by Altenburger et al. (2004). The extrapolation from single species laboratory tests on mixture toxicity to the in situ risk of chemical mixtures for an assemblage of species adds further complexity, and requires development of new methodologies (De Zwart and Posthuma, 2005). An additional challenge is to develop concepts and models for predicting interactions between chemical and physical or biological stressors. As indicated above, it is known empirically that physical and biological entities can modulate chemical toxicity and vice versa, however, few attempts have been made to date to quantitatively analyse and to predict such interactions (Folt et al., 1999; Koppe et al., 2006).

Finally, when talking on "going multiple", we should not only focus on exposure of organisms to multiple stressors, but we need also to consider that one and the same toxicant may induce multiple biological responses. This includes not only the fact that toxicants show a transition in mode of action with exposure duration and concentration (Slikker et al., 2004; Schäfers et al., 2007), but it points also to the fact that a chemical can interact with multiple targets in a biological system, what may result in unexpected toxic effects. For instance, studies on estrogen-active substances focused primarily on their effect on sexual and reproductive parameters; however, estrogens have a series of functions beyond the reproductive system. Studies during recent years have revealed that, for instance, estrogen-active environmental substances can interfere with the arylhydrocarbon receptor pathway (Navas and Segner, 2001; Cheshenko et al., 2007), the growth hormone/insulin-like growth factor system (Berishvili et al., 2006; Filby et al., 2006), with the immune system (Segner et al., 2006) or with the neurosensory system (Kallivretaki et al., 2007) and thereby are able to disrupt a broad array of target systems.

Acknowledgements

The author gratefully acknowledges the financial support by the EU projects EDEN (Endocrine Disrupters: Exploring Novel Endpoints, Exposure, Low Dose and Mixture Effects in Humans, Aquatic Wildlife and Laboratory Animals) and MODELKEY (Models for Assessing and Forecasting the Impact of Environmental Key Pollutants on Marine and Freshwater Ecosystems and Biodiversity).

References

Ahlers J, Diderich R. 1998. Legislative perspective in ecological risk assessment. In: Schüürmann G, Markert B (eds). Ecotoxicology – ecological fundamentals, chemical exposure and biological effects. John Wiley & Sons/Spektrum Akademischer Verlag, New York/Heidelberg. pp. 841–868.

Altenburger R, Nendza M, Schüürmann G. 2003. Miture toxicity and its modeling by quantitative structure activity relationships. Environ Toxicol Chem 22:1900–1915.

Altenburger R, Walter H, Grote M. 2004. What contributes to the combined effect of a complex mixture? Environ Sci technol 38:6353–6362.

Bayne BL, Brown DW, Burns K, Dixon DR, Ivanovici A, Livingstone DR, Lowe DM, Moore MN, Stebbing ARD, Widdows J. 1985. The effects of stress and pollution on aquatic animals. Praeger, New York. 384 pp.

Barnthouse LW, Suter II GW, Rosen AE, Beauchamp JJ. 1987. Estimating responses of fish populations to toxic contaminants. Environ Toxicol Chem 6:811–824.

Behrens A, Segner H. 2005. Cytochrome P4501A induction in brown trout exposed to small streams of an urbanised area: results of a five-year-study. Environ Poll 136:231–242.

Berishvili G, D'Cotta H, Baroiller JF, Segner H, Reinecke M. 2006. Differential expression of IGF-I mRNA and peptide in the male and female gonad during early development of a bony fish, the tilapia Oreochromis niloticus. Gen Comp Endocrinol 146:204–210.

Blanck H, Wängberg SH, Molander S. 1988. Pollution-induced community tolerance. A new ecotoxicological tool. In: Cairns JJ, Pratt JR (eds). Functional testing of aquatic biota for estimating hazards of chemicals. American Society for Testing and Materials, Philadelphia, PA, USA, pp. 219–230.

Boxall AB, Brown CD, Barrett KL, 2002. Highertier laboratory methods for assessing the aquatic toxicity of pesticides. Pest Manag Sci 58:637–648.

Brack W. 2003. Effect-directed analysis: a promising tool for the identification of organic toxicants in complex mixtures. Anal Bioanal Chem 377:397–407.

Bradbury SP, Feijtel TCJ, van Leeuweb NC. 2004. Meeting the scientific needs of ecological risk assessment in a regulatory context. Environ Sci Technol 38:463A–470A.

Braunbeck T, Segner H. 1992. Pre-exposure temperature acclimation and diet as modifying factors for the tolerance of golden ide (*Leuciscus idus melanotus*) to short-term exposure to 4-chloroaniline. Ecotoxicol Environ Safety 24:72–94.

Brown AR, Riddle AM, Cunningham NL, Kedwards TJ, Shillabeer N, Hutchninson TH. 2003. Predicting the effects of endocrine disrupting chemicals on fish populations. Human Ecol Risk Assess 9:761–788.

Burkhardt-Holm P, Scheurer K. 2007. Application of a weight-of-evidence approach to assess the decline of brown trout (*Salmo trutta*) in Swiss rivers. Aquatic Sciences, 69:51–70.

Burkhardt-Holm P, Giger W, Güttinger H, Ochsenbein U, Peter A, Scheurer K, Segner H, Staub E, Suter MJF. 2005. Where have all the fish gone? Environ Sci Technol 39:441A–447A.

Burki R, Krasnov A, Bettge K, Antikainen M, Burkhardt-Holm P, Wahli T, Segner H. 2007. Combined effects of 17beta-estradiol and the parasite *Tetaracapsuloides bryosalmonae* on rainbow trout. submitted.

Cairns J, Cherry DS. 1993. Fresh water multi-species test systems. In: Calow P (ed). Handbook of Ecotoxicology. Blackwell Scientific, Oxford, UK, pp. 101–118.

Calow P, Forbes VE. 2003. Does ecotoxicology inform ecological risk assessment? Environ Sci Technol 37:146A–151A.

Carlson EA, Li Y, Zelikoff JT. 2002. Exposure of Japanese medaka (*Oryzias latipes*) to benzo(a)pyrene suppresses immune function and host resistance against bacterial challenge. Aquat Toxicol 56:289–301.

Carson R. 1962. Silent spring. Houghton-Mifflin Co. Boston, USA.

Cheshenko K, Brion F, Le Page Y, Hinfray N, Pakdel F, Kah O, Segner H, Eggen RIL. 2007. Expression of zebrafish aromatase *cyp19a* and *cyp19b* genes in response to the ligands of estrogen receptor and aryl hydrocarbon receptor. Toxicol Sci, 96:255–267.

Desbrow C, Routledge EJ, Brighty G, Sumpter JP, Waldock M. 1998. Identification of estrogenic chemicals in STW effluent. 1. Chemical fractionation and in vitro biological screening. Environ Sci Technol 32:1549–1558.

De Zwart D, Posthuma L. 2005. Complex mixture toxicity for single and multiple species: proposed methodologies. Environ Toxicol Chem 24:2665–2672.

Downes BJ, Barmuta LA, Fairweather PG, Keough MJ, Faith DP, Lake PS, Mapstone BD, Quinn GP. 2002. Monitoring ecological impacts. Concepts and practice in flowing waters. Cambridge University Press, Cambridge, UK.

Eggen RIL, Segner H. 2003. The potential of mechanism-based bioanalytical tools in ecotoxicological exposure and effect assessment. Anal Bioanal Chem 377:386–396.

Eggen RIL, Suter MJF. 2007. Analytical chemistry and ecotoxicology – tasks, needs and trends. J Toxicol Environ Health, 70:724–726.

Eggen RIL, Behra R, Burkhardt-Holm P, Escher BI, Schweigert N. 2004. Challenges in ecotoxicology. Environ Sci Technol. 38:58A–64A.

Escher BI, Hermens JLM. 2002. Modes of action in ecotoxicology: their role in body burdens, species sensitivity, QSARs and mixture effects. Environ Sci Technol 36:4201–4217.

Filby AL, Thorpe KL, Tyler CR. 2006. Multiple molecular effect pathways of an environmental oestrogen in fish. J Molec Endocrinol 37:121–134.

Folt CL, Chen CY, Moore MV, Burnaford J. 1999. Synergism and antagonism among multiple stressors. Limnol Oceanogr 44:864–877.

Forbes VE, Forbes TL. 1994. Ecotoxicology in theory and practice. Chapman & Hall, London, UK.

Forbes VE, Palmquist A, Bach L. 2006. The use and misuse of biomarkers in ecotoxicology. Environ Toxicol Chem 25:272–280.

Gleason TR, Nacci DE. 2001. Risks of endocrine-disrupting compounds to wildlife: extrapolating from effects on individuals to population response Hum Ecol Risk Assess 7:1027–1042.

Grist EPM, Wells MC, Whitehouse P, Brighty G, Crane M. 2003. Estimating the effects of 17alpha-ethynylestradiol on populations of fathead minnow: are conventional toxicological endpoints adequate? Environ Sci Technol 27:1609–1616.

Gurney WSC. 2006. Modeling the demographic impact of endocrine disruptors. Environ Health Persp 114(suppl 1):122–126.

Hermens J, Canton H, Steyger N, Wegman R. 1984. Joint toxicity of a mixture of 14 chemicals on mortality and reproduction of *Daphnia magna*. Aquat Toxicol 5:315–322.

Heugens EHW, Hendriks AJ, Dekker T, van Straalen NM, Admiraal W. 2001. A review of the effects of multiple stressors on aquatic organisms and analysis of uncertainty factors for use in risk assessment. Crit Rev Toxicol 31:247–284.

Hoffmann-Riem H, Wynne B. 2002. In risk assessment, one has to admit ignorance. Nature 416:223.

Hutchinson TH, Ankley GT, Segner H, Tyler CR. 2006. Screening and testing for endocrine disruption in fish – biomarkers as signposts, not traffic lights, in risk assessment. Environ Health Persp 114(suppl 1):106–114.

Jorgensen SE. 1998. Ecotoxicological research – historical development and perspectives. In: Schüürmann G, Markert B (eds). Ecotoxicology – ecological fundamentals, chemical exposure and biological effects. John Wiley & Sons/Spektrum Akademischer Verlag,, New York/Heidelberg. pp. 3–15.

Kallivretaki E, Eggen RIL, Neuhauss SCF, Segner H. 2007. Knockdown of *cyp19a1* aromatase gene decreases neuromast number in the lateral line organ of zebrafish embryos. submitted.

Kiesecker JM. 2002. Synergism between trematode infection and pesticide exposure: a link to amphibian limb deformities in nature? Proc Natl Acad Sci USA 99:9900–9904.

Köllner B, Wasserrab B, Kotterba G, Fischer U. 2002. Evaluation of immune functions of rainbow trout (*Oncorhynchus mykiss*) – how can environmental influences be detected? Toxicol Lett 131:83–95.

Könnemann WH. 1981. Fish toxicity tests with mixtures of more than two chemicals, a proposal for a quantitative approach and experimental results. Toxicology 19:229–238.

Kooijman SALM. 1998. Process-oriented descriptions of toxic effects. In: Schüürmann G, Markert B (eds). Ecotoxicology – ecological fundamentals, chemical exposure and biological effects. John Wiley & Sons/Spektrum Akademischer Verlag,, New York/Heidelberg. pp. 483–520.

Koppe JG, Bartonova A, Bolte G, et al. 2006. Exposure to multiple environmental agents and their effects. Acta Paed 95 suppl 453:106–113.

Lanno RP, Hickie BE, Dixon DG. 1989. Feeding and nutritional considerations in aquatic toxicology. Hydrobiologia 188/189:525–531.

Mackay ME, Lazier CB. 1993. Estrogen responsiveness of vitellogenin gene expression in rainbow trout (*Oncorhynchus mykiss*) kept at different temperatures. Gen Comp Endocrinol 89:255–266.

McCarthy LS, Ozburn GW, Smith AD, Dixon DG. 1992. Toxicokinetic modelling of mixtures of organic chemicals. Environ Toxicol Chem 11:1037–1047.

Miracle AL, Ankley GT. 2005. Ecotoxicogenomics: linkages between exposure and effects in assessing risks of aquatic contaminants to fish. Reprod Toxicol 19:321–326.

Miracle AL, Toth GP, Lattier DL. 2003. The path from molecular indicators of exposure to describing dynamic biological systems in an aquatic organisms: microarrays and fathead minnow. Ecotoxicology 12:457–462.

Monosson E. 2005. Chemical mixtures: considering the evolution of toxicology and chemical assessment. Environ Health Persp 113:383–390.

Moore MN. 2002. Biocomplexity: the post-genome challenge in ecotoxicology. Aquat Toxicol 59:1–15.

Moriarty F. 1983. Ecotoxicology. The study of pollutants in ecosystems. Academic Press, New York, USA.

Navas JM, Segner H. 2001. Estrogen-mediated suppression of cytochrome P4501A (CYP1A) expression in rainbow trout hepatocytes: role of estrogen receptor. Chemico-Biological Interactions 138:285–298.

Newman MC. 1998. Fundamentals of ecotoxicology. Sleeping Bear/Ann Arbor Press, Chelsea, USA.

Newman MC. 2001. Population ecotoxicology. John Wiley & Sons, New York.

Odum EP. 1985. Trends expected in stressed ecosystems. BioScience 35:419–422.

Peakall DB. 1994. Biomarkers. The way forward in environmental assessment. Toxicol Ecotoxicol News 1:55–60.

Posthuma L, de Zwart D. 2005. Predicted effects of toxicant mixtures are confirmed by changes in fish species assemblages in Ohio, USA, rivers. Environ Toxicol Chem 25:1095–1105.

Posthuma L, Traas TP, Suter II GW (eds). 2002. Species Sensitivity Distributions in Ecotoxicology. Lewis Publishers, Boca Raton, FL, USA.

Ramade F. 1977. Ecotoxicologie. Masson, Paris, France.

Rolland, RM. 2000. Ecoepidemiology of the effects of pollution on reproduction and survival of early life stages in teleosts. Fish and Fisheries 1:41–72.

Schäfers C, Teigeler M, Wenzel A, Maack G, Fenske M, Segner H. 2007. Concentration- and time-dependent effects of the synthetic estrogen, 17alpha-ethynylestradiol, on reproductive capabilities of zebrafish, *Danio rerio*. J Toxicol Environ Health 70:768–779.

Schmitt-Janssen M, Altenburger R. 2005. Predicting and observing responses of algal communities to photosystem II-herbicide exposure using pollution – induced community tolerance and species sensitivity distributions. Environ Toxicol Chem 24:304–312.

Schüürmann G. 1998. Ecotoxic modes of action of chemical substances. In: Schüürmann G, Markert B (eds). Ecotoxicology – ecological fundamentals, chemical exposure and biological effects. John Wiley & Sons/Spektrum Akademischer Verlag, New York/Heidelberg. pp. 665–749.

Safe S. 1990. Polychlorinated biphenyls (PCBs), dibenzo-p-dioxins (PCDDs), dibenzofurans (PCDFs) and related compounds: environmental and mechanistic considerations which support the development of toxic equivalency factors (TEFs). Crit Rev Toxicol 21:51–88.

Segner H. 2003. The need for integrated programs to monitor endocrine active compounds. Pure Appl Chem 75:2435–2444.

Segner H. 2006. Comment on "Lessons from endocrine disruption and their application to other issues concerning trace organics in the aquatic environment". Environ Sci Technol 40:1084–1085.

Segner H, Braunbeck T. 1988. Cellular response profile to chemical stress. In: Schüürmann G, Markert B (eds). Ecotoxicology – ecological fundamentals, chemical exposure and biological effects. John Wiley & Sons/Spektrum Akademischer Verlag, New York/Heidelberg. pp. 521–569.

Segner H, Eppler E, Reinecke M. 2006. The impact of environmental hormonally active substances on the endocrine and immune systems of fish. In: Reinecke M, Zaccone G, Kapoor BG (eds). Fish Endocrinology. Science Publishers, Enfield (NH). pp. 809–865.

Segner H, Chesne C, Cravedi JP, Fauconneau B, Houlihan D, LeGac F, Loir M, Mothersill C, Pärt P, Valotaire Y. 2001. Cellular approaches for diagnostic effects assessment in ecotoxicology: introductory remarks to an EU-funded project. Aquat Toxicol 53:153–158.

Sibly RM, Calow P. 1989. A life cycle theory of responses to stress. Biol J Linnean Soc 37:101–116.

Silva E, Rajapakse N, Kortenkamp A. 2002. Something from "nothing" – eight weak estrogenic chemicals combined at concentrations below NOEC produce significant mixture effects. Environ Sci Technol 15:1751–1756.

Slikker W, Andersen ME, Bogdanffy MS, Bus JS, Cohen SD, Conolly RB, David RM, Doerre
    NG, Dorman DC, Gaylor DW, Hattin D, Rogers JM, Setzer WR, Swenberg JA, Wallace K.
    2004. Dose-dependent transitions in mechanisms of toxicity. Toxicol Appl Pharmacol
    201:203–225.
Sumpter JP, Johnson AP. 2006. Lessons from endocrine disruption and their applications to
    other issues concerning trace organics in the aquatic environment. Environ Sci Technol.
    39:4321–4332.
Truhaut R. 1977. Ecotoxicology: objectives, principles and perspectives. Ecotoxicol Environ
    Safety 1:151–173.
Underwood AJ, Peterson CH. 1988. Towards an ecological framework for investigating
    pollution. Mar Ecol Progr Ser 46:227–234.
Van der Ost R. 2003. Fish bioaccumulation and biomarkers in environmental risk assessment:
    a review. Environ Toxicol Pharmacol 13:57–149.
van Straalen NM. 2003. Ecotoxicology becomes stress ecology. Environ Sci Technol 37:324A–
    330A.
Walker CH, Hopkin SP, Sibly RM, Peakall DB. 1996. Principles of ecotoxicology. Taylor &
    Francis, London, UK.
Williams P, Whitfield M, Biggs J, Fox G, Nicoelt P, Shillabeer N, Sheratt T, Heneghan P,
    Jepson P, Maund S. 2002. How realistic are outdoor microcosms ? A comparison of the
    biota of microcosms and natural ponds. Environ Toxicol Chem 21:143–150.
Winemiller, KO, Rose, KA. 1992. Patterns of life-history diversification in North American
    fishes: implications for population regulation. Can. J. Fish. Aquat. Sci. 49:2196–2218.

CHAPTER 5

A LAYPERSON'S PRIMER ON MULTIPLE STRESSORS

THOMAS G. HINTON* AND KOUICHI AIZAWA

University of Georgia, Savannah River Ecology Laboratory, Aiken, South Carolina, USA

Abstract: This article introduces the concept of multiple stressors. It has been written for the layperson, in terms that do not require a strong scientific background. It has been written to facilitate scientists' communication with the public and funding agencies about multiple stressors. This article briefly explains several major classes of contaminants whose global dispersal and long-term persistence in the environment might cause them to contribute to multiple stressors. Highlighted is our lack of understanding about the potential interactions among multiple stressors and the need for much additional research. Interactions are explained through a simple example of various plausible responses that an organism might exhibit when exposed to both cadmium and radiation. Our current approach for determining human and ecological risks from contaminants is explained such that the reader is aware of why multiple stressor research is needed. This article stresses the need for a coordinated, multinational, multidisciplinary research plan for multiple stressors.

Keywords: interactions; mixed stressors; multiple contaminants; effects

Introduction

Late in 1997, Dr Larry Zobel, the medical director for the 3 M Corporation, was puzzled by some laboratory analyses he had requested (Fisher, 2005). The analyses were on workers that produced a 3 M chemical, perfluorooctane sulfonate (PFOS). PFOS is a chemical used to make SCOTCHGUARD, and is also found in products as dissimilar as GORETEX, TEFLON, power plant pipe linings and jet engine gaskets. PFOS allows two other disparate

---

*To whom correspondence should be addressed. e-mail: thinton@srel.edu

C. Mothersill et al. (eds.), Multiple Stressors: A Challenge for the Future, 57–69.
© 2007 Springer.

chemicals to bond together; a property that gives it wide commercial appeal. What was puzzling Dr Zobel was not that tiny amounts were showing up in the workers' blood samples, which was to be expected, but that the chemical was showing up in the clean blood samples from control individuals. To try to resolve the issue, 3 M contacted biological supply companies and purchased pooled samples from blood donors that represented some 760 random locations within the USA. PFOS was in every sample. This was perplexing. Dr Zobel then went to the Red Cross and asked for samples from 600 more blood donors throughout the United States. Same result, PFOS was in every sample. He then turned to Europe, where the chemical had never been manufactured, and obtained samples from Belgium, the Netherlands and Germany. Same result, PFOS was in every sample. Dr Zobel's lab went on to test over 1,500 more samples, including some 600 children. They found PFOS in every sample but two, with levels in some children scoring above those found in the 3 M workers. Alarmed, 3 M notified the US Environmental Protection Agency of its findings, and 2 years later, 3 M announced it would cease production of PFOS. University researchers, alerted to the problem, began to look for the chemical in non-human samples. They found it everywhere they looked. PFOS was in polar bears of the Canadian Arctic, the blood of Inuit's in Alaska, cormorants in the Sea of Japan. It seems that everything contained trace quantities of PFOS. In the course of a single human generation, we contaminated virtually all of earth's biological systems with PFOS.

Dr Zobel's story illustrates how small and interconnected our world truly is. It is astonishing to imagine all of the biological mechanisms, the physical routes of transport, and the chemical's environmental resilience required to disperse so thoroughly in such a relatively short amount of time. And yet, PFOS is not unique. Mixtures of chemicals are ubiquitous in the air we breathe, the food we eat and the water we drink. Polychlorinated biphenyls (PCBs), pesticides, endocrine disruptors, heavy metals, radiation....the list goes on and on (Muir et al., 2005). Surprisingly, our methods of determining acceptable levels of contaminants and of calculating a pollutant's risk to humans and the environment do a very poor job of considering contaminant mixtures; instead, we largely study contaminants as if they occurred in isolation (Cory-Slechta, 2005). The long-term human and ecological risks from chronic exposures to contaminant mixtures are not known. A lack of knowledge about complex mixtures of pollutants is among the major challenges facing the environmental sciences (Eggen et al., 2004).

This primer (i) introduces the concept of multiple stressors and briefly explores some major classes of pollutants that have the potential to be within complex mixtures; (ii) provides an overview of how human and ecological risk analyses evaluate pollutants and highlights the difficulties of studying multiple stressors; (iii) explains the concept of interactions among multiple stressors and the need for a multinational, consolidated research effort to understand them.

Some Major Players

We live in a chemically sophisticated world. Better living through chemistry is the reality. Humans are masters at combining chemicals in magical ways to produce goods that truly enrich our lives. The price we pay, however, is that complex mixtures of metals, nicotine, and benzene are found in our blood; PCBs, PAHs and POPs settle in our fat; we inhale pesticides that cling to our house dust; fire retardants are found in breast milk; endocrine disruptors are excreted in our urine (Duncan, 2006). All of this occurs while the ice melts in the arctic from global warming. Exposure to multiple stressors is the rule, not the exception.

The global dispersal of PFOS is not unique; as is evident by the occurrence of several contaminants in what was previously thought to be pristine habitats. The arctic environment is a good example (Bard, 1999; AMAP, 2002; Macdonald et al., 2005; Muir et al., 2005). Although thought to be isolated from industrial processes known to cause pollution, mixtures of contaminants are showing up and impacting arctic wildlife. Eagles, sea otters and Steller sea lions in the Aleutian Islands have elevated levels of the pesticide DDT; sea ducks, walrus and caribou have high levels of cadmium; killer whales in the North Pacific are now considered to be the most contaminated mammals on earth (Ayotte et al., 1995). Polar bears with higher concentrations of PCBs have altered immune responses that are likely to increase the animals' susceptibility to infections (Muir et al., 2005). Research on Greenland bears reveal increasing concentrations of DDT, polybrominated diphenyl ethers (PBDE flame retardants), and a banned insecticide called chlordane. The contaminants appear to be influencing reproductive organs. Testis and ovary length, and length of the baculum, a bone that supports a bear's penis, decreased significantly with increasing concentrations of the contaminants (Sonne et al., 2006).

Some of the major classes of contaminants that persist in the environment, readily disperse, and have been shown to be components of mixed contaminants include:

- Persistent organic pollutants (POPs); including industrial chemicals (e.g., PCBs, brominated flame retardants), byproducts of industrial processes (e.g., dioxins and furans, hexachlorobenzene), and pesticides (e.g., DDT, chlordane, atrazine)

- Metals; especially cadmium and mercury, both of which are released by fossil fuel combustion, waste incineration, and in various mining and metallurgical processes

- Radionuclides (e.g., cesium-137, strontium-90, plutonium); primarily from past atmospheric testing of nuclear weapons, accidents such as Chernobyl, releases from nuclear fuel reprocessing plants in Europe, and the dumping and storage of nuclear waste
- Petroleum hydrocarbons; either originating locally as a result of spills and discharges from shipping, pipelines, oil and gas drilling, or transported long distances via the atmosphere

These pollutants tend to have strong geological stability, accumulate in fatty tissues, have relatively long biological half-times within organisms, and increase in concentration at higher levels of the food chains (Wormley et al., 2004). Generally, they have been shown to produce:

- Reproductive effects; reduced ability to conceive and carry offspring, reduce sperm count, and/or feminization of males.
- Immunological effects; decreased ability to fight off disease
- Neurological and developmental effects; reduced growth and permanent impairment of brain function
- Mutations; that lead to cancer or genomic instability

A World of Contaminants

It appears that in mastering the use of chemicals to improve our lives, we have also mastered the fouling of our own nest; indeed, the nests of all living organisms are impacted by our better living through chemistry. And there are a lot of chemicals. As of August 2005, over 26 million substances had been indexed by the American Chemical Society's Abstracts Service (CAS, 2005). One-third of these (nearly 9 million) were commercially available; however, only 240,000 are regulated by government bodies worldwide (Daughton, 2005). Some 82,000 chemicals are registered for commercial use in the USA alone, and an estimated 2,000 new ones are introduced annually for use in everyday items such as foods, personal care products, prescription drugs, household cleaners, and lawn care products (Duncan, 2006). About 10% of these chemicals are recognized as carcinogens, but only a quarter of the 82,000 chemicals in use in the U.S. have ever been tested for toxicity (Suk and Olden, 2005; Duncan, 2006).

The EPA receives approximately 100 applications per month from companies seeking to introduce new chemicals on the market. With each application, the manufacturer supplies information on production volume, use and environmental release rates; but not a word on toxicity, unless the manufacturer happens to have such data. Critical information such as the chemical's effects, physical properties, and health impacts must come from EPA files or public databases. The burden rests with the EPA to prove a problematic chemical should be restricted. Perhaps it comes as no surprise that since 1979, the EPA has forced restrictions on just nine applications. 82,000 chemicals results in a daunting research task when trying to determine potential effects from their innumerable possible combinations. In vivo testing of all the various combinations of mixtures, at all the conceivable dose levels is impossible from an ethical, economical or pragmatic perspective (Cassee et al., 1998).

Rather than requiring a government agency to test for toxicity, the European Union is taking a different approach (Duncan, 2006). Last year they gave initial approval to a measure called REACH—Registration, Evaluation, and Authorization of Chemicals—which would require companies to prove the substances they market or use are safe, or that the benefits outweigh any risks. The chemical industry and the US government oppose the REACH concept.

Interactions

All organisms are exposed to a diverse mix of chemicals, pollutants, and stressors. However, our regulations for deeming when risks are acceptable come largely from assessment protocols based on the unrealistic assumption that pollutants occur in isolation from each other. This approach prevents us from properly evaluating mixtures of stressors; particularly, it prevents a determination of the potential interactions among pollutants or stressors. It is with these interactions that 1+1 can indeed = 3.

Interactions, in this context, can be explained by a brief discussion of the plausible outcomes that an organism might experience when exposed to both radiation and the metal contaminant, cadmium. Radiation is a well-known mutagenic that damages DNA. Ionizing radiation induces DNA strand breaks. The most potent type of radiation-induced DNA lesion is double-strand breaks (DSBs; reviewed in Ward, 1995). When cells detect DSBs they arrest their cell cycle and attempt to reestablish chromosome integrity. If the cell damage is extensive, a cell may program itself to die via apoptosis, perhaps because the cost of repair is too great or to avoid the risk of mis-repair and propagation of damage to subsequent cell generations (reviewed in Zhivotovsky and Kroemer, 2004). Some of the damage caused by radiation is due to the ionization of water within the body and the formation of free radicals. Free radicals are molecular species with unpaired electrons. Free radicals are very reactive and can damage DNA by oxidative reactions. Mammals have evolved very effective methods of repairing DNA damage; and humans often assist the repair process by consuming antioxidants (e.g. vitamins C and E, green tea, red wine) that provide additional defense against free radicals.

Cadmium (Cd) is a metal. When animals are exposed to metals they increase the production of a protein called metallothionein (MT), which attaches to the metal, making it less toxic. MT also has antioxidant characteristics that reduce the impact of free radicals. Additionally, Cd is known to inhibit DNA repair.

What might the organism's response be, then, if these two pollutants act together? Responses could simply be additive, no interaction. Radiation causes damage, Cd causes damage, the damages are additive (1+1 = 2). If, however, Cd induces the upregulation of MT, and if the MT acts as an antioxidant, then it is plausible that the antioxidants produced from Cd exposure also scavenge the free radicals produced by the radiation. In this scenario, the Cd would provide a protective effect, with the interaction being antagonistic (1 + 1 = 1, or perhaps 0). Alternatively, if Cd inhibits DNA repair, it is plausible that Cd would augment the radiation damage by reducing the efficiency of DNA repair mechanisms. In this case a synergistic effect could occur (1 + 1 = 3, or 5, or 35).

Thus several mechanistic reasons exist for interesting interactions to occur when radiation and metals co-occur, and yet, only by researching these contaminants together would you ever be able to discern if any interaction, good or bad, occurs. The two contaminants seldom exist in isolation. Indeed, metals co-occur with radioactive contaminants 99% of the time at the contaminated sites managed by the US EPA's Superfund program (Table 1). This example illustrates the relevance and importance of studying chemicals as they occur in nature – as mixtures.

Sometimes interactions among pollutants can cause effects to significantly magnify. Yang (2004) examined the effects of Kepone, a fire ant pesticide, and carbon tetrachloride in rats. Carbon tetrachloride is used in the production of

TABLE 1. Percent occurrence of the top five contaminant groups occurring on sites in association with radioactive contamination, averaged across all EPA regions. This analysis was done using a database of sites listed as "currently on the final National Priority List" from the Agency for Toxic Substances and Disease Registry (ATSDR). Thus, of all the sites containing radionuclide contamination, 99% also contained metals and 77% also contained VOC

Contaminant	Metals[1]	VOC[2]	Inorganic[3]	PAH[4]	Pesticides[5]
% occurrence with rad contamination	99%	77%	73%	67%	54%

[1]Lead, Arsenic, Cadmium and Zinc; [2]Volatile Organic Compounds (e.g., Acetone, Benzene, Toluene); [3]Inorganic compounds (e.g., Asbestos, Cyanide, Sulfuric acid, Sulfate); [4]Polyaromatic Hydrocarbons (e.g., Fluorine, Anthracene, Diethyl phthalate); [5]Heptachlor, DDT, and Dieldrin.

refrigeration fluid, propellants for aerosol cans, as a pesticide, as a cleaning fluid, in fire extinguishers, and in spot removers. When both chemicals were administered at low, environmentally relevant doses the two synergistically interacted such that effects were magnified 67-fold.

Interactive effects are not only caused from exposure to multiple contaminants, the phenomenon occurs due to exposure to multiple stressors, and stress can come from a myriad of sources. Stress is an unavoidable aspect of life for all populations in differing degrees and manifestations, and thus an inevitable contributor to risk (Cory-Slechta, 2005). Two recent examples concern amphibians, which have been undergoing worldwide population declines. Relyea (2003) examined interactions when amphibians were exposed to carbaryl, a pesticide, in the presence of predators. Pesticide concentrations from short-term acute exposures that would normally not adversely affect growth or survival proved lethal when the exposure occurred in the presence of predatory stress. The chemical stressor was magnified many fold by the non-chemical stressor of the predator cue. Likewise, Teplitsky et al. (2005) reported a greatly enhanced stressor action of the fungicide fenpropimorph to tadpoles when they were developing in the presence of a predator. The combined action of the predatory stress cue and the low-level fungicide resulted in delayed and smaller maturation beyond exposure to either stressor alone.

With 82,000 chemicals in the environment, it becomes quite plausible that they might not all act independently, but instead impacts to organisms could be influenced by exposure to multiple stressors. The interaction of two or more chemicals is determined in part by which mode/mechanism of toxic action is operative, and points to the necessity of doing research at environmentally relevant dose levels (McCarty and Borgert, 2006). The order of exposure also complicates analyses. The response produced by an exposure to chemical A then B may be different from B then A. Additionally, all environmental contaminants are changed to metabolites or conjugates in the body, and these new products may also have biological activity that may or may not be similar to the parent compound. Thus even a single compound may become a functional mixture (McCarty and Borgert, 2006).

Determining Risks and Acceptable Concentrations

How are risks actually determined? Generally, one pollutant at a time! We study mercury, in isolation. We study cadmium, in isolation; we study yet another four-letter-coded, organic contaminant, in isolation. For each, we develop a dose-response curve, from which we determine a no-observable effect level (NOEL); apply safety factors that account for uncertainty, and then derive exposure limits and permissible levels (Dourson and Patterson,

2003; Cormier et al., 2003; Suter et al., 2004). [For a synopsis on the evolution of the ecological risk assessment framework in the USA see Suter et al. (2003) and Suter (2006)]. We then repeat the process for the next contaminant. Each contaminant studied in isolation, and thus with no possibility of detecting interactive effects.

The focus on individual chemical agents has been a significant first step in toxicological/environmental studies. Studying chemicals in isolation provides necessary information on the pollutants mode of action, or the mechanism whereby it causes an effect. However, it means that we lack adequate data, methods and models to assess risks realistically for most mixtures to which people and the environment are routinely exposed (Suk and Olden, 2005).

In 1996, the US EPA was directed to include chemical mixtures in its assessment of risk for pesticides that have a common mode of action. Because the mixtures are limited to those that have the same mode of action, the consequent effect is often one of additivity for mixtures. Thus to date, mixtures of chemicals have been dealt with legislatively by largely restricting them to classes that are chemically related and using an additive approach to risks (Cory-Slechta, 2005). Under these conditions additivity should not be surprising, given that the approach may be little different from simply increasing the dose of a representative agent acting under the same mode of action. Merely to use an effect summation approach, however, has proven to often underestimate risks. For example, Silva et al. (2002) found that simple summation of the individual effects of eight weak estrogenic chemicals, each administered below the No Observable Effect Concentration (NOEC), underestimated observed effects by a factor of 20.

EPA's most recent guidance for mixtures (US EPA, 2000) is for human health risk characterization. It does not recommend any single approach for mixtures, but provides a number of options for the practitioner to consider. Two other US agencies stressing mixture research are the National Institutes of Environmental Health Sciences (NIEHS) and the Agency for Toxic Substances and Disease Registry (ATSDR). Two guidance documents on the assessment of chemical mixtures have been produced by the ATSDR (US DHHS, 2004a, 2004b), and NIEHS is supporting mixture research (Suk and Olden, 2005). The ATSDR has recently undertaken the development of a series of "Interaction Profiles" for substances most commonly found at EPA Superfund Sites. To compensate for current lack of knowledge, ATSDR applies an additional safety factor of 10 for mixtures of non-cancerous chemicals and 100 for cancerous chemicals.

These significant knowledge gaps represent major complications thwarting both academic investigations of and regulatory approaches to the toxicity of chemical mixtures. For example, ATSDR and EPA recommend using data from similar mixtures as surrogates for the mixture of concern if data are lacking. However, despite the promulgation of guidance for assessing mixtures, clear criteria have yet to be developed for determining when two mixtures are sufficiently similar to use one as a toxicological surrogate for the other. Indeed, there is no generally accepted classification scheme for categorizing toxicological effects or modes/mechanisms of toxic action (McCarty and Borgert, 2006).

All agencies involved recognize that substantial enhancements to experimental and risk assessment methods are needed. It is generally believed that improvements can be achieved by using organism-based uptake, distribution, and elimination modeling, coupled with data from well-defined, model-based in vivo and in vitro experiments analyzed with improved statistical and mathematical protocols. Good examples of modeling approaches to assess the ecological effects from multiple stressors while considering spatial and temporal parameters are provided by Hope (2005) and Nacci et al. (2005), while McCarty and Borger (2006) provide an excellent review of chemical mixtures. Mixed-exposure research will require the development and refinement of mathematical and physiological models that can be used to estimate the effects of stressors on whole body systems. In addition to a historical perspective of assessing the effects of chemical mixtures, Yang et al. (2004) highlight the need for physiologically based pharmacokinetic and pharmacodynamic (PBPK/PD) modeling approaches.

To be successful, substantial improvements are needed in our knowledge of biological mechanisms of toxicity, chemical structure function relationships, and dose-response relationships. Such knowledge may in turn lead to the development of biological screening tools and improve our ability to model exposure-effect relationships. Anderson et al. (2006) provide an example of integrating several approaches to contaminant responses. Their method involves quantifying molecular, biochemical, and cellular responses in individual organisms collected from stressed and less-stressed sites, and along gradients within the sites, in conjunction with chemical, organism, and population measures. Ultimately, they use a dynamic-energy-budget model to analyze growth of individuals and potential impacts to the population. However, in the short term the traditional approach of calculating hazard indices and summing cancer risk estimates is likely to remain the predominant from of mixture risk assessment.

Path Forward

Understanding the interactions among chemical mixtures and multiple stressors is one of the most perplexing and difficult areas of science within toxicology and risk assessment (Suk and Olden, 2005). A multinational, multidisciplinary strategic research plan is needed for chemical mixtures that is coordinated, comprehensive and cogent (Suk and Olden, 2005). There has been a lack of federal leadership related to research on chemical mixtures, a situation that has caused chronic funding problems and hindered the development of a broad-based mutually agreed-on and clearly articulated strategic research plan (Sexton et al., 1995). Consequently, despite significant scientific advances, the field of chemical mixture research can generally be characterized as uncoordinated, unsystematic and under funded; problems that are exacerbated by the complexity of mixture-related exposures (Sexton et al., 1995).

The recommendations of Sexton et al. in 1995 are still appropriate. Toxicologic research should proceed along three parallel and complementary tracks: (i) studies of basic interaction mechanism using simple combinations of important chemicals, with the express objective of developing and refining mechanistically based mathematical models; (ii) studies of the toxicity of high-priority, environmentally relevant mixtures with the express objective of reducing critical scientific uncertainty in health risk assessment; and (iii) studies that examine both constituent interactions and whole-mixture toxicity in simplified artificial mixtures (e.g., the 10 most important chemicals impacting the Arctic or Superfund sites).

Any comprehensive framework that seeks to predict and explain the effects of chemical mixtures must take into account the following (McCarty and Borgert, 2006): the mechanisms of toxicity of the component chemicals, the potential points at which these mechanisms interact, the dose-dependence of both the mechanisms of toxicity and the mechanism of interaction, be designed to be used at various levels of biological organization, and account for species-specific difference in both toxicity and interaction.

Intact animals are probably the only model adequate for evaluating mixed stressors (other than chemicals), such as physical stressors (e.g., extreme cold or heat, exercise), personal factors (e.g., nutritional deficiencies, aging, etc.), hormonal changes (e.g., co-exposure to endocrine disruptors, pregnancy), biological stressors (e.g., infectious agents), and psychological stressors. Intact animals are also required to study reproductive (e.g., fertility, teratological), postnatal development and growth phenomena.

Fish, especially zebrafish (*Danio rerio*) and medaka (*Oryzias latipes*), have several features that make them useful models for evaluating mixed stressors. Fish are accepted model vertebrates for studying genetics, developmental biology, toxicology and human disease (reviews: Shima and Mitani, 2004; Hill et al., 2005). The main advantages of using zebrafish and medaka as models over other fish are their small size, ease of husbandry, and prolific breeding capacity. Unlike other fish, such as salmon and trout, the small size of zebrafish and medaka (approximately 2.5–3 cm long) permits reduced breeding space and reduced husbandry cost. These characteristics are important for the large-scale experiments needed to evaluate the effects of contaminants when exposures are at low, environmentally relevant concentrations. In addition, zebrafish and medaka development have been well characterized (Kimmel et al., 1995; Iwamatsu, 2004). Equally important, their eggs have a transparent membrane which allows aberrations to be easily observed, even during early development prior to hatch (Teuschler et al., 2005). A "see-through" strain of medaka has a transparent body in the adult stage as well, such that organ abnormalities can be observed in living adults (Wakamatsu et al., 2001). This mutant, and transgenic mutants that express green fluorescent proteins (GFP), have been used for several toxicity studies (e.g. Hano et al., 2005; Kashiwada, 2006). Such advantages, in addition to the availability of genome information, make these fish models ideal candidates for addressing the difficult questions surrounding multistressors.

The dispersal of so many pollutants beyond national boundaries suggests that solutions will require a long-term development plan of global perspective (McCarty and Borgert, 2006). Research on chemical mixtures should be a broad, multidisciplinary approach that goes beyond the traditional boundaries between academic disciplines, and beyond the traditional boundaries of independent nations. Collaborative funding and calls for joint proposals among several nations will result in the most rapid and efficient research on this most difficult of problems. Until we better understand the potential interactions from chronic exposure to multiple stressors, Suk and Olden (2005) recommend invoking the Precautionary Principle and erring on the side of caution.

Acknowledgments

Compilation of this review was supported in part by the Environmental Remediation Sciences Division of the Office of Biological and Environmental Research, U.S. Department of Energy through the Financial Assistant Award DE-FC09-96SR18546 to the University of Georgia Research Foundation. Suggestions made on a draft version of the manuscript by Yi Yi and Daniel Coughlin of the Savannah River Ecology Laboratory were much appreciated.

References

AMAP, Arctic Monitoring and Assessment Programme. 2002. *Arctic Pollution*. AMAP: Oslo, Norway. 111 pp.
Anderson, S., G. Cherr, S. Morgan, C. Vines, R. Higashi, W. Bennett, W. Rose, A. Brooks, and R. Nisbet. 2006. Integrating contaminant responses in indicator saltmarsh species. Marine Environ. Research 62:S317–S321.
Ayotte, P., E. Dewailly, S. Bruneau, H. Careau, and A. Vezina. 1995. Arctic air pollution and human health: what effects should be expected? Sci. of Total Environ. 161:529–537.

Bard, S. 1999. Global transport of anthropogenic contaminants and the consequences fro the arctic marine ecosystem. Marine Pollution Bulletin 38:356–379.

CAS. Chemical Abstracts Service, American Chemical Society. 2005. The latest CAS registry number 7 and substance count. (accessed Oct. 2006) http://www.cas.org/cgi-bin/regreport.pl

Cassee, F., J. Groten, P. van Bladeren, and V. Feron. 1998. Toxicological evaluation and risk assessment of chemical mixtures. Crit. Rev. Toxicology. 28:73–101.

Cormier, S., S. Norton, and G. Suter, II. 2003. The U.S. Environmental Protection Agency's stressor identification guidance: A process for determining the probably causes of biological impairments. Hum. Eco. Risk Assess. 9:1431–1443.

Cory-Slechta, D. 2005. Studying toxicants as single chemicals: does this strategy adequately identify neurotoxic risk? Neurotoxicology 26:491–510.

Daughton, C. 2005. "Emerging" chemicals as pollutants in the environment: A 21st century perspective. Renewable Resour. J. 23:6–23.

Dourson, M. and J. Patterson. 2003. A 20-year perspective on the development of non-cancer risk assessment methods. Hum. Eco. Risk Assess. 9:1239–1252.

Duncan, D. 2006. The chemicals within us. National Geographic Society Magazine. October: 116–135.

Eggen, R.I.L., R. Behra, P. Burkhardt-Holm, B.I. Escher, and N. Schweigert. 2004. Challenges in ecotoxicology. Environ. Sci. Technol. 38:58a–64a.

Fisher, D. 2005. How we depend on chemicals. (accessed Oct. 2006). http://fluoridealert.org/pesticides/2005/effect.pfos.class.news.133.htm.

Hano, T., Y. Oshima, T. Oe, M. Kinoshita, M. Tanaka, Y. Wakamatsu, K. Ozato, and T. Honjo. 2005. Quantitative bio-imaging analysis for evaluation of sexual differentiation in germ cells of olvas-GFP/ST-II YI medaka (Oryzias latipes) nanoinjected in ovo with ethinylestradiol. Environ. Toxicol. Chem. 24:70–77.

Hill, A., H. Teraoka, W. Heideman, and R. Peterson. 2005. Zebrafish as a model vertebrate for investigating chemical toxicity. Toxicol. Sci. 86:6–19.

Hope, B.K. 2005. Performing spatially and temporally explicit ecological exposure assessments involving multiple stressors. Hum. Eco. Risk Assess. 11:539–565.

Iwamatsu, T. 2004. Stages of normal development in the medaka Oryzias latipes. Mech. Dev. 121:605–618.

Kashiwada, S. 2006. Distribution of nanoparticles in the see-through medaka (Oryzias latipes). Environ. Health Perspect. 114:1697–1702.

Kimmel, C., W. Ballard, S. Kimmel, B. Ullmann, and T. Schilling. 1995. Stages of embryonic development of the zebrafish. Dev. Dyn. 203:253–310.

McCarty, L. and C. Borgert. 2006. Review of the toxicity of chemical mixtures: theory, policy and regulatory practice. Regul. Toxicol. Pharm. 45:119–143.

Macdonald, R., T. Harner, and J. Fyfe. 2005. Recent climate change in the Arctic and its impact on contaminant pathways and interpretation of temporal trend data. Sci. Total Environ. 342:5–86.

Muir, D., R. Shearer, J. Van Oostdam, S. Donaldson and C. Furgal. 2005. Contaminants in Canadian arctic biota and implications for human health: Conclusions and knowledge gaps. Sci. Total Environ. 351: 539–546.

Nacci, D., M. Pelletier, J. Lake, R. Bennett, J. Nichols, R. Haebler,J. Grear, A. Kuhn, J. Copeland, M. Nicholson, S. Watlers, and W. Munn, Jr. 2005. An approach to predict risks to wildlife populations from mercury and other stressors. Ecotoxicology 14:283–293.

Relyea, R. 2003. Predator cues and pesticides: a double dose of danger for amphibians. Ecol. Appl. 13:1515–1521.

Sexton, K., B. Beck, E. Bingham, J. Brian, D. DeMarini, R. Hertzberg, E. O'Flaherty, and J. Pounds. 1995. Chemical mixtures from a public health perspective: the importance of research for informed decision making. Toxicology 105:429–441.

Shima, A. and H. Mitani. 2004. Medaka as a research organism: past, present and future. Mech. Dev. 121:599–604.

Silva, E., N. Rajapakse, and A. Kortenkamp. 2002. Something from "nothing" – Eight weak estrogenic chemicals combined at concentrations below NOECs produce significant mixture effects. Environ. Sci. Technol. 36:1751–1756.

Sonne, C., P. Leifsson, R. Dietz, E. Born, R. Letcher, L. Hyldstrup, F. Riget, M. Kirkegaard, and D. Muir. 2006. Xenoendocrine pollutants may reduce size of sexual organs in East Greenland polar bears (*Ursus maritimus*) Environ. Sci. Tech. 40:5668–5674.

Suk, W. and K. Olden. 2005. Multidisciplinary research: Strategies for assessing chemical mixtures to reduce risk of exposure and disease. Hum. Eco. Risk Assess. 11:141–151.

Suter II, G. 2006. Ecological risk assessment and ecological epidemiology for contaminated sites. Hum. Eco. Risk Assess. 12:31–38.

Suter II, G., S. Norton, and L. Barnthouse, 2003. The evolution of frameworks for ecological risk assessment from the red book ancestor. Hum. Eco. Risk Assess. 9:1349–1360.

Suter II, G., D. Rodier, S. Schwenk, M. Troyer, P. Tyler, D. Urgan, M. Wellman, and S. Wharton. 2004. The U.S. Environmental Protection Agency's generic ecological assessment endpoints. Hum. Eco. Risk Assess. 10:967–981.

Teplitsky, C., H. Phiha, A. Laurila, and J. Merila. 2005. Common pesticide increases costs of antipredator defenses in *Rana temporaria* tadpoles. Environ. Sci. Technol. 39:6979–6085.

Teushcler, L. C. Gennings, W. Hartley, H. Carter, A. Thiyagarajah, R. Schoeny, and C. Cubison. 2005. The interaction effects of binary mixtures of benzene and toluene on the developing heart of medaka (*Oryzias latipes*). Chemosphere 58:1283–1291.

US DHHS. 2004a. Guidance manual for the assessment of joint toxic action of chemical mixtures. U.S. Department of Health and Human Services, Public Health Service, Agency for Toxic Substances and Disease Registry Washington DC.

US DHHS. 2004b. Interaction profile for persistent chemical found in fish (chlorinated dibenzo-p-dioxins, hexachlorobenzene, p,p-DDE, methylmercury and polychlorinated biphenyls). U.S. Department of Health and Human Services, Public Health Service, Agency for Toxic Substances and Disease Registry Washington DC.

US EPA 2000. Supplementary guidance for conducting health risk assessment of chemical mixtures. EPA/630/R-00/-2. U.S. Environmental Protection Agency, Washington, DC.

Wakamatsu, Y., S. Pristyazhnyuk, M. Kinoshita, M. Tanaka, and K. Ozato. 2001. The see-through medaka: a fish model that is transparent throughout life. Proc. Natl. Acad. Sci. USA. 98:10046–10050.

Ward, J. 1995. Radiation mutagenesis – the initial DNA lesions responsible. Radiat. Res. 142:362–368.

Wormley, D., A. Ramesh, and D. Hood. 2004. Environmental contaminant-mixture effects on CNS development, plasticity and behavior. Toxicol. Appl. Pharmacol. 197:49–65.

Yang, R. S. H. 2004. *Toxicology of Chemical Mixtures*. Academic Press, New York.

Yang, R. S. H., H. El-Masri, R. Thomas, I. Dobrev, J. Dennison Jr., D-S. Bae, J. Campain, K. Liao, B. Reisfeld, M. Andersen, and M. Mumtaz. 2004. Chemical mixture toxicology: from descriptive to mechanistic, and going on to in silico toxicology. Environ. Toxicol. Pharmacol. 18:65–81.

Zhivotovsky, B. and G. Kroemer. 2004. Apoptosis and genomic instability. Nat. Rev. Mol. Cell Biol. 5:752–762.

SECTION 2

MULTIPLE EXPOSURE DATA – WHAT RESPONSES ARE SEEN?

CHAPTER 6

EFFECTS OF MULTIPOLLUTANT EXPOSURES ON PLANT POPULATIONS

STANISLAV A. GERAS'KIN*,[1], ALLA A. OUDALOVA[1], VLADIMIR G. DIKAREV[1], NINA S. DIKAREVA[1] AND TATIANA I. EVSEEVA[2]

[1] Russian Institute of Agricultural Radiology and Agroecology, Kievskoe shosse, 109 km, 249020, Obninsk, Russia
[2] Institute of Biology, Komi Scientific Center, Ural Division RAS, Kommunisticheskaya 28, 167982, Syktyvkar, Russia

Abstract: Results of laboratory, "green-house" and long-term field experiments carried out on different plant species to study ecotoxical effects of low doses and concentrations of most common environmental pollutants are presented. Special attention is paid to ecotoxic effects of chronic low dose exposures, synergistic and antagonistic effects of multipollutant exposure. Plant populations growing in areas with relatively low levels of pollution are characterized by the increased level of both cytogenetic disturbances and genetic diversity. The chronic low dose exposure appears to be an ecological factor creating preconditions for possible changes in the genetic structure of a population. A long-term existence of some factors (either of natural origin or man-made) in the plants environment activates genetic mechanisms, changing a population's resistance to exposure. However, in different radioecological situations, genetic adaptation of plant populations to extreme edaphic conditions could be achieved at different rates. The findings presented indicate clearly that an adequate environment quality assessment cannot rely only on information about pollutant concentrations. This conclusion emphasizes the need to update some current principles of ecological standardization, which are still in use today.

Keywords: radioactive and chemical contamination; multipollutant exposure; bioindication; plant populations; environment quality assessment

---

*To whom correspondence should be addressed. e-mail: stgeraskin@gmail.com

C. Mothersill et al. (eds.), Multiple Stressors: A Challenge for the Future, 73–89.
© 2007 *Springer.*

Introduction to Bioindication Approach

Contamination of the environment has become a worldwide problem. Therefore, a clear understanding of all the dangers posed by environmental pollutants to both human health and ecologic systems is needed. Knowledge of the existence of an environmental stress situation is the prerequisite for its solution or amelioration. With this in mind, considerable efforts have been undertaken to develop effective methods for assessing the quality of the environment.

Generally, two approaches are used. The first is based on chemical-physical techniques for laboratory analysis of air, water and soil samples. At this, an evaluation of true exposure characteristics is complicated; however, since most quantification techniques are capable of recognizing just a specific compound or its metabolites. So, this approach gives only a part of the knowledge necessary to evaluate the harmful potential of pollutants.

The other approach is to score biological effects in animals or plants that could be exposed at contaminated sites. An obvious advantage of this method is a demonstration of the direct results from the pollutants' action on living nature. This limits the usefulness of this approach as an analytical method to investigate the total pollutants' burden, but enhances it ability to measure environmental quality. The use of biomarkers could remove much of the uncertainty associated with current ecological risk assessments and provide meaningful indicators of biological damage. In contrast to the specific nature of assessments on exposure, studies of biological effects integrate the impacts of all the harmful agents, including synergistic and antagonistic effects. The biomarkers may also illuminate previously unsuspected chemical or natural stressors in the study area or reveal damage caused by a pollutant that has since degraded and is no longer detectable by residue analysis. Therefore, this approach is particularly useful for assessing unknown contaminants, complex mixtures, or hazardous wastes.

Certainly, it will never be possible to replace direct physical-chemical measurements of pollutant concentrations entirely by a detection of effects in bioindicators. An increased understanding of fate and exposure pathways of harmful substances in various test-systems is also needed for a better prediction of what has happened in the field. It is obvious that a correct estimation of the environment pollution risk needs to be derived from biological tests and pollutant chemical control in ecosystem compartments. Chemical and biological control methods need to be used simultaneously, which allows an identification of the relationships between the pollutant concentrations and the biological effects that they cause. In turn, such relationships may help in an identification of the contribution from specific pollutants to the overall biological effect observed. The knowledge generated makes it possible to limit an effect of unfavorable factors on biota and predict the further ecological alterations in regions submitted to intensive industrial impact.

To assess the quality of the environment, we mostly used plants as test-objects, for several reasons. Plants are higher eukaryotes and essential components of any ecosystem. Plant species are the most important primary producers and most relevant in the food web of the ecosystem. Plants seem to be especially well suited for an environmental assessment since they have fast growth rates and provide a large number of offspring. Owing to their settled nature, plants are constantly exposed to pollution and, therefore, can characterize the local environment in the best way. Plants do not have a predetermined germline (Walbot, 1985); the germ cells are produced during plant development from somatic cells, and, consequently, mutations occurring during somatic development can be inherited. Furthermore, in many cases, plant bioassays are the simplest and most cost effective among test-systems for an environmental assessment.

Effects of contaminants on biota first appear in the cellular level making cellular responses not only the first manifestation of harmful impact, but also suitable tools for an early and reliable detection of exposure. Cellular changes would initially be less obvious than the direct visible effects of pollutants, but in the long run they could be more significant. At ecological risk assessment, basic subjects for protection are populations, communities, and ecosystems, and thus, the biomarker response should be tightly linked to effects in these biological systems. It is becoming increasingly clear (Theodorakis et al., 1997) that cellular alterations may afterwards influence biological parameters important for populations such as health and reproduction. These types of effects are of special concern because they can manifest themselves long after the source of contamination has been eliminated. Therefore, it is genetic test-systems that should be used for an early and reliable demonstration of the alterations resulting from human industrial activity.

Nonlinearity of Dose-Response Relationship

Effects detected in field observations are usually difficult or impossible to relate to specific contaminants or their sources in the environment because of the influence of noncontaminant-mediated factors. In such cases, laboratory studies may sometimes be important stage for establishing a chain of causality. An exposure to low doses and dose-rates of ionizing radiation is one of the inevitable factors in the current environment, and biota, including man, is chronically subjected to low-level radiation. Understanding the risks of low doses of radiation is also important with regard to various issues such as cancer screening, occupational exposure, frequent-flyer risks, and the future of nuclear power. For example, most radiological examinations produce doses in the range from 3 to 30 mSv (Brenner et al., 2003). Since the concept of radiation protection for humans and biota should be based on a clear comprehension of the consequences of low-level exposure, a correct estimation of the effects of low doses is an important topic.

A review of published and own data showed (Geras'kin, 1995) that the regularities of cytogenetic disturbances at low-level radiation are often characterized by a sound nonlinearity and have universal character. To verify this conclusion, a study was undertaken on cytogenetical damage occurrence in meristem cells of irradiated barley seedlings (Geras'kin et al., 2007). Aberrant cells frequency obtained is presented in Fig. 1 in dependence on dose and shows a deviation of linearity. Cytogenetic damage increases the control level at a dose of 50mGy and above, but stays at the same level in the dose range of 50–500mGy.

How important is the deviation of linearity observed in this study? It could be answered from a comparison of linear and non-linear models on their potential to fit the data obtained. At looking for the best model for data approximation, it is important, however, to get an improvement in the goodness-of-fit not merely by means of a model complicating through additional terms, but achieving a mutual conformity between a biological phenomenon and its mathematical model. Figure 1 illustrates results of the data approximation with three of six examined mathematical models (Geras'kin et al., 2007). The piecewise linear (PL) model (2 in Fig. 1) supposes a non-linearity of a dose dependency through including a dose-independent plateau. For the given data set, the plateau limits are calculated as 83.4 and

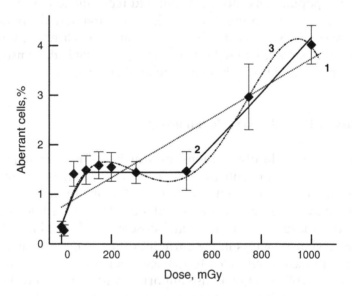

Fig. 1. Frequency of aberrant cells in barley seedlings (mean ± se) exposed to low-radiation doses and approximation of the data with linear (1), piecewise linear (2) and 4th degree polynomial (3) models.

513.7 mGy. The comparison of approximation quality by the most common quantitative criteria (Geras'kin et al., 2007) shows that the PL model statistically surpasses all the other tested variants. In particular, it fits the data significantly better than the linear model, according to the Hayek criteria ($p<5\%$). And this improvement is reached not through model complicating. Indeed, the criterion of structural minimization penalizing a model for any additional free parameter (Geras'kin and Sarapul'tzev, 1993) shows that, despite of five free parameters, the PL model is more advantageous than the linear model with only two parameters.

The current radiological protection practice is based on the linear non-threshold extrapolation of effects into the range of low-level exposures, which has been justified by a lack of reliable data on the effects of low doses. This approach, however, has been widely questioned recently because of many doubts in its consistency with knowledge and last data available. The findings presented here give further evidence that biological response to low dose exposures could essentially deviate of the linear non-threshold dependency. Consequently, it is necessary either scientifically justify the use of the LNT, or, having proved it is not scientifically robust, to develop a new approach that would be fair at different levels of radiation exposure.

Deviations of Additivity at Multipollutant Exposures

Ecosystems will often be polluted with a mixture of pollutants rather than a single pollution. Despite this, and as a result of time and financial constraints, toxicity testing has generally been restricted to studying the effects of single pollutants on a target organism under controlled conditions. Recent research has shown (Howard, 1997) that the synergistic effects among pollutants are much more dramatic than was previously thought. In most environmental situations, potentially harmful substances present at low doses and concentrations; nevertheless, a risk of such impact should not be underestimated as synergetic effects are most often registered at a combination of pollutants at low levels. This was, in particular, demonstrated in our combined-effect studies with a number of common stressors like acute and chronic γ-radiation, heavy metals, pesticides, artificial and heavy natural radionuclides; there were used different plant species such as spring barley, bulb onion, spiderwort and others. Moreover, at certain conditions, nonlinear effects were found to contribute significantly to a plant response (Evseeva et al., 2003a; Geras'kin et al., 2005a).

A study of cytogenetic disturbances induction in intercalary meristem cells of spring barley grown on soil contaminated with radioactive caesium (^{137}Cs) and one of heavy metals, Cd or Pb, can be an example illustrating an importance of non-linear interaction of damage caused by different factors (Geras'kin et al., 2005a). Table 1 presents interaction coefficients calculated

TABLE 1. Interaction coefficient for an endpoint of aberrant cell frequency in root meristem of barley at combined ^{137}Cs + Cd and ^{137}Cs + Pb soil pollution

^{137}Cs, kBq/kg	Cd, mg/kg			Pb, mg/kg		
	2	10	50	30	150	300
4.92	1.84*	1.74**	1.46**	0.82	0.74	0.75
24.6	0.76*	0.63**	0.65	0.61	0.54*	0.53*
49.2	0.36**	0.32**	0.50*	0.52*	0.54**	0.61*

Note: Interaction coefficient differs from 1.
*$p < 5\%$;
**$p < 1\%$.

as a ratio of increments in cytogenetic effect observed at combined pollution, and expected from an additive hypothesis. In a case of ^{137}Cs + Cd soil contamination, the interaction coefficient differs of 1 for eight of nine tested mixtures, so, there are significant deviations of additivity (Table 1). When the lowest ^{137}Cs specific activity of 4.92 kBq/kg was combined with any of the cadmium concentrations, significant synergistic effect was observed. It is of special importance as such ^{137}Cs specific activity occurs in the territories contaminated by the Chernobyl accident. At the ^{137}Cs–Pb combined exposure, the interaction effects are also essential; when these contaminants are applied at high concentrations, the confident antagonisms are registered. These findings show that a forecast of cytogenetic consequences for combined exposures based on an additive model would be incorrect and cause essential deviations from experimentally observed data.

Short-Term Laboratory Studies

The earlier presented examples emphasize that assessments of environmental risks based solely on physical-chemical control methods are often inadequate to an actual situation, and an integrating of bioindication assays could improve the system of ecological monitoring. In most bioindication studies, standard indicator plant species such as *Tradescantia, Allium cepa* or *Vicia faba* were used. Among the test-systems suitable for toxicity monitoring, the *Allium*-test is well known and commonly used in many laboratories. Results from the *Allium*-test have shown a good agreement (Fiskesjo, 1985) with results from other test-systems, eukaryotic as well as prokaryotic. As a genotoxicity test, the *Allium*-based assay of chromosome aberration in anaphase-telophase is for many reasons especially useful for the rapid screening of pollutants posing environmental hazards. In addition, a good toxicity

TABLE 2. *Allium*-test application for the environment quality assessment

Site	Contamination	Results
Radium production industry storage cell territory, Komi Republic, Russia	Heavy natural radionuclides & chemical pollution	All water samples caused a significant increase in the chromosome aberration frequency. Genotoxic effect was a result of chemical toxicity mainly (Evseeva et al., 2003b)
Underground nuclear explosion site, Perm region, Russia	Radionuclides	^{90}Sr significantly contributes to the induction of cytogenetic disturbances (Evseeva et al., 2005)
Radioactive waste storage facility, Obninsk, Russia	Radionuclides & chemicals	All water samples caused a significant increase in the chromosome aberration frequency. Genotoxic effect was a result of chemical toxicity mainly (Oudalova et al., 2006)
Upper Silesian Coal Basin, Poland	Heavy natural radionuclides & chemical pollution	All water and sediment samples caused a significant increase in the chromosome aberration frequency. Ra, Ba, Sr, and Cu contribute significantly to the induction of cytogenetic disturbances
Semipalatinsk Test Site, Kazakhstan	Radionuclides	Data analysis in progress

indicator is given by the mitotic index. In our laboratories, several studies on the environment quality assessment have involved the *Allium*-test, and some results are briefly summarized in Table 2. In all these studies, chemical and biological control methods were applied simultaneously. This helps in identifying relationships between the pollutant concentrations and biological effects they cause.

As an illustration of the *Allium*-test application, the frequency of aberrant cells in root meristem of sprouted bulbs, grown on sediment sampled from the post-mining areas of the Upper Silesia, is presented in Fig. 2. Cytogenetic damage to meristem cells of *Allium cepa* is evident in all non-control variants (1–4) which are significantly higher than the control value of variant 5. So, the clear genotoxic effect of sampled sediment is shown. Also, an important contribution of such severe types of cell damage as chromosome bridges and laggings is demonstrated.

Concentrations of 12 heavy metals and 4 radionuclides are measured in the sediments. All samples contain extremely high concentrations of Ba, from 5 to 50.5 times over the maximum permissible level. In variants 3 and 4, concentrations of Ba, Cu, Mn, and Zn are above the permissible limits. Specific activities of radium nuclides in the samples from the underground gallery (variant 1) and the bank of the Rontok pond (variant 2) are

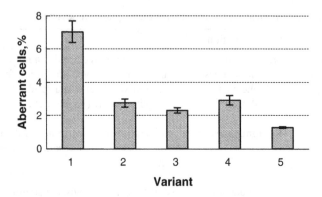

Fig. 2. Aberrant cells frequency in root meristem cells of *Allium cepa*, grown in sediment sampled from the post-mining areas in the Upper Silesia, Poland. Sampling variants: (1) underground galleries, (2–4) settling ponds, (5) control.

over the corresponding values of the minimum significant specific activity. Key pollutants governing biological effect observed are revealed, through an adaptation of mathematical and statistical techniques of multivariate analysis, including correlation analysis, step-by-step inclusion/exclusion of essential predictors, and multivariable linear regression. From preliminary assessments, the best model to describe an aberrant cell's occurrence in root meristem cells of *Allium cepa* germinated on the sediment sampled is

$$Y\ (\%) = (3.32 \pm 0.28) - (0.104 \pm 0.009)[Cu] + (2.10 \pm 0.18)10^{-2}\,[Ba] + (4.94 \pm 0.46)10^{-4}\,[Ra_\Sigma],$$

where [Cu] and [Ba] are concentrations of chemical elements in soil, in mg/kg; $[Ra_\Sigma]$ is total specific activity of Ra isotopes analysed in soil, in Bq/kg; the determination coefficient $R^2 = 60.2\%$; Fisher statistics $F = 49.5$.

Field Studies on Wild Plant Populations and Crops

The other possible way to assess the quality of the environment is the use of plants directly growing in contaminated sites. This approach is particularly useful for assessing long-term ecotoxical effects induced by chronic low dose-rate and multi-pollutant exposure at contaminated sites. Up to now we have known little about responses of plant and animal populations to environmental pollutants in their natural environments. Although radionuclides and heavy metals cause primary damage at the molecular level, there are emergent effects at the level of populations, non-predictable solely from the knowledge of elementary mechanisms of the pollutants' influence. The usefulness of data gathered both in laboratory-based studies and field-based monitoring observations may, therefore, be significantly affected by our present lack of knowledge in this area of environmental research. Previously completed and ongoing field studies on biological effects in different species of wild and agricultural plants are briefly summarized in Table 3.

With each passing year since the Chernobyl accident of 1986, more questions arise about the potential for organisms to adapt to radiation exposure. It is often thought to be attributed to somatic and germline mutation rates in various organisms. Studies into the mechanisms of plant

TABLE 3. Field studies on wild and agricultural plants

Species	Site & Time	Contamination	Assay and/or Endpoints
Winter rye and wheat, spring barley and oats	10 km ChNPP zone (11.7–454 MBq/m²),1986–1989	Radionuclides	Morphological indices of seeds viability, cytogenetic disturbances in intercalar and seedling root meristems (Geras'kin et al., 2003a)
Scots pine, couch-grass	30 km ChNPP zone, (250–2690 µR/h), 1995	Radionuclides	Cytogenetic disturbances in seedling root meristem (Geras'kin et al., 2003b)
Scots pine	Radioactive waste storage facility, Leningrad Region, Russia, 1997–2002	Mixture	Cytogenetic disturbances in needles intercalar and seedling root meristems (Geras'kin et al., 2005b)
Wild vetch	Radium production industry storage cell, Komi Republic, Russia, (73–3300 µR/h), 2003–2006	Heavy natural radionuclides & chemical pollution	Embryonic lethals, cytogenetic disturbances in seedling root meristem (Evseeva et al., 2007)
Scots pine	Sites in the Bryansk Region radioactively contaminated in the Chernobyl accident (451–2344 kBq/m²), 2003–2006	Radionuclides	Cytogenetic disturbances in seedling root meristem, enzymatic loci polymorphism analyses
Scots pine	10 km ChNPP zone (1100 µR/h), 2004	Radionuclides	Morphological modifications in pine needles, cytogenetic disturbances in seedling root meristem
Koeleria gracilis Pers., Agropyron pectiniforme Roem. et Schult.	Semipalatinsk Test Site, Kazakhstan, (74–3160 µR/h), 2005–2006	Radionuclides	Cytogenetic disturbances in seedling root meristem

adaptation to environmental stresses and increased radiation levels still lag far behind many other areas of plant molecular biology. Adaptation is a complex process (Kovalchuk et al., 2004) by which populations of organisms respond to long-term environmental stresses by permanent genetic change. Studies of multiple generations exposed to radiation have rarely been undertaken due to the difficulties of creating a suitable model to study the effects of chronic exposure. While limited in number, these radioecological studies have attracted a great deal of interest in the question of how organisms adapt to ionizing radiation. In 1987–1989, an experimental study on the cytogenetic variability in three successive generations of winter rye and wheat, grown at four plots with different levels of radioactive contamination, was carried out within the 10 km ChNPP zone (Geras'kin et al., 2003a). In the autumn of 1989, aberrant cell frequencies in intercalary meristem of winter rye and wheat of the second and third generations significantly exceeded these parameters for the first generations (Table 4). In 1989, plants of all three generations were maintained in the identical conditions and accumulated the same doses, which is why the most probable explanation of the registered phenomenon relates to a genome destabilization in plants grown from radiation-affected seeds. Such detailed analysis of several generations of plants exposed to radiation provides some insight into possible mechanisms of plant adaptation to chronic radiation exposure. It relates to higher-order ecologic effects, as well as to contaminant-induced selection of resistant phenotypes. From these viewpoints, the results observed in (Geras'kin et al., 2003a) and indicating a threshold character of the genetic instability induction may be a sign of an adaptation processes beginning. That is, the chronic low dose irradiation appears to be an ecological factor creating preconditions for possible changes in the genetic structure of a population.

The adaptation processes in impacted wild plant populations were also investigated within the framework of other studies. The results of these experiments indicate (as an example, see Table 5) that an increased level of cytogenetic disturbance is a typical phenomenon for plant populations growing in areas with relatively low levels of pollution.

In 1949–1963, 116 air and ground-surface explosions for nuclear and hydrogen bomb testing was carried out in the Semipalatinsk Test Site. An ongoing study of *Koeleria gracilis* Pers populations, a typical wild crop for Kazakhstan, has shown that not only did the total frequency of cytogenetic disturbances increase significantly with the dose-rate measured in sampling points (Fig. 3A), but the relative contributions of such severe disturbances as double (chromosome) bridges and tripolar mitoses increased as well (Fig. 3B).

Forest trees have gained much attention in recent years as nonclassical model eukaryotes for population, evolutionary and ecological studies

TABLE 4. Aberrant cells in three successive generations of winter rye and wheat, grown on contaminated plots within the 10km ChNPP zone

Generation	D_{87-88}, Gy	D_{88-89}, Gy	D, Gy	Aberrant cell frequency, %		D_{87-88}, Gy	D_{88-89}, Gy	D, Gy	Aberrant cell frequency, %	
				Rye	Wheat				Rye	Wheat
	Plot 1					Plot 2				
X_1			0.01	20.2 ± 1.8	18.0 ± 1.7			0.09	20.6 ± 1.8	20.6 ± 1.8
X_2		0.11	0.01	23.8 ± 1.9	26.0 ± 2.0*		0.95	0.09	31.0 ± 2.1*	29.2 ± 2.0*
X_3	0.18	0.11	0.01	27.2 ± 2.0	26.4 ± 2.0*	1.62	0.95	0.09	35.8 ± 2.1*	30.0 ± 2.0*
	Plot 3					Plot 4				
X_1			0.06	23.4 ± 1.9	21.2 ± 1.8			0.04	22.8 ± 1.9	25.8 ± 2.0
X_2		0.64	0.06	28.8 ± 2.0	27.0 ± 2.0		4.17	0.04	33.0 ± 2.1*	34.0 ± 2.1*
X_3	1.11	0.64	0.06	30.0 ± 2.0	31.4 ± 2.1*	7.17	4.17	0.04	32.6 ± 2.1*	33.4 ± 2.1

Note: D is the dose, accumulated from planting in autumn 1989 to the sampling time; D_{87-88}, D_{88-89} are the doses, accumulated by parent plants during the whole vegetative period from planting up to harvesting in 1987–1988 and 1988–1989 years, respectively.

X_1-generation – plants grown from intact seeds and accumulated dose D from planting in autumn 1989 to the sampling time;

X_2-generation – parent plants were sown in 1988, harvested in summer 1989 and planted again on the same plots in autumn 1989. Genetical effects are the result of both the ancestral dose D_{88-89} and current exposure D;

X_3-generation – parent plants grew on the same plots in 1987–1988 and in 1988–1989 and accumulated doses D_{87-88} and D_{88-89}, correspondingly. Seeds harvested in 1989 were sown again in autumn 1989 and plants of the X_3 generation got dose D.

*Significance of variation from the level of cytogenetic disturbances in the X_1 generation: $p < 5\%$.

TABLE 5. Aberrant cell frequency in seedling root meristem of Scots pine growing in the Bryansk Region of Russia, radioactively contaminated as a result of the Chernobyl accident

Test site	^{137}Cs contamination density, kBq/m^2	Dose-rate, mGy/year[a]	Aberrant cells (mean ± se), % 2003	2004
Reference	–	0.14	0.90 ± 0.09	0.88 ± 0.09
VIUA	451	7.40	1.47 ± 0.15*	1.59 ± 0.14*
Starye Bobovichy	946	15.3	1.32 ± 0.12*	1.37 ± 0.14*
Zaborie 1	1730	28.3	1.69 ± 0.17*	1.67 ± 0.17*
Zaborie 2	2340	37.8	1.63 ± 0.15*	1.68 ± 0.17*

Note: Seeds were collected in 2003 and 2004;
* Difference from the reference population is significant, $p < 5\%$
[a]Absorbed doses are estimated for γ- and β-radiation

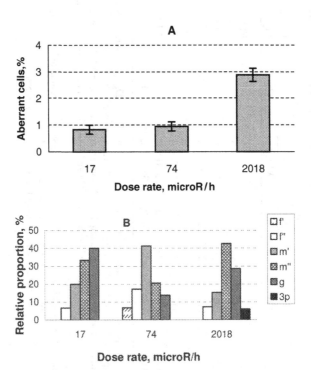

Fig. 3. Frequency of aberrant cells (A) and relative spectrum of aberrations (B) in coleoptiles of *Koeleria* seedlings collected in the Semipalatinsk Test Site, Kazakhstan.

(Gonzalez-Martinez et al., 2006). Because of low domestication, large open-pollinated native populations, and high levels of both genetic and phenotypic variation, they are ideal organisms to unveil the genetic basis of population adaptive divergence in nature. In the field study (Geras'kin et al., 2005b), Scots pine populations were used for an assessment of the genotoxicity originating from an operation of a radioactive waste storage facility. Specifically, frequency and spectrum of cytogenetic disturbances in reproductive (seeds) and vegetative (needles) tissues were studied to examine whether Scots pine trees experienced environmental stress in areas with relatively low levels of pollution. The temporal changes of the cytogenetic disturbances in seedling root meristem from 1997 to 2002 are shown in Fig. 4. There are essential differences between these relationships for the reference and impacted Scots pine populations. Statistical analysis revealed (Geras'kin et al., 2005b) that cytogenetic parameters at the reference site (Bolshaya Izhora) tend to follow cyclic fluctuations in time, whereas in technogenically affected populations ('Radon' LWPE and Sosnovy Bor) this relationship could not be revealed with confidence. Thus, man-made impact in this region is strong enough to destroy natural regularities.

Pollution is a well-documented selective force (Breitholtz et al., 2006), which has been found to induce metal tolerance in plants, pesticide resistance in insects, and pollution tolerance in a variety of aquatic organisms. To study possible adaptation processes in the impacted pine tree populations, a portion of the seeds collected were subjected to an acute γ-ray exposure (Geras'kin et al., 2005b). The seeds from the Scots pine populations experiencing man-made impacts showed (Table 6) a higher resistance than the reference ones. Although the picture of adaptation is far from complete, there is convincing proof (Shevchenko et al., 1992) that the divergence of populations in terms of radioresistance is connected with a selection for changes in the effectiveness of the repair systems.

It is well known (Macnair, 1993; Rajakaruna, 2004) that genetic adaptation in plant populations to extreme environmental conditions can take place quite rapidly, even within a few generations. But this rule does not apply to all the cases. A study of the wild vetch (*Vicia cracca* L.) population for more than 40 years inhabiting a site with an enhanced level of natural radioactivity

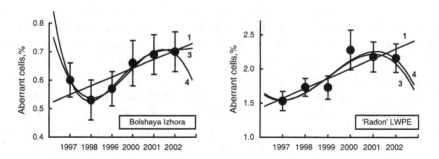

Fig. 4. Aberrant cells percentages in seedling root meristem of Scots pine trees in dependence on year and their approximation by the best models. (1) linear model, (3) and (4) polynomial models of the 3rd and 4th degrees, correspondingly.

TABLE 6. Aberrant ana-telophases frequency in root meristem of seedlings grown from seeds sampled from the reference and impacted Scotch pine populations in the Leningrad Region of Russia and exposed to acute γ-ray dose of 15 Gy

	1999		2000		2001	
Site	Cells total	Aberrant cells, %	Cells total	Aberrant cells, %	Cells total	Aberrant cells, %
1	3536	7.01 ± 0.43	2350	6.94 ± 0.52	1372	7.07 ± 0.69
2	4913	3.44 ± 0.26*	3383	3.55 ± 0.32*	2661	3.53 ± 0.36*
3	4688	3.56 ± 0.27*	2541	3.74 ± 0.38*	3009	3.69 ± 0.34*

Note: 1 is the reference site (Bolshaya Izhora); 2 is the urban area (Sosnovy Bor settlement); 3 is the territory of a radioactive waste storage facility ('Radon' LWPE).
* Difference from the reference site of Bolshaya Izhora is significant; $p < 5\%$

TABLE 7. Embryonic lethal and chromosome aberration frequency in seedling root meristem of wild vetch inhabiting the site with enhanced level of natural radioactivity (the Komi Republic) (Evseeva et al., 2007)

	Dose-rate, μR/h	Chromosome aberrations frequency, %	Embryonic lethals, %
Reference	9	0.80 ± 0.07	24.62 ± 2.12
Plot 1	3300	1.33 ± 0.12*	33.63 ± 23.32*
Plot 2	2400	1.00 ± 0.09	35.38 ± 2.06*
Plot 3	758	0.87 ± 0.10	25.17 ± 2.24
Plot 4	73	0.98 ± 0.10	24.60 ± 2.08

Note: Seeds were collected in 2003;
*difference from the reference population is significant, $p < 5\%$

showed that transformations of genetic structure in the populations are still ongoing (Evseeva et al., 2007). In this site, both low doses of external exposure and incorporated heavy natural radionuclides have significant effects on the variability in plant populations and their potential for an adaptation. As a selection factor, external exposure results in an increasing of the frequency of embryonic lethal mutations (Table 7). In addition, in contrast with data from the pine populations study (Geras'kin et al., 2005b), the acute γ-irradiation demonstrated rather high radiosensitivity of seeds from the impacted populations of wild vetch. The inherited character of the enhanced radiosensitivity was shown in (Alexakhin et al., 1990). At this point, it is possible to suppose that a probable cause of special features found in this site in the Komi Republic relates to a specific radioecological situation within the uranium–radium anomaly, where a decisive role is played by α-emitters. However, the mechanisms involved in this plant response remain to be studied in more detail.

Conclusions

The increasing degradation and pollution of the environment requires the establishment of biological sentinel systems that provide information on adverse effects on ecosystems and on human health. Findings presented here clearly indicate that an adequate environment quality assessment cannot rely only on information about pollutant concentrations. This conclusion emphasizes the need to update some current principles of ecological standardization, which are still in use today. When assessing potential hazards from radionuclide and chemical pollution for ecosystems, a harmonized approach based on ecotoxicology principles should be applied. At the first stage, it is advisable to use biological testing that result in an integral assessment of effects from all substances presented in the environment. If bioassays give positive responses, a more detailed survey should be undertaken including physical-chemical analysis, geochemical barriers, and study of migration parameters of the contaminants for a certain landscape. An effective linking of bioindication screening assays to well-established environmental pollution monitoring is a way of improving and upgrading an existing system of public and environment protection in order to meet requirements of consistency between current scientific knowledge and decision-making processes.

A better understanding of the genetic aspects of population, community and the whole ecosystem responses to toxic agent's exposure is vital to future environmental management programs. The results described here provide evidence that plant populations growing in areas with relatively low levels of pollution are characterized by the increased level of both cytogenetic disturbances and genetic diversity. Man-made pollution may influence an evolution of exposed populations through a contaminant-induced selection process. The long-term existence of some factors (either of natural origin or man-made) in the plants' environment activates genetic mechanisms, changing a population's resistance to exposure. However, in different radioecological situations, genetic adaptation to extreme edaphic conditions in plant populations could be achieved at different rates. Such evolutionary effects are of special concern because they are able to negatively affect population dynamics and local extinction rates. These processes have a genetic basis; therefore, understanding changes at the genetic level should help in identifying more complex changes at higher levels. Recent studies have shown (Whitham et al., 2006) that heritable traits in a single species have predictable effects on community structure and ecosystem processes. Therefore, we can begin to apply the principles of population and quantitative genetics to place the study of complex communities and ecosystems. Finally, in spite of the wealth of information collected so far, much more still remains to be explained in order to fully understand the basis of plant populations' adaptation to a harmful environment.

Acknowledgements

Works presented were supported by the EU-Project WaterNorm (contract No MTKD-CT-2004-003163), Ministry of Science and Innovation of Russian Federation (contract No 02.445.11.7463), Federal Agency on Atomic Energy of Russian Federation (contract No 1.30.06.16/7), and ISTC projects No 3003 and K-1328.

References

Alexakhin, R.M., Arkhipov, N.P., Barhudarov, R.M., Vasilenko, I.Y., Drichko, V.F., Ivanov, U.A., Maslov, V.I., Maslova, K.I., Nikiforov, V.S., Polykarpov, G.G., Popova, O.N., Sirotkin, A.N., Taskaev, A.I., Testov, B.V., Titaeva, N.A., and Fevraleva, L.T., 1990, *Heavy Natural Radionuclides in Biosphere. Migration and Biological Effects on Population and Biocenosis*, Nauka Publishers, Moscow (in Russian).

Breitholtz, M., Ruden, C., Hansson, S.O., and Bengtsson, B.E., 2006, Ten challenges for improved ecotoxicological testing in environmental risk assessment, *Ecotoxicol. Environ. Safety* **63**:324–335.

Brenner, D.J., Doll, R., Goodhead, D.T., Hall, E.J., Land, C.E., Little, J.B., Lubin, J.H., Preston, D.L., Preston, R.J., Puskin, J.S., Ron, E., Sachs, R.K., and Samet, J.M., 2003, Cancer risks attributable to low doses of ionizing radiation: assessing what we really know, *Proc. Natl. Acad. Sci. USA* **100**:13761–13766

Evseeva, T.I., Geras'kin, S.A., and Khramova, E.S., 2003a, The comparative estimation of plant cell early and long-term responses on ^{232}Th and Cd combined short-time or chronic action, *Tsitol. Genet.* **37**:61–66.

Evseeva, T.I., Geras'kin, S.A., and Shuktomova, I.I., 2003b, Genotoxicity and toxicity assay of water sampled from a radium production industry storage cell territory by means of *Allium*-test, *J. Environ. Radioact.* **68**:235–248.

Evseeva, T.I., Geras'kin, S.A., Shuktomova, I.I., and Taskaev, A.I., 2005, Genotoxicity and toxicity assay of water sampled from the underground nuclear explosion site in the north of the Perm region (Russia), *J. Environ. Radioact.* **80**:59–74.

Evseeva, T.I., Majstrenko, T.A., Geras'kin, S.A., and Belych, E.S., 2007, Genetic variance in wild vetch cenopopulation inhabiting site with enhanced level of natural radioactivity, *Radiat. Biol. Radioecol.* **47**:54–62. (in Russian)

Fiskesjo, G., 1985, The Allium test as a standard in environmental monitoring, Hereditas **102**:99–112.

Geras'kin, S.A., 1995, Critical survey of modern concepts and approaches to the low doses of ionizing radiation biological effect estimation, *Radiat. Biol. Radioecol.* **35**:563–571 (in Russian).

Geras'kin, S.A. and Sarapul'tzev, B.I., 1993, Automatic classification of biological objects on the level of their radioresistance, *Automat Remote Control* **54**:182–189

Geras'kin, S.A., Oudalova, A.A., Kim, J.K., Dikarev, V.G., and Dikareva, N.S., 2007, Cytogenetic effect of low dose γ-radiation in Hordeum vulgare seedlings: non-linear dose-effect relationship, *Radiat. Environ. Biophys.*, 46:31–41.

Geras'kin, S.A., Dikarev, V.G., Zyablitskaya, Ye.Ya., Oudalova, A.A., Spirin, Y.V., and Alexakhin, R.M., 2003a, Genetic consequences of radioactive contamination by the Chernobyl fallout to agricultural crops, *J. Environ. Radioact.* 66:155–169.

Geras'kin, S.A., Kim, J.K., Dikarev, V.G., Oudalova, A.A., Dikareva, N.S., and Spirin, Ye.V., 2005a, Cytogenetic effects of combined radioactive (^{137}Cs) and chemical (Cd, Pb, and 2,4-D herbicide) contamination on spring barley intercalar meristem cells, *Mutat. Res.* 586:147–159.

Geras'kin, S.A., Kim, J.K., Oudalova, A.A., Vasiliyev, D.V., Dikareva, N.S., Zimin, V.L., and Dikarev, V.G., 2005b, Bio-monitoring the genotoxicity of populations of Scots pine in the vicinity of a radioactive waste storage facility, *Mutat. Res.* 583:55–66.

Geras'kin, S.A., Zimina, L.M., Dikarev, V.G., Dikareva, N.S., Zimin, V.L., Vasiliyev, D.V., Oudalova, A.A., Blinova, L.D, and Alexakhin, R.M., 2003b, Bioindication of the anthropogenic effects on micropopulations of Pinus sylvestris L. in the vicinity of a plant for the storage and processing of radioactive waste and in the Chernobyl NPP zone, *J. Environ. Radioactivity* 66:171–180.

Gonzalez-Martinez, S.C., Krutovsky, K.V., and Neale, D.B., 2006, Forest-tree population genomics and adaptive evolution, *New Phytologist* 170:227–238.

Howard, H., 1997, Synergistic effects of chemical mixtures – can we rely on traditional toxicology?, *The Ecologist* 27:192–195.

Kovalchuk, I., Abramov, V., Pogribny, I., and Kovalchuk, O., 2004, Molecular aspects of plant adaptation to life in the Chernobyl zone, *Plant Physiol.* 135:357–363.

Macnair, M., 1993, The genetics of metal tolerance in vascular plants, *New Phytol.* 124:541–559.

Oudalova, A.A., Geras'kin, S.A., Dikarev, V.G., Dikareva, N.S., Kruglov, S.V., Vaizer, V.I., Kozmin, G.V., and Tchernonog, E.V., 2006, Method preparation on integrated assessment of ecological situation in the vicinity of the Russian nuclear facilities, In: *Proceedings of the Regional Competition of the Natural-Science Project.,* Issue 9, Kalugian scientific centre, Kaluga, pp. 221–239. (in Russian)

Rajakaruna, N., 2004, The edaphic factor in the origin of plant species, *Intern. Geol. Rev.* 46:471–478.

Shevchenko, V.A., Pechkurenkov, V.L., and Abramov, V.I., 1992, *Radiation Genetics of Natural Populations: Genetic Consequences of the Kyshtym Accident,* Nauka Publishers, Moscow. 221 p. (in Russian)

Theodorakis, C.W., Blaylock, B.G., and Shugart, L.R., 1997, Genetic ecotoxicology I: DNA integrity and reproduction in mosquitofish exposed in situ to radionuclides, *Ecotoxicology* 6:205–218.

Walbot, V., 1985, On the life strategies of plants and animals, *Trends Genet* 1:165–170.

Whitham, T.G., Bailey, J.K., Schweitzer, J.A., Shuster, S.M., Bangert, R.K., LeRoy, C.J., Lansdorf, E.V., Allan, G.J., DiFazio, S.P., Potts, B.M., Fischer, D.G., Gehring, C.A., Lindroth, R.L., Marks, J.C., Hart, S.C., Wimp, G.M., and Wooley, S.C., 2006, A framework for community and ecosystem genetics: from genes to ecosystems, *Nat. Rev. Genet.* 7:510–523.

CHAPTER 7

METHODOLOGY OF SOCIO-ECOLOGICAL MONITORING USING CYTOGENETIC METHODS

ALLA GOROVA AND IRINA KLIMKINA

National Mining University, K. Marks 19, Dnipropetrovs'k, 49006 Ukraine,
e-mail: gorovaa@nmu.org.ua; irina_klimkina@ukr.net

Abstract: Anthropogenic violation of the principles of rational environmental management has resulted in degradation changes in social life and environment. It has caused emergence of abnormal ecosystems and changes in gene pool. For enhancement of socio-ecosystems and improvement of the national gene pool, it is necessary to implement complex systems of socio-ecological monitoring. A methodology of socio-ecological monitoring has been developed. It is based on high-sensitivity indication systems of various levels of organization, and standardized procedures of integral estimation of the state of environmental objects according to toxic and mutagenic background, and level of health of different age groups. All this made it possible to define integral tests, which allow assessing ecological and genetic danger (risks) imposed on biota and human by mutagenic and ecological factors.

Keywords: population health state; environmental state; social-ecological monitoring; ecological risk assessment; damageability index, biological indicators; Dnipropetrovsk region

Introduction

It is known that anthropogenic pressure on the environment in many regions of our planet has come up to the level which threatens the ecosystems state and population health. Therefore, efforts of many countries are directed to the development of the policy, which would provide sustainable ecological, economic and social development of society. The ecological priorities of sustainable development of society were reflected in the following documents singed by the leaders of countries: program document of The International Conference of UN in the Rio-de-Janeiro "The Action Program. Agenda for the XXI century" (1992), Kyoto Protocol to the United Nations Framework

Convention on Climate Change (1997), decisions of The 5th European conference of Environmental Ministers "Environment for Europe" (Kiev, 2003).

The potent base of mineral resources is the feature of Ukraine. There are more than 20,000 deposits of minerals in its bowels; its iron and manganese ore reserves compose 14% and 43% of the world reserves respectively. Ukraine is one of the leading countries in coal, manganese, iron, titan, graphite and kaolin extraction. Development of the coal-mining branch has vital importance for Ukraine because of the priority of coal as the main fuel resource in the twenty-first century. Development of the fuel and energetic complex will allow the level of country energy provision to rise (Pivniak et al., 2003).

However, the long-term development of different branches of industry, agriculture, urbanization etc. in the country has resulted in the high level of technogenic pressure on the environment, accumulation of big amount of waste (including radioactive waste), activation and evolution of dangerous exogenous geological processes, transgressions of hydrogeological conditions etc. Besides, degradation changes in the environment are enforced by the high level of radioactive contamination arising from the Chernobyl accident.

As the result of technogenic pressure on the environment, negative changes in ecosystems and society accrue. Namely biodiversity degrades, natural stability of ecosystems to the action of inimical factors abates, the physical state of human and the state of the gene pool deteriorates (Pivnyak et al., 2002; Gorova et al., 2003).

The high level of air, water, soil and foodstuff pollution with toxic and mutagenic agents has contributed to the rise of population morbidity and deterioration of demographic indices. Diseases of endocrine system, blood and blood-making organs, allergic diseases, neoplasms, chronic diseases of respiratory organs constantly increase their occurrence in Ukraine. The decline in birth-rate and increasing of mortality were observed during last years, and are the main causes for the negative tendencies of changes in the indices of fatality and population reproduction in Ukraine (Serdiuk, 1998; Gorova et al., 2003).

Health of the nation and the environmental state are the important integral parameters of the society civilization level and its social and economic development. It is not by chance that the state of ecosystems and population health are the criteria for life quality and the major priorities of governmental actions in the developed countries (Serdiuk, 1998). Exactly because of this, questions of environmental state and health of nation assessment and management acquire more and more importance in Ukraine.

The absence of systematic approach and methodology of integral assessment of environmental and social changes impedes an adequate complex evaluation of a concrete ecological and social situation at a specific territory. But it can be done on the basis of the implementation of the standardized methodology of social and ecological monitoring. Its main goals are

- Environmental assessment by toxic and mutagenic background on the basis of methods of cytogenetic bioindication;
- Determination of unified assessment criteria for the influence of harmful factors (natural and anthropogenic) upon the population health state;
- Definition of the ecological and genetic danger level for human and biota from the influence of eco-toxicants and mutagens of environment.

Main Methodological Aspects Concerning Assessment of the Population Health

The term "Health" is defined as a state of full social, mental and physical well-being of a human. Health of a population reflects the influence of a complex phenomenon in the environment and depends on the biological, socio-economic, natural and climatic factors.

Figure 1 presents a simplified scheme of the impact of different factors on the human health. As it can be seen from the scheme, the human health is determined by both internal (hereditary) and external (environmental conditions) factors. Health depends on the state of human genome and hereditary constitution of an organism.

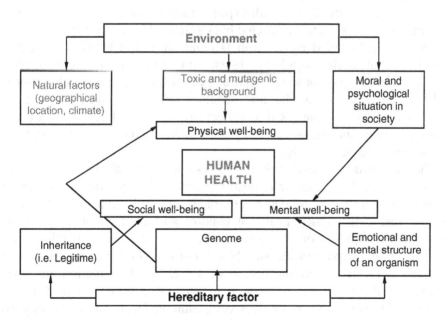

Fig. 1. Structural scheme of the factors influencing human health.

At the same time, environmental pollutants presented in the air, water, food products and other objects also impact on the health condition as well as emotional and mental structure of an organism, and social conditions. Social status and human well-being depend on the material inheritance, which is the result of human's previous work or may be inherited by right of succession from the preceding generation; they also depend on natural and geographical conditions in the place of dwelling that enable to obtain vitally important products with economic effectiveness and ecological expediency.

Not only does the environmental state characterize the ingredient content of pollutants but also toxic and mutagenic background they produce. This background can be determined through methods of bio-indication, among which cytogenetic methods are the most sensitive. They make it possible to evaluate the complex influence of unfavorable factors, allowing for doses and duration of their influence, on living organisms.

Structural Scheme of Complex Social-Ecological Monitoring

Starting from the integral understanding of the term "Health" and influence of environmental factors the structural scheme of social and environmental monitoring has been elaborated (Fig. 2).

As can be seen from the scheme, the top (zero) level characterizes ecological and social state of the integral system of the territory development on the local, regional and national levels. It includes two parameters of lower (first) level: the state of population health (population block) and the state of the environment on toxic and mutagenic background (bio-indication block).

The further detailed elaboration (second structural level) is presented by indices that in the population block "population health" characterize the main demographic processes (natural variation of population), physical health of children and adults, and genetic health. Bio-indication block is characterized by parameters that reflect environmental condition of the territory on toxic and mutagenic background including the state of atmosphere, hydrosphere and pedosphere.

The third structural level presents indices that fill the blocks of the second structural level. In particular natural variation of population is evaluated through the parameters of death-rate, birth-rate and infant mortality rate (up to 1-year old). Physical health of children and adult population is characterized by frequency of the following disease occurrence: infectious and parasitic, endocrine system diseases, blood and hematopoietic organs, mental disorders, nervous system and organs of sense, blood circulation and lung diseases, gastrointestinal diseases, urino-genital diseases, skin diseases, osseous and muscular system diseases, congenital malformations and neoplasm diseases. Genetic health of the population is characterized by such indices

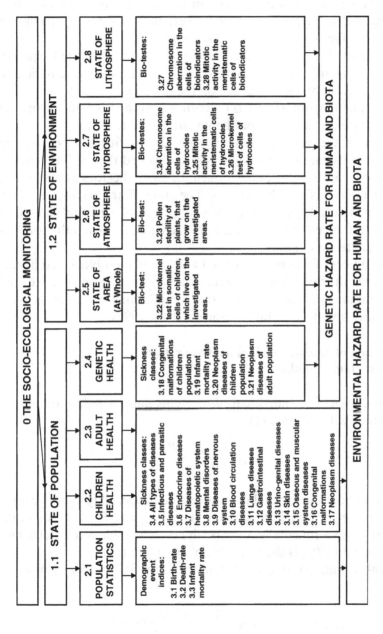

Fig. 2. Structural scheme of socio-ecological monitoring.

like congenital malformations and neoplasm for children and adults and mortality rate of infants (up to 1-year old). Ecological state of environmental objects on toxic and mutagenic background is recommended to evaluate through cytogenetic methods of bio-indication which are the most sensitive, informative and sufficient for appropriate evaluation of associated impact of environmental factors on various biosystems.

The state of the territory as a whole is determined on the frequency of occurrence of microkernel fragments (genetic imbalances) in somatic cells of children of preschool age which live on the investigated territory.

The state of the atmosphere is determined on the level of pollen sterility for indicator plants that grow on the investigated territory.

The state of the hydrosphere is evaluated on the frequency of occurrence of meristem cells with chromosome aberrations and the level of cell mitotic activity of both exogenous and endogenous bio-indicators as well as micro-kernel test in the cells of hydrocoles. The state of the pedosphere (soil) is evaluated on their mutagenicity according to the level of genic and chromosomal mutations in the cells of bio-indicators. Toxic effect is a value of mitotic index in meristematic tissues.

As the indices of the bio-indication and population blocks do have their own units, for the purposes of integral evaluations, it is necessary to reduce them to common non-dimensional form of conditional indices for ecosystems damageability and protectability (Gorova et al., 1996).

Determination of Integral Indices for Bio-systems of Different Damageability

Conditional indices of individual bio-systems damageability are calculated in accordance with formula (1):

$$CI_i = \frac{/P_{comf} - P_{act}/}{/P_{comf} - P_{crit}},\tag{1}$$

CI_i is a conditional index of bio-parameter damageability that is caused by the action of negative environmental factors; P_{comf} and P_{crit} are bio-parameter values that are experimentally (or expertly) estimated for comfortable and for critical conditions; P_{act} is the real value of bio-parameter.

Integral conditional indices of damageability (ICI) are determined using formula (2):

$$ICI_i = \frac{1}{n} \times \sum_{i=1}^{n} ICI_i = \frac{1}{n} \times \sum_{i=1}^{n} \left[\frac{/P_{comf} - P_{act}/}{/P_{comf} - P_{crit}} \right]_i,\tag{2}$$

where ICI_i is one of the integral conditional indices of human heath or environmental state; $P_{comf.}$, $P_{crit.}$, $P_{act.}$ are comfortable, critical and actual value of any n-indices respectively.

Integral index, which characterizes the environmental state by toxic and mutagenic background (ICI_{bioind}), implies the parity of components and is determined using formula (3):

$$ICI_{bioind} = \frac{1}{4} (ICI_1 + ICI_2 + ICI_3 + ICI_4),$$ (3)

where ICI_1, ICI_2, ICI_3, and ICI_4 are integral indices of biological indication of territory as a whole, atmosphere, hydrosphere and lithosphere state respectively.

An integral index, which characterizes total human health (ICI_{popul}), is determined by formula (4) with consideration of expertly weight ratio parameters setting for individual parameter. The larger coefficient values are taken for the most sensitive parameters:

$$ICI_{popul} = 0.30 * ICI_1 + 0.30 * ICI_2 + 0.22 * ICI_3 + 0.18 * ICI_4,$$ (4)

where ICI_1 – demographic events; ICI_2 – physical health of children; ICI_3 – physical health of adults; ICI_4 – genetic heath.

Integral index that characterizes the total ecological danger (ED) for human and biota from environmental pollutant action is calculated using formula (5):

$$ED = 0.60 * ICI_{bioind} + 0.40 * ICI_{popul},$$ (5)

At the same time, the priority of bio-indication block, which parameters are expertly estimated, should be noted.

Integral index that characterizes the genetic danger (GD) for human and biota from environmental mutagen action is calculated using formula (6):

$$GD \quad 0.60 * ICI_{bioind} \quad 0.40 * ICI_{gen.health},$$ (6)

where $ICI_{gen.health}$ is the integral index of population genetic health. In this case, like in previous formula, the priority is given to the data which were experimentally received.

Conditional indices allow comparison and ranking of city, regional and country areas to be made, based on the environmental state and human heath, which is impossible while data are in their natural units. Moreover, using conditional indices is it possible to make ranking of territories by toxical and mutagen background of various environmental objects, and determine the priority classes of population diseases in different regions.

Values of conditional indices of damageability (CI and ICI) change in range from 0 (comfortable conditions for vital functions) to 1 (critical conditions).

As a normative value of damageability, which complies with conditions of sustainable development of territories, a 30% damageability level was taken for all analyzed bio-parameters (i.e. CI_{norm} = 0.300 c.u.). Under these conditions bio-system self-regeneration is possible after negative factor action is stopped. Formulas (7) and (8) are deduced from formula (1). They are used to determine normative values for every actual value.

$$CI_{norm} = 0.3\left(P_{crit} - P_{comf}\right) + P_{comf} \qquad (7)$$

$$CI_{norm} = P_{comf} - 0.3\left(P_{comf} - P_{crit}\right) \qquad (8)$$

Formula (7) is used for values $P_{crit} > P_{comf}$, formula (8) – for values $P_{comf} > P_{crit}$.

We recommend using the uniform scale (Table 1) for the human health and environmental state damageability assessment.

Ecological risks (ER) are based on index that characterizes ecological danger (ED) and are calculated using formula (9). Genetic risks (GR) are based on the index of genetic danger (GD) and are calculated using formula (10):

$$ER = \frac{ED}{CI_{norm}}, \qquad (9)$$

$$GR = \frac{GD}{CI_{norm}}. \qquad (10)$$

The rating scale of the territory state assessment by ecological and genetic risks is presented in Table 2.

TABLE 1. The rating scale of human health and environmental state assessment

Damageability index, numerical values range	Level of human and bio-system damageability	Protection index, numerical values range	Level of protection	Eco-system and human health state	Assessment of ecological situation
0.000–0.150	Low	1.000–0.850	Maximum	Favorable	Comfortable
0.151–0.300	Below average	0.849–0.700	High	To be watched	Satisfactory
0.301–0.450	Average	0.699–0.550	Above average	Conflicted	Unsatisfactory
0.451–0.600	Above average	0.549–0.400	Average	Menacing	Unsatisfactory
0.601–0.750	High	0.399–0.250	Below average	Critical	Catastrophic
0.751 and higher	Maximum	0.249–0.000	Low	Hazardous	Catastrophic

TABLE 2. The rating scale of the area state assessment by ecological and genetic risks

Risk, numerical values range	Risk assessment
0.0–1.0	Risk is absent
1.01–1.5	Low
1.51–2.0	Average
2.01–2.5	High
2.51–3.3	Maximal

Results of the Methodology Approbation

The Socio-ecological monitoring methodology was approbated on those Ukrainian territories which were heavily damaged by industrial activity; among them are Dnipropetrovsk, Donetsk and Lviv regions. The Autonomous Republic of Crimea and Zakarpattia region were chosen to serve as reference territories due to a good environmental state.

Results of the complex social-ecological monitoring of the territories of industrial cites in Dnipropetrovsk region are given on Fig. 2.

According to Table 3, the index of integral ecological danger in Dnipropetrovsk is 0.429, in Zhovti Vody it is 0.591 and on the reference territory of the Autonomous Republic of Crimea (Nikita settlement) its value is 0.036. It means that the level of damageability of bio-indicators and the population heath in Dnipropetrovsk are "average," the state of living organisms is "conflict" and the ecological situation is "unsatisfactory." In Zhovti Vody, which is well-known for its uranium industry, the level of damageability of bio-indication systems is "above average," ecosystems state is "menacing," and the ecological situation is close to "catastrophic." On the contrary, the ecological situation in Nikita settlement is considered to be "comfortable," with "low" levels of damageability of ecosystems and population health, and therefore their "favorable" state.

TABLE 3. Results of the socio-ecological monitoring, ICI (2001–2005)

Dnipropetrovsk		Zhovti vody		Nikita settlement, crimea (reference)	
1.1–0.414	1.2–0.434	1.1–0.732	1.2–0.483	1.1–0.161	1.2–0.099
2.1–0.510	2.5–0.500	2.1–0.674	2.5–0.578	2.1–0.313	2.5–0.072
2.2–0.330	2.6–0.421	2.2–0.790	2.6–0.409	2.2–0.060	2.6–0.097
2.3–0.470	2.7–0.400	2.3–0.787	2.7–0.470	2.3–0.078	2.7–0.110
2.4–0.330	2.8–0.413	2.4–0.701	2.8–0.497	2.4–0.180	2.8–0.117
ED – 0.429		ED – 0.591		ED – 0.036	
GD – 0.390		GD – 0.579		GD – 0.039	

Notation corresponds to the one used in Fig. 2.

A similar tendency is observed with respect to the variation of the genetic danger index, general and genetic heath damageability, as well as of toxic and mutagenic activity of environmental objects. All these are the highest in Zhovti Vody, and are lower in Dnipropetrovsk. Concerning Nikita settlement, it is natural that genetic danger is absent there; toxic and mutagenic activity of environmental objects is the lowest; and population health is the best.

The obtained data made it possible to determine the ecological and genetics risks levels, that are imposed on biota and human (Table 4).

On the territories with highly developed uranium and iron ore industries (Zhovti Vody and Kryvy Rig) the highest values of genetic and ecological dangers can be observed (Table 4). They correspond to "average" and are approaching "high" levels of ecological and genetic risks.

Coal deposits in Donetsk region (Selidovo) have "average" levels of ecological and genetic risks.

TABLE 4. Assessment of eco-genetic danger and risks imposed on biota and humans in mining regions of Ukraine

Deposits	GD	GR	Genetic risks assessment	ED	ER	Ecological risk assessment
Coal deposits						
Pavilograd, Dnipropetrovsk Region	0.399	1.33	Low	0.411	1.37	Low
Selidovo, Donetsk Region	0.480	1.6	Average	0.471	1.57	Average
Chervonograd, Lviv Region	0.420	1.4	Low	0.429	1.43	Low
Iron-ore deposits of Dnipropetrovsk Region						
Kryvy Rig	0.540	1.8	Average	0.561	1.87	Average
Uranium deposits of Dnipropetrovsk Region						
Zhovti Vody	0.579	1.93	Average	0.591	1.97	Average
Manganese deposits of Dnipropetrovsk Region						
Marganets	0.429	1.43	Low	0.480	1.6	Average
Ordjonikidze	0.390	1.3	Low	0.459	1.53	Average
Nikopol	0.443	1.47	Low	0.481	1.60	Average
Granite deposits of Dnipropetrovsk Region						
Dnipropetrovsk	0.390	1.3	Low	0.429	1.43	Low
Monitoring areas						
Nikita, AR Crimea	0.039	0.13	Risk is absent	0.036	0.12	Risk is absent
Polana, Zakarpatian Region	0.039	0.13	Risk is absent	0.036	0.12	Risk is absent

On other coal deposits in Dnipropetrovsk and Lviv regions (Pavlograd and Chervonograd respectively) levels of genetic and ecological risks are "low."

Regions where manganese ore is extracted (Nikopol, Marganets, Ordzhonikidze) show an "average" level of ecological risks, and a "low" level of genetic risks.

"Low" levels of both types of risks groups are observed on granite deposits in Dnipropetrovsk.

Reference territories of Zakarpattia region and the Autonomous Republic of Crimea possess the lowest values of ecological and genetic danger. This is the reason for the absence of both risks groups on these territories.

Conclusion

The analysis has revealed that most centres of mining industry in Ukraine show "average" levels of ecological and genetic risks. This is a source of danger for all living organisms and human, and therefore requires urgent development and realization of the program of rehabilitation of the state of environmental objects and human health (Pivnyak et al., 2002; Gorova et al., 2005).

Implementation of the described methodology of socio-ecological monitoring on the whole territory of Ukraine will allow new ecological and socio-ecological maps to be created, which are necessary for the management of ecological and social processes in the country.

Application of the methodology on territories of other countries will enable a comparative analysis of the state of social and ecological systems to be conducted on an intergovernmental level, as an essential step towards the assessment of the global socio-ecosystem.

Acknowledgments

Authors express their gratitude to the associate professor of the Chair of Ecology (National Mining University) Tatiana Skvortsova, as well as to the assistants of the chair Artem Pavlichenko and Yuriy Buchavyy, Inna Mironova for their collaboration in perfection, approbation and implementation of the methodology of socio-ecological monitoring. We are grateful to assistants of the chair A. S. Kovrov and E. A. Borisovska for their kind help in translation.

Special gratitude is expressed to the associate professor of the chair of Hygiene and Ecology (Medical Academy, Dnipropetrovsk), Candidate of Medical Sciences L.B. Ogir, and to the Head of the Office for Medical Statistics of Dnipropetrovsk region A.A. Sokulsky for the statistical data provided and advices regarding data processing.

In addition, we would like to thank the student of Brandenburg Technical University (Cottbus, Germany) O.O. Balyk for his valuable suggestions and feedback regarding the translation of the manuscript into English.

References

Gorova, A., A. Sokulsky, I. Klimkina, U. Buchavyi. Basic methodological states and evaluation of bad environment influence on human health. Proc. of Intern. Conf. "Anthropogenic changes in Ukraine environment and risk for human health and biota." Kiev:Chernobylinterninform, 2003, 340–353 (in Ukrainian).

Gorova, A., T. Skvortsova, I. Klimkina, A. Pavlichenko. Cytogenetic test use for evaluation of mining industry influence on environment and human health. Proc. of NSU, 7(2) – Dnepropetrovsk, 2003, 522–531 (in Russian).

Gorova, A., T. Skvortsova, I. Klimkina, A. Pavlichenko. Cytogenetic effects humic substances and their use for remediation of polluted environments. NATO Science Series "Use of humic substances to remediate polluted Environments: From theory to Practice, 2005, Springer. Printed in the Netherlands, 311–328.

Gorova, A., L. Bobir, T. Skvortsova, V. Digurko, I. Klimkina, A. Pavlichenko. Methodological aspects of mutagenic background evaluation and genetic risk for human and biota from mutagenic environment//Cytology and Genetics, 1996, 30(6), 78–86. (in Russian).

Pivniak, G., P. Pilov, V. Bondarenko et al. Strategic problems of coal industry development in Ukraine.//Proc. of NSU, 17, V 1. – Dnepropetrovsk, 2003, 5–11 (in Ukrainian).

Pivnyak, G.G., A.I. Gorova, V.A. Dolinsky, V.O. Skvortsov, O.S. Kovrov. To ecological program of mining development of Ukraine. – Proceeding SWEMP 2002, 7-th International Symposium on Environmental Issues and Waste Management in Energy and Mineral Production, R. Ciccu (Ed.), Cagliari, Italy, October 7–10, 2002, pp. 17–19.

Serdiuk, A. Environment and health of Ukraine population.//Environment and health. 4(7), 1998, 2–7 (in Ukrainian).

CHAPTER 8

ROLE OF GENETIC SUSCEPTIBILITY IN ENVIRONMENTAL EXPOSURE INDUCED DISEASES

SOHEIR KORREA

National Center for Radiation Research and Technology, Egypt
email: soheirskorraa@hotmail.com

Abstract: *Inherited susceptibility* due to a defective gene is a factor in a small percentage of people who develop cancer (<5%), while *induced susceptibility*, which is due to the wide variation in individual responses to exogenous agents, is believed to result from the great diversity in responsiveness to risk factors in the environment. Interindividual variations in DNA repair capacity for specific types of DNA damage are documented. Functional polymorphism has been identified primarily at enzymes associated with redox regulation and detoxification, such as glutathione *S*-transferase and cytochrome p450 isozymes. Cancer susceptibility can be the inability to eliminate mutated cells by apoptosis due to mutation in apoptosis regulatory genes and/or induced disruption in gap junction, activation of proto-oncogene and/or inactivation of suppressive genes. Detectable gene mutations and alterations in signal transduction pathways together with modified post-translational proteins offer valuable molecular biomarkers for occupational and environmental human biomonitoring applied for the identification of potentially hazardous exposures before adverse health effects appear and allow the establishment of exposure limits in order to minimize the likelihood of significant health risks. An emerging concept is that the combined action of environmental factors and individual susceptibility determines an individual's likelihood of developing cancer, asthma, diabetes as well as many other aging-associated diseases.

Introduction

During the last decades diseases such as asthma (Eggleston et al., 1999), obstructive lung diseases (Lagorio et al., 2006) diabetes mellitus (Lee et al., 2006), cardiovascular disease (Chen et al., 2005), cancer (Brennan, 2002) atherosclerosis (Wang and Wang, 2005), Alzheimer's disease (Landrigan et al., 2005) and autoimmune disorders (Dooley and Hogan, 2003) are increasing in incidence. These diseases are multifactorial and all are suggested to involve complex interactions between genetic (individual susceptibility) environmental (potentially modifiable) factors. It is recognized that environmental exposures play a key role factor in their propagation.

C. Mothersill et al. (eds.), Multiple Stressors: A Challenge for the Future, 103–123.

Inherited susceptibility due to a defective gene is a factor in a small percentage of people who develop cancer (<5%). Nearly all hereditary diseases are recessive, meaning that both copies of a gene must be mutated in order for the disease to develop (Risch, 2000). Induced susceptibility, which is due to the wide variation in individual responses to exogenous stressors, is believed to result from the great diversity in responsiveness to risk factors in the environment. These variations, known as polymorphism, are caused by sporadic mutations caused by both endogenous and exogenous processes (Elena and de Visser, 2003). In most instances, such mutations, which result in minor changes in the nucleotide sequence of the coding region as well as 5′ and 3′ untranslated regions, are sufficient to alter expression or stability at both the RNA and protein levels (Malkin, 1995). However, there are many instances, where modifications in gene expression do not involve changes in DNA nucleotide sequences. Modifications in gene expression through methylation of DNA and remodelling of chromatin via histone proteins are believed to be the most important events of the epigenetic changes (Verma and Srivastava, 2002).

Potential sources of susceptibility for complex diseases risk include interindividual variation in DNA repair capacity for specific types of DNA damage (Au et al., 1996). Also variation in enzymes, which activate and detoxify procarcinogens and carcinogens (e.g. Phase I enzymes, which catalyze oxidation, reduction, and hydrolysis reactions, and Phase II enzymes, which catalyze conjugation and synthetic reactions) are causes for interindividual susceptibility. Functional polymorphism has been identified primarily at enzymes associated with redox regulation and detoxification, such as glutathione S-transferase (Nakajima et al., 1995) and cytochrome p450 isozymes (Oyama et al., 1997). Cancer susceptibility can be the inability to eliminate mutated cells by apoptosis due to mutations in the apoptosis regulatory genes (Malkin, 1995), and/or induced disruption in gap junction (Trosko et al., 1994), activation of proto-oncogene (You et al., 1989) and/or inactivation of suppressor genes (Weinberg, 1991; Greenblatt et al., 1994).

BIOLOGICAL VARIABILITY IN THE ACTIVITY OF OXIDANT-PRODUCING ENZYMES

Nitric oxide species (NOS) and reactive oxygen species (ROS) regulate multiple cellular functions such as DNA synthesis (Kandacova and Zagrebel'naia, 2004), signal transduction (Ruiz-Ramos et al., 2005), transcription factor activation, (Bove and van der Vilet, 2006), gene expression (Hsu et al., 2004), cell proliferation (Attene-Ramos et al., 2005) and apoptosis (Nair et al., 2004). Numerous various enzymes including: NADPH oxidase (Fialkow et al., 1994) and xanthine oxidase (Weiss, 1986) generate ROS.

Endothelial cells, neutrophils, macrophages and other inflammatory cells generate and release ROS and NOS via an NADPH-oxidase-dependent mechanism that is mediated by membrane receptor activation of protein kinase C and phospholipase C (Cemerski, 2002; Gelinas et al., 2002). One of the major functions of these free radicals is immunological host defence, where they are generated by macrophages and neutrophils and play critical role as bactericidal, antiviral and anti-tumour agent (Wiesman and B. Halliwell, 1996; Guzik et al., 2003). H_2O_2 is considered to activate NF-κB (Siebenlist et al., 1994), which regulates the expression of multiple immune and inflammatory molecules. Generation of such ROS to a level that overwhelms tissue antioxidant defence systems, results in an oxidative stress, whose magnitude depends on the ability of the tissues to detoxify such free radicals (Ames, 1983), and consequently damaging cellular lipids, proteins and DNA inducing lipid peroxides, protein carbonyls and DNA damage (Cerutti, 1985; Henson and Johnston, 1987; Wiesman and Halliwell, 1996). Environmental agents, which generate free radical are numerous and include alcohol, numerous food sources (Ames, 1983), infectious organisms (Freeman and Crapo, 1982), most physical and chemical agents including ionizing radiation (Gisone et al., 2006), dust particles (Fujimura, 2000), asbestos (Dopp et al., 2005), and are also provoked during exercise (Xiao and Li, 2006)

ROS have been implicated in the pathogenesis of most diseased conditions (Favier, 2006). ROS are virtually implicated in every stage of vascular lesion formation, angiotensin II-dependent hypertension (Kazama et al., 2004), hyperhomocysteinemia (Dayal et al., 2006), diabetes (Moore, 2006), metabolic syndrome, (Erdos et al., 2004) inflammation together with ischaemia and reperfusion (Weiss, 1986), subarachnoid haemorrhage (Kim et al., 2002). ROS are also implicated chronic kidney diseases (Shah, 2006), liver diseases (Kouroumalis and Notas, 2006) peripheral arterial disease (Loffredo et al., 2006), Alzheimer's disease (Onyango and Khan, 2006) and many others. Genetic variabilities in intensity of free radicals generation in response to external stressors and environmental stimuli in humans are largely unknown. However, the most striking example is chronic granulomatous diseases of childhood (CGD), which are a group of disorders in which, phagocytic cells are unable to produce superoxide (O^-_2) production from the respiratory burst system, due to defaults in the phagocyte NADPH oxidase, which is a complex system consisting of membrane and cytosolic components that must assemble at the membrane for proper activation. Lack in this system makes children succumb from infection and die at an early age (Nouni et al., 1998).

Nitric oxide (NO) has been identified as a widespread and multifunctional biological messenger molecule in the central nervous system (CNS), with possible roles in neurotransmission, neurosecretion, synaptic plasticity, and tissue injury in many neurological disorders, including schizophrenia. Nitric oxide (NO) has been identified as a widespread and multifunctional biological messenger molecule produced in several types of mammalian cells including CNS, PMNs, macrophages and muscle cells. It participates in a broad range of important physiologic processes, including vasodilatation, neurotransmission, neurosecretion, synaptic plasticity, and host defence (Nathan and Xie, 1994). NO is generated from the amino acid L-arginine by three isoforms of the enzyme NO synthase; the constitutive (cNOS), the endothelial (eNOS) and the inducible (iNOS). The inducible form generates much larger amounts of NO (1,000 times fold than the other two isoforms) and its cellular production continues for many hours (Nathan, 1992). Inducible NOS has been detected in virtually every cell type, and the NO that it produces can perform both beneficial and detrimental actions, where, in physiological amounts, it is the key signal molecule in cell–cell interactions controlling vascular regulation (Clough, 1999) and neuronal communication (Yang and Hatton, 2002), NO can eliminate infiltrating microorganisms (James, 1995), reduce thrombosis, and improve blood supply to injured tissues (Gross and Wolin, 1995). NO can be detrimental, where excess production of NO can cause tissue damage and contribute to the development of a wide spectrum of diseases including septic shock, rheumatoid arthritis, cerebral ischaemia, multiple sclerosis, and diabetes (Nathan, 1992). On the contrary, NO deficiency may contribute to cardiovascular events and progression of kidney damage at end stage renal disease (Boger and Zoccali, 2003). Also loss of endothelial cell-derived nitric oxide (NO) in hypertension is a hallmark of arterial dysfunction as it is associated with decreased arterial vasodilator activity (Thakali et al., 2006). Concurrently, decreased NO levels are closely associated with preeclampsia-related endothelial dysfunction (Var et al., 2003).

Several studies suggest that the nitric oxide synthases gene polymorphism may confer increased susceptibility to several diseases. Increased NO generation has been reported to be caused by a mutation at the C150T iNOS. C150T iNOS polymorphism is associated with the risk of H pylori-related gastric cancer in a Japanese population. And is related to increasing the risk of gastric cancer in Asian countries with the highest rates of gastric cancer (Goto et al., 2006). This polymorphism is also associated with cigarette- and alcohol-induced gastric cancer in Chinese population (Shen et al., 2004). Decreased NO generation is caused by a mutation in eNOS gene promoter T-786C single nucleotide polymorphism. eNOS T-786C SNP has been shown to predict susceptibility to post-subarachnoid haemorrhage vasospasm (Khurana et al., 2004), and rheumatoid arthritis (Melchers et al., 2006). T-786C has been suggested to be an important risk factor in the development of non-arteritic anterior ischaemic optic neuropathy (NAION) among Japanese subjects and is associated with a higher risk of multivessel coronary artery disease in Caucasians (Rossi et al., 2003). It is believed that high salt intake interacts with the T-786C mutation and leads to a significant increase in the risk of hypertension (Miyaki et al., 2005). Severity of carotid atherosclerosis is linked to the eNOS G/T polymorphism (Glu298Asp variant) (Spoto et al., 2005). The endothelial nitric oxide synthase (eNOS) gene is responsible for constitutive nitric oxide synthesis and arterial vasodilatation. 4a allele of the eNOS gene is related to elevated blood pressure levels particularly among type 2 diabetic patients with coronary heart disease (Zhang et al., 2006).

Agents known to induce the expression of iNOS mRNA are numerous and some of them include UV (Artiukhov et al., 2005), ionizing radiation (Chi et al., 2006), Helium neon laser (El Batanouny and Korraa, 2002), ozone (Fakhrzadeh et al., 2004), hypoxia (Lu et al., 2006), fly ash particles (Gursinsky et al., 2006) and asbestos (Sandrini et al., 2006). Morphin (Frenklakh et al., 2006) and dioxin (Cheng et al., 2003; Kuchiiwa, 2003) administration down-regulates NO production, while hydrogen sulphide can inhibit NO production in LPS-stimulated macrophages (Oh et al., 2006). Silymarin, a polyphenolic flavonoid antioxidant, inhibits NO production and iNOS gene expression (Kang et al., 2002) and 2-Chloroethyl ethyl sulphide (CEES) is a sulphur vesicating agent and an analogue of the chemical warfare agent 2,2'-dichlorodiethyl sulphide, or sulphur mustard gas (HD) decreases iNOS expression in murine macrophages (Qui et al., 2006)

BIOLOGICAL VARIABILITY IN NUCLEAR TRANSCRIPTION FACTORS ACTIVITY

Altering gene expression is the fundamental and effective way for a cell to respond to extracellular signals or environmental stresses in short- or long-term responses (D'Angio and Finkelstein, 2000). In the short term, transcription factors are involved in mediating responses to growth factors and a variety of other extracellular signals (Cosma, 2002). Regulation of the signaling responses is governed at the genetic level by transcription factors that bind to control regions of target genes and alter their expression (Alder et al., 1999). Transcription factors are endogenous DNA-binding proteins that enhance the transcription phase proteins by regulating gene expression of a variety of genes and are required for maximal transcription of many cytokines. They are effective in the initiation, stimulation or termination of the genetic transcription process. (Chu and Chang, 1988; Escoubet-Lozach et al., 2002) While in the cytoplasm, the transcription factor is incapable of promoting transcription. The activity of transcription factors is typically regulated by phosphorylation-dependent events that can include the phosphorylation of the transcription factor itself; a signaling event occurs, leading to a change of the state of phosphorylation, followed by protein subunit translocation into the nucleus (Whitmarsh and Davis, 2000).

An example of these nuclear transcription factors is the Hypoxia-inducible factor 1 (HIF-1), which functions as a master regulator of oxygen homeostasis. HIF-1 consists of a constitutively expressed HIF-1β subunit and an oxygen-regulated HIF-1α subunit. Under hypoxic conditions, HIF-1α protein accumulates and translocates to the nucleus where it forms an active complex with HIF-1β, which activates transcription of >60 target genes important for the adaptation and survival under hypoxia (Semenza, 2003). HIF-1 target genes encode proteins that increase oxygen delivery, such as angiogenic factors (Tanimoto et al., 2003), as well as proteins that mediate adaptive responses to oxygen deprivation in ischaemic tissue, such as glucose transporters and glycolytic enzymes (Semenza, 2000). Genetic variations in HIF-1α genotype have been reported. HIF-1α may influence development of coronary artery collaterals in patients with significant coronary artery disease (Kelly et al., 2003), where the development of collateral circulation plays an important role in protecting tissues from ischaemic damage. Clinical observations have documented substantial differences in the extent of collateralization among patients with coronary artery disease, with some individuals demonstrating marked abundance and others showing nearly complete absence of these vessels (Resar et al., 2005). Mutation in two nucleotide sequence variants in exon 12 of the human *HIF1A* gene that affect the coding sequence of HIF-1α were lately reported to be present in patients with renal cell carcinoma (Clifford et al., 2001).

Hypoxia-inducible factor 1 (HIF-1) is affected by external stressors, where smoking damages the human placenta by altering the expression of HIFs, which play a key role in enhancing mediators of placental development (Genbacev et al., 2003). Carbon monoxide suppresses the activation of HIF-1 by hypoxia in a dose-dependent manner (Huang et al., 1999) by decreasing the binding of HIF-1 to its enhancer as exhibited by nuclear proteins isolated from CO-treated cells (Lui et al., 1998). HIF-1α is also overexpressed in the vast majority of patients with squamous cell cancer of the oropharynx and the degree of its expression has predictive and prognostic significance in individuals undergoing curative radiation therapy (Aebersod et al., 2001).

Another nuclear transcription factor is the nuclear transcription factor κB (NF-κB). It designates a group of critical transcription factors involved in a variety of immunologic and/or inflammatory processes and apoptosis in response to external stressors in many cell types. The predominant complex of NF-κB in most mammalian cells is p50/p65. NF-κB is required for maximal transcription of many cytokines, including tumour necrosis factor

(TNF-), interleukin-1 (IL-1), IL-2, IL-6, and IL-8, which are thought to be important in the generation of acute inflammatory responses (Siebenlist et al., 1994).

NF-κB activation was shown to be stimulated by alcohol consumption (Jaruga et al., 2004), bacterial endotoxin, chemical mitogens, viral proteins (Parsonnet, 1995), and certain chemical agents including ozone (Fakhrzadeh et al., 2004), arsenic and chromium (Ding et al., 2000; Dong, 2002) and asbestos (Faux and J. Howden, 1997). Excessive activation of NF-κB in leucocytes is stimulated by Short wavelength UV (Li and Karin, 1998; Wu et al., 2004) and ionising radiation in a dose-response pattern (Iarilin, 1999). There is an increasing body of evidence suggesting a role for NF-κB in carcinogenesis (Baldwin, 1996). For example, NF-κB is implicated in signaling tumour promoter-induced transformation and is activated by viral-transforming proteins (Dahr et al., 2002). The importance of NF-κB cannot be overstated, as failure in any of the mechanisms leading to NF-κB activation can have serious consequences for the cell. Impaired ability to signal and activate specific gene transcription through NF-κB has been directly linked to immunodeficiency (Uzel, 2005). Agents such as hydrogen sulphide can inhibit NO NF-κB activation in LPS-stimulated macrophages (Oh et al., 2006). Silymarin, a polyphenolic flavonoid antioxidant (Kang et al., 2002), and mustard gas analogue (Qui et al., 2006) individually inhibits NF-κB activation.

Concurrently, the constitutive activation of NF-κB has been linked with a wide variety of human diseases, including asthma, atherosclerosis, AIDS, rheumatoid arthritis, diabetes, osteoporosis, Alzheimer's disease, and cancer. Several agents are known to suppress NF-κB activation, including Th2 cytokines (IL-4, IL-13, and IL-10), interferons, endocrine hormones (LH, HCG, MSH and GH), phytochemicals, corticosteroids, and immuno-suppressive agents. Because of the strong link of NF-κB with different stress signals, it has been called a "smoke-sensor" of the body (Ahn and Aggarwal, 2005). Polymorphism in the promoter region of the human NFKB1 gene was found to be associated with susceptibility to ulcerative colitis (Borm et al., 2005). Hippocampal pyramidal neurons in mice lacking the p50 subunit of NF-κB (p50$^{-/-}$) exhibit increased damage after exposures to excitotoxins (Yu et al., 1999; Kassed et al., 2002).

Antioxidant Enzymes, encoded by numerous genes in mammalian systems, have been shown to be responsive to oxidants, although a systematic mechanism for gene regulation by oxidative stress has not been elucidated. Oxidative stress has been shown to alter the expression of mammalian antioxidant enzymes including superoxide dismutase (SOD), glutathione peroxidase (GPx), α;-glutamylcysteine synthetase, catalase, glutathione S-transferase and quinone reductase (Amstad and Cerutti, 1990). SOD catalyses the dismutation of superoxide radicals into hydrogen peroxide that glutathione peroxidase and catalase break down into water (Halliwell, 1994). Induction of antioxidants by oxidative stress may both function in intracellular signalling and serve to protect cells from further oxidant injury. Accordingly, an imbalance in antioxidant mechanisms may influence cellular sensitivity to free-radical damage and alter susceptibility to disease (Ames, 1983).

Antioxidant enzymes especially the inducible SOD enzyme has been shown to be elevated in individuals at risk of exposure to low doses of various stresses. Superoxide dismutases (SODs) are the major antioxidant enzymes that inactivate superoxide and thereby control oxidative stress as well as redox signaling.

There are three types of mammalian SODs: manganese SOD (MnSOD) on the mitochondria, copper-zinc SOD on the cytosol and extracellular SOD in extracellular compartments (Zelko et al., 2002). Cigarette smokers (Kanehira et al., 2006), asbestos exposed workers (Kamal et al., 1992), radiotherapy exposed patients (Vucic et al., 2006), and athletes have higher levels of SOD enzymes compared to controls (Elosua et al., 2003). Asthmatic Chinese patients were shown to have elevated erythrocyte SOD activities in comparison with healthy controls (Mak et al., 2006). Also the antioxidant enzyme SOD in samples from patients with malignant tumours revealed up to 45-fold greater than that of controls (Yoshii et al., 1999; Soini et al., 2006).

MnSOD locus has been linked to the atherogenic lipoprotein phenotype, i.e. the excess of small dense LDL in humans (Allayee et al., 1998). Overexpression of MnSOD has been shown to protect transgenic mice against myocardial ischaemia (Cheng et al., 1998) and in rabbits to reverse vascular dysfunction in carotid arteries without atherosclerotic changes, but not in vessels with atherosclerotic plaques (Zanetti et al., 2001). Overexpression of MnSOD inhibits in vitro oxidation of LDL by endothelial cells (Fang et al., 1998) and ox-LDL is able to induce the expression of MnSOD in macrophages (Kinscherf et al., 1997). The apoE-deficient mice lacking MnSOD had more severe atherosclerosis compared to the apoE-deficient mice (Ballinger et al., 2002). In addition, the signal sequence polymorphism of the MnSOD gene has been associated with non-familial dilated cardiomyopathy in Japanese subjects (Hiroi et al., 1999), but it has not been investigated earlier in human atherosclerosis. It is suggested that MnSOD has a protective role for in retinal capillary cell death and, ultimately, in the pathogenesis of diabetes induced retinopathy (Kowluruet et al., 2006).

Extracellular SOD (ECSOD or SOD 3) is a major extracellular antioxidant enzyme. It distributed in the extracellular matrix of many tissues and especially blood vessels (Strålin et al., 1995). A fundamental property of ECSOD is its affinity, through its heparin-binding domain (HBD), for heparan sulphate proteoglycans located on cell surfaces and in extracellular matrix (Sandström et al., 1992, 1993). It has been demonstrated that vascular effects of ECSOD require an intact HBD (Fattman et al., 2003). A common genetic variant with a substitution in the HBD (ECSOD(R213G) was reported recently to be associated with ischaemic heart disease. is a major extracellular antioxidant enzyme and it is suggested that human beings carrying ECSOD(R213G) are predisposed to vascular diseases (Chu et al., 2005)

Substitution of arginine-213 by glycine (R213G), which results from a C-to-G transversion at the first base of codon 213, is a common human gene variant in the HBD of ECSOD (Sandström et al., 1994). Plasma concentrations of ECSOD are increased greatly in the 2–5% of the population that carries $ECSOD_{R213G}$ (Adachi et al., 1996). This alteration in the HBD reduces affinity for heparin but does not affect the enzymatic activity of ECSOD. ECSOD affected individuals in Sweden, who did not have major phenotypic abnormalities, but there was a trend for increased triglycerides and body weight (Marklund et al., 1997). A recent large study in Denmark suggested a 2.3-fold increase in risk of ischaemic heart disease in heterozygotes carrying $ECSOD_{R213G}$, with a 9-fold increase after adjustment for plasma levels of ECSOD (Busse, 2001).

Glutathione peroxidase (GPX1), is an intracellular selenium-dependent enzyme that is ubiquitously expressed and detoxifies hydrogen and lipid peroxides plays a significant role in protecting cells from the oxidative stress induced by ROS. GPX1 levels are particularly responsive to fluctuations in selenium levels compared with other selenoproteins. Mice null for Gpx1 and GPx2 exhibit severe ileocolitis at a young age and develop microflora-associated cancers in the lower gastrointestinal tract (Chu et al., 2004).

GPX1 is polymorphic at codon 198, resulting in either a proline or a leucine at that position, and the frequency of the leu allele is strongly associated with an increase in the risk for lung (Ratnasinghe et al., 2000), and possibly breast cancer (Hu et al., 2004). The identity of the amino acid at codon 198 (proline or leucine) has functional consequences with regard to level of enzyme activity in response to increasing levels of selenium provided to cells in culture (Hu and Diamond, 2003). Loss of heterozygosity (LOH) occurs at the *GPX1* locus during the development of several cancer types, including those occurring in lung, breast, and head and neck (Moscow et al., 1994). In the case of head and neck cancers, *GPX1* allelic loss was shown to occur in histopathologically normal tissue adjacent to tumours, indicating that loss at this locus may be an early event in cancer evolution (Hu et al., 2004).

Meanwhile, GPX1 codon 198 polymorphism was associated with an increased risk of lung cancer and individuals carrying the Pro/Leu or Leu/Leu genotype of GPX1 were at a higher risk for lung cancer and were shown to have high urinary 8-OH-dG concentrations compared to the individuals with the GPX1 Pro/Pro genotype. On the other hand, the polymorphism of the hOGG1 gene did not affect the lung cancer risk and the oxidative DNA damage (Lee et al., 2006)

Catalase enzyme is an endogenous antioxidant enzyme that neutralizes ROS by converting H_2O_2 into H_2O and O_2. A $-262C \rightarrow T$ polymorphism in the promoter region of the gene (CAT) is associated with risk of several conditions related to oxidative stress. It is plausible that the endogenous variability associated with this polymorphism plays a role in the host response to oxidative stress and progression to breast cancer (Ambrosone, 2000). Asthma patients due to polymorphism in the C allele of catalase gene at C-262T had elevated erythrocyte CAT activities in comparison with healthy controls in Hong Kong (Mak et al., 2006).

BIOLOGICAL VARIABILITY IN DNA REPAIR CAPACITY

DNA repair processes restore the normal nucleotide sequence and DNA structure after damage. It assists in maintaining genomic integrity by removing inappropriate bases and other possible deleterious lesions from DNA. Overlap among these pathways exists in terms of the types of damage removed by each. The complex series of DNA repair pathways employ many different proteins. Numerous DNA repair mechanisms have been identified: (i) site-specific repair (ii) nucleotide excision repair (NER), (iii) base excision repair, (iv) mismatch repair (MMR), (v) direct reversal of the damage, in which no incision is made in the backbone of the DNA (Bohr et al., 1987). An increased incidence of neoplasia is correlated with a defect in the repair or replication of damaged DNA in some human genetic diseases. Examples of such hereditary disorders include xeroderma pigmentosum, ataxia telangiectasia, Fanconi's anaemia, and Bloom's syndrome (Fuss and Cooper, 2006). Several types of cancer have been linked with defects in all types of DNA repair pathways. For example, hereditary nonpolyposis colon cancer results from defects in MMR genes, and hereditary breast cancer is caused by mutations affecting the breast cancer associated proteins BRCA1 or BRCA2 that play a role in DSB repair by homologous recombination (Fuss and Cooper, 2006). Patients with multiple sclerosis were shown to exhibit reduced DNA excision reparation capacity in their peripheral blood lymphocytes which correlated with the disease severity but not with its duration (Moskaleva et al., 1988).

The biological consequences of unrepaired or misrepaired DNA damage depend on the precise locations of the lesions. DNA lesions at specific sites in the mammalian genome can lead to mutation, recombination, gene amplification, translocation, and other chromosomal abnormalities. These changes in turn may result in malignant transformation, faulty differentiation patterns, or cell death. Thus, it has become clear that damage to DNA at particular loci can cause activation of the protooncogenes and inactivation of tumour suppressor genes that may be implicated in subsequent tumourigenesis and age-related diseases (1). There is increasing evidence that human atherosclerosis is associated with damage to the suppressor gene p53 of both circulating cells, and cells of the vessel wall. DNA damage produces a variety of responses, including cell senescence, apoptosis and DNA repair. Decreased endogenous levels of p53 promotes plaque formation in vascular muscle smooth cells and stromal cells by promoting apoptosis, while inhibiting apoptosis in macrophages, leading to atherosclerosis development (Mercer et al., 2005). Similarly, in rheumatoid arthritis, oxidative damage caused by inflammation appears to cause *p53* mutations in synovium Most of the p53 mutations in RA are characterized by transition base changes (Inazuka et al., 2000). Furthermore, certain p53 mutations in RA are dominant negative and can suppress endogenous wild-type p53 function (Han et al., 1999). Inactivation of p53 protein can recapitulate many of the phenotypic changes observed in RA, such as increased proliferation and invasion of synovial cells (Aupperle et al., 1998; Pap et al., 2001). Elevated levels of DNA alkylation damage have been detected in schistosome-infected bladders and are accompanied by an inefficient capacity of DNA repair mechanisms. Consequently, high frequency of $G \rightarrow A$ transition mutations were observed in the H-ras gene and at the CpG sequences of the p53 tumour suppressor gene (Badawi, 1996). It is suggested that the excess of transitions at CpG dinucleotides in squamous cell carcinoma induced by Bilharzial infections results from nitric oxide (NO) produced by the inflammatory response provoked by schistosomal eggs. NO could produce such mutations directly, by deamination of 5-methylcytosine, and indirectly, following conversion to nitrate, bacterial reduction to nitrite and endogenous formation of urinary N-nitroso compounds. These produce O6-alkylguanines in DNA, leading to very high rates of $G:C \rightarrow A:T$ transitions, a process possibly augmented by inefficient repair of alkylated bases at CpG dinucleotides (Warren et al., 1995). DNA repair capacity decreases by ageing (Cabelof et al., 2006) giving clue to the increased incidence of ageing associated diseases.

INDUCTION OF APOPTOSIS

Apoptosis or programmed cell death is a gene-regulated process in which a coordinated series of morphological changes such as nucleus and chromatin condensation, cell membrane blebbing and fragmentation of the cell into membrane-bound apoptotic bodies occurs, resulting in cell death (Barazzone and White, 2000). It is accepted that morphological changes observed during programmed cell death are the consequence of an activation of caspases cascade (Green, 1998). At least two main signaling pathways have been postulated to participate in this process. The first one involves membrane receptors called death receptors (i.e. TNF receptor-1 and Fas/Apo-1) (Mak and Yeh, 1999; Hengartner, 2000), and the second one relies on the cell's ability to sense changes in the ratio between the protein levels of the members of the Bcl-2 family. Bcl-2 prevents apoptosis induced by a wide range of stimuli, suggesting that different pathways of transduction signals converge at this point (Adams and Cory, 1998; Zornig et al., 2001).

Membrane receptors include FAS (also known as TNFSF6, CD95, or APO-1), which is a cell surface receptor that plays a central role in apoptotic signaling in many cell types (Nagata and Goldstein, 1995). This receptor interacts with its natural ligand FASL (also known as CD95L), a member of the tumour necrosis factor superfamily, to initiate the death signal cascade, which results in apoptotic cell death (Reichmann, 2002). An immuno-privileged status for tumours is established via the FAS-mediated apoptosis of tumour-specific lymphocytes (Nagata and Goldstein, 1995). Decreased expression of FAS and/or increased expression of FASL favors malignant transformation and progression [for a review, see (Muschen et al., 2000). In addition, functional germline and somatic mutations in the FAS gene and perhaps also in the FASL gene that impair apoptotic signal transduction are associated with a high risk of cancer. Thus, the FAS/FASL system appears to have a role in the development and progression of cancer (Lee et al., 1999).

Mitochondria play a key role in the apoptotic pathway through the release of several factors from the intermembrane space to the cytoplasm, such as cytochrome C (Liu et al., 1996). It has been suggested that this pathway could be regulated by the relative levels and subcellular distribution of Bcl-2 family proteins. The antiapoptotic members (i.e. Bcl-2 or Bcl-X_L) are mostly associated to the outer membrane of mitochondria and inhibit cytochrome C release. On the other hand, the proapoptotic molecules such as Bax, Bad, or Bid are cytosolic proteins; they translocate to the mitochondria and trigger cytochrome C release on apoptosis induction Several authors have identified a variety of proteins related with Bcl-2, such as Bax, Bak, Bid, and the different Bcl-X isoforms, which can either promote or prevent apoptosis (Cory, 1995).

Apoptosis plays an important role in sculpting the developing organism and eliminating unwanted or potentially dangerous cells throughout life. Abnormal regulation of apoptosis is associated with a variety of diseases. Cells that should die but do not can cause cancer and autoimmune diseases, whereas cells that should not die but do can cause stroke and neurodegenerative disorders (Thompson, 1995). The adaptive increase in apoptosis that accompanies the oncogene-activated dysregulation in proliferation selectively eliminates poten-tially preneoplastic cells in hyperplastic foci. Acquired resistance to apoptosis appears to be a pivotal event in the transition to malignancy (Schulte-Hermann et al., 1995). The homeostatic balance between cell proliferation and apoptosis in the maintenance of constant cell numbers may provide a hormetic effect by minimizing the consequences of proliferation-related mutagenesis during tumour promotion (McDonnell, 1993; Meikrantz and Schlegel, 1995).

In conclusion, it has been long established that when organisms or cells are exposed to low levels of specific harmful physical or chemical agents, a beneficial physiologic effect is observed. Concurrently; exposure to sublethal challenges of stress may rejuvenate the cell by repairing damage before the challenge and may provide transient protection against further damage from subsequent sublethal or lethal challenges with a different otherwise harmful physical or chemical stressor. Due to the wide variation in individual responses to exogenous agents is believed to result from the great diversity in responsiveness to risk factors in the environment. Detectable gene mutations and alterations in signal transduction pathways together with modified post-translational proteins offer valuable molecular biomarkers for occupational and environmental human biomonitoring applied for the identification of potentially hazardous exposures before adverse health effects appear and allow the establishment of exposure limits in order to minimize the likelihood of significant health risks. Concurrently, most of these elicited gene expressions are also expressed in tissue transformation and progression of tumours and are similar to the hallmarks used for cancer prognosis. Some of these proteins represent protective mechanisms against different environmental stresses, while others amplify adaptation to environmental conditions. The resultant balance between protective proteins and adaptive proteins seem to determine an individual likelihood to develop diseases.

References

Adachi, T., Yamada, H., Yamada, Y., Morihara, N., Yamazaki. N., Murakami, T., Futenma, A., Kato, K., Hirano, K., 1996, Substitution of glycine for arginine-2.13 in extracellular-superoxide dismutase impairs affinity for heparin and endothelial cell surface. *Biochem. J.* 313: 235–239.

Adams, J. and Cory, S., 1998, The Bcl-2 protein family: arbiters of cell survival. *Science* 281: 1322–1326.

Aebersod, D., Burri, P., Beer, K., Laissue J., Djonov, V., Greiner, R., and Semenza G. L., 2001, Expression of hypoxia-inducible factor-1alpha: a novel predictive and prognostic parameter in the radiotherapy of oropharyngeal cancer. *Cancer Res.* 61(7): 2911–2916.

Ahn, K., and Aggarwal, B., 2005, Transcription Factor NF-{kappa}B: A Sensor for Smoke and Stress Signals. *Ann. NY Acad. Sci.* 1056: 218–233.

Alder, V., Yin, Z., Tew, K. D., and Ronai, Z., 1999, Role of redox potential and reactive oxygen species in stress signaling. *Oncogene* 18: 6104–6111.

Allayee, H., Aouizerat, B., Cantor, R., Dallinge-Thie, G., Krauss, R., Lanning, C., Rotter, J., Lusis, A., and de Briun, T., 1998, Families with familial combined hyperlipidemia and families enriched for coronary artery disease share genetic determinants for the atherogenic lipoprotein phenotype. *Am. J. Hum. Genet.* 63(2): 577–585.

Ambrosone, C., 2000, Oxidants and antioxidants in breast cancer. *Antioxid. Redox Signal* 2: 903–917.

Ames, B., 1983, Dietary carcinogens and anticarcinogens, *Science* 121: 1250–1264.

Amstad, P., and Cerutti, P., 1990, Genetic modulation of the cellular antioxidant defense capacity, *Environ. Health Perspect.* 88: 77–82.

Artiukhov, V., Gusinskaia, V., and Mikhileva, E., 2005, Level of nitric oxide and tumor necrosis factor-alpha production by human blood neutrophils under UV-irradiation. *Radiat. Biol. Radioecol.* 45(5): 576–580.

Attene-Ramos, M., Kitiphongspattana, K., Ishii-Schrade, K., and Gaskins, H., 2005, Temporal changes of multiple redox couples from proliferation to growth arrest in IEC-6 intestinal epithelial cells, *Am. J. Physiol. Cell Physiol.* 289(5): C1220–C1228.

Au, W., Wilkinson, G., Tyring, S., Legator, M., El Zein, R., Hallberg, L., and Heo, M., 1996, Monitoring populations for DNA repair deficiency and for cancer susceptibility. *Environ. Health Perspect.* 104(Suppl 3): 579–584.

Aupperle, K., Boyle, D., Hendrix, M., Seftor, E., Zvaifler, N., Barbosa, M., and Firestein, G., 1998, Regulation of synoviocyte proliferation, apoptosis, and invasion by the p53 tumor suppressor gene. *Am. J. Pathol.* 152: 1091–1098.

Badawi, A., 1996, Molecular and genetic events in schistosomiasis-associated human bladder cancer: role of oncogenes and tumor suppressor genes. *Cancer Lett.* 105(2): 123–138.

Baldwin A., 1996, The NF-κB and IκB Proteins: New Discoveries and Insights. *Ann. Rev. Immunol.* 14: 649–681.

Ballinger, S., Patterson, C., Knight-Lozano, C., Burow, D. et al., 2002, Mitochondrial integrity and function in atherogenesis. *Circulation* 106: 544–549.

Barazzone, C. and White, C., 2000, Mechanisms of cell injury and death in hyperoxia. Role of cytokines and Bcl-2 family proteins. *Am. J. Resp. Cell Mol. Biol.* 22: 517–519.

Brennan, P., 2002, Gene–environment interaction and aetiology of cancer: what does it mean and how can we measure it? *Carcinogenesis* 23(3): 381–387.

Boger, R. and Zoccali, G., 2003, ADMA: a novel risk factor that explains excess cardiovascular event rate in patients with end-stage renal disease. *Atheroscler. Suppl.* 4(4): 23–28.

Bohr, V., Phillips, D., and Hanawalt, P., 1987, Heterogeneous DNA damage and repair in the mammalian genome. *Cancer Res.* 47: 6426–6436.

Borm, M., van Bodegraven, A., Mulder, C., Kraal, G., Bouma, G., 2005, A NFKB1 promoter polymorphism is involved in susceptibility to ulcerative colitis. *Int. J. Immunogenet.* 32(6): 401.

Bove, P. and van der Vilet, A., 2006, Nitric oxide and reactive nitrogen species in airway epithelial signaling and inflammation. *Free Rad. Biol. Med.* 41(4): 515–527.

Busse, W. and Lemanske, R., 2001, Asthma. *N. Engl. J. Med.* 344: 350–362.

Cabelof. D., Raffoul, J., Ge, Y., Van Remmen, H., Matherly, L., and Hedari, A., 2006, Age-related loss of the DNA repair response following exposure to oxidative stress. *J. Gerontol. A Biol. Sci. Med. Sci.* 61(5): 427–434.

Cemerski, S., Cantagrl, A., Van Meerwijki, J., and Romagnoli, P., 2002. Immediate and delayed VEGF-mediated NO synthesis in endothelial cells: role of PI3K, PKC and PLC pathways. *Br. J. Pharmacol.* 137(7): 1021–1030.

Cerutti, P., 1985, Prooxidant states and tumor promotion. *Science* 227: 375–381.

Chen, L., Knutsen, S., Shavlik, D., Beeson W., Petersen, F., Ghamsary, M., and David, A., 2005, The association between fatal coronary heart disease and ambient particulate air pollution: are females at greater risk? *Environ. Health. Perspect.* 113 (12): 1723–1729.

Cheng, S., Kuchiiwa, S., Ren, X., Gao, H., Kuchiiwa, T., and Nakagawa, S., 2003, Dioxin exposure down-regulates nitric oxide synthase and NADPH-diaphorase activities in the hypothalamus of Long-Evans rat. *Neurosci. Lett.* 345(1): 5–8.

Cheng, S., Chen, Z., Siu, B., Ho, Y., Vincent R et al., 1998, Overexpression of MnSOD protects against myocardial ischemia/reperfusion injury in transgenic mice. *J. Mol. Cell. Cardiol.* 30: 2281–2289.

Cory, S., 1995, Regulation of lymphocyte survival by the bcl-2 gene family. *Annu. Rev. Immunol.* 13: 513–543.

Chi, C., Ozawa, T., and Anazi, T., 2006, In vivo nitric oxide production and iNOS expression in X-ray irradiated mouse skin. *Biol. Pharm. Bull.* 29(2): 348–353.

Chu, G. and Chang, E., 1988, Xeroderma Pigmentosum Group E Cells Lack a Nuclear Factor that Binds to Damaged DNA, *Science* 242: 564–567.

Chu, F., Esworthy, R., Chu, P., Longmate, J., Huycke, M., Wilczynski, S., and Doroshow, J., 2004, Bacteria-induced intestinal cancer in mice with disrupted Gpx1 and Gpx2 genes. *Cancer Res.* 64: 962–968.

Chu, Y., Alwahdani, A., Iida, S., Lund, D., Faraci, F., and Heistad, D., 2005, Vascular effects of the human extracellular superoxide dismutase R213G variant. *Circulation* 112(7): 1047–1053.

Clifford, S., Astuti, D., Hooper, L., Maxwell, P., Ratcliffe, P., and Maher E., 2001, The pVHL-associated SCF ubiquitin ligase complex: molecular genetic analysis of elongin Band C, Rbx1, and HIF-1α in renal cell carcinoma. *Oncogene* 20: 5067–5074.

Clough, G., 1999, Role of nitric oxide in the regulation of microvascular perfusion in human skin in vivo. *J. Physiol.* 516: 549–557.

Cosma, M., 2002, Ordered recruitment: gene-specific mechanism of transcription activation. *Mol. Cell* 10(2): 227–236.

D'Angio, C., and Finkelstein, J., 2000, Oxygen regulation of gene expression: a study in opposites. *Mol. Genet. Metab.* 71: 371–380.

Dahr, A., Young, M., and Colbum, N., 2002, The role of AP-1, NF-kappaB and ROS/NOS in skin carcinogenesis: the JB6 model is predictive. *Mol. Cell Biochem.* 234–235(1–2): 185–193.

Dayal, S., Wilson, K., Leo, L., Aming, E., Bottiglieri, T., and Lentz, S., 2006, Enhanced susceptibility to arterial thrombosis in a murine model of hyperhomocysteinemia. *Blood* 108(7): 2237–2243.

Ding, M., Shi, X., Castranova, V., and Vallyathan, V., 2000, Predisposing factors in occupational lung cancer: inorganic minerals and chromium. *J. Environ. Pathol. Toxicol. Oncol.* 19(1–2): 129–138.

Dong, Z., 2002, The molecular mechanisms of arsenic-induced cell transformation and apoptosis. *Environ. Health Perspect.* 110(Suppl 5): 757–759.

Dooly, M. and Hogan, S., 2003, Environmental epidemiology and risk factors for autoimmune disease. *Curr. Opin. Rheumatol.* 15(2): 99–103.

Dopp, E., Yadav, S., Ansari, F., Bhattacharya, K., von Recklinghausen, U., Rauen, U., Rodelsperger, K., Shokouhi, B., Geh, S., and Rahman Q., 2005, ROS-mediated genotoxicity of asbestos-cement in mammalian lung cells in vitro. *Part. Fibre Toxicol.* 6: 2–9.

Eggleston, P., Buckley, T., Breysse, P., Wills-Karp, M., Kleeberger, S., and Jaakkola, J., 1999, The environment and asthma in U.S. inner cities. *Environ. Health Perspect.* 107(Suppl 3): 439–450.

Erdos, B., Snipes, J., Miller, A., and Busija, D., 2004, Cerebrovascular dysfunction in Zucker obese rats is mediated by oxidative stress and protein kinase C. *Diabetes* 53: 1352–1359.

El Batanouny, M. and Korraa, S., 2002, Effects of low intensity laser on the activity and expression of nitric oxide synthase in human polymorphonuclear leucocytes in vitro, *Arab J. Lab. Med.* 28(3): 289–297.

Elena, S. and de Visser, J., 2003, Environmental stress and the effects of mutation. *J. Biol.* 2(2): 12–17.

Elosua, R., Molina, L., Fito, M., Arquer, A., Sanchez-Quesada, J., Covas, M., Ordonez-Llanos, J., and Marrugat, J., 2003, Response of oxidative stress biomarkers to a 16-week aerobic physical activity program, and to acute physical activity, in healthy young men and women. *Atherosclerosis* 167(2): 327–334.

Escoubet-Lozach, L., Glass, C. K., and Wasserman, S. I, 2002, The role of transcription factors in allergic inflammation. *Allergy Clin. Immunol.* 110(4): 553–564.

Fakhrzadeh, L., Laskin, J., and Laskin, D., 2004, Ozone-induced production of nitric oxide and TNF-alpha and tissue injury are dependent on NF-kappaB p50. *Am. J. Physiol. Lung Cell Mol. Physiol.* 87(2): L279–L285.

Fang, X., Weintraub, N., Rios, C., Chappell D., et al., 1998, Overexpression of human superoxide dismutase inhibits oxidation of low-density lipoprotein by endothelial cells. *Circ. Res.* 82: 1289–1297.

Fattman, C., Schaefer, L., and Oury, T., 2003, Extracelluar superoxide dismutase in biology and medicine. *Free Radic. Biol. Med.* 35: 236–256.

Faux, S. and Howden, J., 1997, Possible Role of Lipid Peroxidation in the Induction of NF-κB and AP-1 in RFL-6 Cells by Crocidolite Asbestos: Evidence following Protection by Vitamin E. *Environ. Health Perspect.* 105(Suppl 5): 1127–1130.

Favier, A., 2006, Oxidative stress in human diseases. *Ann. Pharm. Fr.* 64(6): 390–396.

Fialkow, L., Chan, C., Rotin, D., Grinstein, S., and Downey, G., 1994, Activation of the mitogen-activated protein kinase signaling pathway in neutrophils. Role of oxidants. *J. Biol. Chem.* 269(49): 31234–31242.

Freeman, B., and Crapo, J., 1982, Biology of disease: free radicals and tissue injury. *Lab. Invest.* 47: 412–426.

Frenklakh, L., Bhat, R., Bhaskaran, M., Sharma, S., Sharma, M., Dinda, A., and Singhal, P., 2006, Morphine-induced degradation of the host defense barrier role of intestinal mucosal injury. *Dig. Dis. Sci.* 51(2): 318–325.

Fujimura, N., 2000, Pathology and pathophysiology of pneumoconiosis. *Curr. Opin. Pulm. Med.* 6: 140–144.

Fuss, J. and Cooper, P., 2006, DNA repair: dynamic defenders against cancer and aging. *PLoS. Biol.* 4(6): e203.

Genbacev, O., McMaster, M., Zdravkovic, T., and Fischer, S., 2003, Disruption of oxygen-regulated responses underlies pathological changes in the placentas of women who smoke or who are passively exposed to smoke during pregnancy. *Reprod. Toxicol.* 17(5): 509–518.

Gelinas, D., Bernatchez, P., Rollin, S., Bazan, N., and Sirois, M., 2002, Reactive oxygen species differentially affect T cell receptor-signaling pathways. *J. Biol. Chem.* 277(22): 19585–19593.

Gisone, P., Robello, E., Sanjurjo, J., Dubner, D., Perez Mdel R., Michelin, S. and Puntarulo, S., 2006, Reactive species and apoptosis of neural precursor cells after gamma-irradiation. *Neurotoxicology* 27(2): 253–259.

Goto, Y., Ando, T., Naito, M., Goto, H., and Hamajima N., 2006, Inducible nitric oxide synthase polymorphism is associated with the increased risk of differentiated gastric cancer in a Japanese population. *World J. Gastroenterol.* 12(39): 6361–6365.

Green, D., 1998, Apoptotic pathways: the roads to ruin. *Cell* 94: 695–698.

Greenblatt, M., Bennett, W., Hollstein, M., and Harris, C., 1994, Mutations in the *p53* tumor suppressor gene: clues to cancer etiology and molecular Pathogenesis. *Cancer Res.* 54(18): 4855–4878.

Gross, S. and Wolin, M., 1995, Nitric oxide: pathophysiological mechanisms. *Ann. Rev. Physiol.* 57: 737–769.

Gursinsky, T., Ruhs, S., Friess, U., Diabate, S., Krug, H., Silber, R., and Simm, A., 2006, Air pollution-associated fly ash particles induce fibrotic mechanisms in primary fibroblasts. *Biol. Chem.* 387(10–11): 1411–1420.

Guzik, T., Korbut, R., and Adamek-Guzik, T., 2003, Nitric oxide and superoxide in inflammation and immune regulation. *J. Physiol. Pharmacol.* 54(4): 469–487.

Halliwell, B., 1994, Free radicals, antioxidants, and human disease: curiosity, cause, or consequence? *Lancet* 344: 721–724.

Han, Z., Boyle, D. L., Shi, Y., Green, D., and Firestein, G., 1999, Dominant-negative p53 mutations in rheumatoid arthritis. *Arthritis Rheum.* 42: 1088–1092.

Henson, P. and Johnston, R., 1987, Tissue injury in inflammation, oxidants, proteinases and cationic proteins. *J. Clin. Invest.* 79: 669–674.

Hengartner, M., 2000, The biochemistry of apoptosis. *Nature* 407: 770–776.

Hiroi S., Harada, H., Nishi, H., Satoh, M., Nagai, R., and Kimura, A., 1999, Polymorphisms in the SOD2 and HLA-DRB1 genes are associated with nonfamilial idiopathic dilated cardiomyopathy in Japanese. *Biochem. Biophys. Res. Commun.* 261: 332–339.

Hsu, Y., Chen, J., Chang, C., Chen, C., Liu, J., Chen, T., Jeng, C., Chao, H., and Chen, T., 2004, Role of reactive oxygen species-sensitive extracellular signal-regulated kinase pathway in angiotensin II-induced endothelin-1 gene expression in vascular endothelial cells. *J. Vasc. Res.* 41(1): 64–74.

Hu, Y. and Diamond, A., 2003, Role of glutathione peroxidase 1 in breast cancer: loss of heterozygosity and allelic differences in the response to selenium. *Cancer Res.* 63: 3347–3351.

Hu, Y., Dolan, M., Bae, R., Yee, H., Roy, M., Glickman, R., Kiremidjian-Schumacher, L., and Diamond, A., 2004, Allelic Loss at the GPx-1 Locus in Cancer of the Head and Neck. *Biol. Trace Elem. Res.* 101: 97–106.

Huang, L., Willmore, W., Gu, J. Goldberg, M., and Bunn, H., 1999, Inhibition of hypoxia-inducible factor 1 activation by carbon monoxide and nitric oxide, Implications for oxygen sensing and signaling. *J. Biol. Chem.* 274(13): 9038–9044.

Iarilin, A., 1999, Radiation and immunity, interference of ionizing radiation with key immune processes. *Radiat. Biol. Radioecol.* 39(1): 181–189.

Inazuka, M., Tahira, T., Horiuchi, T., Harashima, S., Sawabe, T., Kondo, M., Miyahara, H., and Hayashi, K., 2000, Analysis of p53 tumour suppressor gene somatic mutations in rheumatoid arthritis synovium. *Rheumatology* 39: 262–266.

James, S., 1995, Role of nitric oxide in parasitic iInfection. *Microbiol. Rev.* 59(4): 533–547.

Jaruga, B., Hong, F., Kim, W., Sun, R., Fan, S., and Gao, B., 2004, Chronic alcohol consumption accelerates liver injury in T cell-mediated hepatitis: alcohol disregulation of NF-{kappa}B and STAT3 signaling pathways. *Am. J. Physiol. Gastrointest. Liver Physiol.* 87(2): G471–G47.

Kamal, A., Al Khafif, M., Massoud, A., and Korraa, S., 1992, Plasma lipid peroxide and blood superoxide dismutase among asbestos exposed workers. *Am. J. Indust. Med.* 21(31): 341–352.

Kandacova, N. and Zagrebel'naia, G.V., 2004, The influence of peroxide Radicals & nitric oxide on DNA synthesis in tumour cells. *Biomed. Khim.* 50: 566–575.

Kanehira, T., Shibata, K., Kashiwazaki, H., Inoue, N., and Morita, M., 2006, Comparison of antioxidant enzymes in saliva of elderly smokers and non-smokers. *Gerodontology.* 23(1): 38–42.

Kang, J., Jeon, Y., Kim, H., Hans, S., Han, S., and Yang, K., 2002, Inhibition of inducible nitric-oxide synthase expression by silymarin in lipopolysaccharide-stimulated macrophages. *J. Pharmacol. Exp. Ther.* 302(1): 138–144.

Kassed, C., Willing, A., Garbuzova-Davis, A., Sanberg, P., Pennypacker, K., 2002, Lack of NF-κB p50 exacerbates degeneration of hippocampal neurons after chemical exposure and impairs learning. *Exp. Neurol.* 176: 277–288.

Kazama, K., Anrather, J., Zhou, P., Girouard, H., Frys, K., Milner, T., and Iadecola, C., 2004, Angiotensin II impairs neurovascular coupling in neocortex through NADPH oxidase-derived radicals. *Circ. Res.* 95(10): 1019–1026.

Kelly, B., Hackett, S., Hirota, K., Oshima, Y., Cai, Z., Berg-Dioxin, S., Rowan, A., Yan, Z., Campochiaro, P., and Semensa, G., 2003, Cell type-specific regulation of angiogenic growth factor gene expression and induction of angiogenesis in nonischemic tissue by a constitutively active form of hypoxia-inducible factor 1. *Circ. Res.* 93: 1074–1081.

Khurana, V., Sohni, Y., Mangrum, W., McClelland, R., O'Kane, D., Meyer, D. and Meisser, I., 2004, Endothelial nitric oxide synthase gene polymorphisms predict susceptibility to aneurysmal subarachnoid hemorrhage and cerebral vasospasms. *J. Cereb. Blood Flow. Metab.* 24(3): 291–297.

Kim, D., Suh, Y., Lee, M., Kim, K., Lee J., Lee, H., Hong, K., and Kim, C., 2002, Vascular NAD(P)H oxidase triggers delayed cerebral vasospasm after subarachnoid hemorrhage in rats. *Stroke* 33: 2687–2691.

Kinscherf, R., Deigner, H., Usinger, C., Pill, J., Wagner, M., Kamencic, H., Hou, D., Chen, M., Schmiedt, W., Schrader, M., Kovacs, G., Kato, K., and Metz, J., 1997, Induction of mitochondrial manganese superoxide dismutase in macrophages by oxidized LDL: its relevance in atherosclerosis of humans and heritable hyperlipidemic rabbits. *FASEB J.* 11: 1317–1328.

Kouroumalis, E. and Notas, G., 2006, Pathogenesis of primary biliary cirrhosis: a unifying model. *World J. Gastroenterol.* 12(15): 2320–2327.

Kowluruet, R., Atasi, I., and Ho, Y., 2006, Role of mitochondrial superoxide dismutase in the development of diabetic retinopathy. *Invest. Ophthalmol. Vis. Sci.* 47(4): 1594–1599.

Kuchiiwa, S., Ren, X., Gao, H., Kuchiiwa, T., and Nakagawa, S., 2003, Dioxin exposure down-regulates nitric oxide synthase and NADPH-diaphorase activities in the hypothalamus of Long-Evans rat. *Neurosci. Lett.* 345(1): 5–8.

Lagorio, S., Forastiere, F., Pistelli, R., Iavarone, I., Michelozz, P., Fano, V., Marconi, A., Ziemacki, G., and Ostro, B., 2006, Air pollution and lung function among susceptible adult subjects: a panel study. *Environ. Health.* 5: 11–16.

Landrigan, P., Sonawane, B., Butler, R., Trasande, T., Richard, Callan, R., and Droller, D., 2005, Early environmental origins of neurodegenerative disease in later life. *Environ. Health Perspect.* 113(9): 1230–1233.

Lee, C., Lee, K., Choe, K., Hong, Y., Noh, S., Eom, S., Ko, Y., Zhang, y., Yim, D., Kang, J., Kim, H., and Kim Y., 2006, Effects of oxidative DNA damage and genetic polymorphism of the glutathione peroxidase 1 (GPX1) and 8-oxoguanine glycosylase 1 (hOGG1) on lung cancer. *J. Prev. Med. Pub. Health* 39(2): 130–134.

Lee, D., Lee, I., Song, K., Steffes, M., Toscano, W., Baker, B., and Jascobs, D., 2006, A strong dose-response relation between serum concentrations of persistent organic pollutants and diabetes: results from the National Health and Examination Survey 1999–2002. *Diabetes Care* 29(7): 1638–1644.

Lee, S. H., Shin, M. S., Park, W. S., Kim, S. Y., Kim, S. H., Han, J. Y., et al., 1999, Alterations of Fas (Apo-1/CD95) gene in non-small cell lung cancer. *Oncogene* 18: 3754–3760.

Li, N. and Karin, M., 1998, Ionizing radiation and short wavelength UV activate NF-κB through two distinct mechanisms. *Cell Biol.* 95(22): 13012–13017.

Liu, X., Kim, C., Yang, J., Jemmerson, R., and Wang, X., 1996, Induction of apoptotic program in cell-free extracts: requirement for dATP and cytochrome c. *Cell* 86: 147–157.

Loffredo, L., Pignatelli, P., Cangemi, R., Andreozzi, P., Panico, M., Meloni, V., and Violi, F., 2006, Imbalance between nitric oxide generation and oxidative stress in patients with peripheral arterial disease: effect of an antioxidant treatment. *J. Vasc. Surg.* 44(3): 525–530.

Lu, D., Liou, H., Tang, C., and Fu, W., 2006, Hypoxia-induced iNOS expression in microglia is regulated by the PI3-kinase/Akt/mTOR signaling pathway and activation of hypoxia inducible factor-1alpha. *Biochem. Pharmacol.* 72(8): 992–1000.

Lui, Y., Christou, H., Morita, T., Laughner, E., Semenza, G., and Kourembanas, S., 1998, Carbon monoxide and nitric oxide suppress the hypoxic induction of vascular endothelial growth factor gene via the 5 enhancer. *J. Biol. Chem.* 273(24): 15257–15262.

Mak, T. and Yeh, W., 1999, Genetic analysis of apoptotic and survival signals. *Cold Spring Harb. Symp. Quant. Biol.* 64: 335–342.

Mak, J., Leung, Ho, S., Ko, F., Cheung, A., Ip, M., and Chan-Yeung, M., 2006, Polymorphisms in manganese superoxide dismutase and catalase genes: functional study in Hong Kong Chinese asthma patients. *Clin. Exp. Allergy* 36(4): 440–447.

Malkin, D., 1995, Age-specific oncogenesis: the genetics of cancer susceptibility. *Environ. Health Perspect.* 103(Suppl 5): 45–48.

Marklund S., Nilsson, P., Israelsson, K, Schampi, I., Peltonen. M., Asplund. K., 1997, Two variants of extracellular-superoxide dismutase: relationship to cardiovascular risk factors in an unselected middle-aged population. *J. Intern. Med.* 242: 5–14.

McDonnell, T., 1993, Cell division versus cell death: a functional model of multistep neoplasia. *Mol. Carcinog.* 8: 209–213.

Meikrantz, W., and Schlegel, R., 1995, Apoptosis and the Cell Cycle. *J. Cell Biochem.* 58: 160–174.

Melchers, I., Blaschke, S., Hecker, M., and Cattaruzza, M., 2006, The -786C/T single-nucleotide polymorphism in the promoter of the gene for endothelial nitric oxide synthase: insensitivity to physiologic stimuli as a risk factor for rheumatoid arthritis. *Arthritis. Rheum.* 54(10): 3144–3151.

Mercer, I., Blashke, S., Hecker, M., and Cattaruzza, M., 2005, Endogenous p53 protects vascular smooth muscle cells from apoptosis and reduces atherosclerosis in ApoE knockout mice. *Circ. Res.* 96(6): 667–674.

Miyaki, K., Tohyana, S., Murata, M., Kikuchi, H., Takei, I., Watanabe, K., and Omae K., 2005, Salt intake affects the relation between hypertension and the T-786C polymorphism in the endothelial nitric oxide synthase gene. *Am. J. Hypertens.* 18(12 Pt 1): 1556–1562.

Moore, K., 2006, Glucose fluctuations and oxidative stress. *JAMA* 296(14): 1730.

Moscow, J., Schmidt, L., Ingram, D., Gnarra, J., Johnson, B., Cowan, K., 1994, Loss of heterozygosity of the human cytosolic glutathione peroxidase I gene in lung cancer. *Carcinogenesis* 15: 2769–2773.

Moskaleva, Elu., Tobolov, I., Shumilov, V., Demina, T., and Gorbunov, E., 1988, Decreased capacity of the peripheral blood lymphocytes for excisional DNA repair in disseminated sclerosis. *Zh. Nevropatol. Psikhiatr. Im S. S. Korsakova.* 88(7): 87–88.

Muschen, M., Warskulat, U., and Beckmann, M. W., 2000, Defining CD95 as a tumor suppressor gene. *J. Mol. Med.* 78: 312–325.

Nair, V., Yuen, T., Olanow, C., and Sealfon, S., 2004, Early single cell bifurcation of pro- and antiapoptotic states during oxidative stress. *J. Biol. Chem.* 279(26): 27494–27501.

Nagata, S. and Golstein, P., 1995, The Fas death factor. *Science* 267: 1449–1456.

Nakajima, T., Elovaara, E., Okino, T., Gelboin, H., Klockars, M., Riihimaki,V., Aoyama, T., and Vainio, H., 1995, Different contributions of cytochrome P450 2E1 and P450 2B1/2 to chloroform hepatotoxicity in rat. *Toxicol. Appl. Pharmacol.* 133(2): 215–222.

Nathan, C., 1992, Nitric oxide as a secretory product of mammalian cells, *FASEB J.* 6: 3051–3064.

Nathan, C. and Xie, Q., 1994, Nitric oxide synthetases: roles, tolls and controls, *Cell* 78: 915–918.

Nouni, H., Rotrosen, D., Gallin, J., and Malech, H., 1998, Two forms of autosomal chronic granulomatous disease lack distinct neutrophil cytosol factors. *Science* 242(4883): 1298–1301.

Oh, G., Pae, H., Lee, B., Kim, B., Kim, J., Jeon, S., Jeon, W., Chae, H., and Chung, H., 2006, Hydrogen sulfide inhibits nitric oxide production and nuclear factor-kappaB via heme oxygenase-1 expression in RAW264.7 macrophages stimulated with lipopolysaccharide. *Free Radic. Biol. Med.* 41(1): 106–119.

Onyango, L. and Khan, S., 2006, Oxidative stress, mitochondrial dysfunction, and stress signaling in Alzheimer's disease. *Curr. Alzheimer Res.* 3(4): 339–349.

Oyama, T., Kawamoto, T., Mizoue, T., Sugio, K., Kodama, Y., Mitsudomi, T., and Yasumoto, K., 1997, Cytochrome P450 2E1 polymorphism as a risk factor for lung cancer: in relation to p53 gene mutation. *Anticancer Res.* 17: 583–588.

Pap, T., Aupperle, K. R., Gay, S., Firestein, G., and Gay, R., 2001, Invasiveness of synovial fibroblasts is regulated by p53 in the SCID mouse in vivo model of cartilage invasion. *Arthritis Rheum.* 44: 676–681.

Parsonnet, J., 1995, Bacterial infection as a cause of cancer. *Environ. Health Perspect.* 103(Suppl 8): 263–268.

Qui, M., Paromov, V., Yang, H., Smith, M., and Stone, W., 2006, Inhibition of inducible nitric oxide synthase by a mustard gas analog in murine macrophages. *BMC Cell Biol.* 30: 7–39.

Ratnasinghe, D., Tangrea, J., Andersen, M., Barrett, M., Virtamo, J., Taylor, P., and Albanes, D., 2000, Glutathione peroxidase codon 198 polymorphism variant increases lung cancer risk. *Cancer Res.* 60: 6381–6383.

Reichmann, E., 2002, The biological role of the Fas/FasL system during tumor formation and progression. *Semin. Cancer Biol.* 12: 309.

Resar, J., Roguin, A., Voner, J., Nasir, K., Hennebry, T., Miller, J. M., Ingersoll, R., Kasch, L., and Semenza, G., 2005, Hypoxia-inducible factor 1alpha polymorphism and coronary collaterals in patients with ischemic heart disease. *Chest* 128(2): 787–791.

Risch, N., 2000, Searching for genetic determinants in the new millennium. *Nature* 405: 847–856.

Ruiz-Ramos, R., Cebrian, M., and Garrido, E., 2005, Benzoquinone activates the ERK/MAPK signaling pathway via ROS production in HL-60 cells. *Toxicology* 209(3): 279–287.

Rossi, G., Cesari, M., Zanchetta, M., Colonna, G., Pedon, L., Cavalin, M., Maiolino, P., and Pession, A., 2003, The T-786C endothelial nitric oxide synthase genotype is a novel risk factor for coronary artery disease in Caucasian patients of the GENICA study. *J. Am. Coll. Cardiol.* 41(6): 930–937.

Sandrini, A., Johnson. A., Thomas, P., and Yates, D., 2006, Fractional exhaled nitric oxide concentration is increased in asbestosis and pleural plaques. *Respirology* 11(3): 325–329.

Sandström, J., Carlsson, L., Marklund, S., and Edlund, T., 1992, The heparin-binding domain of extracellular superoxide dismutase C and formation of variants with reduced heparin affinity. *J. Biol. Chem.* 267: 18205–18209.

Sandström, J., Karlsson, K., Edlund, T., and Marklund, S., 1993, Heparin-affinity patterns and composition of extracellular superoxide dismutase in human plasma and tissues. *Biochem. J.* 294: 853–857.

Sandström, J., Nilsson, P., Karlsson, K., and Marklund, S., 1994, Ten-fold increase in human plasma extracellular superoxide dismutase content caused by a mutation in heparin-binding domain. *J. Biol. Chem.* 269: 19163–19166.

Schulte-Hermann, R., Bursch, W., Gras-kraupp, B., Torok, L., and Ellinger, A., 1995, Role of active cell death (apoptosis) in multi-stage carcinogenesis. *Toxicol Lett.* 82–83: 143–148.

Semenza, G. L., 2000, Surviving ischemia: adaptive responses mediated by hypoxia-inducible factor 1. *J. Clin. Invest.* 106: 809–812.

Semenza, G. L., 2003, Targeting HIF-1 for cancer therapy. *Nat. Rev. Cancer* 3: 721–732.

Shah, S., 2006, Oxidants and iron in progressive kidney disease. *J. Ren. Nutr.* 16(3): 185–189.

Shen, J., Wang, R., Wang, L., Xu, Y., Wang X., 2004, A novel genetic polymorphism of inducible nitric oxide synthase is associated with an increased risk of gastric cancer. *World J. Gastroenterol.* 10: 3278–3283.

Siebenlist, U., Franzoco, G., and Brown, K., 1994, Structure, regulation and function of NF-κB. *Annu. Rev. Cell Biol.* 10: 405–455.

Soini, Y., Kallio, J., Hirvikoski, P., helin, H., Kellokumpu-Lehtinen, P., Tammela, T., Peltoniemi, M., Martikainen, P., and Kinnula, L., 2006, Antioxidant enzymes in renal cell carcinoma. *Histol. Histopathol.*, 21(2): 157–165.

Spoto, B., Benedetto, F., Testa, E., Tripepi, G., Mallamaci, F., Maas, R., Boeger, R., Zoccali, C., Parlongo, R. and Pisano, A., 2005, Atherosclerosis and the Glu298Asp polymorphism of the eNOS gene in white patients with end-stage renal disease. *Am. J. Hypertens.* 18(12 Pt 1): 1549–1555.

Strålin, P., Karlsson, K., Johansson, B., and Marklund, S., 1995, The interstitium of the human arterial wall contains very large amounts of extracellular superoxide dismutase. *Arterioscler. Thromb. Vasc. Biol.* 15: 2032–2036.

Tanimoto, K., Yoshiga, K., Eguchi, H., Kaneyasu, M., Ukon, K., Kumazak, T., Oue, N., Yasui, W., Imai K., Nakachi, K., Poellinger, L., Nishiyama, M., 2003, Hypoxia-inducible factor-1α′ polymorphisms associated with enhanced transactivation capacity, implying clinical significance. *Carcinogenesis* 24: 1779–1783.

Thakali, K., Lau, Y., Fink, G., Gallican, J., Chen, A. and Watts, S., 2006, Mechanisms of hypertension induced by nitric oxide (NO) deficiency: focus on venous function. *J. Cardiovasc. Pharmacol.* 47(6): 742–750.

Thompson, C., 1995, Apoptosis in the pathogenesis and treatment of disease. *Science* 10: 1456–1462.

Trosko, J., Chang, C., Madhukar, B., 1994, The role of modulated gap junctional intercellular communication in epigenetic toxicology. *Risk Anal.* 14(3): 303–312.

Uzel, G., 2005, The range of defects associated with nuclear factor kappaB essential modulator. *Curr. Opin. Allegy. Clin. Immunol.* 5(6): 513–518.

Wang, X. and Wang, J., 2005, Smoking-gene interaction and disease development: relevance to pancreatic cancer and atherosclerosis. *World J. Surg.* 29(3): 344–353.

Warren, W., Biggs, P., El-Baz, M., Ghoneim, M., Stratton, M., and Vebit, s., 1995, Mutations in the p53 gene in schistosomal bladder cancer: a study of 92 tumours from Egyptian patients and a comparison between mutational spectra from schistosomal and non-schistosomal urothelial tumours. *Carcinogenesis* 16(5): 1181–1189.

Weinberg, R., 1991, Tumor suppressor genes. *Science* 254: 1138–146.

Weiss, S., 1986, Oxygen, ischemia and inflammation. *Acta Physiol. Scand. Suppl.* 548: 9–37.

Whitmarsh, A. J. and Davis, R. J., 2000, Regulation of transcription factor function by phosphorylation. *Cell Mol. Life Sci.* 57: 1172–1183.

Wiesman, M. and Halliwell, B., 1996, Damage to DNA by reactive oxygen species and nitrogen species: role of inflammatory disease and progression to cancer. *Biochem. J.* 313: 17–29.

Wu, S., Tan, M., Hu, Y., Wang, J., Scheuner, D., and Kaufman, R., 2004, Ultraviolet light activates NFkappaB through translational inhibition of IkappaBalpha synthesis. *J. Biol. Chem.* 279(33): 34898–34902.

Var, A., Yi;dirim, Y., Onur, M., Kuscu, E., Uyanik, B., Goktalay. K., and Guvenc, Y., 2003, Endothelial dysfunction in preeclampsia. Increased homocysteine and decreased nitric oxide levels. *Gynecol. Obstet. Invest.* 56(4): 221–224.

Verma, M. and Srivastava, S., 2002, Epigenetics in cancer: implications for early detection and prevention, *Lancet Oncol.* 3(12): 755–763.

Vucic, V., Isenovic, E., Adzic, M., Ruzdijic, S., and Radojcic, M., 2006, Effects of gamma-radiation on cell growth, cycle arrest, death, and superoxide dismutase expression by DU 145 human prostate cancer cells. *Braz. J. Med. Biol. Res.* 39(2): 227–236.

Xiao, G. and Li., H., 2006, Effects of inhalation of oxygen on free radical metabolism and oxidative, antioxidative capabilities of the erythrocyte after intensive exercise. *Res. Sports Med.* 14(2): 107–115.

Yang, Q. and Hatton, G., 2002, Histamine H(1)-receptor modulation of inter-neuronal coupling among vasopressinergic neurons depends on nitric oxide synthase activation, *Brain Res.* 955(1–2): 115–122.

Yoshii, Y., Saito, A., Zhao, D., and Nose, T., 1999, Copper/zinc superoxide dismutase, nuclear DNA content, and progression in human gliomas. *J. Neurooncol.* 42(2): 103–108.

You, M., Candrian, U., Maronpot, R., Stoner, G., and Anderson, M., 1989, Activation of the Ki-*ras* protooncogene in spontaneously occurring and chemically induced lung tumors of the strain A mouse, *Proc. Natl. Acad. Sci. USA* 86: 3070–3074.

Yu, Z., Zhou, D., Bruce-Keller, A., Kindy, M., and Mattson, M., 1999, Lack of the p50 subunit of nuclear factor- B increase the vulnerability of hippocampal neurons to excitotoxic injury. *J. Neurosci.* 19: 8856–8865.

Zanetti, M., Sato, J., Jost, C., and Gloviczki, P., 2001, Gene transfer of manganese superoxide dismutase reverses vascular dysfunction in the absence but not in the presence of atherosclerotic plaque. *Hum. Gene Ther.* 12: 1407–1416.

Zelko, N., Mariani, J., and Folz, J., 2002, Superoxide dismutase multigene family: a comparison of the CuZn-SOD (SOD1), Mn-SOD (SOD2), and EC-SOD (SOD3) gene structures, evolution and expression. *Free Radic. Biol. Med.* 33: 337–349.

Zhang, C., Lopez-Ridaura, R., Hunter, D., Rifai, N., and Hu, F., 2006, Common variants of the endothelial nitric oxide synthase gene and the risk of coronary heart disease among U.S. diabetic men. *Diabetes* 55(7): 2140–2147.

Zornig, M., Hueber, A., Baum, W., and Evan, G., 2001, Apoptosis regulators and their role in tumorigenesis. *Biochim. Biophys. Acta* 1551:F1–F37.

SECTION 3

MULTIPLE STRESSOR DATA: LONG-TERM EFFECTS

CHAPTER 9

POST-RADIATED AND POST-STRESSED VOLATILE SECRETIONS: SECONDARY IMMUNE AND BEHAVIORAL REACTIONS IN GROUPS OF ANIMALS

B.P. SURINOV, A.N. SHARETSKY, D.V. SHPAGIN, V.G. ISAYEVA, AND N.N. DUKHOVA

Medical Radiological Research Center of the Russian Academy of Medical Sciences, 249036 Obninsk, Kaluga Region, ul. Koroliova, 4, Russia.
e-mail: surinov@mrrc.obninsk.ru

Abstract: It was shown that the irradiated mice gave off with urine immunosuppressive components which possessed of high volatility. With their help even the one irradiated individual can induce in intact mice the disturbance of immunity and alteration of number of blood cells. Moreover, the predominant mouse from exposed group of mice also was capable to elicit disturbance of immunity in next group of intact animals. The same effects could be received by transfer of the urine samples from irradiated mice or from intact mice exposed with urine of irradiated animals to the box with intact individuals. It was established that ionizing radiation (4–6 Gy) of male mice increase their scent attractiveness to intact singenic male conspecifics. The carried out experiments have shown existence of the mediated by volatile chemosignals mechanism of multiplication second post-radiated disturbances of immunity in the groups of animals. The immunosuppressive volatile components were induced also by stress and some immunodepressants (dexametasone and cyclophosphamide).

Keywords: immunosuppressive volatile components; irradiation; stress; immunode-pressants; olfactory behavior of mice

Introduction

It is known that in physiological conditions animals give off secretion including urine volatiles (pheromones). These substances cause specific physiological behavioral reactions in the recipients and provide them with information about the social, sexual and reproductive status of other individuals within the species (Halpin, 1986; Penn and Potts, 1998; Yamazaki et al., 1999; Moshkin

C. Mothersill et al. (eds.), Multiple Stressors: A Challenge for the Future, 127–138.

et al., 2002). There are practically no findings suggesting that animals possess any chemical signals of pathologic conditions. The exception is provided by some reports on alterations in the scent attractiveness of female mice to chemosignals of infected male mice (Penn and Potts, 1998; Moshkin et al., 2002), "the smell of fear" produced by stressed animals (Zalcman et al., 1991).

Previous experiments have shown that mice and rats exposed to radiation or stress can induce some disorders of the immune reactivity and the content of formed elements of blood in intact individuals (Surinov et al., 1998). Such effects have been designated as communicative. Subsequently, they were found to be conditioned by volatile components (VC) produced with urine (Surinov and Dukhova, 2004). These VC appear to function to attract intact individuals.

These VC can cause not only mediated secondary post-radiation reactions not associated with direct exposure of animals to radiation but also contribute, as proved, to their multiplication and diffusion (Surinov et al., 2005). This effect is similar to the bystander effect, which is well known in radiobiology and manifests itself as attenuation of the genome of intact cells adjoining the irradiated cell (Mothersill and Seymour, 1998–2001).

Materials and Methods

IMMUNOLOGICAL INVESTIGATIONS

CBA and C57Bl/6 (B6) and F1 (CBAxC57Bl6) strains of male mice (weighted 22–24 g) were used in this study. They were maintained under vivarium conditions, provided standard food, water, and kept under a usual light/dark cycle. Before the experiment, mice were housed in tens in standard plastic boxes for 2 weeks. They were given whole-body irradiation of 4, 6 Gy using a 60Co source with a dose-rate of 3.2 mGy/sec on "Gamma-Cell-220" equipment (Atomic Energy Canada Limited, Canada).

Mice were exposed to stress by one-time swim within 1 h at 30°C.

Immunodepressants (dexametasone or cyclophosphamide) were injected introperitoneal.

In 1–2 days after irradiation, stress or injection of immunodepressants to mice, on bottom of box put a sheet of filter paper (bedding). Access to this bedding was limited by the second perforated bottom, lifted above basic bottom on 0.5 cm. Such paper bedding, containing of absorbed 24 hrs urine was transferred in box with individuals (recipients) also under perforated bottom. In 24 hs after exposition with bedding animals immunized by sheep red blood cells (SRBC) in dose 1×10^8 cells/mouse. In 4 days, mice were decapitated of under ester narcosis. The number of nuclear cells and antibody-forming cells (AFCs) was determined by Cunningham's method. Statistic analysis of the results of the study was performed with the use of Student's t-test.

BEHAVIORAL INVESTIGATIONS

We used also a choice assay to test the VC preferences of irradiated mice. For this purpose, we used a modified T-maze (Bures et al., 1983) with a "field of choice" represented as an open plastic 30 × 35 cm cage with a wall height of 35cm. On its external opposite sides, there were two "hiding boxes" (lightproof plastic 10 × 10 × 5 cm boxes), where mice (testers) could freely go out through the holes in the "field" walls. At a distance of 0.5 m above the "field", there was an electrical halogen 50 wt lamp. Male mice used as testers (10 individuals) were individually placed six times each in the middle of the "field" before irradiation (several series of observations at 1–2 days intervals) and at different intervals after irradiation to record which of "hiding boxes" will be chosen by a given individual. Within first 1–3 min, the testers went round the "field" and alternately entered both "hiding boxes". Finally, the most testers spent more than 0.5 min inside one of the boxes. The latter was considered as preference for VC cues outgoing from bedding containing a certain urine sample. Each series of observations was conducted within the first half of the light day and contained 60 preference assessments.

The results were expressed in the percentage of the preferences for compared samples (number of the preferences for one sample in relation to the total number of assessments in a series). We used Wilcoxon Signed-Ranks test to assess the significance of the VC preferences by one of the groups in comparison with another group in each period of observations. All data were subjected to Student's t-test to assess the significance of the differences between preference rates.

Results and Discussion

IMMUNOLOGICAL INVESTIGATIONS

As a result of investigation of influence 24 h bedding from the irradiated in doses 4 Gy mice on immunological parameters of intact singenic mice were obtained the following data (Table 1).

The bedding from a cage with irradiated animals transferred to a box with intact ones and then these mice immunized by the SRBC. As control used mice which transferred bedding from intact singenic conspecifics. In 5 days at experimental animals defined a content of AFC in a spleen. Immunosuppressive effect was observed in that case when the bedding was from mice in 3–7 days after an irradiation. In other periods of research of essential suppression, antibody genesis was not defined. According to the numerous data just for 3–7 days the greatest postradiation disturbance of immunity takes place.

TABLE 1. Immunosuppressive activity ($M \pm m$) of volatile components of urine of mice CBA in different periods after irradiation (4 Gy)

Animals–donors of VC	Time after irradiation, days	Spleen		Thymus
		Number of cells, 1×10^6	AFC, 1×10^3	Number of cells, 1×10^6
Control	1	123 ± 12.6 (100 ± 10.3)	178 ± 12.9 (100 ± 7.2)	53.0 ± 3.9 (100 ± 7.3)
Irradiated		100 ± 9.5 (81.0 ± 7.7)	151 ± 10.5 (84.8 ± 5.7)	39.3 ± 1.8* (74.2 ± 3.4)
Control	2	125 ± 14.8 (100 ± 11.9)	173 ± 21.0 (100 ± 12.0)	39.0 ± 1.3 (100 ± 3.3)
Irradiated		122 ± 17.5 (97.6 ± 14.0)	122 ± 8.5* (70.6 ± 4.9)	32.5 ± 7.1 (83.0 ± 18.0)
Control	3	103 ± 6.7 (100 ± 6.5)	187 ± 11.2 (100 ± 6.0)	35.5 ± 1.5 (100 ± 4.2)
Irradiated		100 ± 12.2 (97.0 ± 11.8)	126 ± 13.7* (67.7 ± 7.3)	36.1 ± 7.3 (101 ± 20.6)
Control	7	162 ± 9.2 (100 ± 5.7)	148 ± 13.6 (100 ± 9.0)	45.2 ± 3.8 (100 ± 8.4)
Irradiated		133 ± 5.4* (82.0 ± 3.3)	111 ± 6.0* (75.0 ± 4.0)	38 ± 2.2 (84 ± 4.9)
Control	14	130 ± 9.0 (100 ± 7.0)	143 ± 22.0 (100 ± 15.0)	53.2 ± 4.3 (100 ± 8.1)
Irradiated		170 ± 19.3 (131 ± 14.8)	135 ± 20.0 (94.4 ± 14.0)	31.6 ± 4.5* (59.4 ± 8.5)
Control	21	130 ± 7.1 (100 ± 5.5)	260 ± 34.0 (100 ± 13.0)	44.0 ± 6.3 (100 ± 14.3)
Irradiated		120 ± 10.5 (92.0 ± 8.0)	191 ± 41.2 (73.5 ± 15.8)	38.4 ± 4.2 (87.3 ± 9.5)

Note: In brackets – % to control
*Significant different ($p < 0.05$) from control.

Thus mice subjected to single action of ionizing radiation in sublethal doses secreted with urine of volatile components possessing by immunosuppressive activity reference to intact animals. The time of appearance of such components coincided with period the greatest postradiation disturbance of immunity.

For comparison post-radiated and post-stressed dynamics of a secretion of volatile components with their immunosuppressing properties the following research was carry out. A 24 h bedding from a box with stressed mice transferred to a box with the intact singenic recipients just as was made in experiences with an irradiation.

As has appeared within the first 2 days the stressed animals produced volatile components, which twice reduced the immune response to SRBS at intact individuals (Table 2).

TABLE 2. Immunological parameters ($M \pm m$) at intact male mice F1(CBA×C57Bl6) after transfer to them of bedding received in different time from stressed mice

Animals–donors of VC	Time after stress, days	Number of cells in spleen, 1×10^6	AFC in spleen, 1×10^3
Control	2	146 × 10 (100 ± 6.8)	86.6 ± 7.8 (100 ± 9)
Stress		109 ± 3.1 (74.7 ± 2.1)*	44 ± 5.6 (50.8 ± 6.5)*
Control	7	111 ± 15.6 (100 ± 14)	63.6 ± 4.7 (100 ± 7.4)
Stress		113 ± 11.7 (102 ± 10.5)	54.4 ± 7.7 (85.5 ± 12.1)
Control	14	112 ± 7.5 (100 ± 6.7)	78.4 ± 5.1 (100 ± 6.5)
Stress		125 ± 5.7 (112 ± 5.1)	51 ± 4 (65.1 ± 5.1)*
Control	21	125 ± 12.6 (100 ± 11)	85.7 ± 4.6 (100 ± 5.4)
Stress		128 ± 15.2 (102±12.2)	55 ± 4 (64.2 ± 4.7)*
Control	30	110 ± 12.3 (100 ± 12)	132 ± 14 (100 ± 11)
Stress		106 ± 11.6 (96.4 ± 10.5)	95.6 ± 14.1 (72.4 ± 10.7)

Note: In brackets – % to control
*Significant different ($p < 0.05$) from control.

By 7 days after a stressing, the suppressing activity of bedding was minor. By 14 days the suppressing activity of components to some extent was restored and by 30 days it was again reduced.

Described earlier dynamics of immunosuppressing activity of a bedding from stressed mice practically completely coincident with dynamics of alteration of the antibodygenesis at mice subjected single stress, which was investigated and reported by us earlier. The greatest suppression of the immune response took place just for 1–2 days after a stress.

Thus a post-stressed dynamics of secretion with urine of mice volatile components coincide with dynamics of post-stressed suppression of immunity.

Ability to induce secretion of immunosuppressive volatile components possessed also the synthetic glucocorticoid dexametasone. Under its direct influence since a dose 4 mg/kg in mice-donors there was a dose-dependent reduction of the number of AFC in spleen and nuclear cells.

At the recipients the reduction of antibody formation was observed only in that case when the donors of volatile components received high doses of the drug (16 and 32 mg/kg). The immunosuppressive state elicited by direct action of dexametasone at the donors was replicated somewhat with the help volatile secrets at the mouse-recipient. However, at the recipients of VC the inhibition of the immune response was less expressed and was not accompanied by decrease of number of cells in lymphoid organs.

For research of dynamics of secretion of volatile immunosuppressive components the samples of urine of donors obtained through 1–28 days after injection of dexametasone in dose 16 mg/kg exposed to the recipients. It was showed that the suppressive activity of volatile components is observed at least within 3 days after injection of the drug (Table 3).

TABLE 3. Secretion of immunosuppressive components by mice-donors after injection of dexametasone (16 mg/kg)

Groups of mice-recipients	Time after stress, days	Number of cells in spleen, 1×10^6	AFC in spleen, 1×10^3	Number of cells in Thymus, 1×10^6
Control	1	151.9 ± 10.8 (100 ± 7.1)	76.3 ± 4.8 (100±6.3)	66.2 ± 2.6 (100±3.9)
Dexametasone		138.6 ± 12.4 (91.2 ± 8.2)	51.0 ± 4.7 (66.8 ± 3.2)*	62.8 ± 5.1 (94.9 ± 7.7)
Control	3	152.3 ± 9.5 (100 ± 6.2)	129.2 ± 11.7 (100 ± 9.1)	82 ± 9.5 (100 ± 11.6)
Dexametasone		150.0 ± 14.5 (98.5 ± 9.5)	89.1 ± 3.8 (69.7 ± 2.9)*	62.2 ± 7.5 (75.9 ± 9.1)
Control	7	137.2 ± 9.0 (100 ± 6.6)	107.2 ± 10.0 (100 ± 9.3)	92.2 ± 6.1 (100 ± 6.6)
Dexametasone		154.4 ± 10.5 (112.5 ± 7.7)	134.9 ± 15.6 (125.8 ± 14.6)	95.8 ± 8.3 (103.9 ± 9.0)
Control	15	173.4 ± 13.8 (100 ± 7.9)	44.0 ± 6.3 (100 ± 14.3)	43.0 ± 3.0 (100 ± 7.1)
Dexametasone		143.3 ± 10.4 (82.6 ± 6.0)	41.1 ± 9.9 (93.4 ± 22.5)	40.5 ± 3.3 (94.2 ± 7.7)
Control	21	227.8 ± 24.5 (100 ± 10.8)	39.7 ± 3.7 (100 ± 9.3)	63 ± 1.0 (100 ± 1.6)
Dexametasone		170 ± 14.2 (74.6 ± 6.2)	36.4 ± 1.4 (88.3 ± 3.5)	63.5 ± 3.4 (100.8 ± 5.4)
Control	28	172.2 ± 12.6 (100 ± 7.3)	53.1 ± 8.5 (100 ± 16.0)	49.8 ± 2.11 (100 ± 4.2)
Dexametasone		165.6 ± 9.6 (96.2 ± 5.6)	52.2 ± 3.0 (98.3 ± 5.7)	48.7 ± 2.1 (97.8 ± 4.2)

Note: In brackets – % to control
*Significant different ($p < 0.05$) from control.

By ability to induce secretion of the immunosuppressive volatile components except for a synthetic glucocorticoid dexametasone had also the cytostatic cyclophosphamide. However, the effect of this drug took place only at the large doses.

According to results of present research at mice-donors under influence of cyclophosphamide in dose 300 mg/kg took place deep suppression of the immune response up to 24.0 ± 2.1% which accompanied by a reduction of a number of cells in a spleen and thymus up to 35.7 ± 3.2% and 73.6 ± 9.3% respectively. The volatile secretions of these animals elicited in the recipients a decrease in number of AFC in a spleen up to 63.4 ± 6.3% and a number of cells in a thymus up to 69.0 ± 8.6%. A number of cells in spleen remained without an alteration.

Earlier we formulated the supposition about a possibility of multiplication and spreading of mediated secondary post-radiated disturbances in groups of animals with the help of volatile components. As the basic model the co-content in the same cage of irradiated and intact mice or rats in the ratio 5:5 was used. The presence in this cage among intact animals only of one irradiated individual did not ensure steady effect. We have assumed that the effect of one individual can depend upon its hierarchical position in group. The experiments with an irradiation of the dominant mouse were carried out. For this purpose from group of outbred mice have chosen individuals with dominant behavior according to described in the literature procedure (Bures et al., 1983). The research have shown that the dominant mouse returned in the group after an irradiation induced at intact mice augmentation of number of neutrophilic leucocytes in peripheral blood up to $181 \pm 28\%$ as compared with control (in control group the dominant individual did not subject an irradiation). Then from groups of animals 1 and 2 took out the next dominant individual (having contact with irradiated mouse) and put it for 24h in group of other intact mice (Table 4). After taking out of this dominant individual at the stayed 4 mice the decrease in number of blood cells in particular of erythrocytes ($44.0 \pm 8.7\%$) and leucocytes ($77.0 \pm 5.0\%$) was observed.

TABLE 4. Effect of dominant individual-inductor irradiated in dose 4Gy or dominant individual-inductor exposed by irradiated one on parameters of peripheral blood of intact outbred mice

Experiments	Number of cells in 1 mL of peripheral blood			
	Erythrocytes, 1×10^6	Leucocytes, 1×10^3	Lymphocytes, 1×10^3	Neutrophils, 1×10^3
Control 1 – contained with intact individual	8.72 ± 0.8 (100 ± 9.2)	6.2 ± 0.8 (100 ± 13.0)	4.3 ± 0.7 (100 ± 15.9)	0.9 ± 0.1 (100 ± 14.6)
Experiment 1 – contained with irradiated individual	8.3 ± 0.4 (95.1 ± 4.9)	6.8 ± 0.6 (110.0 ± 9.2)	3.9 ± 0.6 (92.5 ± 12.8)	1.7 ± 0.3 $(181 \pm 26.8)^*$
Control 2 – contained with individual from group of control 1	8.6 ± 0.7 (100 ± 8.1)	7.9 ± 0.5 (100 ± 6.8)	5.7 ± 0.7 (100 ± 11.9)	1.33 ± 0.1 (100 ± 9.4)
Experiment 2 – contained with exposed individual from group of experiment 1	3.8 ± 0.7 $(44.0 \pm 8.7)^*$	6.1 ± 0.4 $(76.9 \pm 5.0)^*$	4.1 ± 0.4 (71.6 ± 7.7)	1.0 ± 0.1 (77.0 ± 8.9)
Control 3 – contained with individual from group of control 2	8.1 ± 0.2 (100 ± 2.9)	9.8 ± 0.6 (100 ± 6.1)	7.1 ± 0.5 (100 ± 7.1)	1.4 ± 0.2 (100 ± 11.2)
Experiment 2 – contained with exposed individual from group of experiment 1	8.0 ± 0.2 (99.6 ± 2.7)	7.9 ± 0.9 (81.0 ± 9.4)	5.2 ± 0.5 $(72.8 \pm 6.8)^*$	1.3 ± 0.1 (93.0 ± 7.0)

Note: In brackets – % to control
*Significant different ($p < 0.05$) from control.

A repetition of the earlier described procedure – translocation of one by one dominant individual from exposed groups 3 and 4 in next groups of intact mice – group 5 and 6 – also induced a decrease in number of blood cells mainly of lymphocytes (72.8 ± 7.0%).

The sequential "transmission" of disturbances from one group of animals in other one can be realized not only with the help of separate individuals but also with the help of samples of urine absorbed in a paper bedding within 24 h.

So transfer of bedding from a box with the irradiated singenic mice into box with group of intact mice (male F1(CBAxC57Bl6) resulted in reduction of their ability to immune response up to 74.8 ± 4.0% versus control (bedding from a box with intact animals). When from these exposed by urine of mice received a bedding saturated with their own urine then it also caused an immunosuppression in other intact animals. The ability to the immune response was reduced practically up to the same level (73.0 ± 7.9%) as well as in the previous group. Thus even one irradiated mouse can cause of spreading of secondary indirect disturbance of an immune reactivity and decrease a number of blood cells in group contacting with its individuals.

The absence at individual, group (it is showed in the present work), specific restrictions for reactions on post-radiated secretions obviously can have ecological consequences.

The phenomena found by us remind the bystander effect which is widely considered in the literature with only one difference that in the later case there is a disturbance of a genome stability in intact cells under influence of the factors outgoing from irradiated cells (Mothersill and Seymour, 1998; 2000; 2001). Whereas, we study a distant action of one organism to others in groups of animals. Namely influence of pheromone-like volatile components induced by damaging actions on immunity of intact animals.

BEHAVIORAL INVESTIGATIONS

One of the important problems at study of properties of post-radiated volatile secretion is their similarity or difference from post-stressed volatile secretion which also have immunosuppressive activity (Surinov and Dukhova, 2004)

The comparative research of dynamics of the reaction of intact testers on post-radiated and post-stressed volatile secretions in test on olfactory preference-avoidance has shown the following. At comparison secretion of the intact and irradiated mice the intact mice-testers preferred post-radiated secretions. At comparison secretion of the intact and stressed mice the same mice-testers preferred volatile secretions of stressed individuals. Hence, as post-radiation secretions and post-stressed these of mice CBA have the high scent attractiveness for intact mice. These attractive properties of the volatile components were found only during first week after an irradiation or stressing action. Later such attractiveness was not found.

In the same experiments the comparative research of reaction of preference-avoidance by intact testers of the post-radiated and post-stressed volatile secretions was carried out (Table 5).

It was supposed that in the case of identity of compared secretions the mice-testers cannot distinguish. However testers with the greater frequency chose cover with bedding from stressed individuals than from irradiated ones.

Thus obtained data do not exclude that post-irradiated and post-stressed secretions are not identical.

In our other experiment, intact male mice significantly preferred VC produced by individuals of the congenial genotype while choosing between hiding boxes containing VC of CBA and B6 males. Before irradiation, intact CBA males preferred VC of syngeneic males with an average daily rate of 59.67 ± 0.48%. Males of the B6 strain preferred VC of B6 males with an average daily rate of 59.50 ± 0.50%. This olfactory behavior of male mice has been previously designated as homing, i.e. yearning for their males (Surinov and Dukhova, 2004; Surinov et al., 2005) unlike the reproductively conditioned attractiveness to females with a strange genotype (Yamazaki et al., 1999; Moshkin et al., 2002).

After irradiation (in dose of 4 Gy), testers of both strains showed changes in the short-term attractiveness to syngeneic chemosignals when choosing hiding boxes with VC of syngeneic or allogeneic male mice as compared with data before irradiation. The attractiveness of CBA testers was significantly

TABLE 5. Comparative evaluation of preference rate (%) by intact testers (CBA-mice) of volatile secretions of intact irradiated (4Gy) or stressed mice CBA.

Time after irradiation, days	CBA male testers		CBA male testers		CBA male testers	
	VC control	VC irradiated	VC control	VC stressed	VC irradiated	VC stressed
1	40 ± 4.62	60 ± 4.62*	44 ± 2.78	56 ± 2.78	46 ± 4.34	54 ± 4.34*
2	43 ± 4.34	57 ± 4.34*	39 ± 4.45	61 ± 4.45*	44 ± 2.78	56 ± 2.78*
3	38 ± 3.56	62 ± 3.56*	42 ± 4.34	58 ± 4.34*	42 ± 3.56	58 ± 3.56*
4	40 ± 4.45	60 ± 4.45*	42 ± 2.78	58 ± 2.78*	46 ± 3.69	54 ± 3.69
7	43 ± 4.62	57 ± 4.62*	42 ± 4.62	58 ± 4.62*	44 ± 3.56	56 ± 3.56
9	48 ± 4.34	52 ± 4.34	46 ± 3.69	54 ± 3.69	46 ± 4.62	54 ± 4.62
11	48 ± 3.56	52 ± 3.56	44 ± 4.62	56 ± 4.62*	48 ± 3.56	52 ± 3.56
14	48 ± 4.62	52 ± 4.62	50 ± 4.34	50 ± 4.34	46 ± 4.34	54 ± 4.34
16	45 ± 4.45	55 ± 4.45	48 ± 3.56	52 ± 3.56	44 ± 4.45	56 ± 4.45
18	45 ± 4.34	55 ± 4.34*	46 ± 4.62	54 ± 4.62	54 ± 3.69	46 ± 3.69
21	48 ± 2.78	52 ± 2.78	48 ± 3.69	52 ± 3.69	46 ± 4.34	54 ± 4.34

*Significantly (according to Wilcoxon) preferred odortype.

TABLE 6. Preference rate ($M \pm m$, %) in CBA and B6 male testers exposed to 4 Gy irradiation for volatile secretions produced by intact syngeneic and allogeneic males

Period after irradiation, days	CBA male testers		B6 male testers	
	VC of CBA	VC of B6	VC of B6	VC of CBA
Before irradiation	65.00 ± 9.61*	34.00 ± 9.61	62.38 ± 2.14*	37.62 ± 2.14
1	68.33 ± 4.66*	31.67 ± 4.66	71.67 ± 3.56**	28.33 ± 3.56
2	73.33 ± 3.69***	26.67 ± 3.69	71.67 ± 4.34*	28.33 ± 4.34
3	70.00 ± 4.16*	30.00 ± 4.16	80.00 ± 5.44***	20.00 ± 5.44
4	65.00 ± 4.62*	35.00 ± 4.62	76.67 ± 4.45***	23.33 ± 4.45
7	73.33 ± 5.67*	26.67 ± 5.67	68.33 ± 4.62*	31.67 ± 4.62
8	73.33 ± 4.45*	26.67 ± 4.45	76.67 ± 4.45***	23.33 ± 4.45
10	78.33 ± 5.00*	21.67 ± 5.00	68.33 ± 5.24*	31.67 ± 5.24
11	73.33 ± 4.45*	26.67 ± 4.45	73.33 ± 3.69***	26.67 ± 3.69
14	68.33 ± 5.24*	31.67 ± 5.24	70.00 ± 4.84*	30.00 ± 4.84
17	68.33 ± 3.89*	31.67 ± 3.89	71.67 ± 4.34*	28.33 ± 4.34
19	66.67 ± 4.30*	33.33 ± 4.30	75.00 ± 3.73***	25.00 ± 3.73
23	60.00 ± 2.72*	40.00 ± 2.72	75.00 ± 2.78***	25.00 ± 2.78
28	61.67 ± 4.34*	38.33 ± 4.34	68.33 ± 2.99*	31.67 ± 2.99
32	61.67 ± 3.56*	38.33 ± 3.56	68.33 ± 3.89*	31.67 ± 3.89

*Significantly (according to Wilcoxon) preferred odortype;
**Significant differences (according to Student, $p < 0.05$) as compared with the attractiveness of testers before the exposure to radiation.

enhanced on the second day (Table 6). Interestingly, this effect lasted considerably longer (up to 23 days after irradiation) in irradiated B6 males.

Our studies indicate that within short times after irradiation in sublethal dose the attractiveness of CBA and B6 male mice to volatile secretions of syngeneic males changes and exceeds the physiological level, showing no disorder of the olfactory sensitivity. Together, these findings support the idea that male mice experience selective changes in chemosignals-induced behavioral reactions. It might be caused by disturbances of metabolic and regulatory processes suggestive of a "syndrome of disease" (Barnett, 1975).

The group of animals with the absorbed dose of 6 Gy showed some different patterns (Table 7).

Within a short period after exposure to radiation, the animals lost their olfactory selectivity in respect to volatile cues from syngeneic and allogeneic individuals. For the CBA strain, that time period was 2 days and for the B6 strain – 4 days. The average of the preferences of CBA testers for volatile cues from syngeneic individuals achieved the physiological level on the third day, and the attractiveness in B6 testers regenerated up to the background level on the seventh day. Subsequently, testers of both strains showed a temporary enhancement of the attractiveness versus intact mice. In CBA testers, the attractiveness enhanced significantly over the period of 11 days, and in B6 testers – up to 14 days.

TABLE 7. Relative preference rate ($M \pm m$, %) in CBA and B6 male testers exposed to 6 Gy irradiation for volatile secretions produced by intact syngeneic and allogeneic males

Period after irradiation, days	CBA male testers		B6 male testers	
	VC of CBA	VC of B6	VC of CBA	VC of B6
Before irradiation	59.67 ± 0.48*	40.33 ± 0.48	40.5 ± 0.50	59.5 ± 0.50*
1	48.33 ± 2.99	51.67 ± 2.99	48.33 ± 2.99	51.67 ± 2.99
2	51.67 ± 2.99	48.33 ± 2.99	48.33 ± 5.24	51.67 ± 5.24
3	55.00 ± 2.55*	45.00 ± 2.55	46.67 ± 5.44	53.33 ± 5.44
4	73.33 ± 2.72***	26.67 ± 2.72	50.00 ± 3.51	50.00 ± 3.51
7	76.67 ± 2.72***	23.33 ± 2.72	46.67 ± 2.22	53.33 ± 2.22*
8	75.00 ± 2.78***	25.00 ± 2.78	28.33 ± 2.55	71.67 ± 2.55***
9	78.33 ± 3.56***	21.67 ± 3.56	26.67 ± 2.72	73.33 ± 2.72***
10	71.67 ± 4.34***	28.33 ± 4.34	31.67 ± 3.89	68.33 ± 3.89***
11	65.00 ± 1.67***	35.00 ± 1.67	31.67 ± 5.24	68.33 ± 5.24*
14	65.00 ± 2.99*	35.00 ± 2.99	28.33 ± 2.55	71.67 ± 2.55***
16	56.67 ± 2.72*	43.33 ± 2.72	38.33 ± 3.56	61.67 ± 3.56*
18	58.33 ± 2.78*	41.67 ± 2.78	36.67 ± 3.33	63.33 ± 3.33*
21	60.00 ± 2.72*	40.00 ± 2.72	36.67 ± 4.16	63.33 ± 4.16*

*Significantly (according to Wilcoxon) preferred odortype;
**Significant differences (according to Student, $p < 0.05$) as compared with the attractiveness of testers before the exposure to radiation.

These studies have provided experimental evidence illustrating a clear relationship between olfactory conditioned behavioral reactions of male mice of both strains and the absorbed dose of ionizing radiation.

Conclusions

Thus, the data obtained in this study provide evidence that, at early stages after exposure to sublethal doses of ionizing radiation, stress and administration of immunodepressants the animals secrete volatile substances with urine, which attract intact animals, inducing immune disturbances in them. Apparently, they may cause the development of secondary, reactive disturbances in groups of animals that were but contacted with damaging animals. The ability of volatile substances to induce the distribution of them in groups of animals may probably have ecological consequences.

References

Barnett, S.A., 1975, The rat: A study in behavior. University of Chicago Press, Chicago, USA.
Bures, J., Buresova, O., and Huston, J.P., 1983, Techniques and basic experiments for the study of brain and behavior. Amsterdam, New York.
Halpin, Z.T., 1986, Individual odors among mammals origins and function. *Adv. Stud. Behav.* **16**: 39–70.

Moshkin, M.P., Gerlinskaya, L., Morosova, O., Bakhvalova, V., and Evsikov, V., 2002, Behaviour, chemosignals and endocrine functions in male mice infected with tick-borne encephalitis virus. *Psychoneuroendocrinology* **27**: 603–608.

Mothersill, C. and Seymour, C.B., 1998, Cell-cell contact during gamma-irradiation is not required to induce a bystander effect in normal human keratinocytes: evidence for release of a survival controlling signal into medium. *Radiat. Res.* **149**: 256–262.

Mothersill, C. and Seymour, C.B., 2000, Genomic Instability, bystander effects & Radiation Risks: Implications for the development of Protection Strategies for man & the environment. *Radiat. Biol. Radioecol.* **40**(5): 615–620 (in Russian).

Mothersill, C. and Seymour, C.B., 2001, Radiation-induced bystander effects: past history and future directions. *Radiat. Res.* **155**: 759–767. Review.

Penn, D.J. and Potts, W.K., 1998, Chemical signals and parasite-mediated sexual selection. *Trends Ecol. Evol.* 13: 391–396.

Surinov, B.P. and Dukhova, N.N., 2004, Postradiation mice's volatile secretions which are attractive for intact individuals. *Radiat. Biol. Radioecol.* **44**(6): 662–665 (in Russian).

Surinov, B.P., Isaeva, V.G., and Dukhova, N.N., 2005, Post-radiation immunosuppressing and attractive volatile secretions: "Bystander effect" or allelopathy in groups of animals. *Proc. Russian Acad. Sci.* **400**(5): 711–713 (in Russian).

Surinov, B.P., Karpova, N.A., Isayeva, V.G., and Kulish, Y.S., 1998, Natural secretions in the post-radiation period and contact induction of immunodeficiency. *Radiat. Biol. Radioecol.* **38**(1): 9–13 (in Russian).

Yamazaki, K., Beauchamp, G.K., Singer, A.J., and Boyse, E.A., 1999, Odortypes: their origin and composition. *Proc. Natl. Acad. Sci. USA* **96**:1522 – 525.

Zalcman, S., Kerr, L., Anisman, H., 1991, Immunosuppression elicited by stressors and stressor-related odors. *Brain Behav. Immun.* **5**(3): 262–273.

CHAPTER 10

RADIATION-INDUCED GENOMIC INSTABILITY IN THE OFFSPRING OF IRRADIATED PARENTS

YURI E. DUBROVA

Department of Genetics, University of Leicester, Leicester LE1 7RH, UK. e-mail: yed2@le.ac.uk

Abstract: So far, mutation induction in the germline of directly exposed parents has been regarded as the main component of the genetic risk of ionising radiation. However, recent data on the delayed effects of exposure to ionising radiation challenge for the existing paradigm. The results of some publications imply that exposure to ionising radiation results in elevated mutation rates detectable not only in the directly irradiated cells, but also in their non-irradiated progeny. Here I review the data on transgenerational instability showing that radiation-induced instability in the germline of irradiated parents manifests in their offspring, affecting their mutation rates and some other characteristics. This paper summarises the data on increased cancer incidence and elevated mutation rates in the germline and somatic tissues of the offspring of irradiated parents. The possible mechanisms of transgenerational instability are discussed.

Keywords: instability; mutation; radiation; germline; genetic risk; mouse

Introduction

The effort to predict the long-term genetic effects of ionising radiation for humans has certainly been one of the most important issues of radiation biology in the past years. However, despite the vast amount of experimental data describing the phenomenon of mutation induction in the directly exposed somatic and germ cells (UNSCEAR, 2001), the results of some recent studies clearly show that the genetic risks of ionising radiation may be far greater than previously thought. For example, the results of numerous in vitro studies have shown that mutation rates in the progeny of irradiated cells remains elevated over a considerable period of time after the initial exposure

C. Mothersill et al. (eds.), Multiple Stressors: A Challenge for the Future, 139–154.
© 2007 *Springer.*

(Morgan, 2003a). The manifestation of radiation-induced genomic instability has also been reported in vivo (Morgan, 2003b). It should be stressed that these data challenge the existing paradigm in radiation biology which regards mutation induction in the directly exposed somatic and germ cells as the main component of genetic risk for humans (UNSCEAR, 2001). As mutation rates in the non-exposed progeny of irradiated cells remain considerably elevated over many cell divisions after irradiations, radiation-induced genomic instability can be regarded as a risk factor for radiation-induced carcinogenesis. It is well established that carcinogenesis is a multistep process in which somatic cells acquire mutations in a specific clonal lineage (Loeb et al., 2003). However, the pattern of accumulation of multiple mutations in the irradiated cells over a clinically relevant time period still remains unclear. It was therefore suggested that ongoing genomic instability could result in the accumulation of mutations over a certain period of time after irradiation which, together with mutations directly induced in the irradiated cells, may significantly enhance radiation carcinogenesis (Little, 2000; Goldberg, 2003; Huang et al., 2003).

The issue of the delayed effects of radiation is also highly relevant to the understanding of the mechanisms underlying a therapy-induced second malignancy (Goldberg, 2003; Sigurdson and Jones, 2003). The development of effective radiotherapy and chemotherapy regimes for the treatment of cancer has recently resulted in a dramatically increased number of long-term survivors. Given that many treatments used for cancer, including ionising radiation, are mutagenic, the impressive increase in cure and survival rates has been accompanied by a worrisome increase in the incidence of therapy-related second malignancies (Garwicz, 2000; UNSCEAR, 2000). With respect to cancer patients, the tissue at risk for the induction of secondary malignancy is normal tissue that has been exposed to ionising radiation, but not killed. It appears that therapy-related genomic instability can follow radiation therapy or chemotherapy and may thus contribute to the development of cancer in normal tissue. However, the extent to which genomic instability plays a role in the development of therapy-induced second malignancy still remains unknown.

Apart from the studies on mutation rates in somatic cells, considerable progress has been made in the analysis of radiation-induced instability in the mammalian germline, where the effects of radiation exposure were investigated among the offspring of irradiated parents (reviewed in Dubrova, 2003; Nomura 2003; Barber and Dubrova, 2006). These transgenerational studies were designed to test the hypothesis that radiation-induced instability in the germline of irradiated parents could manifest in the offspring, affecting their mutation rates and some other characteristics. The aim of this paper is to review a number of publications addressing transgenerational instability in mice and other laboratory animals. Given that a considerable number of publications have characterised the transgenerational changes using a variety of phenotypic traits, this review mainly describes the progress made in the analysis of transgenerational changes in cancer predisposition and mutation rates.

Cancer Predisposition

The in-depth analysis of the incidence of cancer in the offspring of irradiated parents was initiated by the findings showing clustering of childhood leukaemia in the vicinity of the Sellafield nuclear reprocessing plant (Gardner et al., 1990) and a substantial increase in the incidence of tumours in the non-exposed first-generation offspring (F_1) of male mice exposed to X-rays or uretane (Nomura, 1982). It should be noted that the Nomura's data were not confirmed in the later studies (Cattanach et al., 1995, 1998), the results of which indicated that this phenomenon could partially be attributed to the seasonal variation in tumour incidence in mouse colonies. However, the results of some publications suggest that although the incidence of cancer among the offspring of exposed parents does not exceed that of the control, the morphology of the tumours in the offspring differs. For example, it has been shown that the mean number of lung adenomas per tumour in the offspring of parents pre-natally exposure to benz(a)pyrene remains persistently elevated over several generations (Turusov et al., 1990).

A substantial contribution to the analysis of transgenerational cancer predisposition has been made in more recent studies where the offspring of irradiated parents were exposed to carcinogens (Nomura, 1983; Vorobtsova, 1993; Lord et al., 1998; Hoyes et al., 2001). In contrast to the data obtained on the non-treated offspring, the results of these studies showed an elevated incidence of cancer among the carcinogen-challenged offspring of irradiated males (Fig. 1a).

The pattern of malignancy among the treated offspring of irradiated males was also modified (Lord et al., 1998a, b). Thus the treatment shortened latent period for the leukaemia and resulted in a switch from the predominant thymic lymphoma in the controls to a predominance of leukaemia in the offspring of irradiated males.

Somatic Mutation Rates

Given carcinogenesis is a process of accumulation of mutations in somatic cells (Loeb et al., 2003), the data showing a substantially elevated cancer risk in the offspring of irradiated parents indicates that somatic mutation rates in these animals may also be increased. Indeed, the results of some studies provide a strong evidence for transgenerational increases in somatic mutation rates.

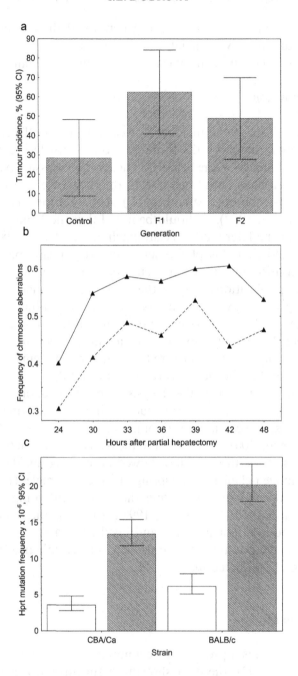

Fig. 1. Transgenerational changes in somatic tissues of the offspring of irradiated parents. (a), Promotion of skin tumours by 12-*O*-tetradecanoylphorbol-13-acetate (TPA) in the offspring of irradiated male mice (data from Vorobtsova et al., 1993). (b), Frequency of chromosome aberrations in liver of controls (dashed line) and F_1 offspring (solid line) of irradiated male rats (data from Vorobtsova, 2000). (c), Frequency of *hprt* mutations in controls (open bars) and F_1 offspring of irradiated male mice (hatched bars, data from Barber et al., 2006). The 95% confidence intervals (CI) are shown.

To date, the frequency of somatic mutations in the offspring of irradiated parents has been analysed using a variety of endpoints. Thus, the study by Vorobtsova (2000) provided a convincing evidence for elevated frequency of chromosome aberrations in the liver of F_1 offspring of irradiated male rats (see Fig. 1b). These data were later confirmed by studying the frequency of chromosome aberrations in the liver and as well as in some other rat tissues (Kropacova, 2002; Slovinska et al., 2004; Sanova et al., 2005). An elevated frequency of chromosome aberrations was also found in the bone marrow tissue of F_1 offspring of irradiated male mice (Lord et al., 1998a, b). These results are further supported by the data showing an elevated frequency of micronuclei in the offspring of irradiated male mice (Fomenko et al., 2001). Given the micronuclei contain the fragments of mis-repaired or damaged chromosomes (Fenech et al., 1999), the observed increase is therefore attributed to chromosomal instability.

Transgenerational changes in somatic mutation rates were also observed by studying the frequency of mutations at some protein-coding genes. Thus, Luke et al. (1997) showed an elevated frequency of mutations at the *lacI* transgene in the F_1 offspring of irradiated male mice. The most compelling data addressing somatic instability in the F_1 offspring of irradiated male mice were obtained from the analysis of somatic reversions of the pink-eyed unstable mutation (p^{un}). The mouse p^{un} mutation is caused by DNA sequence duplication within the *pink-eyed dilute* locus and results in reduced coat and eye coloration. This duplication is highly unstable and spontaneous reversions caused by deletion of one of the duplicated sequences are quite frequent (Brilliant et al., 1991; Gondo et al., 1993). Using this endpoint, Shiraishi et al., 2002 demonstrated that the frequency of p^{un} somatic reversions in the first-generation offspring of irradiated male mice was significantly elevated. The authors studied transgenerational effects in the first-generation offspring of two reciprocal matings of irradiated fathers – $\male p^{un} \times \female p^j$ and $\male p^j \times \female p^{un}$ – and found that the frequency of somatic reversions was equally elevated in the offspring of both reciprocal matings. Given that reversions only occur at the p^{un} allele, the data showing elevated frequency of somatic reversions in the offspring of $\male p^j \times \male p^{un}$ mating therefore demonstrate that transgenerational instability is manifested at the alleles derived from both irradiated and non-irradiated parents. These results were supported in recent studies on somatic mutation in the offspring of irradiated fish (Shimada and Shima, 2004; Shimada et al., 2005).

We have recently analysed the frequency of thioguanine-resistant mutations at the hypoxanthine guanine phosphoribosil transferase (*hprt*) locus in the spleenocytes of F_1 offspring of irradiated male mice (Barber et al., 2006). A highly significant ~3.5-fold increase in the frequency of *hprt* mutations was found in both strains of mice (Fig. 1c). Given that in the male offspring of irradiated males the X-linked *hprt* locus is transmitted from the non-exposed mothers, these data further confirm the abovementioned conclusion that transgenerational changes in mutation rates equally affect both alleles derived from the irradiated fathers and the unexposed mothers, thus implying a genome-wide destabilisation after fertilisation.

Germline Effects

The first evidence for the transgenerational increases in germline mutation rates was obtained by Luning et al., 1976. By analysing the frequency dominant lethal mutations in the germline of directly irradiated male mice and their first-generation offspring, the authors demonstrated that mutation rates in the germline of directly exposed parents and their non-irradiated offspring were equally elevated. Similar data were later obtained from the analysis of the F_1 offspring of male rats treated by cyclophosphamide (Hales et al., 1992). The results showing decreased proliferation of early embryonic cells and increased frequency of malformations in the F_2 offspring of irradiated parents are also consistent with these observations (Wiley et al., 1997; Pils et al., 1999). It should be noted that further analysis of transgenerational instability requires a sensitive technique capable of detecting relatively modest changes in mutation rate.

We have previously developed a new sensitive technique for monitoring mutation induction in the mouse germline by ionising radiation and chemical mutagens (Dubrova et al., 1993, 1998, 2000; Vilarino-Guell, 2003). This technique employs highly unstable expanded simple tandem repeat (ESTR) loci which consist of homogenous arrays of relatively short repeats (4–6 bp) and show a very high spontaneous mutation rates both in germline and somatic cells (Kelly et al., 1989; Gibbs et al., 1993; Bois et al., 1998; Yauk et al., 2002).

In our early studies, we have used this technique to evaluate ESTR mutation rates in the germline of F_1 and F_2 offspring of irradiated male mice (Dubrova et al., 2000; Barber et al., 2002). The analysis of the F_1 offspring of a male mouse exposed to fission neutrons showed that their germline mutation rates did not return to the mutation rates seen in unexposed individuals, but remained similar to those observed in directly exposed males (Dubrova et al., 2000). The increase was observed in most F_1 offspring and was in part attributable to increased mutational mosaicism in the germline, therefore indicating that transgenerational destabilisation should occur either immediately after fertilisation or on the very early stages of the developing F_1 germline.

To verify these results and to gain some insights on the mechanisms of transgenerational instability in the mouse germline, we later analysed ESTR mutation rates in the germline of first- and second-generation offspring of inbred male CBA/H, C57BL/6 and BALB/c mice exposed to either high-LET fission neutrons or low-LET X-rays (Barber et al., 2002). Figure 2a presents the main result of this study, showing that paternal exposure to ionising radiation results in increased mutation rates in the germline of two subsequent generations of all inbred strains, demonstrating that transgenerational instability is not restricted to one particular inbred strain of mice. Our data also revealed inter-strain differences in the transgenerational effects, demonstrating that ESTR mutations rates in the F_1 and F_2 germline of BALB/c and CBA/H mice were significantly higher than in those of C57BL/6 mice. These data are consistent with the results of previous studies showing that that BALB/c and CBA mice are significantly more radiosensitive, and display higher levels of radiation-induced genomic instability in somatic cells than C57BL/6 mice (Roderick, 1963; Watson et al., 1997; Mothersill et al., 1999; Ponnaiya et al., 1997). The high level of radiation-induced genomic instability observed in BALB/c mice could potentially be explained by the strain-specific amino-acid substitutions affecting the activity of the p16[ink4a] cyclin-dependent kinase inhibitor and the catalytic-subunit of the DNA-dependent protein kinase (Zhang et al., 1998; Yu et al., 2001). Given the wide range of inherited variation in DNA repair capacity in humans (Mohrenweiser et al., 2003), there is potential that the same phenomenon may also exist in humans.

In this study we also compared the transgenerational effects of paternal exposure to high-LET fission neutrons and low-LET X-rays. It is well established that high-LET radiation produces highly complex and localised initial DNA damage, which is different to the sparse damage produced by low-LET radiation, resulting in the unique final biological effects of these different radiation sources (Goodhead, 1988). However, it appears that exposure to both types of radiation is capable of inducing genomic instability in somatic cells, though some studies have failed to detect the effects of low-LET exposure (Limoli et al., 2000).

It should be noted that our results indicated that most of the offspring of irradiated males showed elevated mutation rates in their germline, therefore providing important evidence for the involvement of epigenetic mechanisms in transgenerational instability. However, these data were obtained using a pedigree-based approach, which has low statistical power for the detection of mutation rate heterogeneity between individuals due to the relatively small litter size in mice. Our study also raised the possibility of transgenerational increases in ESTR somatic mutation rates. To establish whether ESTR mutation rates are equally elevated in the germline and somatic tissues of first-generation (F_1) offspring of irradiated males, we have used a single-molecule PCR approach which allows the recovery of large numbers of de novo mutants from a single individual and hence provides robust estimates of individual mutation rates (Yauk et al., 2002).

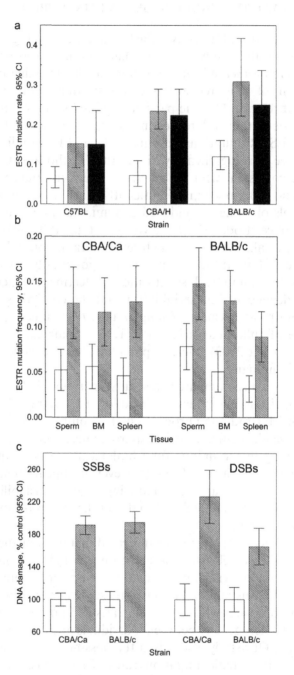

Fig. 2. Transgenerational changes in the germline and somatic tissues. (a), ESTR mutation rate in the germline of controls (open bars), F_1 (hatched bars) and F_2 (black bars) offspring of irradiated male mice (data from Barber et al., 2002). (b), ESTR mutation frequencies in the germline and somatic tissues of controls (open bars) and F_1 offspring (hatched bars) of irradiated males (data from Barber et al., 2006). (c), Endogenous DNA damage (single-strand, SSBs and double-strand, DSBs DNA breaks) in controls (open bars) and F_1 offspring (hatched bars) of irradiated male mice males (data from Barber et al., 2006).

The frequency of ESTR mutation was established in DNA samples prepared from sperm, bone marrow (BM) and spleen from the same animal (Barber et al., 2006). A statistically significant increase in the mean mutation frequency was found in all tissues of the offspring of irradiated males (Fig. 2b). These data confirmed our previous results obtained in the germline of F_1 offspring of irradiated males, using the more traditional pedigree-based approach (Dubrova, 2000; Barber, 2002) and showed that transgenerational genomic instability at ESTR loci was also manifested in somatic tissues. Most importantly, we observed that the frequency of ESTR mutation was elevated in the germline and somatic tissues of all the offspring of irradiated males (Barber et al., 2006).

Mechanisms

Cellular response to ionising radiation is a multistep process which includes the recognition of DNA damage, its repair, cell cycle arrest and apoptosis (Friedberg et al., 2006; Sancar et al., 2004; Bakkenist and Kastan, 2004). To date, the majority of pathways involved in the mammalian cellular response to radiation have been characterised. It should be stressed that the data on the delayed effects of exposure to ionising radiation represent a serious challenge to the existing paradigm. Although the mechanism(s) underlying the phenomenon of radiation-induced genomic instability still remain unknown, the results of some publications show that the ability of cells to exhibit elevated mutation rates cannot be ascribed to the conventional mechanisms of mutator phenotype and is most likely related to the epigenetic events (Lorimore et al., 2003; Morgan, 2003a,b). Such a conclusion is based on two sets of experimental data, showing that: (i) radiation-induced genomic instability persists over a long period of time after the initial exposure; (ii) the number of cells/organisms manifesting radiation-induced genomic instability is too high to be explained by the conventional mechanisms of direct targeting of DNA-repair and related genes. Our data showing that transgenerational changes affect the majority of the offspring of irradiated parents (Barber et al., 2006), together with the results of recent publication on transgenerational effects of paternal exposure to endocrine disruptors (Anway et al., 2005), further support this hypothesis.

As already mentioned, the data on transgenerational instability at the p^{un} and *hprt* loci showed that somatic mutation rates in the F_1 offspring are equally elevated at the alleles derived from the irradiated fathers and non-irradiated mothers (Shiraishi et al., 2002; Barber et al., 2006). Similar transgenerational data were obtained by studying the mouse ESTR loci (Dubrova et al., 2000; Barber et al., 2002; Niwa and Kominami, 2001). Taken together, these results show that an increased mutation rate in the offspring of irradiated males results from a genome-wide elevation of mutation rate.

In a number of publications the issue of stage-specificity of transgenerational changes has been addressed. These data provide some important clues onto the mechanisms of radiation-induced genomic instability. Thus, our data were obtained on the descendants conceived 3 and 6 weeks after the initial paternal exposure to ionising radiation (Barber et al., 2002, 2006; Dubrova et al., 2000). Given that these stages of the mouse spermatogenesis are transcriptionally active, their exposure to radiation could result in an accumulation of certain classes of RNA in the paternal germ cells which, being transmitted to the fertilised egg, may affect gene expression and stability in the developing embryo. If transgenerational instability is attributed to the zygotic transfer of RNA (Rassoulzadegan et al., 2006), then the offspring conceived just few days after paternal irradiation, from transcriptionally inert sperm cells (Rousseaux et al., 2005), should be genetically stable.

However, several recent publications report transgenerational changes in the offspring of male mice irradiated during the late post-meiotic stages of spermatogenesis, where gene expression is practically shut down (Vorobtsova et al., 1993, 2000; Shiraishi et al., 2002). Given that pre-mutational radiation-induced lesions in sperm DNA are effectively recognised and repaired within a few hours of fertilisation (Matsuda and Tobari, 1989; Derijck et al., 2006), it would therefore appear that radiation-induced damage to sperm DNA could trigger a cascade of events in the zygote, including profound changes in the expression of DNA repair genes in the pre-implantation embryo (Harrouk et al., 2000; Shimura et al., 2002) and alterations in DNA methylation and histone acetylation (Barton et al., 2002). The presence of such dramatic changes at fertilisation could also result in delayed effects, which may influence the stability of the developing embryo. The results showing an unusually high level of mutational mosaicism in the germline and somatic tissues of F_1 mice (Wiley et al., 1997; Niwa and Kominami, 2001) suggest that the destabilisation could occur at the very early stages of development.

Given that the results of transgenerational studies, together with the data on radiation-induced genomic instability in vitro clearly indicate that these phenomena are attributed to epigenetic events, we and other hypothesised that DNA methylation may be regarded as a strong candidate for such an epigenetic signal resulting in transgenerational mutagenesis (Wiley et al., 1997; Anway et al., 2005). DNA methylation and histone modification represent the main mechanisms by which DNA is epigenetically marked (Jones and Baylin, 2002). Methylation is known to survive the reprogramming of DNA methylation during spermatogenesis and early development (Holliday, 1997; Reik and Walter, 2001; Rakyan et al., 2001) and can to be transmissible through many cell divisions (Roemer et al., 1997). Alterations in the pattern of DNA methylation might affect genes responsible for maintaining genomic integrity and influence the recognition of DNA damage or its repair. For example, promoter methylation switches off the transcription of the *hMLH1* mismatch repair gene colorectal carcinomas and results in microsatellite instability (Jones and Baylin, 2002). The transgenerational increases can also be attributed to the change in the expression patterns of genes involved in DNA repair in the offspring of irradiated males. Indeed, recent data showed persistently altered pattern of expression of some genes in the offspring of irradiated male mice (Nomura et al., 2004; Baulch et al., 2001; Vance et al., 2002).

It should be stressed that the altered expression of DNA repair genes cannot explain the transgenerational increases in mutation rates detected across a number of endpoints, including protein-coding genes, ESTR loci and chromosome aberration. Given that the mechanisms of spontaneous and induced mutation at these systems substantially differ, these observations imply that the efficiency of multiple DNA repair pathways should be simultaneously compromised in the offspring of irradiated parents. The presence of such highly coordinated changes appears to be highly unlikely. Indeed, the results of our recent study show that the efficiency of repair of some DNA lesions (detected by the alkaline Comet assay; Kassie et al., 2000) in the offspring of irradiated males is not compromised (Barber et al., 2006).

The results of our study (Barber et al., 2006) also demonstrate that the multiplicity of transgenerational changes is most likely attributed to an abnormally high level of DNA damage in F_1 offspring of irradiated parents (Fig. 2c). Given that in tissues with a high mitotic index, such as bone marrow and spleen, the life-span of cells containing deleterious lesions such as single- and double-strand breaks is restricted since these types of DNA damage are not compatible with replication, these data clearly demonstrate that transgenerational instability is an ongoing process occurring in multiple adult tissues. As unrepaired/uncorrected double- and single-strand breaks are known to be highly mutagenic and thus may result in tumour development and/or progression, our data therefore provide a plausible explanation for elevated predisposition to cancer among the offspring of irradiated parents. Given that destabilisation of genome occurs in multiple adult tissues, it is likely that transgenerational carcinogenesis may be due, in part, to this phenomenon.

The high level of DNA damage could be attributed to an oxidative stress/inflammatory response. The involvement of inflammatory-type processes in the delayed increases in mutation rates in the progeny of irradiated cells has long been suspected (Morgan, 2003a,b; Lorimore et al., 2003). Reactive oxygen species are the major source of endogenous DNA damage, including single- and double-strand breaks, abasic sites, and a variety of nucleotide modifications (Jackson and Loeb, 2001). The diversity of DNA alterations detected in the offspring of irradiated parents could therefore be explained by this mechanism. However, according to our data, the F_1 offspring of irradiated males did not show any increases in the somatic tissue level of oxidatively damaged nucleotides. This being the case, then transgenerational instability could be attributed to replication stress. Indeed, the results of recent studies suggest that in human precancerous cells the ATR/ATM-regulated checkpoints are activated through deregulated DNA replication, which leads to the multiplicity of DNA alterations (Bartkova et al., 2005; Gorgoulis et al., 2005). It has also been shown that radiation-induced chromosome instability *in vitro* could be attributed to the long-term delay in chromosome replication (Breger et al., 2004). Given that our previous results suggest that the mechanism of ESTR mutation is most probably attributed to replication slippage (Yauk et al., 2002; Barber et al., 2004; Dubrova, 2005), delayed/stalled replication can therefore provide a plausible explanation for the transgenerational increases in mutation rate at these loci, as well as the multiplicity of DNA alterations detected in the offspring of irradiated parents.

In summary, the results of studies reviewed here provide strong evidence for long-term increases in mutation rate, observed in the offspring of irradiated parents. However, being unequivocally established in laboratory animals, the phenomenon of radiation-induced transgenerational instability in humans still awaits its final clarification. Future work should address these important issues and provide experimentally based estimates of the delayed effects of radiation in humans.

Acknowledgements

This work was supported by grants from the Wellcome Trust, the Department of Energy and the European Commission and the Medical Research Council.

References

Anway, M. D., A. S. Cupp, M. Uzumcu, and M. K. Skinner, Epigenetic transgenerational actions of endocrine disruptors and male fertility, *Science*, 308, 1466–469 (2005).

Bakkenist, C. J. and M. B. Kastan, Initiating cellular stress response, *Cell*, 118, 9–17 (2004).

Barber, R. and Y. E. Dubrova, The offspring of irradiated parents, are they stable? *Mutat. Res.*, 598, 50–60 (2006).

Barber, R. C., L. Miccoli, P. P. W. van Buul, K. L. Burr, A. van Duyn-Goedhart, J. F. Angulo and Y. E. Dubrova, Germline mutation rates at tandem repeat loci in DNA-repair deficient mice, *Mutat. Res.*, 554, 287–295 (2004).

Barber, R. C., P. Hickenbotham, T. Hatch, D. Kelly, N. Topchiy, G. Almeida, G. G. D. Jones, G.E. Johnson, J.M. Parry, K. Rothkamm, and Y.E. Dubrova, Radiation-induced transgenerational alterations in genome stability and DNA damage, *Oncogene*, 25, 7336–7342 (2006).

Barber, R., M. A. Plumb, E. Boulton, I. Roux, and Y. E. Dubrova, Elevated mutation rates in the germ line of first- and second-generation offspring of irradiated male mice, *Proc. Natl. Acad. Sci. USA*, 99, 6877–882 (2002).

Bartkova, J., Z. Horejsi, K. Koed, A. Kramer, F. Tort, K. Zieger, P. Guldberg, M. Sehested, J. M. Nesland, C. Lukas, T. Orntoft, J. Lukas, and J. Bartek, DNA damage response as a candidate anti-cancer barrier in early human tumorigenesis, *Nature*, 434, 864–870 (2005).

Barton, T. S., B. Robaire, and B. F. Hales, Epigenetic programming in the preimplantation rat embryo is disrupted by chronic paternal cyclophosphamide exposure, *Proc. Natl. Acad. Sci. USA*, 102, 7865–870 (2005).

Baulch, J. E., O. G. Raabe, and L. M. Wiley, Heritable effects of paternal irradiation in mice on signaling protein kinase activities in F3 offspring, *Mutagenesis*, 16, 17–23 (2001).

Bois, P., J. Williamson, J. Brown, Y.E. Dubrova, and A.J. Jeffreys, A novel unstable mouse VNTR family expanded from SINE B1 element, *Genomics*, 49, 122–128, (1998).

Breger, K. S., L. Smith, M. S. Turker and M. J. Thayer, Ionizing radiation induces frequent translocations with delayed replication and condensation, *Cancer Res.*, 64, 8231–238 (2004).

Brilliant, M. H., Y. Gondo, and E. M. Eicher, Direct molecular identification of the mouse pink-eyed unstable mutation by genome scanning, *Science,* 252, 566–569 (1991).

Cattanach, B. M., D. Papworth, G. Patrick, D. T. Goodhead, T. Hacker, L. Cobb, and E. Whitehill, Investigation of lung tumour induction in C3H/HeH mice, with and without tumour promotion with urethane, following paternal X-irradiation, *Mutat. Res.*, 403, 1–12 (1998).

Cattanach, B. M., G. Patrick, D. Papworth, D. T. Goodhead, T. Hacker, L. Cobb, and Whitehill, E., Investigation of lung tumour induction in BALB/cJ mice following paternal X-irradiation, *Int. J. Radiat. Biol.*, 67, 607–615 (1995).

Derijck, A. A. H. A., G. W. van der Heijden, M. Giele, M. E. Philippens, C. C. A. W. van Bavel CC, and P. de Boer, γH2AX signalling during sperm chromatin remodelling in the mouse zygote, *DNA Repair,* 5, 959–971 (2006).

Dubrova, Y. E., A. J. Jeffreys, and A. M. Malashenko, Mouse minisatellite mutations induced by ionizing-radiation, *Nat. Genet.*, 5, 92–94 (1993).

Dubrova, Y. E., M. Plumb, B. Gutierrez, E. Boulton, and A. J. Jeffreys, Genome stability - Transgenerational mutation by radiation, *Nature*, 405, 37–37 (2000).

Dubrova, Y. E., M. Plumb, J. Brown, E. Boulton, D. Goodhead, and A. J. Jeffreys, Induction of minisatellite mutations in the mouse germline by low-dose chronic exposure to gamma-radiation and fission neutrons, *Mutat. Res.*, 453, 17–24 (2000).

Dubrova, Y. E., M. Plumb, J. Brown, J. Fennelly, P. Bois, D. Goodhead, and A. J. Jeffreys, Stage specificity, dose response, and doubling dose for mouse minisatellite germ-line mutation induced by acute radiation, *Proc. Natl. Acad. Sci. USA*, 95, 6251–255 (1998).

Dubrova, Y. E., Radiation-induced mutation at tandem repeat DNA loci in the mouse germline: spectra and doubling doses, *Radiat. Res.*, 163, 200–207 (2005).

Dubrova, Y. E., Radiation-induced transgenerational instability, *Oncogene*, 22, 7087–7093 (2003).

Fenech, M., N. Holland, W.P. Chang, E. Zieger, and S. Bonassi, The HUman MicroNucleus Project – An international collaborative study on the use of the micronucleus technique for measuring DNA damage in humans, *Mutat. Res.*, 428, 271–283 (1999).

Fomenko, L. A., G. V. Vasil'eva, and V. G. Bezlepkin, Micronucleus frequency is increased in bone marrow erythrocytes from offspring of male mice exposed to chronic low-dose gamma irradiation, *Biol. Bull.,* 28, 419–423 (2001).

Friedberg, E. C., G. C. Walker, W. Siede, R. D. Wood, R. A. Schultz, and T. Ellenberger, *DNA Repair and Mutagenesis* (ASM Press, Washington, 2006).

Gardner, M. J., M. P. Snee, A. J. Hall, C. A. Powell, S. Downes, and J. D. Terrell, Results of case-control study of leukaemia and lymphoma among young people near Sellafield nuclear plant in West Cumbria, *Br. Med. J.*, 300, 423–429 (1990).

Garwicz, S., H. Anderson, J. H. Olsen, H. Dollner, H. Hertz, G. Jonmundsson, F. Langmark, M. Lanning, T. Moller, R. Sankila, and H. Tulinius, Second malignant neoplasms after cancer in childhood and adolescence: a population-based case-control study in the 5 Nordic countries, *Int. J. Cancer,* 88, 672–678 (2000).

Gibbs, M., A. Collick, R. G. Kelly, and A. J. Jeffreys, A tetranucleotide repeat mouse minisatellite displaying substantial somatic instability during early preimplantation development, *Genomics*, 17, 121–128 (1993).

Gondo, Y., J. M. Gardner, Y. Nakatsu, D. Durham-Pierre, and S. A. Deveau, C. Kuper and M. H. Brilliant. High-frequency genetic reversion mediated by a DNA duplication: the mouse pink-eyed unstable mutation, *Proc. Natl. Acad. Sci. USA*, 90, 297–301 (1993).

Goodhead, D. T., Spatial and temporal distribution of energy, Health Phys., 55, 231–240 (1988).

Gorgoulis, V. G., L. V. Vassiliou, P. Karakaidos, P. Zacharatos, A. Kotsinas, T. Liloglou, M. Venere, R. A. Ditullio, Jr., N. G. Kastrinakis, B. Levy, D. Kletsas, A. Yoneta, M. Herlyn, C. Kittas, and T. D. Halazonetis, Activation of the DNA damage checkpoint and genomic instability in human precancerous lesions, *Nature*, 434, 907–913 (2005).

Hales, B. F., K. Crosman, and B. Robaire, Increased postimplantation loss and malformations among the F2 progeny of male rats chronically treated with cyclophosphamide, *Teratology*, 45, 671–678 (1992).

Harrouk, W., A. Codrington, R. Vinson, B. Robaire, and B. F. Hales, Paternal exposure to cyclophosphamide induces DNA damage and alters the expression of DNA repair genes in the rat preimplantation embryo, *Mutat. Res.*, 461, 229–241 (2000).

Holliday, R., The inheritance of epigenetic defects, *Science*, 238, 163–170 (1987).

Hoyes, K. P., B. I. Lord, C. McCann, J. H. Hendry, and I. D. Morris, Transgenerational effects of preconception paternal contamination with (55)Fe, *Radiat. Res.*, 156, 488–494 (2001).

Huang, L., A. R. Snyder, and W. F. Morgan, Radiation-induced genomic instability and its implications for radiation carcinogenesis, *Oncogene*, 22, 5848–854 (2003).

Jackson, A. L. and L. A. Loeb, The contribution of endogenous sources of DNA damage to the multiple mutations in cancer, *Mutat. Res.*, 477, 7–21 (2001).

Jones, P. A. and S. B. Baylin, The fundamental role of epigenetic events in cancer, Nat. Rev. Genet., 3, 415–428 (2002).

Kassie, F., W. Parzefall, and S. Knasmuller, Single cell gel electrophoresis assay: a new technique for human biomonitoring studies, *Mutat. Res.*, 463, 13–31 (2000).

Kelly, R., G. Bulfield, A. Collick, M. Gibbs and A. J. Jeffreys, Characterization of a highly unstable mouse minisatellite locus: evidence for somatic mutation during early development, *Genomics*, 5, 844–856 (1989).

Kropacova, K., L. Slovinska, and E. Misurova, Cytogenetic changes in the liver of progeny of irradiated male rats, *J. Radiat. Res.*, 43, 125–133 (2002).

Limoli, C. L., B. Ponnaiya, J. J. Corcoran, E. Giedzinski, M. I. Kaplan, A. Hartmann, and W. F. Morgan, Genomic instability induced by high and low LET ionizing radiation, *Adv. Space. Res.*, 25, 2107–117 (2000).

Little, J. B., Radiation carcinogenesis, *Carcinogenesis*, 21, 397–404 (2000).

Loeb, L. A., K. R. Loeb, and J. P. Anderson, Multiple mutations and cancer, *Proc. Natl. Acad. Sci. USA*, 100, 776–781 (2003).

Lord, B. I., L. B. Woolford, L. Wang, D. McDonald, S. A. Lorimore, V. A. Stones, E. Wright, G. and D. Scott, Induction of lympho-haemopoietic malignancy: impact of preconception paternal irradiation, *Int. J. Radiat. Biol.*, 74, 721–728 (1998).

Lord, B. I., L. B. Woolford, L. Wang, V. A. Stones, D. McDonald, S. A. Lorimore, D. Papworth, E. G. Wright, and D. Scott, Tumour induction by methyl-nitroso-urea following preconceptional paternal contamination with plutonium-239, *Br. J. Cancer*, 78, 301–311 (1998).

Lorimore, S. A., P. J. Coates, and E. G. Wright, Radiation-induced genomic instability and bystander effects: inter-related nontargeted effects of exposure to ionizing radiation, *Oncogene*, 22, 7058–7069 (2003).

Luke, G. A., A. C. Riches, and P. E. Bryant, Genomic instability in haematopoietic cells of F1 generation mice of irradiated male parents, *Mutagenesis*, 12, 147–152 (1997).

Luning, K. G., H. Frolen, and A. Nilsson, Genetic effects of ^{239}Pu salt injections in male mice, *Mutat. Res.*, 34, 539–542 (1976).

Matsuda, Y. and I. Tobari, Repair capacity of fertilized mouse eggs for X-ray damage induced in sperm and mature oocytes, *Mutat. Res.*, 210, 35–47 (1989).

Mohrenweiser, H. W., D. M. Wilson, and I. M. Jones, Challenges and complexities in estimating both the functional impact and the disease risk associated with the extensive genetic variation in human DNA repair genes, *Mutat. Res.,* 526, 93–125 (2003).

Morgan, W. F., Non-targeted and delayed effects of exposure to ionizing radiation: I. Radiation-induced genomic instability and bystander effects in vitro, *Radiat. Res.*, 159, 567–580 (2003a).

Morgan, W. F., Non-targeted and delayed effects of exposure to ionizing radiation: II. Radiation-induced genomic instability and bystander effects in vivo, clastogenic factors and transgenerational effects, *Radiat. Res.*, 159, 581–596 (2003b).

Mothersill, C. E., K. J. O'Malley, D. M. Murphy, C. B. Seymour, S. A. Lorimore, and E. G. Wright, Identification and characterization of three subtypes of radiation response in normal human urothelial cultures exposed to ionizing radiation, *Carcinogenesis*, 20, 2273–2278 (1999).

Niwa, O. and R. Kominami, Untargeted mutation of the maternally derived mouse hypervariable minisatellite allele in F1 mice born to irradiated spermatozoa, *Proc. Natl. Acad. Sci. USA*, 98, 1705–710 (2001).

Nomura, T., H. Nakajima, H. Ryo, L.Y. Li, Y. Fukudome, S. Adachi, H. Gotoh, and H. Tanaka, Transgenerational transmission of radiation- and chemically induced tumors and congenital anomalies in mice: studies of their possible relationship to induced chromosomal and molecular changes, *Cytogenet. Genome Res.*, 104, 252–260 (2004).

Nomura, T., Parental exposure to x rays and chemicals induces heritable tumours and anomalies in mice, *Nature*, 296, 575–577 (1982).

Nomura, T., Transgenerational carcinogenesis: induction and transmission of genetic alterations and mechanisms of carcinogenesis, *Mutat. Res.*, 544, 425–432 (2003).

Nomura, T., X-ray-induced germ-line mutation leading to tumors. Its manifestation in mice given urethane post-natally, *Mutat. Res.*, 121, 59–65 (1983).

Pils, S., W-U. Muller, and C. Streffer, Lethal and teratogenic effects in two successive generations of the HLG mouse strain after radiation exposure of zygotes - association with genomic instability?, *Mutat. Res.*, 429, 85–92 (1999).

Ponnaiya, B., M. N. Cornforth, and R. L. Ullrich, Radiation-induced chromosomal instability in BALB/c and C57BL/6 mice: the difference is as clear as black and white, *Radiat. Res.*, 147, 121–125 (1997).

Rakyan, V. K., J. Preis, H. D. Morgan, and E. Whitelaw, The marks, mechanisms and memory of epigenetic states in mammals, *Biochem. J.*, 356, 1–10 (2001).

Rassoulzadegan, M., V. Grandjean, P. Gounon, S. Vincent, I. Gillot, and F. Cuzin, RNA-mediated non-Mendelian inheritance of an epigenetic change in the mouse, *Nature,* 241, 469–474 (2006).

Reik, W. and J. Walter, Genomic imprinting: parental influence on the genome, *Nat. Rev. Genet.,* 2, 21–32 (2001).

Roderick, T. H., The response of twenty-seven inbred strains of mice to daily doses of whole body X-irradiation, *Radiat. Res.*, 20, 631–639 (1963).

Roemer, I., W. Reik, W. Dean, and J. Klose, Epigenetic inheritance in the mouse, *Current Biol.*, 7, 277–280 (1997).

Rousseaux, S., C. Caron, J. Govin, C. Lestrat, A. K. Faure, and S. Khochbin, Establishment of male-specific epigenetic information, *Gene,* 345, 139–153 (2005).

Sancar, A., L. A. Lindsey-Boltz, K. Unsal-Kacman, and S. Linn, Molecular mechanisms of mammalian DNA repair and the DNA damage checkpoints, *Annu. Rev. Biochem.,* 73, 39–85 (2004).

Sanova, S., S. Balentova, L. Slovinska, and E. Misurova, Effects of preconception gamma irradiation on the development of rat brain, *Neurotoxicol. Teratol.*, 2005, 27, 145–151 (2005).

Shimada, A. and A. Shima, Transgenerational genomic instability as revealed by a somatic mutation assay using the medaka fish, *Mutat. Res.*, 552, 119–124 (2004).

Shimada, A., H. Eguchi, S. Yoshinaga, and A. Shima, Dose-rate effect on transgenerational mutation frequencies in spermatogonial stem cells of the medaka fish, *Radiat. Res.*, 163, 112–114 (2005).

Shimura, T., M. Inoue, M. Taga, K. Shiraishi, N. Uematsu, N. Takei, Z-M. Yuan, T. Shinohara, and O. Niwa, p53-dependent S-phase damage checkpoint and pronuclear cross talk in mouse zygotes with X-irradiated sperm, *Mol. Cell. Biol.*, 22, 2220–228 (2002).

Shiraishi, K., T. Shimura, M. Taga, N. Uematsu, Y. Gondo, M. Ohtaki, R. Kominami, and O. Niwa, Persistent induction of somatic reversions of the pink-eyed unstable mutation in F1 mice born to fathers irradiated at the spermatozoa stage, *Radiat. Res.*, 157, 661–667 (2002).

Sigurdson, A. J. and I. M. Jones, Second cancers after radiotherapy: any evidence for radiation-induced genomic instability?, *Oncogene*, 22, 7018–7027 (2003).

Slovinska, L., A. Elbertova, and E. Misurova, Transmission of genome damage from irradiated male rats to their progeny, *Mutat. Res.*, 559, 29–37 (2004).

Turusov, V. S., T. V. Nikonova, and Yu. D. Parfenov, Increased multiplicity of lung adenomas in five generations of mice treated with benz(a)pyrene when pregnant, *Cancer Lett.*, 55, 227–231 (1990).

UNSCEAR, *Hereditary Effects of Radiation* (United Nations, New York, 2001).

UNSCEAR, *Sources and Effects of Ionizing Radiation. Annex I. Epidemiological evaluation of radiation-induced cancer*. Vol. 2. (United Nations, New York, 2000).

Vance, M. M., J. E. Baulch, O. G. Raabe, L. M. Wiley and J. W. Overstreet, Cellular reprogramming in the F3 mouse with paternal F0 radiation history, *Int. J. Radiat. Biol.*, 78, 513–526 (2002).

Vilarino-Guell, C., A. G. Smith, and Y. E. Dubrova, Germline mutation induction at mouse repeat DNA loci by chemical mutagens, *Mutat. Res.*, 526, 63–73 (2003).

Vorobtsova, I. E., Irradiation of male rats increases the chromosomal sensitivity of progeny to genotoxic agents, *Mutagenesis*, 15, 33–38 (2000).

Vorobtsova, I. E., L. M. Aliyakparova, and V. N. Anisimov, Promotion of skin tumors by 12-O-tetradecanoylphorbol-13-acetate in two generations of descendants of male mice exposed to X-ray irradiation, *Mutat. Res.*, 287, 207–216 (1993).

Watson, G. E., S. A. Lorimore, S. M. Clutton, M. A. Kadhim, and E. G. Wright, Genetic factors influencing alpha-particle-induced chromosomal instability, *Int. J. Radiat. Biol.*, 71, 497–503 (1997).

Wiley, L. M., J. E. Baulch, O. G. Raabe, and T. Straume, Impaired cell proliferation in mice that persists across at least two generations after paternal irradiation, *Radiat. Res.*, 148, 145–151 (1997).

Yauk, C. L., Y. E. Dubrova, G. R. Grant, and A. J. Jeffreys, A novel single molecule analysis of spontaneous and radiation-induced mutation at a mouse tandem repeat locus, *Mutat. Res.*, 500, 147–156 (2002).

Yu, Y., R. Okayasu, M. M. Weil, A. Silver, M. McCarthy, R. Zabriskie, S. Long, R. Cox, and R. L. Ullrich, Elevated breast cancer risk in irradiated BALB/c mice associates with unique functional polymorphism of the Prkdc (DNA-dependent protein kinase catalytic subunit) gene, *Cancer Res.*, 61, 1820–824 (2001).

Goldberg, Z. Clinical implications of radiation-induced genomic instability, *Oncogene*, 22, 7011–7017 (2003).

Zhang, S., E. S. Ramsay, and B. A. Mock, Cdkn2a, the cyclin-dependent kinase inhibitor encoding p16INK4a and p19ARF, is a candidate for the plasmacytoma susceptibility locus, Pctr1, *Proc. Natl. Acad. Sci. USA*, 95, 2429–434 (1998).

CHAPTER 11

EVOLUTION PROCESSES IN POPULATIONS OF PLANTAIN, GROWING AROUND THE RADIATION SOURCES: CHANGES IN PLANT GENOTYPES RESULTING FROM BYSTANDER EFFECTS AND CHROMOSOMAL INSTABILITY

V.L. KOROGODINA* AND B.V. FLORKO

*Joint Institute for Nuclear Researches (JINR), 141980
Dubna, Moscow region, Russia.
e-mail: korogod@jinr.ru*

Abstract: The viability of seeds growing around the nuclear power plant (NPP) can decrease up to 20–30%. We consider the appearance of both multiple secondary cells and chromosomes with abnormalities. We used the ideas of adaptation to explain these phenomena. **The aim** was the statistical analysis of the appearances of cells and chromosomes with abnormalities in dependence on radiation factor around the NPP and seeds' antioxidant status (AOS). **Methods**. The chromosome bridges and acentric fragments were registered as chromosomal abnormalities in root meristems of plantain seeds collected in tested populations. For sites within a 30 km radius of the NPP, the annual γ-radiation dose rates, ^{137}Cs soil concentrations and NPP fallouts were standard and did not exceed norms. Seeds were collected in years with normal and extreme high summertime temperatures. **Results**. The modelling showed that the appearances of a number n cells (or chromosomes) with abnormalities would be described by the formulas G_n, P_n, $G_n + P_n$, where P – Poisson and G – geometrical regularities. The parameters of distributions are the AOS- and NPP fallouts dose rate dependent, especially communicative G-component. Due to communication mechanisms, some of cells and seedlings accumulated abnormalities and many others died near the NPP. **Conclusions**. (i) statistical biomarker of stress effects is intensification of the communicative processes; (ii) dose-dependent microevolution process is observed (an appearance of new genotypes and their selection) in indigenous populations under the combined effect of NPP atmosphere fallouts and higher summer temperatures; (iii) the communicative processes in meristem are AOS-dependent; (iv) strategies of survival of populations are different for middle-lethal doses and those which do not exceed norms of their ecological niche (stresses).

---

*To whom correspondence should be addressed.

C. Mothersill et al. (eds.), Multiple Stressors: A Challenge for the Future, 155–170.

Keywords: low dose irradiation; seeds; cells with abnormalities; chromosomes with abnormalities; survival; adaptation, statistical analyses

Introduction

In the last decade, many investigations were devoted to radiation stress effects reviewed in (Mothersill and Seymour, 2005; Little, 2006). As a basic feature, researchers mark out the genetic instability including spatial intracellular (Mothersill and Seymour, 2005) and intercellular (Arutyunyan et al., 2001) spreading of the DNA changes. These phenomena could be described as the non-randomly processes of the appearance of multiple secondary cells with abnormalities (Korogodina et al., 2005) and chromosomal abnormalities (Chebotarev, 2000). At present, some biologists suggest an idea of an adaptation to explain the genetic instabilities (Mothersill and Seymour, 2005).

Earlier, the effects of radiation stress were studied on the plantain seeds collected in populations growing around the radiation sources in years with normal and extreme high summertime temperatures (Korogodina et al., 2000; 2004). The biological values prevented from the simple conclusion of the radiation effects on seeds and the mechanisms of damaging (Korogodina et al., 2004). We used an idea of adaptation to analyze structure of distributions of both seeds on the numbers of cells with chromosomal abnormalities (CCAs) in root meristem of seeds (Korogodina et al., 2005; 2006) and meristem cells on the numbers of the chromosomal abnormalities (CAs).

Our aims were to analyse statistically the mechanisms of the CCAs and CAs appearances in seedlings root meristems and to study changes in plant genotypes resulting from the intra- and intercellular communication processes in dependence on radiation factor around the nuclear power plant (NPP) and seeds' antioxidant status (AOS).

Materials and Methods

SEED

The plantain seeds (*Plantago major*) were used in the natural experiment (1998, 1999). For this kind of seeds, the reported quasi-threshold radiation dose, which corresponds to the inflection of the survival curve from a shoulder to mid-lethal doses, is 10–20 Gy (Preobrazhenskaya, 1971). The plantain populations were located in sites within 80 km of the Balakovo NPP and in Chernobyl trace area (Saratov region), and near the accelerator facilities in Moscow region. In 1999 the temperatures during daylight hours reached 30–32 °C in the Moscow region and 38–40 °C in the Saratov region (they are extreme for both provinces), and the seeds experienced elevated temperatures during the maturation period in nature (SCEPSR, 2000). The plantain populations were chosen in similar biotopes. Seeds were collected at the end of August in 1998 and 1999 from 20–30 plants. The seeds were refrigerated until the following April at $T = 3$–$4\,°C$ and relative humidity $= 13$–14%.

Seeds of all populations were germinated on wet filter paper in petri dishes at $23\,°C$ until seedling roots reached $3.5 \pm 2\,mm$, a length corresponding to the first mitoses before which seedling growth is due only to swelling without cell division. After this point, a number of proliferated cells fluctuates at a constant level. Seedlings were fixed in ethyl alcohol and acetic acid (3:1) and stained with acetoorcein. Seedlings $<1.5\,mm$ after 13 days were scored as non-surviving $(1$–$S)$ because too few cells reached the first mitosis (the mitotic index in the shoot zone is already known to increase in parallel with the size of the seedling (Gudkov, 1985). The first fixation was started when approximately 1/3 part of whole seeds' population was germinated. After 13 days we ended the fixation. Prolonging the germination period (up to 6 weeks) increases S by only 2–3%; some rootlets occurred too small and brown. The methods of seed sprouting and fixing have been described (Korogodina et al., 1998; 2004). Antioxidant activities were studied with a photochemiluminescence method (Korogodina et al., 2000). An amount of seed infusion, which inhibited chemiluminescence by 50% $(C_{1/2})$, was adopted as a measure of AOS.

Ana-telophases were scored for CCAs containing CAs. The chromosome bridges and acentric fragments were registered as CAs in ana-telophases. The mitotic activity (MA) of cells in meristem was scored as the number of cells in ana-telophases.

CHARACTERISTICS OF SITES AND WEATHER CONDITIONS

To investigate the possibility of radiation stress effects the sites near the NPP in Saratov region and accelerator facilities in Moscow region were selected. In Saratov region, a site in Chernobyl trace territory was also studied.

In 30-km zone of the NPP two sources of radioactivity are placed, which could be an influence on plantain populations: the atomic station (P2–P6 sites) (MAPRF, 1998) and the phosphogypsum dump (P8–P10 sites) (SCEPSR, 2000) (Fig. 1).The sites were chosen in the view of the wind rose. For the most part, the NPP atmospheric fallouts influence on populations P2–P6 resulting from the direction of the winds in summer. Perhaps the populations P7–P10 experienced the effects of phosphogypsum dump, which can influence a soil contamination in this area (Korogodina et al., 2000). Therefore, these populations (P7–P10) did not studied in 1999. The site P1 was at the left bank of the Volga (~80 km from NPP). The population P11 was chosen at the right bank of the Volga (100 km from the NPP) on a Chernobyl radioactivity-deposition track with well-characterized ^{137}Cs soil

Fig. 1. Locations of the selected plantain populations in the vicinity of the Balakovo NPP: 2–10 (P2–P10). Populations P1, P11 are located 80 km and 100 km from NPP; P12 one is placed in the Moscow region.

contaminations (average concentration ~30 Bq/kg, EC, 1998). The Moscow region site P12 was selected in Moscow region to know in detail both radiation exposures from the accelerator facilities and soil pollution.

The annual rainfall near the Volga is 1.5 times higher than in steppe, therefore the microclimate of P1–P6 sites is damper than that of P7–P10 (SCEPSR, 2000). In 1999, the summertime high temperatures in the European part of Russia averaged 2–3 °C above normal (SCEPSR, 2000).

DETERMINATION OF RADIOACTIVITY

For sites within a 100 km radius of the NPP, the annual γ-radiation dose rates (DR) and ^{137}Cs soil concentrations (C_{Cs}) varied little from the ranges ~0.10–0.15 μSv/h and ~5–10 Bq/kg reported in independent radiological surveys (MAPRF, 1998; SCEPSR, 2000). In site P11 DR is ~0.10–0.15 μSv/h (SCEPSR, 2000) and C_{Cs} is 30 Bq/kg (EC, 1998). DR is ~0.10–0.12 μSv/h and C_{Cs} is ~5–10 Bq/kg in site P12 (Moscow region) (Zykova et al., 1995; Alenitskaja et al., 2004). These values (excluding the concentrations in site P11) do not exceed the average radiation values over the Saratov and Moscow regions (Zykova et al., 1995; SCEPSR, 2000; Alenitskaja et al., 2004).

We examined the upper 10–12 cm soil in tested sites. Measurements were carried out using the low-background γ-spectrometers with a NaI(Tl) crystal as well as a Ge one, which were described in (Alenitskaja et al., 2004). The

TABLE 1. Soil contamination of ^{137}Cs and ^{40}K (Bq/kg)

	P1	P2	P3	P4	P5	P6	P7	P8	P9	P10	P11	P12
^{137}Cs	9	33	–	5	–	–	10	8	8	6	39	9
	5	4	3	5	5	5	–	5	–	–	15	10
^{40}K	36	33	–	55	–	–	47	45	46	70	45	58
	0	0	40	0	23	50	0	0	0	0	0	0
	40	34	0	41	0	0	–	41	–	–	32	60
	0	0		0			0				0	0

errors of detection efficiency of γ-quanta did not exceed 7%. Total errors of radioactivity determination for different isotopes were 20–40%. The artificial isotope ^{137}Cs soil contamination did not differ significantly in 1998 and 1999 (Table 1), although usually the fluctuations can be observed in the same site (Alenitskaja et al., 2004). The data on C_{Cs} agree with published values (Zykova et al., 1995; EC, 1998; MAPRF, 1998; SCEPSR, 2000; Alenitskaja et al., 2004) and do not correlate with NPP fallout. The accumulated doses were calculated using the Brian-Amiro model (Amiro, 1992) and were estimated to be ~1–3 cGy for seeds P1–P6. We accounted the secondary wind rising (SCEPSR, 2000) in P7–P10 populations by means of (Gusev and Beljaev, 1986). The results did not differ significantly from 1998 to 1999 (for accumulated radiation doses and soil concentrations) in tested sites.

CALCULATION OF THE NPP FALLOUT IRRADIATION OF SEEDS

Plantain seeds experienced the NPP fallout irradiation in nature (annual fallouts on isotopes: Kr ~2.5 TBq; Xe ~2.5 TBq, and I ~4.4 TBq (MAPRF, 1998), the dose rates are controlled by NPP administration). Distribution of the particulate emissions and gases were estimated according to the Smith-Hosker model (Hosker, 1974) based on NPP characteristics (MAPRF, 1998) and winds in summer near the ground in the NPP region (SCEPSR, 2000). The isotopes fallouts result in γ-irradiation mainly (mean energy ~1.1 MeV/ γ-quanta (Ivanov et al., 1986)). The relative fallout dose rates (RFRD) values were calculated in the ratio to the dose in site P1. The RFRD value in P7 site is higher than in P2, P3, P8–P10 sites due to short half life of I isotopes, which do not reach the populations P2, P3, P8–P10. This fallout irradiation is not chronic and depends on location of populations. Intensity of irradiation is shown in Table 2. We used the γ-quanta' LEP-dependence on their energy (Ivanov et al., 1986) and the NPP characteristics (MAPRF, 1998) to calculate a mean γ-quanta energy deposition per plant cell nucleus, which is 1.4 keV. In our calculations, each seedling' meristem experiences an

TABLE 2. Relative daily Balakovo NPP fallout (Kr, Xe, I) dose rates (RFRD), experienced by populations, calculated in the ratio to that in the site P1

Populations	Relative dose rate RFRD	Intensity of γ-quanta per cell nucleus per $\times 10^{-7}$	Intensity of γ-quanta per cell nucleus per min for 3 months
P1	1	1.9×10^{-4}	2.5×10^{-6}
P2	80	1.7×10^{-2}	2.2×10^{-4}
P3	80	1.7×10^{-2}	2.2×10^{-4}
P4	560	0.11	1.4×10^{-3}
P5	5700	65	0.85
P6	1350	0.26	3.4×10^{-3}
P7	340	6.7×10^{-2}	8.6×10^{-4}

influence of even one g-quantum per 3 months. The expected irradiation dose was calculated after the accelerator facilities operation in the P12 site. In 1998 the calculated neutron dose level was 1 mSv (for two months), and the neutron dose rate level was 0.8 µSv/h (the neutron background dose rate ~ 9.3 nSv/h (Wiegel et al., 2002). In 1999, the neutron irradiation did not increase. The averaged neutron energy was 5.5 MeV (MCNP calculations).

STATISTICAL ANALYSIS

The data were processed using standard statistical methods (Van der Vaerden, 1957) and statistical criteria (Akaike, 1974; Schwarz, 1978; Glotov et al., 1982; Geras'kin and Sarapul'tzev, 1993). The maximum-likelihood method was used for approximations. For estimated regression the following criteria were used: R^2_{adj}- determination coefficient adjusted on the number of range of discretion; criterion AIC (Akaike criterion); criterion BIC (choice of the most probable model from an ensemble of ones on the assumption of their prior equal probability). The details of the approximation procedure are described in (Florko and Korogodina, 2007).

Hypotheses

PROCESSES IN MERISTEMS AND IN CELLS

At germination of seeds, the adaptations of rootlet meristems as well as meristem cells are required by the environment, which can result in process of some changes (reconstructions) in meristems and cells. Irradiation of meristems can be over by stage of first mitoses that can be considered as an "adapted stage of seedling" (Fig. 2A), or their death. Some of the irradiated cells come into mitosis, which can be considered also as "adapted stage of cells" (Fig. 2B), and the others die. The reconstructions, which are necessary for adaptation, can be provided by communication processes ($I^{com}/I^{p2} > R_{ad} > I^{rad}/I^{p1}$) or primary

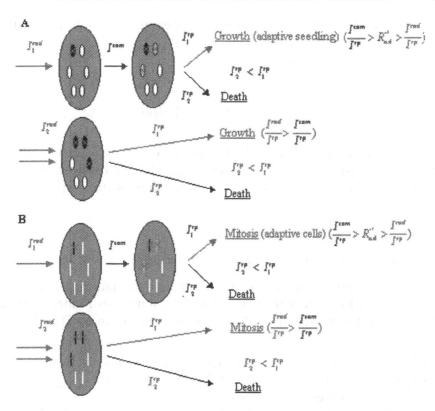

Fig. 2. The processes in meristems (A) and cells (B). I^{rad} – intensity of irradiation; I^{com} – intensity of communications; I^{rp} – intensity of repair. Primary damaged cells (A) or chromosomes (B) are crossed black, communicative ones – crossed gray, and undamaged ones – white colored.

damages ($I^{\mathrm{rad}}/I^{\mathrm{rp1}} > I^{\mathrm{com}}/I^{\mathrm{rp2}}$), where I^{rad} – intensity of irradiation; I^{com} – intensity of communications; I^{rp1}, I^{rp2} – intensities of repair. The abnormalities lead to death of seedling or cell if repair systems are not effective.

THE ADAPTATION HYPOTHESES IN MATHEMATICAL TERMS

- Primary damages of cells or chromosomes are rare and independent events and can be described by Poisson distribution;
- The communication processes could induce the appearance of the secondary abnormalities. This increases the Poisson parameter a;
- These reconstructions in meristems and cells provide selection of "adapted" seedlings and cells. A waiting period of adaptation is exponentially distributed that results in geometrical distribution on the reconstruction numbers.

Therefore, we conclude that an appearance of number n abnormalities could include Poisson and geometrical components (Florko and Korogodina, 2007).

A probability of a number n CCAs (CAs) appearance in adapted meristem (cell) could be described by the formulas

$$T_n = P_n + G_n$$

$$P_n = \frac{a^n}{n!}e^{-a}; G_n = G(1-\theta)\theta^n,$$

where P_n and G_n are probabilities of a number n CCAs (CAs) appearance in P, G-distributions; parameter a is the sample mean of the P-distributions; q is a part of seedlings (cells) without CCAs (CAs); G – value of this subpopulation.

Results

BIOLOGICAL VALUES OF SEEDS AND SEEDLINGS

Table 3 presents non-survival of seeds and values that characterize processes in seedlings meristems and their cells. For the populations located near the NPP (P2–P6), the non-survival of seeds grew high (up to 80%) in 1999 in

TABLE 3. Antioxidant status and non-survival of seeds, frequencies of both cells with abnormalities in meristem and chromosomes with abnormalities in cells

Seeds	Number of seeds	Number of ana-telofases	AOS $(C_{1/2})$	Non-survival 1-S,%	CAs frequency	CAs frequency	Mitotic activity
				1998			
P1	167	726	0.22	10.8	2.5 ± 0.5	0,05	9.7 ± 0.9
P2	152	942	0.16	34.2	3.1 ± 0.8	0,03	6.0 ± 0.6
P4	156	518	0.20	19.9	2.5 ± 0.6	0,04	6.3 ± 0.7
P7	149	763	0.50	13.4	1.3 ± 0.3	0,02	10.9 ± 1.0
P8	148	1047	0.80	31.8	3.2 ± 0.8	0,03	7.8 ± 0.8
P9	167	528	0.76	65.3	4.6 ± 1.4	0,07	6.1 ± 1.0
P10	153	231	0.66	29.4	3.2 ± 0.7	0,05	9.7 ± 3.0
P11	153	342	0.83	55.6	4.4 ± 0.9	0,08	7.3 ± 1.0
P12	148	1805	0.28	12.3	1.2 ± 0.3	0,01	14.9 ± 0.9
				1999			
P2	500	2228	0.16	72.6***	3.2 ± 0.4**	0,04	17.8 ± 1.2***
P3	500	3827	0.33	32.6	5.4 ± 0.6	0,06	21.9 ± 1.3
P4	500	1035	0.22	83.6***	6.8 ± 0.9***	0,09	17.5 ± 1.4***
P5	500	2209	0.25	67.8	6.3 ± 0.6	0,07	15.0 ± 0.9
P6	500	2385	0.25	71.6	5.1 ± 0.4	0,07	17.9 ± 1.1
P11	500	2220	0.50	43.6*	5.5 ± 0.5*	0,07	9.8 ± 0.5**
P12	200	832	0.31	37.5***	5.6 ± 0.8***	0,01	8.0 ± 0.6***

Note: Standard error of the CAs frequency does not exceed 0.01
Comparing 1998 and 1999 data: *$p>0.1$; **$p>0.5$; ***$p<0.001$.

comparison with 1998 and with those in control populations. The mitotic activity is higher than the same value over all studied populations in 1998 ($p < 0.05$, ~ threefold), and higher than in P11, P12 plants (F, $p < 0.05$, ~ twofold) in 1999 (Korogodina et al., 2006).

The correlations of both frequencies of cells with abnormalities and chromosomes with abnormalities with non-survival of seeds were examined. The correlation between these values could mean the predominance of one mechanism influencing the non-survival of seeds. If such correlation is absent, we can suspect that some mechanisms have approximately equal rights. In 1998, a strong correlation of the (1–S) value with both frequencies of cells with chromosomal abnormalities ($|r_{1-S, CCA}| = 0.92$, df = 7, $p < 0.001$) (Korogodina et al., 2006) and the CAs ($|r_{1-S, CA}| = 0.76$, df = 7, $p < 0.02$) was observed. In 1999, the correlation between the (1–S) value and CCA frequency disappeared ($|r| = 0.02$), as well as with CA frequency was non-confidence statistically ($|r_{1-S, CA}| = 0.51$, df = 5). These data indicated that some mechanisms acted in seeds collected near the NPP in 1999. We can think that observed effects were induced by radiation and heat factors because the radiation factors were the same both years and the temperatures were extreme high in all sites in 1999.

The hypothesis of the combined effects of radiation and heat stresses is verified by our laboratory experiments on pea seeds (the methods are the same as in natural experiments) (Korogodina et al., 2005). Figure 3 presents a non-viability of seeds, a frequency of CCAs in root meristem of seedlings and a frequency of CAs in meristem cells at γ-irradiation (^{60}Co) with 7 cGy and dose rates 0.3; 1.2; 19.1 cGy/h. The first group of seeds stored 8 months at 4°C and then was tested. The second one stored 8 months in the same conditions and was stressed during the outdoor storage of seeds (2 months) at extreme high summertime temperatures (32–34°C in Moscow region) (Korogodina et al., 2005). The non-viability of irradiated seeds of the second group increased up to 28–62%, whereas frequencies of CCAs at dose rates 1.2; 19.1 cGy/h as well as CAs at dose rate 19.1 cGy/h differed

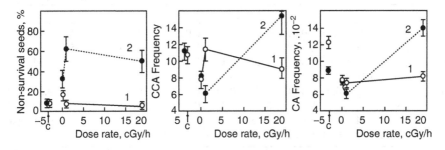

Fig. 3. The non-viability (A) of pea seeds, frequency of CCAs (B) and frequency of CAs (C) in seedlings meristem in dependences on dose rate lab irradiation. Explanation in text.

significantly in comparison with the first group (F-criterion) (Fig. 3). That is not surprising because high temperatures can induce DNA damages and apoptosis (Rainwater et al., 1996).

THE STATISTICAL MODELING

The statistical modeling was used to investigate mechanisms in plant seedlings and their dependence on irradiation. Table 4 presents the parameters of distributions of seeds on the number of cells with abnormalities and cells on the number of chromosomes with abnormalities in 1998 and 1999.

All distributions of seeds include Poisson and geometrical components, and the Poisson one predominated in 1998. In 1999, the P-value decreased and averaged sample means increased. It is expected because the heat stress (1999) increases an appearance of the reactive oxidative species (ROS)

TABLE 4. Parameters of distributions of plantain seeds on cells with abnormalities and distributions of meristem cells on chromosomes with abnormalities

Seeds	Distribution of seeds on cells with abnormalities				Distribution of cells on chromosomes with abnormalities					
	mG	G	mP*	P*	mG1*	G1*	mG2*	G2*	mP*	P*
					1998					
P1	1.10 ± 0.05	0.17 ± 0.03	0.22	0.41	0.01	0.42	0.30	0.07		
P2	0.05 ± 0.05	0.10 ± 0.03	0.17	0.45	0.03	0.26				
P4	0.07 ± 0.04	0.06 ± 0.02	0.17	0.64	0.04	0.38				
P7	0.14 ± 0.06	0.27 ± 0.04	0.19	0.38	0.02	0.52				
P8	0.10 ± 0.06	0.07 ± 0.02	0.26	0.39	0.03	0.34				
P9	2.40 ± 0.09	0.01 ± 0.01	0.36	0.19	0.07	0.39	0.28	0.04		
P10	2.30 ± 0.06	0.04 ± 0.02	0.32	0.46	0.03	0.35				
P11	0.70 ± 0.10	0.09 ± 0.02	0.45	0.21	non-confidence statistically data					
P12	0.06 ± 0.04	0.12 ± 0.03	0.21	0.69					0.01	0.9
av.	**0.77 ± 0.32**	**0.10 ± 0.03**	**0.26**	**0.42**						
					1999					
P2	1.05 ± 0.17	0.13 ± 0.01	0.36	0.11	0.02	0.43	0.14	0.12		
P3	1.85 ± 0.13	0.30 ± 0.02	1.20	0.37	0.02	0.40	0.11	0.55		
P4	2.01 ± 0.46	0.02 ± 0.01	1.26	0.12	0.03	0.25			0.53	0.03
P5	0.08 ± 0.31	0.04 ± 0.01	1.20	0.26	0.06	0.50	0.23	0.05		
P6	2.45 ± 0.42	0.04 ± 0.01	0.96	0.21	0.02	0.36			0.22	0.12
P11	0.32 ± 0.13	0.10 ± 0.01	0.84	0.33	0.02	0.36			0.20	0.17
P12	1.81 ± 0.34	0.12 ± 0.02	0.36	0.39	0.02	0.19			0.35	0.03
av.	**1.36 ± 0.34**	**0.11 ± 0.02**	**0.89**	**0.26**						

*Standard errors of the parameters do not exceed 20–30% (the sample means) and 10–15% (the relative values)

(Rainwater et al., 1996), which induce the DNA damages (Janssen et al., 1993) and apoptosis (Davies, 2000). The ROS increase the communicative processes (Burlakova et al., 2001) that result in the decreasing of the P-subpopulations also.

In 1998, some populations of cells are geometrical-distributed on the number of CAs. Two distributions of cells include the first and the second geometrical components (their plantain populations located in high chemical-polluted sites (Korogodina et al., 2000)). One cell population is P-distributed (these plants were growing near the accelerator). In 1999, all distributions are compound of two-geometrical or geometrical plus Poisson components. The Poisson component is observed in populations growing at the border of sanitary zone, in Chernobyl trace territory and near the accelerator.

It is interesting that cells' distributions on the number of CAs can be geometrical whereas seeds' distributions on the number of CCAs could have Poisson component. We explain this fact by more intensive irradiation of meristem as a whole than each separate cell. It suggests that irradiation effects are more pronounced in meristems than in cells.

THE AOS- AND DOSE-DEPENDENCE OF MODELS' PARAMETERS

The relative NPP fallouts dose is the same for populations P2 (AOS_{50} ~0.16) and P3 (AOS_{50} ~0.33), therefore it is possible to compare their characteristics in dependence on AOS of seeds. Figure 4 shows the AOS- and dose-dependence of distributions of seeds on the number of cells with abnormalities. The sample means and values of both P and G-distributions of seeds increase with seeds AOS ($p < 0.05$) (Fig. 4A, B). G-graphs split on two ones according to the sites distance from the NPP (AOS_{50} of seeds growing at the border and inside of sanitary zone are ~0.22–0.25). By means of interpolation, the distributions parameters for the "conventional" population in Balakovo (with AOS_{50} ~0.23) were determined. Near the NPP, the non-linear dose-dependence of the P- and G-parameters is observed (Figs. 4C, D). It is stronger for G-parameters: the sample mean is increased at the border of sanitary zone (not significantly) and decreased inside it ($p < 0.05$); the G-value is decreased dramatically for those populations ($p < 0.001$) in comparison with P2, P3 ones. P-parameters are dependent on NPP fallouts in this dose interval to a lesser degree, not significantly.

Figure 5 shows the parameters of distribution of cells on the number of chromosomes with abnormalities in dependence on seeds' AOS and NPP fallouts relative dose. The G1-sample mean is not dependent significantly on seeds AOS and dose irradiation (Fig. 5A, C). It can be suggested that G1-cells are more stable, in contrast with G2- and P-ones, which have fast changed genotype. Cells in meristem can be divided into two groups: the first - (G1) subpopulation and the second – (G2 + P) one.

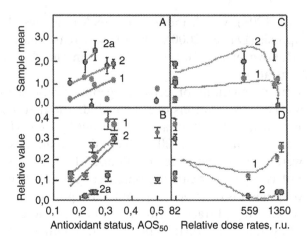

Fig. 4. The dependence of parameters of distributions of seeds on CCA numbers on antioxidant status (A, B) and the relative NPP fallout dose (C, D) in 1999. The parameters of subpopulations: P – red (1), G – green (2). The regressions, A: for P2, P3 $y = -0.48 + 5.04x$ ($p < 0.05$) (1), $y = 0.25 + 4.80x$ ($p < 0.05$) (2); for P4, P6 $y = -0.92 + 12.96x$ ($p < 0.05$) (2a); B: for P2, P3 $y = -0.15 + 1.43x$ ($p < 0.05$) (1), $y = -0.04 + 1.02x$ ($p < 0.05$) (2); for P4, P6 $y = -0.10 + 0.53x$ ($p < 0.05$) (2a); C: for P2–P6 polynomial fits (1, 2); D: for P2–P6 polynomial fit (1); $y = 100/x^{1.4} + 0.027$ (2).

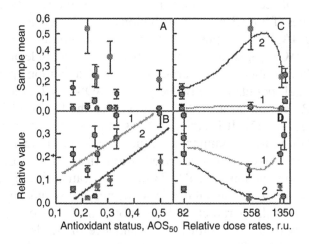

Fig. 5. The dependence of parameters of distributions of meristem cells on CCA numbers on antioxidant status (A, B) and the relative NPP fallout dose (C, D) in 1999. The parameters of subpopulations: P –red, G1 –light green; G2 –dark green. Subpopulation G1 – curve 1, G2 + P ones – 2. The regressions, B: for all populations $y = 0.73x + 0.02$ ($p < 0.05$) (1), $y = 0.77x - 0.12$ ($p < 0.05$) (2); C: for P2–P6 polynomial fits (1, 2); D: for P2–P6 polynomial fit (1); $y = 100/x^{1.4} + 0.027$ (2).

The distribution parameters of these two subpopulations of cells show the similar regularities as parameters of seeds distributions. Their values increase with AOS ($p < 0.05$) (Fig. 5B). It is observed a partition of "$G2_{CA} + P_{CA}$" graphs into two ones. Figure 5C, D demonstrates a non-linear dependence of distribution parameters on NPP fallouts dose. The values of subpopulation ($G2_{CA} + P_{CA}$) are decreased in populations growing at the border and inside of sanitary zone ($p < 0.001$) (Fig. 5D). The dose-decrease of the subpopulations of cells with fast changed genotype is correlated with that of the subpopulation of seeds with cell-to-cell communication ($|r| = 0.97$; df $= 3$; $p < 0.01$).

Discussion

TWO EVOLUTION STRATEGIES OF SURVIVAL

In Poisson subpopulation, repair systems decrease the primary damages and mutations. So, survival is negative connected with a number of abnormalities and their frequency. In geometrical subpopulation, the communicative processes increase a number of abnormalities, which are accumulated. If values of geometrical and Poisson subpopulations are congruent quantities, survival become positive connected with a number of abnormalities and their frequency. Figure 6 shows correlations between survival of seeds and frequency of cells with abnormalities in root meristem. In 1998, a negative correlation is observed (Fig. 6A). In 1999, a correlation is positive for populations P2, P3, P11, P12 and negative for P4, P5, P6 ones (Fig. 6B). Table 3 shows that P-distribution is predominated in all populations in 1998 and at the border and inside of sanitary zone (P4, P5, P6) in 1999.

We can conclude that two strategies of survival are observed: at middle-lethal irradiation (Poisson distribution) and stressed low-intensity one (geometrical distribution). In the first case cells rid oneself of DNA damages

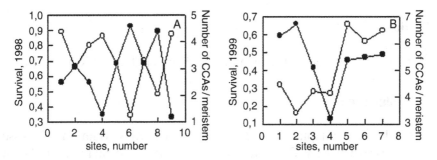

Fig. 6. Negative and positive connections of the numbers of CCAs in meristem with seeds survival. Survival marked in white and numbers of CCAs/meristem in black colors.

by protection systems. In the second one, stresses activate the repair systems, which maintain DNA changes. These repair systems are homologous to mismatch repair one of *E. coli* that influences multiple aspects of genetic stability, including genome rearrangement (reviewed in (Longerich et al., 1995)). In both cases the numbers of abnormalities increase whereas old genotype eliminate, especially at looking for new adaptive genotypes. In 1939 N.W. Timofeeff-Ressovsky pointed out for the first time (Timofeeff-Ressovsky, 1939) that microevolution process includes necessary both the increasing of the material for evolution and decreasing of a part of old population. Then Eigen and Schuster (1979) have shown that elimination of organisms with old genotypes is necessary for survival of new ones in the view of living resources of adapted population.

Conclusions

- Irradiation factor ~1.4 keV/nucleus influences the communicative processes;
- The irradiation effects are more pronounced in meristems than in cells;
- The communicative processes in meristem are AOS-dependent;
- Dose-dependent microevolution process is observed (an appearance of new genotypes and their selection) in indigenous populations under the combined effect of NPP atmosphere fallouts and higher summer temperatures;
- Statistical biomarker of stress effects is intensification of the communicative processes;
- Strategies of survival of populations are different for middle- lethal doses and those, which are not exceed norms of their ecological niche (stresses). The first maintains old genotypes, and the second look for new adapted ones.

Dedication

This article is dedicated to the memory of Prof. Vladimir I. Korogodin.

Acknowledgements

The hypothesis of adaptation was evolved by V.I. Korogodin and discussed with V.B. Priezzhev. G.A. Ososkov suggested some statistical methods. We are grateful to these researchers, who contributed in these investigations. We would like to thank for discussions Yu.A. Kutlakhmedov and E.B. Burlakova. We are grateful to I.V. Kronstadtova for technical assistance.

References

Akaike, H., 1974, A new look at the statistical model identification. *IEEE Trans. Automatic Control*, AC-19, 716–723.

Alenitskaja, S.I., Bulah, O.E., Buchnev, V.N., Zueva, M.V., Kargin, A.N., Florko, B.V., 2004, Results of long-term supervision over the environment radioactivity in area of JINR arrangement. *PAPAN Lett.* 1, 5(122), 88–96.

Amiro, B.D., 1992, Radiological dose conversion factors for generic non-human biota. Used for screening potential ecological impacts. *J. Environ. Radioact.* 35(1), 37–51.

Arutyunyan, R., Neubauer, S., Martus, P., Durk, T., Stumm, M., and Gebhart, E., 2001, Intercellular distributions of aberrations detected by means of chromosomal painting in cells of patients with cancer prone chromosome instability syndromes. *Exp. Oncol.* 23, 23–28.

Burlakova, E.B., Mikhailov, V.F., and Mazurik V.K., 2001, The redox homeostasis system in radiation-induced genomic instability. *Radiat. Biol. Radioecol.* 4(5), 489–499 (in Russian).

Chebotarev, A.N., 2000, A mathematical model of origin multi-aberrant cell during spontaneous mutagenesis, *Dokl. Biol. Sci.* 371, 207–209.

Davies, K.J., 2000, Oxidative stress, antioxidant defenses, and damage removal, repair, and replacement systems. *IUBMB Life*, 50(4–5), 279–289.

EC, 1998, Atlas of cesium contamination in Europe after the Chernobyl accident European Commission. *DG XII. Safety Programme for Nuclear Fission (Protection against Radiation)*.

Eigen, M. and Schuster, P., 1979, The Hypercycle, Berlin-Heidelberg-NY: Springer-Verlag, p. 270.

Florko, B.V., and Korogodina, V.I., 2007, Analysis of the Distribution Structure as Exemplified by One Cytogenetic Problem. *Physics of Particles and Nuclei Letters*, 4(4), 331–338.

Geras'kin, S.A. and Sarapul'tzev, B.I., 1993, Automatic classification of biological entities by the level of radiation stability. *Avtomatika & telemechanika*, 2, 183–189 (in Russian).

Glotov, N.V., Zhivotovskjy, L.A., Khovanov, N.V. and Khromov-Borisov, N.N., 1982, Biometrics, Leningrad: Leningrad State University.

Gudkov, I., 1985, Cell mechanisms of postradiation repair in plants, *Kiev: Naukova Dumka*, 224 (in Russian).

Gusev, N.G. and Beljaev, V.A., 1986, Radionuclide releases into Biosphery. *M: EnergoAtomIzdat*, (in Russian).

Hosker, R.P. Jr., 1974, Estimates of dry deposition and plume depletion over forests and grasslands. *Proceedings of Physical behavior of radioactive contaminants in the atmosphere. IAEA*, 291–309.

Ivanov, V.I., Lyszov, V.N., and Gubin, A.T., 1986, Handbook on microdosimetry. *M: AtomIzdat*, (in Russian).

Janssen, Y.M., Van Houten, B., Borm, P.J., Mossman, B.T., 1993, Cell and tissue responses to oxidative damage. *Lab. Invest.* 69(3), 261–274.

Korogodina, V., Bamblevskij, V., Grishina, I., Gustova, M., Zabaluev, S., Korogodin, V., Kuraeva, T., Lozovskaja, E., and Maslov, O., 2000, Antioxidant status of plant (*Plantago major* L.) seeds of the pollution in the Balakovo NPP and chemical enterprisers region. *Radiats. Biologija. Radioecologija*, 40, 334–338 (in Russian).

Korogodina, V., Bamblevsky, V., Grishina, I., Gustova, M., Florko, B., Javadova, V., Korogodin, V., Lozovskaya, E., Malikov, A., Maslov, O., Melnikova, L., Shlyakhtin, G., Stepanchuk, V., and Zueva, M., 2004, Evaluation of the consequences of stress factors on plant seeds growing in a 30-km zone of Balakovo NPP, *Radiats. Biologija. Radioecologija* 44(1), 83–90 (in Russian).

Korogodina, V.L., Florko, B.V., and Korogodin, V.I., 2005, Variability of seed plant populations under oxidizing radiation and heat stresses in laboratory experiments. *IEEE Trans. Nucl. Sci.* 52(4), 125–144.

Korogodina V. L., Bamblevsky C.P., Florko B.V., and Korogodin V.I. 2006. Variability and viability of seed plant populations around the nuclear power plant, In: Cigna A.A. and Durante (eds.) *Impact of radiation risk estimates in normal and emergency situations. Proceedings of a NATO advanced research workshop Yerevan, Armenia, 8–11 Sept. 2005, Springer*: 271–282.

Korogodina, V.L., Panteleeva, A., Ganicheva, I., Lazareva, G., Melnikova, L., Korogodin, V., 1998, Influence of the weak gamma-irradiation dose rate on mitosis and adaptive response in meristem cells of pea seedlings. *Radiats. Biologija. Radioecologija,* 38, 643–649, (in Russian).

Little, J.B., 2006, Cellular radiation effects and the bystander response. *Mutat. Res.* 11; 597(1–2),113–118.

Longerich, S., Galloway, A.M., Harris, R.S., Wong, C., and Rosenberg, S.M., 1995, Adaptive mutation sequences reproduced by mismatch repair deficiency. *Proc. Natl. Acad. Sci. USA* 92, 12017–20.

MAPRF, 1998, *The General Information on Balakovo NPP,* The Ministry of Atomic Power of the RF, Rosenergoatom concern, Balakovo NPP: 3–5 (in Russian).

Mothersill, C. and Seymour, C., 2005, Radiation-induced bystander effects: are they good, bad or both? *Med. Confl. Surviv.,* 21(2), 101–110.

Preobrazhenskaya, E. 1971, Radioresistance of Plant Seeds. *M: Atomizdat,* (in Russian).

Rainwater, D.T., Gossett, D.R., Millhollon, E.P., Hanna, H.Y, Banks, S.W., and Lucas, M.C., 1996, The relationship between yield and the antioxidant defense system in tomatoes grown under heat stress. *Free Radical Res.,* 25(5), 421–435.

Schwarz, G., 1978, Estimating the dimension of a model. *The Annals of Statistics,* 6, 461–464.

State Committee on the Environment Protection in the Saratov Region (SCEPSR), 2000, *Conditions of the environment in the Saratov region in 1999.* Acvarius, Saratov: 1–193 (in Russian).

Timofeeff-Ressovsky, N.W., 1939, Genetik und Evolution. *Z. und Abst. Vererbl.,* 76(1–2), 158–218.

Van der Vaerden, B.L., 1957, Mathematische statistik. *Berlin-Göttingen-Heidelberg, Springer-Verlag.*

Wiegel, B., Alevra, A.V., Matzke, M., Schrewe, U.J., and Wittstock, J., 2002, Spectrometry using the PTB neutron multisphere spectrometer (NEMUS) at flight altitudes and at ground level. *Nucl. Instr. Meth. A,* 476 (1–2), 52–57.

Zykova, A.S., Voronina, T.F., and Pakulo, A.G., 1995, Radiation situation in the Moscow and Moscow region caused by fallouts in 1989–1993. *Gigiena i Sanitarija,* 2, 25–27 (in Russian).

CHAPTER 12

CLASTOGENIC FACTORS, BYSTANDER EFFECTS AND GENOMIC INSTABILITY IN VIVO

SERGEY MELNOV*, PAVEL MAROZIK,
AND TATIANA DROZD

*Department of Environmental and Molecular Genetics,
International Sakharov Environmental University,
Dolgobrodskaya 23, 220009, Minsk, Belarus
e-mail: sbmelnov@rambler.ru*

Abstract: For the last 15 years, we have investigated low dose radiation genetic effects on human populations affected by the Chernobyl accident. Cytogenetic longitudinal investigations showed that amount of radiation markers for clean-up workers remained at the elevated level and had trend to grow up with the time. A dynamic profile of the amount of aberrations confirms that this group has symptoms of the genomic instability. State of the genomic instability correlates with accumulation of clastogenic factors, responsible for increased genomic instability in clean-up workers peripheral blood. As a model for clastogenic activity testing, we used human keratinocyte cell line with blocked 1st check point of cell cycle. Our results confirm that cytogenetic and molecular effects of irradiation can be fixed even 20 years after the Chernobyl accident.

Keywords: Chernobyl accident; liquidators; affected populations; clastogenic factors; bystander effect; cytogenetic and molecular effects; low dose

Introduction

Today problem of genomic instability is crucial point of modern radiobiology. In a lot of publications they have shown that people suffered from additional irradiation manifest the stable long time elevation of chromosomal aberrations (Melnov, 2004). Mechanism of this event is not clear until now.

From the other side, in a last few years phenomenon of radiation-induced bystander effect has been fixed (Nagasawa and Little, 1992; Mothersill and Seymour, 1997).

C. Mothersill et al. (eds.), Multiple Stressors: A Challenge for the Future, 171–182.
© 2007 *Springer.*

In our understanding, bystander effect could be a key point of the genomic instability formation in vivo. In such situation, clastogenic bystander inducers may be spreaded out via blood serum in body. Possibility of spreading of such factors via liquid phase previously has been shown in vitro (Mothersill and Seymour, 1997).

Basing on this idea, we investigated the clastogenic effect of the serum samples from the patients affected by radiation from different sources and compared it with serum samples from patients suffered from acute and chronic virus diseases.

Materials and Methods

CELL CULTURES

HPV-G cells

The HPV-G cell line is a human keratinocyte line, which has been immortalised by transfection with the HPV virus, rendering the cells *p53* null. They grow in culture to form a monolayer, display contact inhibition and gap junction intracellular communication.

HPV-G cells were cultured in Dulbecco's MEM: F12 (1:1) medium supplemented with 10% fetal bovine serum, 1% penicillin-streptomycin, 1% L-glutamine and 1 µg/mL hydrocortisone. The cells were maintained in an incubator at 37°C, with 95% humidity and 5% carbon dioxide and routinely subcultured every 8–10 days. When 80–100% confluent, the medium was poured from the flask and replaced with 1:1 solution of versene (1nM solution) to trypsin (0.25% in Hank's Balanced Salt Solution) (Gibco, Irvine, UK) after washing with sterile PBS. The flask was placed in the incubator at 37°C for about 11 min until the cells started to detach. The flask was then shaken to ensure that all cells had been removed from the base of the flask. The cell suspension was added to an equal volume of DMEM F12 medium to neutralize the trypsin. From this solution new flasks could be seeded at the required cell quantity.

Human peripheral blood lymphocytes

Blood samples were collected by standard venous puncture and stored with heparin (20 units/mL) no longer than 2–3 h before treatment. Human peripheral blood lymphocytes were cultured in RPMI-1640 medium (100 mL), including fetal bovine serum (20%), and gentamycin (0,2%). 4.5 mL of mixtures were aliquoted into 50 mL culture flasks and frozen at −20°C, before cultivation.

SAMPLE DONORS

The victims of the Chernobyl accident included three main categories: Chernobyl liquidators 1986–1987, workers from Polessky State Radiation and Environment Reserve (PSRER workers) and people living in territories of Gomel region contaminated by radionuclides. The control group were clinically healthy people corresponding to other groups in age and sex aspects.

BLOOD SERUM EXTRACTION

The blood samples were taken and placed in Vacutainers for serum extraction (Becton Dickinson), centrifuged at 2,000g for 10 min, and the serums were frozen and stored at $-20°C$ before use. Before freezing the serum was filtered through Nalgene 0.22 µm filters in order to remove all residual cell components of the blood.

ROUTINE CYTOGENETIC TEST

At the day of analysis flasks with medium were defrosted, 0.5 mL of blood was added and 0.13 mL of PHAM (Sigma, 1 mg/1 mL). Flasks were cultivated in incubators for 48 h at 37°C. About 3 h before the end of cultivation 30 µL/ flacks (standard stock solution) of colchicine was added. Lymphocytes were centrifuged for 5 min at 1,500 rpm, supernatant was removed and cells are washed with warm 0.55% KCl solution (at 37°C) for 20–25 min at incubator. Cells were fixed using Karnua solution (1 part of glacial acetic acid and 3 parts of methanol, 5 mL of solution per flask). Cold fixator was changed three times for 10, 20 and 10 min.

Suspensions were mounted on microscope glasses (3–4 drops of sample per glass) and dried at 40–42°C. Cells were stained using 10% Giemsa solution for 8 min and after drying analyzed under light microscope. Only cells with the chromosomes number between 44 and 47 were analyzed.

MICRONUCLEI TEST

After seeding human keratinocytes, cells were left at 37°C in the CO_2 incubator to attach for 12 h. The blood serum from affected populations was added to 24 cm² flasks (NUNC, USA) (6000 cells per flask) 1–2 days after seeding, and cells were cultivated in the incubator for 1–2 h. Then cytochalasine B was added (7 µg/mL concentration) and the cells were incubated for 24 h, additionally.

After this the cell-culture medium was removed, the cells were washed with PBS and fixed with chilled Karnua solution (1 part of glacial acetic acid and 3 parts of methanol, 10–15 mL three times for 10–20 min). Later flasks were dried and stained with 10% Giemsa solution.

The micronuclei count was carried out under inverted microscope (×400). Micronuclei were counted only in binucleated cells (1,000 binucleated cells per flask). All the data is calculated as the micronuclei number recorded per 1,000 binucleated cells (micronuclei were analyzed only in binucleated cells).

Results and Discussions

Results of the dynamic investigation of the main group (liquidators) has been analyzed (Marozik et al., 2007) and presented on Figs. 1 and 2.

Shortly summarizing this data, one can conclude that with time the sum of aberrations is growing up constantly, and maximal level has been fixed in 2004–2005 (Fig. 1). At the same time, we did not fix prominent dynamics for markers (Fig. 2), dicentric and ring chromosomes – main irradiation specific cytogenetic markers.

It means that the main increase of the total number of aberrations took place because of the elevated levels of unspecific aberrations (predominantly single and double fragments). In other words, investigated group manifested the presence of genomic instability symptoms in somatic cells.

This conclusion is absolutely reliable in comparison with the same data for control group (Figs. 3 and 4), were dynamics of the same parameters are absolutely different.

In control group statistically reliable dynamics for both total number of aberrations and for the level of marker aberrations were absent.

So, basing on this one can conclude that even 20 years after the Chernobyl accident liquidators had the cytogenetic effects in peripheral blood lymphocytes.

This situation was typical not only for liquidators, but also for children-prominent residents of contaminated areas.

Our results presented on Fig. 5 show the results of the investigation of the children from the special areas of Brest region where situation is characterized by quick migration of radionuclides along the food chains, resulting in their high accumulation in human body (children – "accumulators").

This group has been investigated three times with interval 2–3 years (whole period – 8 years).

Analyzing data from Fig. 5, we were able to confirm that dynamics of the investigated parameters of cytogenetic status in this group is very close to liquidator situation – practically all types of aberrations increased with the

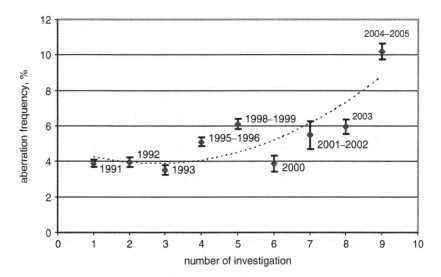

Fig. 1. Dynamics of the total number of aberrations in liquidators lymphocytes.

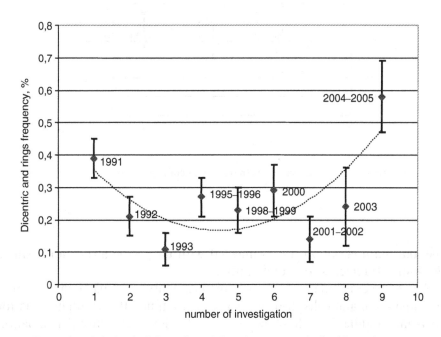

Fig. 2. Dynamics of the level of dicentric and ring chromosomes in liquidators lymphocytes.

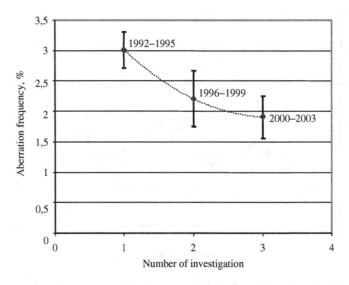

Fig. 3. Dynamics of aberrations frequency in control group.

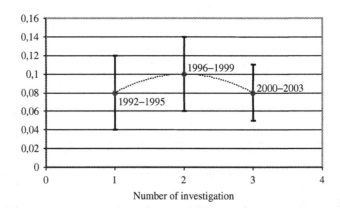

Fig. 4. Dynamics of dicentrics and rings frequency in control group.

time, but main elevation was connected with unspecific aberrations (single and double fragments) predominantly.

So we can conclude that the genomic instability is a typical phenomenon for human somatic cells in delayed period of time after irradiation as for short-time middle dose (liquidators) so as for low dose chronic irradiation (children from contaminated areas).

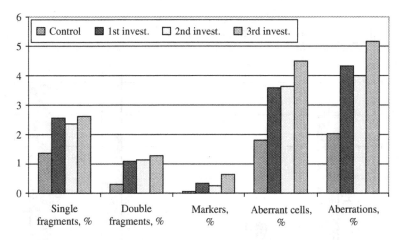

Fig. 5. Dynamics of cytogenetic status of children – "accumulators".

During the investigation we collected blood samples from four groups of patients: liquidators (short-time irradiation), workers of Polessky State Radiation and Environmental Reserve (PSRER, 30 km area around Chernobyl station – low dose chronic irradiation), people from contaminated territories of Gomel region suffered with hepatitis C and children from Gomel region with acute virus infection (flu).

Results of the incubation of keratinocytes in the presence of serum samples from liquidators are summarized in Fig. 6.

Basing on collected data, we can conclude that liquidators serum samples showed real clastogenic effect – the amount of induced micronuclei (in comparison with control group serum samples) increased approximately in three times (273.7 ± 22.4 and 91.8 ± 12.4, correspondingly; $p < 0.01$).

Specially important that these serums induced elevated levels of polymicronuclei cells (e.g., in control group cells with three micronuclei were absent absolutely).

Similar situation took place for workers of PSRER (Fig. 7).

Again the discrepancies between main and control groups were statistically reliable (the total micronuclei frequency 260.3 ± 18.3 vs. 91.8 ± 12.4; frequency of cells with micronuclei 232.2 ± 13.1 vs. 84.2 ± 11.5; $p < 0.05$). At the same time, the discrepancy between levels of micronuclei for liquidators and PSRER workers was statistically unreliable, but at the same time, for PSRER workers induced levels were constantly a little bit lower than for liquidators.

Similar situation has been fixed for two other investigated groups – for patients with chronic (hepatitis C) and for children with the acute (flu) virus infection (Figs. 8 and 9). Maximal induced level has been fixed for patient with acute virus infections (Fig. 9).

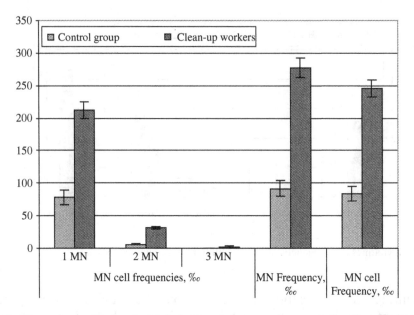

Fig. 6. Comparison of micronuclei induction by serums from people of control group and liquidators.

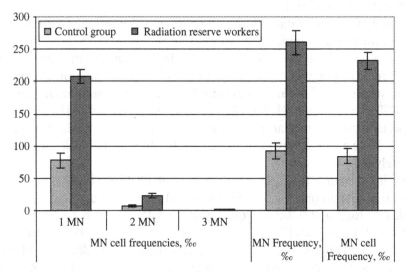

Fig. 7. Comparison of the levels of micronuclei induction by serums from patients of control group and PSRER workers.

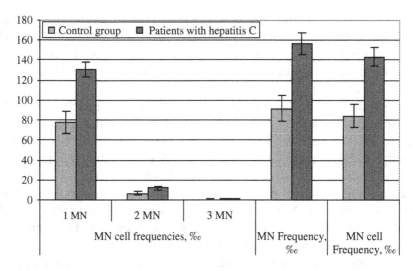

Fig. 8. Comparison of micronuclei induction by serum samples from patients of control group and with hepatitis C.

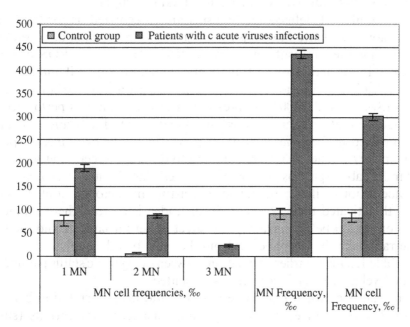

Fig. 9. Comparison of micronuclei induction by serum samples from patients of control group and with acute virus infection.

But we need to mention that 2 weeks later after virus infection the level of clastogenic factors for patients recovered from virus infection moved to norm, and serums did not induced additional micronuclei. In the serum samples from patients with chronic virus infection (hepatitis C) these factors continued to exist.

The similarity of clastogenic factor profiles for people affected by radiation and virus infection could be rooted in the similar role and higher induction of free radicals. Their effect for radiation effects is of common knowledge today. At the same time, it is well known that free radicals are generated in excess in a diverse array of microbial infections. Free-radical induced pathogenicity in virus infections is of great importance, because evidence suggests that NO and oxygen radicals such as superoxide are key molecules in the pathogenesis of various infectious diseases and have practically radio mimetic effect (Kash et al., 2004). Although oxygen radicals and NO have an antimicrobial effect on bacteria and protozoa, they have opposing effects in virus infections such as influenza virus pneumonia and several other neurotropic virus infections. The unique biological properties of free radicals are further illustrated by recent evidence showing accelerated viral mutation by NO-induced oxidative stress (Akaike, 2001).

Neutrophil host defense mechanisms are categorized as oxidative and non-oxidative. Oxidative mechanisms rely upon the production of superoxide, primarily by the multisubunit enzyme NADPH oxidase. ROS are also implicated in the activation of transcriptional factors (NF-κB and activator protein 1) leading to the transcription of genes that accentuate the inflammatory process (Swaun et al., 2004). Expression microarray analysis performed on lung tissues isolated from the infected animals showed activation of many genes involved in the inflammatory response, including cytokine, apoptosis and lymphocyte genes that were common to different infection groups.

Our data above gave us possibility to suggest that elevated level of chromosome aberrations will be stimulated by clastogenic factors and at the same time with activated their synthesis. If such idea is correct, there should be direct correlation between clastogenic effect of serum samples and the level of aberrations in the lymphocytes from the same blood samples.

In Table 1, shown earlier correlation between cytogenetic status parameters and the level micronuclei induction is presented.

Our results confirm that there is stable and deep correlation between the number of induced micronuclei and unspecific aberrations (single fragments $r = 0.50$; double fragments $r = 0.49$; $p < 0.005$) and total number of aberrations and aberrant cells ($r = 0.76$ and $r = 0.65$, correspondingly; $p < 0.005$ for both). In our opinion, it means that there is direct correlation between genomic destabilization and accumulation of clastogenic factors.

TABLE 1. Correlation matrix (by Spearman) for cytogenetic parameters compared to induced effect

Parameters of comparison, %	Number of cells with MN, ‰	Total MN number, ‰
Single fragments	0.50*	0.62*
Double fragments	0.49*	0.55*
Dicentric and ring	0.07**	0.08**
Atypic chromosomes	−0.24**	−0.24**
Polyploid cells	0.13**	0.10**
Number of aberrant cells	0.65*	0.71*
Total number of aberrations	0.76*	0.84*

$*p < 0.005; **p > 0.1$

Conclusions

Shortly summarizing our results, we can confirm the next:

- After irradiation destabilized genome initiate the cell production of bystander factors, able to stimulate the increase of the genomic destabilization of other somatic cells, transforming this situation in long time effect, possibly responsible for delayed health effects;

- Virus infections, which are able to increase the level of free radicals, can stimulate the same effect. But duration of its manifestation is depended upon the type of virus: if the decease is temporal, such situation will take place only during the acute phase; in the chronic situation (hepatitis C) it is a prominent effect.

Shortly speaking, our data confirm that radiation-induced bystander effect, induced in vivo, is a standard biological phenomena, responsible for delayed health effect of radiation.

References

Akaike, Takaaki, 2001, Role of free radicals in viral pathogenesis and mutation, *Rev Med Virol* 11(2):87–101.

Kash, J.C., Basler, C.F., Garcia-Sastre, A., Carter, V., Billharz, R., Swayne, D.E., Przygodzki, R.M., Taubennerger, J.K., Katze, M.G., and Tumpey, T.M., 2004, Global host immune response: pathogenesis and transcriptional profiling of type A influenza viruses expressing the hemagglutinin and neuraminidase genes from the 1918 pandemic virus, *J Virol* **78** (17):9499–511.

Marozik, P., Mothersill, C., Seymour, C.B., Mosse, I., and Melnov, S., 2007, Bystander effect induced by serum from the survivors of the Chernobyl accident, *Exp Haem*, **35**:55–63.

Melnov, S.B., 2004, *Molecular and Genetic Effects of Ecological Trouble (Possibilities of Flow Cytometry)*, Committee "Chernobyl children", Minsk, pp. 10–32.

Mothersill, C. and Seymour, C., 1997, Medium from irradiated human epithelial cells but not human fibroblasts reduces the clonogenic survival of unirradiated cells, *Int J Radiat Biol* **71**:421–427.

Nagasawa, H. and Little, J.B., 1992, Induction of sister chromatid exchanges by extremely low doses of alpha-particles, *Cancer Res* **52**:6394–396.

Swaun, S.D., Wright, T.W., Degel, P.M., et al., 2004, Neither neutrophils nor reactive oxygen species contribute to tissue damage during Pheumocystis pneumonia in mice, *Infect Immun* **72**(10):5722–732.

SECTION 4

MULTIPLE STRESSORS: MECHANISMS

CHAPTER 13

MULTIDISCIPLINARY ASPECTS OF REGULATORY SYSTEMS RELEVANT TO MULTIPLE STRESSORS: AGING, XENOBIOTICS AND RADIATION

C. DAVID ROLLO*

Department of Biology, McMaster University, Hamilton, Ontario, Canada L8S 4K1

Abstract: Free-radical biology, which is central to the fields of radiation, aging and xenobiotics, has shifted from a paradigm highlighting damage to a paradigm emphasizing the role of free radicals in regulatory processes. A unified approach is possible since multiple stressors tend to activate a coordinated set of common mechanisms. These include antioxidant defenses, metal chelators, DNA repair systems, heat-shock proteins, xenobiotic efflux transporters, protein degradation systems, cell survival and apoptosis pathways and detoxification systems. Nearly all MAPK signal transduction pathways employ oxidative signaling, largely generated via membrane-bound NADPH oxidase systems. These regulate most cellular stress, growth and apoptotic responses. A new global perspective highlighting "Electroplasmic Cycles" incorporates numerous cellular aspects of control including free radicals, protein and histone modifications, nuclear–cytoplasmic transport, and ion channels. Aspects critical to multiple stressors include complex interactions related to apoptosis-necrosis, immunological responses (Toll-like receptors), bystander effects, chaperone proteins, and multiple xenobiotic efflux proteins. The synthesis suggests that a systems approach to multiple stressor impacts is required since understanding requires holistic appreciation of integrated regulatory circuitry.

Keywords: radiation; xenobiotics; aging; free radicals; signal transduction; redox; ion channels; growth; chaperone proteins; multiple drug resistance; apoptosis; clearance; Toll-like receptors; auto-immunity; nuclear–cytoplasmic transport

---

*To whom correspondence should be addressed. e-mail: rollocd@mcmaster.ca

C. Mothersill et al. (eds.), Multiple Stressors: A Challenge for the Future, 185–224.
© 2007 *Springer.*

The Regulatory Paradigm

Free-radical and radiation biology are rapidly expanding from a paradigm emphasizing damage, protection and repair (see McBride et al., 2004) to encompass the pervasive regulatory impacts of reactive oxygen, nitrogen, and peroxidation species (e.g. Fornace, 1992; Herrlich et al., 1999; Schmidt-Ullrich et al., 2000; Bauer, 2002; Droge, 2002, 2005; Dent et al., 2003; Finkel, 2003; Esposito et al., 2004; Balaban et al., 2005; Cai, 2005; Fedoroff, 2006; Meng et al., 2006). This is globally applicable to growth, development, organismal and cellular function, aging, stress, xenobiotics, radiation, immunology, transformation and numerous pathologies. Regulatory impacts of radiation and xenobiotics trace largely to generation of free radicals from water by radiation and xenobiotic detoxification (e.g. p450 monooxygenases). Metals are particularly important pollutants as they synergize conversion of relatively benign reactive oxygen species such as superoxide and hydrogen peroxide to the more harmful hydroxyl radical (e.g. iron and Fenton chemistry). Very small changes in transition metals can have large impacts such as alterations in p53 function (Meplan et al., 2000). Moreover, antioxidants, particularly the glutathione (GSH) system, importantly regulate cellular redox status. All of these aspects are also incorporated in the prevailing free-radical theory of aging which predicts that chronic production of endogenous free radicals from mitochondria, NADPH oxidases (NOX) and other processes results in damage to DNA, proteins and lipids. Accumulating damage impacts cellular constituents, organelles, extracellular matrix, tissues and organ systems resulting in an aging organismal phenotype.

Radiation, xenobiotics and aging are unified by their joint linkage to free-radical biology. In fact, the free-radical theory of aging was derived directly from radiation biology (Harman, 1956). Diverse stressors (e.g. free radicals, radiation, heat, toxins, ionic disruption, dietary restriction, and even severe sleep deprivation) invoke a generalized and coordinated upregulation of numerous defensive systems (Jazwinski, 1996; Martin et al., 1996; Rollo, 2002; Gems and McElwee, 2005). These include antioxidants, metal chelators, heat-shock proteins, DNA repair systems, protein degradation systems, P450 monooxygenases, glutathione-S-transferases, apoptotic pathways and multiple drug resistance (efflux) proteins. Most models of extended longevity express stress-resistant phenotypes, and selection for multiple-stress resistance can yield extended maximal longevities (Wang et al., 2004; Frankel and Rogina, 2006; Stuart and Brown, 2006; Wang and Tissenbaum, 2006).

Cysteine–Methionine Residues, NADPH Oxidase and Redox Regulation

Advances in free-radical-mediated signaling have involved parallel increases in understanding NOX and redox-sensitive residues of diverse regulatory peptides (Scmidt-Ullrich et al., 2000; Bauer, 2002; Droge, 2002, 2005; Dent et al., 2003; Cai, 2005; Kuroda et al., 2005). Mitochondria are generally considered the main source of cellular free radicals (see Balaban et al., 2005) but NOX isoforms are highlighted more in signal transduction. Regardless, mitochondrial function and NOX are linked via mutual impingement on NAD–NADH and the associated cellular redox environment (Rollo, 2006). Redox state regulates numerous signal transduction and transcription factors largely via reversible oxidation or reduction of sensitive cysteine and methionine residues. General oxidative inhibition of tyrosine phosphatases means that cell signal transduction via antagonistic phosphatases and kinases is redox dependent. Tyrosine phosphatases can also recognize phosphorylated serine/threonine residues (Meng et al., 2006).

Redox balance and associated cysteine residue status are mainly modulated by the glutathione antioxidant system. Thioredoxin and glutaredoxin augment GSH reduction of critical residues in signaling molecules (NOX-dependent processes) that control ion channels, the cell cycle, apoptosis, mitochondrial functioning, cytoplasmic-nuclear shuttling (e.g. RNA and transcription factors), stress, repair and degradation systems, secretion, excretion and nutrient uptake. Coordinated temporal cycling of numerous important functions is closely linked to shifts between oxidative and reducing redox status (with the reduced state being basal). This includes cycles involving membrane polarization, intracellular pH, calcium and potassium levels. Complications include localized alterations in redox status within various cellular compartments. Oxidative stress from multiple stressors is likely to trigger multiple mechanisms and pathways simultaneously (see Dent et al., 2003). Although some aspects may allow protective compensation, general upregulation of redox-sensitive systems is likely to be disruptive.

The co-repressor, carboxyl-terminal binding protein (CtBP) regulates numerous aspects of development, the cell cycle and transformation via binding to transcriptional repressors. Such binding is positively linked to NADH, allowing CtBT to serve as a redox sensor (Zhang et al., 2002). Furthermore, some NOX isoforms are localized to the nucleus where they generate SOR. In endothelial cells nuclear NOX4 activity increases the transcription of genes containing the MARE response element that coordinates part of the oxidative stress response (Kuroda et al., 2005). Since cellular redox state and NOX function both hinge on NADH, Kuroda et al. (2005) suggest that NOX4 also functions as a nuclear redox sensor. Silencing of NOX4 prevents induction of the "unfolded protein" stress response (Fedoroff, 2006). NOX4 may also contribute to the cellular senescence phenotype via alterations which may include impacts on telomeres.

SOME FEATURES REGULATED BY REVERSIBLE OXIDATION/ NITROSYLATION OF PROTEIN CYSTEINE AND METHIONINE RESIDUES

Pathways: (MAPK-ERK, PI3K, JAK-STAT, p38, JNK)
Growth factors, cytokines
Mitochondrial complex I
Creatine kinase
Caspases
Tyrosine phosphatases (e.g. PTP1B, PTN, SHPs, MKPs)
Ion channels (many)
Nuclear transport (import and export recognition sequences)
Transcription factors (e.g. AP-1, p53, NFκB, HIF-1α, glucocorticoid receptor)
Keap-Nrf2-ARE phase 2 genes
Protease inhibitors
Proteases
Glutathione, thioredoxin, peroxiredoxin
Thiol peroxidase
Chaperones

Ion Channels, Free Radicals and Radiation

Whereas the free-radical theory of aging has dominated gerontological and radiation research, in the realm of neurodegenerative disease and aging, a paradigm highlighting Ca^{2+} dysregulation has also predominated for at least 25 years. An ion channel perspective may in fact be expanded to subsume free-radical biology. Presumably ion channels were already crucially deployed in cells living in ancient reducing environments, with redox regulation being elaborated for protection against radiation (before the ozone layer) and as atmospheric oxygen increased. The basal reducing state of cells likely reflects this early heritage, which is at least as ancient as the preservation of cellular ionic conditions resembling seawater. Many pesticides impact ion channels and can alter numerous features including brain neurochemistry (particularly the dopamine system) (Richardson et al., 2006). Most antibiotics (like Rapamycin) impact ion channels. Furthermore, nearly all evolved toxins and venoms target channels, and at least one third of medicinal drugs act on channels (Pardo et al., 2005). The association of NOX with ion channels links free-radical production directly to ion channel activities, and a voluminous literature documenting the involvement of NOX isoforms in obesity, diabetes, atherosclerosis, cancer, neurotransmission, long-term potentiation and memory, most neurodegenerative diseases, growth factor signaling, cellular pH and redox state, immunological activities and aging.

Radiation deposition of energy into cells causes some direct damage but it also lyses water to form free radicals (including damaging hydroxyl radicals). Low-level γ-irradiation activates outward K^+ channels within seconds. This is associated with generation of free radicals by radiation, and the response is inhibited by N-acetyl cysteine. K^+ channel activation is implicated in other stress responses such as to heat or hydrogen peroxide (Kuo et al., 1993). K^+ channels are regulated by reversible oxidation of sensitive cysteine and methionine residues (Ruppersberg et al., 1991; Chen et al., 2000). This can modulate neuronal excitability. Reactive oxygen species generally allow Ca^{2+} entry and disrupt Ca^{2+} homeostasis in a variety of cell types (Huang et al., 2003). ROS generally impair and antioxidants often protect ion transport proteins underlying transmembrane signal transduction. Kourie (1998) reviews such impacts for (i) ion channels such as those for Ca^{2+} (including voltage-sensitive L-type Ca^{2+} currents, dihydropyridine receptor voltage sensors, ryanodine receptor Ca^{2+}-release channels, and D-*myo*-inositol 1,4,5-trisphosphate receptor Ca^{2+}-release channels), K^+ channels (such as Ca^{2+}-activated K^+ channels, inward and outward K^+ currents, and ATP-sensitive K^+ channels), Na^+ channels, and Cl^- channels; (ii) ion pumps, such as sarcoplasmic reticulum and sarcolemmal Ca^{2+} pumps, Na^+-K^+-ATPase (Na^+ pump), and H^+-ATPase (H^+ pump); (iii) ion exchangers such as the Na^+/Ca^{2+} exchanger and Na^+/H^+ exchanger; and (iv) ion cotransporters such as K^+–Cl^-, Na^+–K^+–Cl^-, and P_i–Na^+ cotransporters.

Many of the features listed earlier for free-radical processes have very broad impacts on cell functions in their own right. Consider ion channels:

MAJOR FUNCTIONS REGULATED BY ION CHANNELS

- Cell size
- Osmotic pressure
- Electrical polarization, depolarization
- Nutrient uptake (including amino acids)
- Neuroendocrine secretion and re-uptake
- Regulation of pH
- Excretion, detoxification
- Muscle contraction
- Redox control
- Aquaporins transport (including hydrogen peroxide)

All of the earlier circumstances highlight a critical linkage between free-radical processes, ionic alterations, channel functions and electrical properties. Consequently, the temporal organization of cells might well reflect "Electroplasmic Cycles." Many critically important cell transduction-transcription factor pathways are redox-ion regulated and sensitive to the impacts of radiation (particularly ionizing radiation) and xenobiotics (see later). A global perspective reveals the existence of redox-associated cycles that dichotomously coordinate major organismal and cellular functions. Probably the most obvious resolution of such cycling is circadian (24 h) rhythmicity. This is associated with endogenous clocks and the antagonistic functioning of the waking versus sleeping states. These states in turn represent opposing watersheds of regulation of various functions by the stress hormone axis in waking versus the growth hormone axis in sleeping (The clockwork genome hypothesis – Rollo, 2007). Although the circadian rhythm is the most obvious expression of endogenous rhythmicity, it is actually composed of and emerges from shorter-duration cycles. Of these, the ultradian rhythms of foraging-feeding and sleep (period lengths of 3–4 h) are well established. Even finer resolution of cycling can be found; however, some examples of nuclear shuttling being remarkably fast. This may allow both nuclear import and export to be carried out within relatively short time frames. Rhythmicity is critical for a multiple stressor paradigm since biological systems respond differently at different cycle phases (e.g. the effectiveness of drugs and detoxification systems shows strong circadian variation).

Yin–Yang Cycles of Redox and Cell Functions (The Electroplasmic Cycle)

Oxidizing	Reducing
Elevated reactive oxygen species	Lower reactive oxygen species
Elevated reactive nitrogen species	Lower reactive nitrogen species
Increasing GSSG/GSH ratio	Lower GSSG/GSH ratio
NAD(P)H oxidase activity Reduced	NOX activity
Acidification	Alkalization
Catabolism	Anabolism
Activity, output	Inactivity, sleep (stimulated by GSSG)
Depolarization	Hyperpolarization
Elevated intracellular Ca^{2+}	Reduced intracellular Ca^{2+}
Decreased intracellular K^+	Increased intracellular K^+
Kinase activity	Phosphatase activity
Acetylation (p300/CBP)	Deacetylation (HDAC, SIRTs)
Oxidized cysteine residues	Reduced cysteine residues
Oxidized methionine residues	Reduced methionine residues
Nuclear import (e.g. AP-1, NF-κB)	Nuclear export (AP-1, NF-κB)

Spatial Compartments

Besides temporal aspects, organ, tissue and intracellular compartmentalization is crucial to specialized functioning. Spatial compartmentalization of glutathione, particularly cytosolic versus mitochondrial pools, may be highly relevant to radiation. Lack of clear dose-response relationships for low dose radiation is an outstanding quandary for radiation biology. One factor may be the presence of a large effective buffer of reduced glutathione such that discernable impacts require 40–50% depletion of mitochondrial GSH (Han et al., 2003). Initial depletion of mitochondrial GSH has little effect but a dose-response relationship in mitochondrial H_2O_2 production emerges beyond a threshold of 50% GSH depletion. A further compensation to redox stress involves the anti-apoptotic function of bcl-2. Besides raising cellular levels of GSH, bcl-2 redistributes GSH from the cytosol into the nucleus. Whereas the nucleus normally sequesters ~30% of cellular GSH, bcl-2 upregulation increases this to ~75% (Voehringer et al., 1998). This mechanism would probably antagonize radiation impacts, particularly at low doses, and like GSH buffering in mitochondria, yield a threshold response to increasing radiation dosage. Ubiquitination is also partially regulated by spatial localization. The ubiquitin-conjugating enzyme is transported into the nucleus if charged with ubiquitin (Sommer and Jarosch, 2005). Regardless of the impacts of cytosolic redox status, the DNA binding of many transcription factors (e.g. AP-1, p53, NF-κB) require reduction of sensitive cysteine residues in the DNA binding region. The co-activator, Ref-1 serves to reduce such cysteines and synergizes transcription (Meplan et al., 2000). Such activity could act in concert with the import of GSH by bcl-2, creating some dichotomy between cytosolic and nuclear regulation.

Cytoplasmic–Nuclear Transport

Of critical importance to both spatial and temporal regulatory impacts is cytoplasmic–nuclear transport of materials into or out of the nucleus (Macara, 2001). This applies to histone protein H1, actin, ribosomal proteins, Cyclin B1, p53, Smad, PKA, PKC, FOXO, MEK1, ERK, c-fos, protein kinase inhibitor α, p38, STAT1, helicase, STAT5, androgen receptor, parathyroid hormone-related protein, fibroblast growth factor, p53, human telomerase, NF-AT4, histone deacetylase2/6, tRNA, and mRNA (Kudo et al., 1999a,b; Verdel et al., 2000; Strom and Weis, 2001; Poon and Jans, 2005; Wiedlocha et al., 2005). Many of these elements shuttle between the cytoplasm and nucleus, highlighting the existence of (electroplasmic) regulatory cycles that appear to be correlated to redox. Among elements that are subject to shuttling via importing family peptides are molecular components of the circadian clock such as mammalian Cryptochrome 2 and mammalian Period-2. Transport processes are an essential aspect of the transcriptional oscillations and auto-regulatory feedback by translated proteins that derive clock function (Sakakida et al., 2005). Clocks in turn diversely regulate cellular and organismal functions, and disruption can lead to cardiovascular disease, psychiatric disorders and cancer (Sakakida et al., 2005).

Cytoplasmic retention of particular proteins can trace to heat-shock proteins like HSP90 (Jans, 1995; Poon and Jans, 2005). Transport proteins belong to the large importin-β-family (at least 22 members) (Strom and Weis, 2001). The GTPase, Ran, binds importin in the nucleus, inducing release of cargo. Gradients of Ran may also regulate directionality of transport. Importin-β proteins contain sequences for recognizing Ran, cargo and nucleoporins (Kudo et al., 1999a; Strom and Weis, 2001; Petosa et al., 2004; Fahrenkrog, 2006). The nuclear pore complexes may be the largest protein structures in eukaryotic cells (Macara, 2001).

Importin-β-family proteins contain sensitive cysteine residues that are crucial for functioning. Various transport recognition sequences may also have such cysteine residues, and yet consideration of regulation by reversible oxidation or glutathionation appears relatively unexplored. Regardless, there are several suggestive examples of reversible oxidation of cysteine residues as key modulators of nuclear–cytoplasmic transport. Pap-1, a fission yeast homologue of Jun is maintained in the nucleus by oxidative stress, and this traces to oxidation of critical cysteine residues. Pap-1 is involved in free-radical responses, metal detoxification and multiple drug resistance acting via catalase, thioredoxin, thioredoxin reductase, glutathione reductase and ABC transporters (Multiple drug resistance proteins) (Kudo et al., 1999b). The yeast multiple drug transporters are homologues of human p-glycoprotein encoded by the MDR1 gene that is associated with JNK activity (Toone et al., 1998). Yap1p is also homologous to mammalian AP-1 (Jun) in budding yeast and regulates responses to oxidative stress and cadmium toxicity. The nuclear export signal of Yap1p is regulated by redox modulation of reversible disulfide bond formation in critical cysteine residues. Kuge et al. (2001) suggest that nuclear localization may be regulated by thioredoxin. Leptomycin B, an antifungal antibiotic, suppresses the function of the key exportin protein CRM1 and suppresses virtually all nuclear export by interfering with a conserved cysteine residue. This also blocks the cell cycle (Kudo et al., 1999a) which is strongly regulated by shuttling of various cyclin proteins (Jackman et al., 2002). The deacetylase mHDAC6 is actively maintained in the cytoplasm and nuclear import also arrests cell proliferation (Verdel et al., 2000). Similarly, oxidative stress mediates nuclear localization of HSF1 and its binding to HSP gene promoters (Fedoroff, 2006).

Although the involvement of a conserved cysteine in exportin1 is suggestive from a redox perspective (i.e., could oxidation directly inhibit nuclear export?) general wisdom currently favors that control of cytoplasmic–nuclear transport resides in access to the import or export signatures of cargo. This allows for differential transport of cargos that may have antagonistic actions or that are relevant to different functions or electroplasmic states. Redox control of some cytoplasmic–nuclear transport may also arise from the pervasive involvement of kinase-phosphatase activities (Jans, 1995; Poon and Jans, 2005). Furthermore, all of the MAPK pathways that regulate nuclear transport are also upregulated by oxidizing conditions and inhibition of tyrosine phosphatases. All of this suggests that a major impact of radiation and xenobiotic interactions will involve transport and localization of numerous critical regulatory peptides.

Most transcription factors relevant to the multiple stressor paradigm are regulated by redox-sensitive transport. The nuclear pore complex and nuclear transport also critically regulate apoptosis. Cytochrome c, caspase-2, caspase-3, caspase-6 and apoptosis inducing factor are all transported into the nucleus during apoptosis, whereas acetylated histone H2A and inhibitors of apoptosis are exported (Fahrenkrog, 2006). Heat-shock proteins can inhibit import of apoptosis-inducing factor. Export of the glucocorticoid receptor and PKA from the nucleus involves the Ca^{2+}-binding protein, calreticulin, suggesting potential linkages between calcium homeostasis and transport activities. Furthermore the Ca^{2+}-responsive phosphatase, calcineurin binds NFAT4, masking its nuclear export signal (Poon and Jans, 2005). Ca^{2+} elevations are generally associated with the oxidative phase of electroplasmic cycles. Fahrenkrog (2006) suggests that reduced nuclear import, as seen in aging cells or following heat shock, may in fact represent a stress response. Indeed, severe oxidative stress disrupts nuclear protein import (Kodiha et al., 2004).

Release of the glucocorticoid receptor (a key effector of the stress hormone axis) from HSP-90 and subsequent nuclear import are inhibited by oxidative conditions. This traces to Cys-481 (Okamoto et al., 1999). Alternatively, nuclear export and transcriptional inhibition were enhanced when the glucocorticoid receptor was phosphorylated by activated JNK (Itoh et al., 2002). The AP-endonuclease (=Redox factor (1) functions in DNA repair and activation of oxidatively induced transcription factors like AP-1. It also alters drug resistance of cancers. This factor is translocated to the nucleus following oxidative stress in some cell types or may be maintained

in the nucleus by inhibition of export (Jackson et al. 2005). The nuclear localization signal for NF-κB is masked by IκB until IκB is phosphorylated and
degraded. IκB itself also shuttles between the nucleus and cytoplasm (Macara,
2001). Nuclear export of telomerase reverse transcriptase traces to phosphorylation of tyrosine 707, and this is effected by H_2O_2 (Haendeler et al., 2003).

This mechanism reduces the anti-apoptotic actions of the telomerase,
and has strong implications for telomere maintenance and cell senescence.
Antioxidants can reverse these impacts (Furumoto et al., 1998) suggesting
that free-radical stress associated with radiation and xenobiotics could accelerate cellular senescence, particularly in humans.

The transcription factor Nrf2 is essential for activating phase II detoxification and oxidative stress defensive genes acting via antioxidant response
elements and Maf recognition elements (ARE/MARE). Nrf2 is inhibited
when bound in the cytoplasm by Keap1. Previous models envisioned that
oxidative stress frees Nrf2 and allows its transport to the nucleus (Hoshino
et al., 2000). Velichkova and Hasson (2005) present evidence that Keap1
has a nuclear export motif which is neutralized by oxidative modification
of crucial cysteine residues. This then allows the complex to move to the
nucleus via a nucleus localization motif on Nrf2. Nrf2/Keap1 are considered to act as a sensor for cytoplasmic oxidative stress. The Bach2 transcription factor is related to Nrf2, but has opposite actions on induction by
MARE. Conditional nuclear export of Bach2 is also sensitive to oxidative
stress, perhaps via a cysteine residue in the cytoplasmic localization motif
(Hoshino et al., 2000).

Growth, Metabolism, PI3K and Aging

Insulin, IGF-1 and all growth factor signaling requires endogenous production
of free radicals (mainly by membrane-bound NOX systems). Free radicals
and low dose radiation can substitute for receptor ligands to activate
NOX, induce growth, or at higher levels, growth arrest and apoptosis.
Mitogen-activated protein kinase pathways involved include MAPK-ERK
(growth, apoptosis, long-term potentiation), PI3K (metabolism, aging,
growth, apoptosis), JAK-STAT (growth hormone, leptin, cytokines), JNK
(immunity, stress), and p38 (stress) (see Schmidt-Ullrich et al., 2000; Dent
et al., 2003). Of these, radiation biology has tended to emphasize JNK
and p38 as stress response pathways, but MAPK-ERK and (especially)
PI3K modulate aging processes and maximal longevity. Important impacts
of PI3K relating to radiation and xenobiotics include NOX (Frey et al., 2006),
the target of rapamycin (mTOR), the multiple drug resistance proteins
(e.g. p-glycoprotein and MRP), the chaperone protein gp96, ATM (the
mutation causing the progeroid syndrome, ataxia telangiectasia), oxidation of

phospholipids like phosphatidylserine and the dietary-restriction-regulating deacetylase, SIRT1. ATM also induces high sensitivity to radiation. Our previous research demonstrated an inverse relationship between adult body size and intra-specific longevity within rats and mice, and curtailed maximal longevity of giant mice (Rollo et al., 1996; Rollo, 2002). Giant transgenic growth hormone mice, which are a model of upregulated free-radical processes (Lemon et al., 2003; Rollo, 2007) were also more sensitive to radiation (Lemon et al., unpublished).

Mutations in the PI3K pathway that extend longevity, and the longevity enhancement associated with dietary restriction and SIRT1, are tightly regulated by activation of forkhead transcription factors. Activation of FOXO4 (=AFX) is suppressed by PI3K activity and phosphorylation by PKB/Akt. FOXO4 is activated following nuclear import but CRM1 exports the peptide. Phosphorylation of FOXO4 by PKB actually occurs in the nucleus and inhibits return transport of the exported transcription factor (Brownawell et al., 2001). Phosphorylation of FOXO4 by PKB/Akt also creates two 14-3-3 binding motifs. Binding of 14-3-3 to both of these sites prevents FOXO4 DNA binding. 14-3-3 peptides can mask transport signaling motifs on target proteins. Examples include CDC25 phosphatases, telomerase, histone deacetylase and FOXO transcription factors (Obsil et al., 2003). For FOXO1, binding by 14-3-3 in response to phosphorylation masks a nuclear localization signal, preventing nuclear import (Zhao et al., 2004).

DNA Modifications, Repair and Regulation

A new wrinkle in redox regulation was the identification of an oxidative "hotspot" in the promoter regions of the vascular endothelial growth factor gene associated with iron residues. Specific modification of guanine in the hypoxia inducible factor-1 response element enhanced incorporation of Hif-1 and Ref-1 into the transcription complex and doubled transcription of a reporter gene (Ziel et al., 2005). This implies that DNA sequences may be adaptive regulatory targets of oxidative signaling which is potentially reversible via DNA repair systems. Such a mechanism resembles a re-writable memory device.

The General Proteomic Code

Regulatory proteins are tuned by a multiplicity of mechanisms, including chaperones, ligands, dimerization, scaffolds, acetylation (p300/CBP, SIRT, HDAC), phosphorylation (kinases, phosphatases), methylation, oxidation, nitrosylation, ubiquitination-degradation, glycosylation, pH, voltage (voltage-gated channels), and transport-compartmentalization. Some of these mechanisms have been hailed as comprising a new "Histone code." In fact this extends to regulatory peptides throughout the cell and should consequently be considered as a more general "Proteomic Code."

Consider p53 regulation. Redox regulation of p53 is well established, but complex. Radiation and oxidative stress can induce transport and p53 activity, but this may partially be an indirect response to DNA damage. Depletion of GSH and treatment with H_2O_2 can inhibit p53 whereas hypoxia is activating (Meplan et al., 2000). A number of antioxidants increase p53 activity and induce apoptosis in several types of cancer (Brash and Havre, 2002). Oxidative status of multiple cysteine residues modifies p53 binding to response elements. When activated by DNA damage, reversible oxidation of the p53 residue Cys277 specifically modulated binding of p53 to response elements of the GADD45 promoter, but did not change p21WAF1/CIP1 binding (Buzek et al., 2002). The ATM protein (PI3K pathway) phosphorylates p53 and it is required for induction of p53 activity by radiation (Meplan et al., 2000). Some variability in p53 binding may also trace to variations in the p53 response element binding motif. Selenium (seleno-L-methionine) induced redox-factor-1 (Ref1) and p53 proteins and specifically upregulated the DNA repair pathway mediated by Ref-1 – p53. The p53 cysteine residues 275 and 277 were implicated. This was protective against induction of DNA damage by UV radiation.

Ref-1 itself has cysteine reducing abilities (Carrero et al., 2000) and it is recycled by thioredoxin (Fedoroff, 2006). Ref-1 functions to reduce sensitive cysteine residues in the DNA binding domain of p53, which is essential for transcription (Meplan et al., 2000). Brca1 was also required for selenium induction of DNA repair pathways (Seo et al., 2002; Fischer et al., 2006). Differential response element binding by p53 could impact major functions such as DNA repair, apoptosis, and cell-cycle regulation. A nuclear export motif occurs in the NH2-terminal region of p53. Phosphorylation following DNA damage can mask this signal (Zhang and Xiong, 2001a). Tetramerization of p53 in the nucleus also masks the nuclear export signal, ensuring nuclear retention until tetramere dissociation (Stommel et al., 1999). Ref-1 promotes p53 tetramere formation and stability. MDM2 is also a key regulator of p53, promoting ubiquitination and degradation in the cytoplasm, and inhibiting transcription in the nucleus. It may also interfere with acetylation by p300/CBP (Zhang and Xiong, 2001b). Alternatively, SIRT1 deacetylases p53 (Finkel and Cohen, 2005). Binding of p53 by 14-3-3 peptide also enhances activity (Meplan et al., 2000). Understanding of p53 regulation remains incomplete, but there is clearly great complexity entailing numerous interacting mechanisms.

Like p53, the critical transcription factor FOXO3 is acetylated at 5 different lysine residues, and phosphorylated at eight serine or threonine residues

(Brunet et al., 2004). Acetylation can involve SIRT1 (Finkel and Cohen, 2005). Nuclear transport proteins may also be both phosphorylated and acetylated (Wang et al., 2004; Poon and Jans, 2005). Interestingly, direct phosphorylation and acetylation (via p300) of Importin α1 was mediated by AMP-activated protein kinase (AMPK). AMPK can mediate a stress signal of low energy supply (Wang et al., 2004). Thus, the numerous mechanisms impinging on various signaling peptides appear to provide mechanisms to obtain combinatorial specificity across a potentially broad spectrum of targets. Multiple stressors have enormous potential to rattle this cage.

Histone Deacetylases (e.g. SIRT) – Histone Acetylases (p300/CBP) and Regulation of Chromatin Structure, Epigenetics and Gene Transcription

Intense interest in sirtuins (NAD^+-dependent deacetylases) emerged with the recognition that sirtuins (SIRT) regulate responses to dietary restriction associated with extended longevity. Actions of SIRT involve the forkhead (FOXO) transcription factor that is upregulated when Akt/PKB (i.e. PI3K pathway) is downregulated. Besides acting on chromatin histone proteins to inactivate gene transcription (via formation of heterochromatin), SIRT deacetylases critical regulatory factors including Ku70, p53, NFκB, PCG1-α1, and PPARγ (Finkel and Cohen, 2005; Frankel and Rogina, 2006). Sirtuins extend longevity of yeast, nematodes and *Drosophila* (Wood et al., 2004; Frankel and Rogina, 2006). Resveratrol and some other flavonoids upregulated SIRT, mimicking the DR response (Wood et al., 2004). Decreasing another deacetylase that is not NAD^+ dependent, upregulated SIRT and extended fly longevity (Frankel and Rogina, 2006).

Despite intense interest in SIRT there is little reciprocal gerontological discussion of acetylases that must be equally important in modulating longevity and the DR response. Indeed p300/CBP are important acetylases that are likely antagonists of SIRT1 and other HDAC. Like SIRT1, p300/CBP also modify important transcription factors like p53 and NFκB. Oxidative conditions may promote histone acetylation, perhaps acting via these pathways (Gilmour et al., 2003). Interestingly, the longevity of *Drosophila* was extended by ~40–50% by pharmacological inhibition of deacetylases which leads to increased histone acetylation (Kang et al., 2002). Feeding adult flies 4-phenylbutyrate (PBA), a histone deacetylase inhibitor, significantly extended longevity without impacting locomotor vigor, reproduction or stress resistance. Histone protein acetylation was globally elevated and gene expression was complexly altered (including upregulation of SOD, GSH-S-TR, Cytochrome P450, 3 chaperones. Downregulated genes included glyceraldehyde-3-phosphate dehydrogenase, NADH: ubiquinone reductase, cytochrome oxidase subunit

VIb, fatty acid synthetase and cyclin-dependent kinase (Kang et al., 2002). These authors suggest (like Droge, 2005), that extension of longevity may require optimal balance between expression and repression of various alternative gene sets regulated by changes in chromatin structure.

Compensation for environmental or genetic stress (e.g. toxins, temperature, mutations) may involve developmental, molecular, physiological, behavioural or maternal adjustments, even within isogenic inbred strains (Crawley, 1996; Gingrich and Hen, 2000). Chromatin can transmit heritable non-genetic variation via alterations in DNA methylation and associated histone protein structure. Such mechanisms can fix the lifetime metabolic character of young mice (Cooney et al., 2002; Jaenisch and Bird, 2003; Rollo, 2006) and may even suppress transgenic insertions via local induction of heterochromatin (Morgan et al., 1999). Richardson et al. (2006) suggest that exposure of neonatal mice to dieldrin may cause epigenetic fixation of elevated NURR1, a transcription factor that regulates dopamine transporters, resulting in greater susceptibility of the dopamine system to various toxins in later life.

Demethylating actions of radiation have some potential to release transposable elements or viruses that are locked down by embedding them in heterochromatin. Indeed, inhibition of histone deacetylases can induce expression of bovine leukaemia virus (Merezak et al., 2002). Release of transposons theoretically could explain some genomic instability induced by radiation.

Heat-Shock Proteins: Roles in Stress, Immunity, Development and Cell Function Relative to Radiation and Multiple Stressors

Chaperone proteins function in housekeeping activities including protein folding, protein transport across membranes, normal protein turnover, and transcriptional protein assembly. They also critically contribute to important signaling mechanisms, including those involved in cell senescence and cell death (Soti et al., 2003; Bagatell and Whitesell, 2004). Numerous heat-shock/chaperone proteins are upregulated as part of the generalized stress response which includes induction by heat-shock and oxidative conditions (e.g. HSP70) (Callahan et al., 2002). Fedoroff (2006) differentiates between the cytoplasmic stress response and the endoplasmic reticulum "unfolded protein response." The latter stress response is triggered by the presence of misfolded or unfolded proteins. This involves elevated Protein disulfide isomerase activity (which acts on disulfide bonds) and associated GSH depletion. The actions and effectiveness of HSPs determines cell function, recovery and survival. Thus, HSPs are critical players in a multiple stressor paradigm. HSPs increase xenobiotic resistance, including resistance to drugs, and many (e.g. HSP70 and HSP90) are strongly anti-apoptotic (Soti et al., 2003; Bagatell and Whitesell, 2004). HSP70 may be activated by glutathionylation at sensitive cysteine residues and the coordinating transcription factor HIF1 is directly regulated by redox (Fedoroff, 2006).

Profound effects of stress proteins on immune function have long been appreciated, but mechanisms have only resolved recently (Pennell, 2005). Gp96 (=glucose-regulated protein 94), a member of the HSP90 family, is a chaperone protein that occurs in the endoplasmic reticulum and cell membrane. It functions in binding and processing antigens and proteins and effectively presenting them to the immune system (e.g. Arnold-Schild et al., 1999; Gidalevitz et al., 2004; Binder and Srivastava, 2005; Srivastava, 2005; Biswas et al., 2006). A major source of proteins transported to the endoplasmic MHC system are appropriately-sized fragments derived from ubiquitin-proteosome function. These are transported into the endoplasmic reticulum by the "transporter associated with antigen processing" (TAP). A "presentasome" complex involving the proteosome, TAP and heat-shock proteins is suggested. MHC I molecules loaded with protein are then transported to the cell surface by way of the golgi (Castellino et al., 2000; Binder et al., 2001; Norbury et al., 2004). Gp96 elicits $CD8^+$ T-cell responses against its bound peptides, which requires access to the MHC cross-presentation pathway of antigen presenting cells (dendritic cells, macrophages, B lymphocytes). Neutrophils and monocytes that encounter gp96 that is shed following apoptosis bind the protein at the site of apoptosis (Radsak et al., 2003).

Antigen-presenting cells likely migrate to lymphatic organs to prime T-cells. This may involve release of HSP such as HSP70 (Kumaraguru et al., 2003). HSP may also exit normal cells and induce potent immunological effects (Asea et al., 2002). This can involve secretion of small membrane vesicles (exosomes) that are produced by many cell types. Many other proteins are also found in exosomes (Lancaster and Febbraio, 2005). Toll-like receptors (TLR) are implicated in many of these phenomena, and CD14 is a required link between HSPs and TLR Asea et al., 2002). Mobility of gp96 in the cell membrane may be regulated in part by cytokines (Stolpen et al., 1988). Radiation can induce an inflammatory environment, including up-regulation of heat-shock proteins (Friedman, 2002). Gp96 increases after irradiation in treatment of cervical cancer, and resistance to radiotherapy was associated with its expression (Kubota et al., 2005). This suggests that inhibition or suppression of this chaperone might enhance radiation effectiveness. Alternatively, radiation and vaccination with gp96 from viral-infected cells killed by radiation may synergistically inhibit tumor cell growth (Liu et al., 2005b).

Barley leaf shows large increases in the mRNA for a heat-shock protein of the HSP90 family after exposure to powdery mildew. The protein shows remarkable similarity to vertebrate glucose–regulated protein 94 (=gp96) (Walther-Larsen et al., 1993). As in mammals this protein is involved in immunological responses, including the hypersensitivity response of plants (which involves an NOX and closely resembles the mammalian bystander phenomenon).

Barley leaf mRNA showed strong increases in the mRNA of a heat-shock protein of the HSP90 family, gp94, in response to infection with powdery mildew. The encoded protein closely resembles vertebrate glucose-regulated protein 94 (=gp96) (Walther-Larsen et al., 1993).

Thus, the association of gp96, NOX and immunity is phylogenetically ancient. In mammals, HSP receptors include TLR, CD91 and scavenger receptors type A, LOX-1, CD94, CD40, CD36 and CD 14 receptors (Binder and Srivastava, 2004, 2005; Quintana and Cohen, 2005; Biswas et al., 2006; Warger et al., 2006). CD91 is considered the major sensor of danger (McBride et al., 2004), and high CD91 expression is the only known marker for HIV resistance (Stebbing et al., 2005). A role for CD91 was contested (Berwin et al., 2002), but the weight of the evidence strongly supports a role of CD91 as a primary receptor for gp96 (Binder and Srivastava, 2004; Srivastava, 2005). Peptides introduced to the cytosol were processed 100 times more efficiently (to major histocompatability complex I molecules) when bound by gp96 or HSP70 (as opposed to free peptides) (Binder et al., 2001). A cytokine function of gp96 was questioned based on the possibility that earlier experiments were performed with materials contaminated with microbial products (e.g. Tsan and Gao, 2004a,b). HSPs, in fact, may have little independent ability to activate the immune system but they bind diverse ligands (including polyliposaccharide) and greatly enhance detection and signaling of immunological receptors (Quintana and Cohen, 2005; Warger et al., 2006). Regardless, preparations from systems unlikely to have microbial contamination support some role of HSP to stimulate macrophages and dendritic cells (Quintana and Cohen, 2005). Thus, gp96 enhances immune responses in a wide variety of situations.

Remarkably, mice vaccinated with gp96 derived from tumor-, virus or bacteria-infected cells developed T-cellular immune responses with corresponding specificities (Rapp and Kaufmann, 2004; Demine and Walden, 2005). In some patients, tumor-derived gp96 peptide complexes strongly suppressed metastatic melanoma (Belli et al., 2002). Specific immunity to tumors induced by gp96 (and little such function in HSP90), may trace to the ATPase activity of gp96, which may be required to transfer proteins to acceptor molecules (Udono and Srivastava, 1994). Anti-tumor immunity provided by gp96 isolated from sarcoma cells was lost of the CD91 receptor was down regulated (Binder et al., 2004). Thus, gp96 is of great interest as a contributing component of vaccines (Liu et al., 2005b; Srivastava, 2005). Given a strong role in immunity and recognition of foreign elements, it is significant that gp96 is closely associated with the membrane-bound NAD(P)H oxidases that serve in immunological responses in mammals. A common denominator for mechanisms inducing HSP70 is oxidation of the cytosol. Altered protein structure by modification of cysteine residues has been proposed for HSP70 as this mechanism was also found in HSP33. Changes in activity of heat-shock proteins in response to oxidative conditions suggests that immunogenicity may also be redox sensitive (Callahan et al., 2002).

Cellular stress responses may be considered with respect to various compartments, and the endoplasmic reticulum is strongly targeted by stress because it is the site where protein folding essential to proper enzyme functions takes place. Disruption results in accumulation of misfolded proteins (Fathallah-Shaykh, 2005). Glioma cells that are resistant to oxidative stress show upregulation of a connected system of genes that includes gp96 and many associated with NOX activity (e.g. GSH-S transferase pi, peroxiredoxin 1, thioredoxin reductase 1, GADD34). Important connections in this system include Keap1, Nrf2, and ARE response elements (Fathallah-Shaykh, 2005). This process involves the actions of various chaperone proteins. Chaperone proteins rarely act alone but are normally associated with larger protein complexes (Bagatell and Whitesell, 2004). Thus, gp96 is closely associated with the NOX /oxidoreductase complex on mammalian cell membranes (Scarlett et al., 2005).

Robert et al. (1999) detected membrane surface gp96 on several types of cancer cells and on various immunocytes across vertebrate phylogenies. Surface expression of many HSP, including HSP 60, HSP70 and HSP90 are abundantly expressed in various cancers (Robert et al., 1999; Shin et al., 2003). Robert et al. (1999) suggest it acts as a "danger" signal. A possible wider distribution appears likely if gp96 is commonly associated with NOX. Given the association of NOX with responses to various pathogens and parasites (perhaps best explored in the plant hypersensitivity response) it seems possible that gp96 serves as a component of a sensing and recognition system? In plants, HSP90 members have indeed been added to the list of proteins involved in surveillance and resistance protein-triggered immunity (Schulze-Lefert, 2004). Most interest has focused on cytosolic HSP90.2. The developing paradigm envisions a complex that engages ATP-dependent modifications of a steroid protein, making it accessible to activating ligand.

Important clients of HSP90 include receptor tyrosine kinases (including IGF-1$_r$), SRC family kinases, serine/threonine kinases (including Akt/PKB), cell cycle G2 checkpoint kinases, steroid hormone receptors (glucocorticoid, androgens, estrogen, progesterone), and transcription factors (including p53, NFκB) (Bagatell and Whitesell, 2004). HSP90 also serves many other clients functioning during development. Consequently, deflection of HSP90 service to manage stress responses (i.e., protect proteins from misfolding) may de-stabilize development. HSP90 can serve as a buffer against even inherent genetic variation to provide greater canalization of the phenotype (Rutherford and Lindquist, 1998; Queitsch et al., 2002; Rutherford, 2003). Stress that impacts HSP90 function may consequently release hidden genetic variation that would facilitate evolutionary responses. It can also release expression of mutations that are otherwise silenced by chaperones (Soti et al., 2003). Functioning of HSP90 is dependent on ATP, so energy depletion, which is reliably associated with cellular stress, would also be destabilizing. Soti et al. (2003) describe the "protein homeostasis hypothesis" which proposes that cellular homeostasis depends on an appropriate match between levels of damaged proteins and other clients versus the complement of chaperones available to serve them. Multiple stressors can be expected to disrupt such balance.

Alternatively, heat-shock proteins may facilitate tumorigenesis by acting as a buffer against the genetic and functional instability associated with cancer (Bagatell and Whitesell, 2004). Under stress this could allow expression of genetic variation that would allow cellular adaptation of cancer lineages. This could extend to genetic instability induced by radiation or xenobiotics. Heavy metals also elicit heat-shock responses and this might be expected to be synergized by radiation (Bagatell and Whitesell, 2004). HSP90 is also of interest in cancer since it stabilizes clients that might be targets for therapy. In breast cancer these include mutant p53, estrogen receptor, and Akt (PKB) (Beliakoff and Whitesell, 2004). Regulatory mechanisms impacted by HSP90 extend to regulation of chromatin structure (Sangster et al., 2003; Sollars et al., 2003) suggesting consequences on release of silenced genes, transposons, viruses or epigenetic alterations.

Professional antigen-presenting cells (monocytes and dendrites) show specific binding of gp96 and specific receptor-mediated internalization of the HSP and bound proteins. Internalized protein colocalizes with surface MHC class I molecules (Arnold-Schild et al., 1999). Gp96 greatly increases the efficiency of presentation of associated peptides to T-cells. As little as 1–2 ng of peptides complexed to gp96 are sufficient to elicit a cytotoxic +independently of binding proteins, and this was associated with release of TNF-α and IFN-γ (Baker-LePain et al., 2004). The powerful role of gp96 in delivering proteins and mediating uptake by professional antigen-presenting cells, makes it especially important that stressed or otherwise defective cells that overexpress gp96 are effectively cleared by engulfing macrophages, since the contents of such cells represent a complex soup of bystander signals to nearby cells and the immune system. Gp96 (and HSP70) remains functional in vitro and complexes formed with peptides in cell-free serum are immunologically active with respect to generating anti-tumor and CD8$^+$ cytolytic T lymphocyte responses (Blachere et al., 1997;

Baker-LePain et al., 2004). In vivo, failure to clear such cells may mediate strong local inflammatory states with the possibility of self-perpetuating vicious feedback cycles. Heat-shock proteins (including HSP10, HSP70, HSP90, calneticulin and gp96) that are released by necrotic cell rupture, are recognized as "danger signals" that may promote inflammation and contribute to bystander cell death. HSP in serum promote maturation of dendritic cells, activation of NF-κB, and subsequent release of inflammatory cytokines (Basu et al., 2000; Soti et al., 2003).

The potential importance of such processes is likely to be overlooked in cell culture approaches to phenomenon like the bystander effect. Indeed, there appear to be two broad potential classes of bystander mechanisms. Activation of the immune system can impact organ to organismal-level responses, particularly via cytokine feedback to central regulation. An endocrinological triumvirate has been proposed as the key organismal regulatory framework. This involves the antagonistic regulation of the stress hormone (waking) and growth hormone (sleeping) axes largely integrated in the hypothalamus, and feedback of status in the periphery from what amounts to a distributed endocrine axis.... the immune system. The immune system is the only other tissue that produces nearly all hypothalamic and pituitary regulatory peptides. The fact that radiation induces the stress hormone axis suggests that peripheral free-radical processes may be sensed and relayed to the hypothalamus (see Rollo, 2002). Even in a cell-free context, gp96 can bind proteins can possibly enhance immunological recognition and activation. Radiation or xenobiotic impacts that overwhelm cell clearance mechanisms could induce a secondary immunological response, mediated partially by release of gp96 and other chaperones that constitute "danger signals" (Soti et al., 2003). This extends to radionucleotide-induced bystander effects as well (Xue et al., 2002). Such a mechanism qualifies as a type of bystander response. Plant responses to pathogens also likely trace to diverse foreign molecule recognition by ligand–receptor interactions and transduction of recognition to responses, including the hypersensitive response (Nimchuk et al., 2003).

Alternatively, local mechanisms that activate associated bystander cells surrounding sites of impact may be relatively independent of the immune system. Thus, P450 monooxidase enzymes involved in detoxification can actively generate free radicals in cell-free media (Clejan and Cederbaum, 1992) and would likely synergize actions of active NOX components in conveying oxidative bystander signals. ... in particular the activation of NOX on cell membranes of intact bystander cells (Rollo, 2006). A crucial property of the NOX is that it is activated by free radicals and consequently can maintain free-radical generation via an auto-stimulatory feedback loop (Cai, 2005). Chaperones may also convey direct signals to surrounding cells that sensitize stress and cell death programs (Soti et al., 2003). A remarkable phenomenon resembling a bystander effect is associated with cells expressing a senescent phenotype. The cell senescence phenotype can be derived by numerous mechanisms including free-radical stress (i.e., related to radiation, organismal aging or xenobiotics). Senescent fibroblasts secrete numerous materials including cytokines, growth factors, degradative enzymes and extracellular matrix. Some factors secreted from senescent cells stimulate the growth of premalignant and malignant (but not normal) epithelial cells (Krtolica et al., 2001). Thus, radiation, pollutants and aging may predispose to both transformation and senescence, and senescence may not provide a defense against cancers as such cells accumulate with age.

The Phosphatidylserine "EAT ME" Signal and Clearance of Defective Cells

A critical aspect of gp96 (and other chaperones) related to radiation, is that heat-shock proteins and their bound client proteins are released or shed from apoptotic cells (Basu et al., 2000). Shedding of active chemical species extends to other important factors such as protein kinase C, phosphatidylserine, arachidonic acid, ATP, cytochrome c, lipid peroxidation products and oxidized sterols. Hydrogen peroxide generated by apoptotic cells is sufficient to induce apoptosis in surrounding cells (Milan et al., 1997; Reznikov et al., 2000; Cusato et al., 2003) and has been suggested as a major bystander candidate (Pletjushkina et al., 2005, 2006). Extracellular ATP can promote cell death acting via purinergic receptors and Ca^{2+} elevations. Ca^{2+} itself is implicated in bystander impacts (Budd and Lipton, 1998; Vinken et al., 2006). Growth factors and cytokines released into serum also have particularly potent inflammatory actions (Zheng et al., 1991; Proskuryakov et al., 2002). Uric acid produced from catabolism of DNA and RNA from dying cells is also a very strong danger signal and immunological activator (Shi et al., 2003). Excitotoxic neurotransmitters can also stress neurons adjacent to those undergoing cell rupture. There is clearly considerable complexity. Add to this that Bauer (2002) distinguishes bystander impacts mediated by gap-junctions versus inter-cellular signals that did not require gap junctions.

Removal of apoptotic cells by phagocytic engulfment may serve to prevent release of cellular constituents, including self antigens that could induce autoimmune responses (Proskuryakov et al., 2002; Fadeel, 2003; Fadeel and Xue, 2006). Clearance of apoptotic cells may induce little or an altered immune response compared to cell rupture (Mevorach et al., 1998). Thus, clearance constitutes an essential aspect of tissue homeostasis which usually highlights only mitosis and apoptosis. Indeed accelerated apoptosis and impaired clearance of apoptotic material exacerbates autoimmune responses in mice (Denny et al., 2006). Furthermore, strong exposure to irradiation-induced apoptotic cells generates auto-antibody production, including anti-cardiolipin and anti-double strand DNA antibodies (Mevorach et al., 1998).

Cardiolipin is a critical and predominant constituent of the mitochondrial membrane. Its oxidation during apoptosis is important with respect to associated cytochrome c (Tyurina et al., 2006). In T-cell lineages, ionizing radiation induced apoptosis in association with elevated levels of mRNA and the Apo2 ligand (also called TRAIL). This also activated the AposL death receptor 5 (also called KILLER) associated with apoptosis and externalization of phosphatidylserine. Bax, induced by p53 is also a mediator of apoptosis in irradiated cells (Gong and Almasan, 2000).

Impaired engulfment of apoptotic cells was found in mice with defective lactadherin (=milk fat globule epidermal growth factor 8), a factor produced by activated macrophages that mediates phosphatidylserine recognition. Injection of the mutated protein also induced autoimmunity, probably by masking recognition of phosphatidylserine (Asano et al., 2004; Fadeel and Xue, 2006). Autoimmunity was also obtained in mice overexpressing gp96 on the cell surface, and this involved the MyD88 adaptor protein crucial to TLR signaling (Liu et al., 2003, 2005a). Massive apoptosis was detected in lymphoid organs in a mouse model of multiple sclerosis (with depletion of B and T cells). Injection of apoptotic thymocytes exacerbated demyelination and disease progression (Tsunoda et al., 2005). Activation of NOX by phorbol myristate in HL-60 cells generated SOR, depleted intracellular GSH and peroxidized all three major classes of membrane phospholipid (phosphatidylcholine, phosphatidyl-ethanolamine, and phosphatidylserine). Radiation induced similar changes in rodent brain (Richards and Budinger, 1988). Enteropathogenic infection by *Escherichia coli* also induces externalization of both phosphatidylserine and PKC (crucially involved in NAD(P)H oxidase activation) (Fadeel and Xue, 2006). Activation of PKC involves movement to the cell surface and binding with phosphatidylserine. PKC externalization was also induced by phorbol myristate acetate and ultraviolet light (~25% of cell content) (Crane and Vezina, 2005). Both of these factors also induce NAD(P)H oxidase. Phosphatidylserine also induces NOX, in cell-free media, although its oxidation state was not considered (McPhail et al., 1993).

Apoptosis is associated with externalization of oxidized phosphatidylserine that serves as a signal for ingestion of cells by macrophages (Arroyo et al., 2002). HSP70 and Hsc70 interact with phosphatidylserine, and accelerate apoptosis, an effect exacerbated by ATP. HSP are also able to induce ion channels (Arispe et al., 2004) which could contribute to NADPH channel associations or apoptotic processes. Thus NOX is associated with both apoptotic and clearance functions. Extracellular ATP can induce thymocyte apoptosis via purinoceptor activation. This included early phosphatidylserine externalization, mediated by activation of the cationic PsX7 (=P2Z) channel. Phosphatidylserine movement required ATP-induced Ca^{2+} and/or Na^+ influx (Courageot et al., 2004). Apoptotic cells express elevated levels of

oxidized phospholipids that function as immunological and pro-inflamma-
tory signals. Fluorescence-conjugated annexin V specifically binds externalized phosphatidylserine, making it a reliable biomarker for apoptosis (Fadeel
and Xue, 2006).

Apoptotic cells induced production of T helper cell Th1 and Th2 cytokines,
and induced monocyte adhesion in endothelial cells. The latter was inhibited
by an antibody specific to oxidized phosphatidylcholine (Chang et al., 2004).
Lysophosphatidylcholine can act as a chemoattractant that guides macrophages to sites of apoptotic lesions (Fadeel and Xue, 2006). Oxidized
phospholipids induced monocyte adhesion in endothelial cells via MAPK
pathways and activation of cytosolic phospholipase A_2 and 12- lipoxygenase
(Huber et al., 2006). Externalization and oxidation of phosphatidylserine in
the cell membrane is a critical signal to macrophages for engulfing apoptotic
cells (Kagan et al., 2003). Oxidized phosphatidylserine externalized during
apoptosis is recognized by specific macrophage receptors. Selective oxidation
of phosphatidylserine precedes externalization and oxidation is required for
engulfment of apoptotic cells. H_2O_2 was effective in oxidation and externalization of phosphatidylserine (Tyurina et al., 2004). Lipid antioxidants (Tyurina
et al., 2004), NOX inhibitors, superoxide dismutase and catalase protected
all phospholipids from oxidation and externalization of phosphatidylserine
(Arroyo et al., 2002). Radiation induces caspase 3 and caspase 8 (Gong and
Almasan, 2000). Caspase 3 is redox-regulated and appears pivotal in determining
whether cell death proceeds by apoptosis or necrosis. Inhibition of caspase
3 yields necrosis and failure to externalize oxidized phosphatidylserine. Both
forms of cell death may be elicited by radiation (Coelho et al., 2000). A complication is that NOX ROS production could inhibit activity of redox-sensitive
caspase 3 (Arroyo et al., 2002).

Despite considerable evidence that failure to clear apoptotic cells may
induce autoimmunity, and that radiation can exacerbate this process, radiation
is not reliably associated with autoimmune responses. The immune system is
particularly impacted by radiation, such that various autoimmune diseases
can actually benefit from irradiation. Suppression of antigen-presenting
dendritic cells and inhibition of the proteosome (which produces peptide
fragments of appropriate size for major histocompatability complex processing)
are crucial mechanisms of radiation-induced hypo-immunity (McBride
et al., 2004). Dosage is likely a crucial factor. Regardless, cell death, clearance
and immune responses remain potentially critical for a radiation and a multi-
stressor paradigm because high doses or interactions that are additive or multiplicative may shift apoptosis and effective clearance to necrosis. Xenobiotic
impacts synergized by relatively low levels of radiation may derive massive
cell death without the immuno-suppressive impacts of high dose radiation.
This would likely result in clearance failures, immunological infiltration and

inflammation. The enhanced induction of immunological sensitivity (e.g. by gp96) extends a linkage between multiple stressors (including xenobiotics) to allergenic and autoimmune responses.

Multiple Drug Resistance Proteins (MDRP)

Multiple drug resistance proteins (members of the ATP-binding cassette (ABC) transporter superfamily) such as multiple drug resistance protein-1 (MDR-1 gene, P-glycoprotein, Pgp) and multiple resistance protein-1 (MRP-1) export cellular detoxification products, xenobiotics and various conjugates of waste products or toxins. Powell and Abraham (1993) suggest they also excrete growth factors that can mediate strong immunological and stress responses (e.g. FGF, TNF, TGF-α). Steroid hormones may also be exported (Tatsuta et al., 1992). MDR-1 is transcriptionally upregulated by heat stress, heavy metals, drug treatment and UV radiation (Toone et al., 1998). At least 50 family members have been identified (Minier et al., 1999). They can also inhibit apoptosis induced by TNF, serum withdrawal, Fas ligand, and UV radiation (Baker and El-Osta, 2004). MDRP are intensely studied because they can export antibiotics and cancer drugs, resulting in resistance to diverse chemotherapeutic agents. They can also mediate pesticide resistance (Minier et al., 1999). Thus, most medical research aims to inhibit MDRP activity (e.g. Tan et al., 2000). Although MDRPs may be detrimental to cancer therapy and pest control, their activities would be highly advantageous for resistance to multiple pollutants. Unlike MDR-1, MRP-1 mainly transports anionic Phase II-conjugates, and this is highly dependent on high intracellular GSH levels. There are at least 6 MRP homologues (Renes et al., 2000). Multiple drug resistance proteins, in conjunction with p450 enzyme activity, may provide xenobiotic barriers in the kidney, buccal cavity, gastrointestinal tract, endothelium, liver, placenta, blood brain barrier (where they may exclude antidepressants) and choroid plexus (Tatusta et al., 1992; Minier et al., 1999; Renes et al., 2000; Uhr et al., 2000; Miller et al., 2002; Leggas et al., 2004). MDR1 may play a role in host-bacterial interactions in the intestinal epithelium (Ho et al., 2003).

Many elements exported by MDRPs are conjugates of GSH (catalyzed by glutathione-S-transferases), or even GSH itself, indicating a crucial role in detoxification and function of the GSH system. High levels of GSH can downregulate expression of MRP-1, although GSH depletion is not necessarily upregulating (Renes et al., 2000). Zaman et al. (1995) found that GSH depletion impacted MRP-1 more than MDR-1. Renes et al. (2000) suggests that a critical level of GSH depletion may be necessary before MRP-1 transcription is affected. This is consistent with my argument that the lack of obvious dosage response for low dose radiation is that it may require >50% depletion of GSH to reach a critical threshold for redox vulnerability (Rollo, 2006). In addition, export of oxidized glutathione (GSSG) by MRP-1 (Leier et al., 1996) suggests a mechanism for regulating intracellular redox balance.

Export of ATP by MDRP has potentially profound impacts, including control of endothelial function, apoptosis, the cell cycle, signal transduction, thiol protection and calcium levels. MDRP transport of ATP may be an initial response to radiation. ATP was promptly released by 0.1 Gy of radiation (Powell and Abraham, 1993). Extra-cellular ATP fuels diverse cell membrane and extra-cellular processes, including ion channel and NOX activities. ATP can act on a variety of purine and adenosine receptors (e.g. P1, P2, A1, A2) which may have bystander impacts or derive intracellular auto-stimulatory feedback loops (Powell and Abraham, 1993). The adenosine neurotransmitter system has widespread important impacts, especially in brain. Further, adenosine receptors A_{2B} interact with melatonin via adenylate cyclase to entrain expression of the circadian clock gene Period, thus coordinating the circadian clocks of peripheral tissues like the pituitary with the hypothalamic suprachiasmatic nucleus (von Gall et al., 2002). The circadian rhythm may constitute a large amplitude electroplasmic cycle.

Overexpression of COX-2 (which generates prostaglandins from arachidonic acid) enhances expression of MDR-1 (Sorokin, 2004). NOX (also activated by arachidonic acid) also upregulates MDR-1, as does radiation and heat shock (Minier et al., 1999). Alternatively, MDR is associated with increased expression and activity of CYP3A (a major drug detoxifying cytochrome) (Schuetz et al., 2000; Baron et al., 2001). The association with the membrane NOX is noteworthy as this oxidase is upregulated by radiation (Narayanan et al., 1997; Azzam et al., 2002) and regulates MAPK signaling. Moreover, the oxidase may be part of a larger NADPH oxidoreductase complex that functions to maintaining intracellular ionic and oxidative homeostasis via regulation of general cation and K^+ channels (Scarlett et al., 2005). The MRP-1 gene contains an antioxidant response element and multiple drug resistance is strongly upregulated by free radicals and oxidation (Renes et al., 2000; Sorokin, 2004). Mechanisms generating cellular free radicals induce gene expression for MRP-1 and the rate-limiting enzyme for GSH synthesis, γ-glutamylcysteine (Renes et al., 2000). MRP-1 modulates dauer (diapausing larvae) formation in *C. elegans*, a mechanism associated with modulation of longevity and linked to the insulin/IGF-1 PI3K pathway (Yabe et al., 2005). This pathway has been highlighted in modulation of aging and stress resistance across broad phylogenies spanning yeast, nematodes, insects and vertebrates.

Although MRP-1 is upregulated by free radicals, it does not directly provide protection from radiation or other sources of free-radical damage.

However, drug resistance transporters can export materials that would otherwise exacerbate free-radical stress. A crucial question then is to what extent low-level radiation might upregulate MDR-1/MRP-1 and the shedding of pollutants, and at what dosages possible benefits exceed the possible costs induced by radiation exposure. Demonstrated expression of MDRPs in filtering species such as sponges and some bivalves (gill) suggests a sensitive biomarker for general environmental pollution (Minier et al., 1999) and possibly associated radiation background. The antibody for human P-glycoprotein is effective across broad phylogenies. Interestingly, heavy metal export is one target of these transporters, and export of such materials could reduce free-radical damage (including that induced by radiation). It should also be noted that one client of MRP-1 is leukotriene C_4, a glutathione-conjugated organic anion that functions as an inflammatory mediator (Leier et al., 1996; Renes et al., 2000) and that may be a candidate bystander signal. Alternatively, leukotriene B_4 stimulates release of arachidonic acid which promotes free-radical generation by NOX. This also occurs in cell-free serum (Brash, 2001; Cherny et al., 2001). In fact, activated phagocytes export arachidonic acid, and this can cause free-radical generation on other cells (DeCoursey, 2002). Thus, arachidonic acid qualifies as a bona fide bystander signal in its own right. Arachidonic acid also impacts a number of ion channels (Brash, 2001). TNF-α also upregulates MRP-1, possibly via its ability to generate free radicals (Renes et al., 2000). TNF is also upregulated by radiation and NAD(P)H: quinone oxidoreductase-1 (NOQ1) is a critical component of the TNF signaling cascade to NF-κB. NQO1 is also known to be induced by xenobiotics, oxidants and radiation (McBride et al., 2004; Ahn et al., 2006).

The promoter region of MDR-1 contains a CpG methylation sequence, and the gene is importantly regulated by epigenetic alterations of chromatin structure. It is possible that this gene may be included in early-life epigenetic modifications that effect life-long changes in stress responses. Chromatin regulation involves methyl-CpG-binding protein-2 (MeCP2) and its recruitment of the corepressors, HDAC1 and HDAC2. Gene regulation requires both demethylation of DNA and hyperacetylation of associated histones (Baker and El-Osta, 2004). Since radiation is generally associated with demethylation, this could represent one mechanism of radiation-induced MDR-1 activation.

Toll-like Receptors

TLRs underpin the ability of innate immunity to discriminate between highly diverse pathogens and self. TLRs activate macrophages in response to pathogens which in turn present pathogen-derived peptides to T-helper cells. Of relevance to the free-radical elevations associated with radiation and p450 xenobiotic detoxification systems, TLR expression (and activation of NF-κB) can also be upregulated by free radicals, including those generated by NOX (Fan et al., 2003; Asehnoune et al., 2004; Qureshi et al., 2006). Expression of TLR4 even confers resistance to hyperoxia-induced lung injury and apoptosis (Zhang et al., 2005; Qureshi et al., 2006). The mechanism of hyperoxia resistance involved the ability to express Bcl-2, a critical element in PI3K-mediated resistance to apoptosis. Lipopolysaccharide induction of free radicals and activation of NF-κB requires direct interaction of TLR4 and NOX4 (Park et al., 2004a). Loss of the TLR signaling element, MyD88, diminished NOX function and killing of gram-negative bacteria (Laroux et al., 2005). TLR4 may also be involved in susceptibility to ozone-induced lung injury (Kleeberger et al., 2000).

TLRs can also act cooperatively in clusters or co-activate via other microbial recognition receptors. Numerous immunocytes employ distinct arrays of TLRs (e.g. neutrophils, natural killer cells, mast cells, B-cells, eosinophils and monocytes). Structure and signaling of TLR closely resemble the IL-1 receptor, activating TRAF6 and NF-κB (Underhill, 2003). Until the 1990s how organisms recognized pathogens was virtually unknown. Since 1991 at least 13 mammalian Toll paralogues (11 expressed in humans) have been found. Each responds to different phylogenetically invariant components of microbial lineages and may have cytosolic or membrane-bound distributions. Like the NOX, TLRs are particularly associated with tissues and locations which face potential pathogenic contacts, including vascular endothelium, bronchial, gastrointestinal and urogenital epithelium and blood-brain barrier (Hopkins and Sriskandan, 2005). Interestingly, most of these are also regions expressing multiple drug resistance proteins. TLR are phylogenetically conserved across plants to insects to vertebrates (Lescot et al., 2004).

TLR2 and TLR4 were implicated as gp96 receptors on immunocytes (Baker-Lepain et al., 2004). HSP, including gp96, complex with TLRs and cooperate in binding and signaling diverse ligands (Underhill, 2003). Earlier concerns that microbial contaminants may have artifactually mediated the apparent gp96 interaction with TLRs have been offset by careful attention to uncontaminated preparations. HSP may have little independent immunogenicity, but even low amounts of gp96 greatly (up to 100-fold) amplified bacterial product-induced TLR-mediated signaling relevant to activation of T cells and dendritic cells (Warger et al., 2006). Membrane expressed TLR2 and TLR4 bind HSP (HSP60, HSP70, HSP90/gp96), suggesting they may serve as immunocyte receptors for HSP. This also suggests the possibility that HSP may be closely connected to surveillance systems more generally. With respect to bacterial lipopolysaccharide, this may be first bound by CD14 associated with TRL4, and then transferred to a HSP70-HSP90 complex (Underhill, 2003).

One outcome of TLR activation is engagement of the NADPH oxidative burst, and ultimately, adaptive immunity (Hopkins and Sriskandan, 2005). Lipopolysaccharide from gram negative bacteria mediates direct activation of NOX4 via TLR4. This is known to involve Rac1. Lipopolysaccharide from *Helicobacter pylori* also induced NOX1 (~gp91[phox]) activity and mRNA via a PI3K pathway activation of Rac1 in gastric mucosal cells. Expression of NOXO1 (~p47[phox]) mRNA was also enhanced. Flagella of *Salmonella* also activated NOX1. Detection of lipopolysaccharide and flagella involved TLR4 and TLR5 respectively (Geiszt and Leto, 2004; Kawahara et al., 2005). Signaling of TLR4 (and many other TLRs) via MyD88 to IL-1R associated kinase to TRAF6 (tumor necrosis factor-associated factor-6) can activate NF-κB (Tsan and Gao, 2004a; Park et al., 2004b). NF-κB activation may also be mediated by free radicals generated by NOX4, since H_2O_2 activates NF-κB (Park et al., 2004a, b). In the colon epithelium, TLR5 mediated free-radical production via NOX1 in response to stimulation by flagellin from *Salmonella enteritidis* or by lipopolysaccharide from *Helicobacter pylori*. In addition to activation of NF-κB, TLRs induce AP-1, TNF-α, IFN-β, IL-8 (involved in chemoattraction of immunocytes) and TGF-β-activated kinase 1 (Yamamoto et al., 2003; Kawahara et al., 2004; Kawai and Akira, 2006).

The IL-1 receptor associated kinase is a key element in TLR signaling, linking receptor activation to TRAF6 and subsequent activation of MAPK (e.g. ERK1/2, p38), NF-κB, TNF-α, IL-1and IL-8. A complex containing HSP90 and Cdc37 regulate the folding and possible level of activation of IL-1R associated kinase, providing a further linkage of HSP90 family members to immunological recognition. In fact, heat-shock proteins are also involved in regulation of many important signaling elements including Raf-1, p53, and IκB (De Nardo et al., 2005). Overall, this discussion identifies complex interactions among NOX, Rac, TLR, HSP90, and gp96. Although TLRs recognize specific pathogen derivatives, they may also respond to other factors. They may monitor endogenous damage, including signals generated by radiation (McBride et al., 2004). TLR9 is activated by unmethylated CpG didioxynucleotides. Mammals have few such sequences, and most are methylated, providing discrimination of bacterial and mammalian host DNA (Hemmi et al., 2000; Takeshita et al., 2001; Chuang et al., 2002).

Although such signals may be associated with recognition of pathogens, it is interesting that demethylation of mammalian CpG islands may release endogenous viral activity, and radiation (and cancer) are generally associated with demethylation. Growing evidence implicates inappropriate responses to host DNA in autoimmune diseases. T cells from lupus patients are hypomethylated, and mice receiving CD4 + T cells that have been chemically demethylated developed a lupus-like condition (Blank and Shoenfeld, 2005; Kaplan et al., 2005). More than 100 genes sensitive to methylation status have been detected. Upregulation of such genes (particularly performin) may contribute to autoimmunity by altering the regulation of apoptosis and clearance (Kaplan et al., 2005). HSP90 is also involved in signaling induced by CpG motifs (Quintana aand Cohen, 2005). Immunization of mice with DNA from activated lymphocytes induced antibodies against double-stranded DNA, and others characteristic of lupus (Qiao et al., 2005).

TLR9, which is the well-characterized DNA recognition receptor is high-lighted in serious conditions such as lupus and multiple sclerosis (Anders, 2005; Means et al., 2005; Prinz et al., 2006). Strong signaling via TLR9 can induce apoptosis and even organ failure (Yi et al., 2006). For lupus, DNA from nucleosomes may be most critical, highlighting the need for efficient clearance of defective cells (Radic et al., 2004; Blank and Shoenfeld, 2005; Ishii and Akira, 2005). Remarkably, nucleosomes may be released from the nucleus to the cell surface during apoptosis. The presence of nucleosome core particles in apoptotic membrane blebs may provide an additional signal for phagocytosis (Radic et al., 2004). Immunization of mice with nucleo-somes triggers Th1-type autoimmune T cells (Mohan et al., 1993; Blank and Shoenfeld, 2005). Other host factors may also serve as endogenous ligands, particularly those associated with cell damage or dysfunction (e.g. heat-shock proteins, fibrinogen, hyaluronan, heparin sulfate and mRNA) (Tsan and Gao, 2004a; De Nardo et al., 2005). An important question, however, is whether some such findings are confounded by the presence pathogen contaminants (Tsan and Gao, 2004a,b).

TLR appear to be involved in recognition of DNA damage and NF-κB-associated responses. The cancer-treatment drug "taxol" activates NF-κB and may be a ligand for TLR4. This receptor may also be required to medi-ate NF-κB activation in response to chemically mediated DNA damage, possibly linking radiation-induced DNA damage and NF-κB activation to TLR (Park et al., 2004a,b). In this regard, low dose radiation (~2Gy/d) was beneficial in follicular lymphoma, and besides upregulation of p53, elevated expression of TLR4 was part of the "immune signature." Induction of specific immune modulators by radiation was suggested to likely impact death and clearance of tumor cells (Knoops et al., 2005).

Specific immunity to tumors by gp96 (and little such function in HSP90), may trace to the ATPase activity of gp96, which may be required to transfer proteins to acceptor molecules (Udono and Srivastava, 1994).

Conclusions

Despite great complexity, some unity amidst diversity can be found in considerations of diverse stressors, including aging, radiation, xenobiotics and many others. This is possible because stressors share a strong basis in free-radical-mediated regulation and coordination of biological stress response functions. A unifying perspective highlighting integrated electroplasmic cycling brings together diverse mechanisms associated with free-radical processes, growth and development, signal transduction, ion channels, chaperone proteins, immunological activities, clocks, and transcription. Regardless, there remains bewildering complexity, and the intricacies of regulatory mechanisms continue to unfold at a great pace. There is a serious need for a computer modeling and systems approach to provide a concrete focus for research into regulatory circuitry, and some ability to predict the consequences of multiple stressors on ecosystems and the people who inhabit them.

References

Ahn, K.S., Sethi, G., Jain, A.K., Jaiswal, A.K., and Aggarwal, B.B. 2006. Genetic deletion of NAD(P)H:quinone oxidoreductase 1 abrogates activation of nuclear factor-κB, IκBα kinase, C-Jun N-terminal kinase, Akt, p38, and p44/42 mitogen-activated protein kinases and potentiates apoptosis. J. Biol. Chem. 281: 19798–808.

Anders, H.J. 2005. A Toll for lupus. Lupus 14: 417–422.

Arispe, N., Doh, M., Simakova, O., Kurganov, B., and De Maio, A. 2004. Hsc70 and Hsp70 interact with phosphatidylserine on the surface of PC12 cells resulting in decrease of viability. FASEB J. 18: 1636–645.

Arnold-Schild, D., Hanau, D., Spehner, D., Schmid, C., Rammensee, H.G., de la Salle, H., and Schild, H. 1999. Cutting edge: Receptor-mediated endocytosis of heat shock proteins by professional antigen-presenting cells. J. Immunol. 162: 3757–760.

Arroyo, A., Modriansḱ, M., Serinkan, F.B., Bello, R.I., Matsura, T., Jiang, J., Tyurin, V.A., Tyurina, Y.Y., Fadeel, B., and Kagan, V.E. 2002. NADPH oxidase-dependent oxidation and externalization of phosphatidylserine during apoptosis in Me$_2$SO-differentiated HL-60 cells: role in phagocytic clearance. J. Biol. Chem. 277: 49965–975.

Asano, K., Miwa, M., Miwa, K., Hanayama, R., Nagase, H., Nagata, S., and Tanaka, M. 2004. Masking of phosphatidylserine inhibits cell engulfment and induces autoantibody production in mice. J. Exp. Med. 200: 459–467.

Asea, A., Rehli, M., Kabingu, E., Boch, J.A., Bare, O., Auron, P.E., Stevenson, M.A., and Calderwood, S.K. 2002. Novel signal transduction pathway utilized by extracellular HSP70. J. Biol. Chem. 277: 15028–5034.

Asehnoune, K., Strassheim, D., Mitra, S., Kim, J.Y., and Abraham, E. 2004. Involvement of reactive oxygen species in Toll-like receptor 4-dependent activation of NF-κB. J. Immunol. 172: 2522–529.

Azzam, E.I., de Toledo, S.M., Spitz, D.R., and Little, J.B. 2002. Oxidative metabolism modulates signal transduction and micronucleus formation in bystander cells from α-particle-irradiated normal human fibroblast cultures. Cancer Res. 62: 5436–442.

Bagatell, R. and Whitesell, L. 2004. Altered Hsp90 function in cancer: a unique therapeutic opportunity. Mol. Cancer Ther. 3: 1021–1030.

Baker, E.K. and El-Osta, A. 2004. Epigenetics – normal control and deregulation in cancer: MDR1, chemotherapy and chromatin remodeling. Cancer Biol. Ther. 3: 819–824.

Baker-LePain, J.C., Sarzotti, M., and Niccitta, C.V. 2004. Glucose-regulated protein 94/glycoprotein 96 elicits bystander activation of CD4 + T cell Th1 cytokine production in vivo. J. Immunol. 172: 4195–203.

Balaban, R.S., Nemoto, S., and Finkel, T. 2005. Mitochondria, oxidants and aging. Cell 120: 483–495.

Baron, J.M., Goh, L.B., Yao, D., Wolf, R., and Friedberg, T. 2001. Modulation of P450 CYP3A4-dependent metabolism by P-glycoprotein: implications for P450 phenotyping. J. Pharmacol. Exp. Ther. 296: 351–358.

Basu, S., Binder, R.J., Suto, R., Anderson, K.M., and Srivastava, P.K. 2000. Necrotic but not apoptotic cell death releases heat shock proteins, which deliver a partial maturation signal to dendritic cells and activate the NF-κB pathway. Int. Immunol. 12: 1539–546.

Bauer, G. 2002. Signaling and proapoptotic functions of transformed cell-derived reactive oxygen species. Prostagl. Leukot. Essen. Fatty Acids 66: 41–56.

Beliakoff, J. and Whitesell, L. 2004. Hsp90: an emerging target for breast cancer therapy. Anti-Cancer Drugs 15: 651–662.

Belli, F., Testori, A., Rivoltini, L., Maio, M., Andreola, G., Sertoli, M.R., Gallino, G., Piris, A., Cattelan, A., Lazzari, I., Carrabba, M., Scita, G., Santantonio, C., Pilla, L., Tragni, G., Lombardo, C., Arienti, F., Marchiano, A., Queirolo, P., Bertolini, F., Cova, A., Lamaj, E., Ascani, L., Camerini, R., Corsi, M., Cascinelli, N., Lewis, J.J., Srivastava, P., and Parmiani, G. 2002. Vaccination of metastatic melanoma patients with autologous tumor-derived heat-shock protein gp96-peptide complexes: clinical and immunologic findings. J. Clin. Oncol. 20: 4169–180.

Berwin, B., Hart, J.P., Pizzo, S.V., and Nicchitta, C.V. 2002. CD91-independent cross-presentation of GRP94(gp96)-associated peptides. J. Immunol. 168: 4282–286.

Binder, R.J. and Srivastava, P.K. 2004. Essential rosle of CD91 in re-presentation of gp96-chaperoned peptides. PNAS 101: 6128–133.

Binder, R.J. and Srivastava, P.K. 2005. Peptides chaperoned by heat-shock proteins are a necessary and sufficient source of antigen in the cross-priming of CD8+ T cells. Nat. Immunol. 6: 593–599.

Binder, R.J., Blachere, N.E., and Srivastava, P.K. 2001. Heat shock protein-chaperoned peptides but not free peptides introduced into the cytosol are presented efficiently by major histocompatability complex I molecules. J. Biol. Chem. 276: 17163–171.

Biswas, C., Sriram, U., Ciric, B., Ostrovsky, O., Gallucci, S., and Argon, Y. 2006. The N-terminal fragment of GRP94 is sufficient for peptide presentation via professional antigen-presenting cells. Intern. Immunol. 18:1147–157.

Blachere, N.E., Li, Z., Chandawarkar, R.Y., Suto, R., Jaikaria, N.S., Basu, S., Udono, H., and Srivastava, P.K., 1997. Heat shock protein-peptide complexes, reconstituted in vitro, elicit peptide-specific cytotoxic T lymphocyte response and tumor immunity. J. Exp. Med. 186: 1315–322.

Blank, M. and Shoenfeld, Y. 2005. Experimental models of systemic lupus ertythematosus: anti-dsDNA in murine lupus. Rheumatology 44: 1086–1089.

Brash, A.R. 2001. Arachidonic acid as a bioactive molecule. J. Clin. Invest. 107: 1339–345.

Brash, D.E. and Havre, P.A. 2002. New careers for antioxidants. PNAS 99: 13969–971.

Brownawell, A.M., Kops, G.J.P.L., Macara, I.G., and Burgering, B.M.T. 2001. Inhibition of nuclear import by protein kinase B (Akt) regulates the subcellular distribution and activity of the forkhead transcription factor AFX. Mol. Cell. Biol. 21: 3534–546.

Brunet, A., Sweeney, L.B., Sturgill, J.F., Chua, K.F., Greer, P.L., Lin, Y., Tran, H., Ross, S.E., Mostoslavsky, R., Cohen, H.Y., Hu, L.S., Cheng, H.L., Jedrychowski, M.P., Gygi, S. P., Sinclair, D.A., Alt, F.W., and Greenberg, M.E. 2004. Stress-dependent regulation of FOXO transcription factors by the SIRT1 deacetylase. Science 303: 2011–2015.

Budd, S.L. and Lipton, S.A. 1998. Calcium tsunamis: do astrocytes transmit cell death messages via gap junctions during ischemia? Nat. Neurosci. 1: 431–432.

Buzek, J., Latonen, L., Kuri, S., Peltonen, K., and Laiho, M. 2002. Redox state of tumor suppressor p53 regulates its sequence-specific DNA binding in DNA-damaged cells by cysteine 277. Nucleic Acids Res. 30: 2340–348.

Cai, H. 2005. NAD(P)H oxidase-dependent self-propagation of hydrogen peroxide and vascular disease. Circ. Res. 96: 818–822.

Callahan, M.K., Chaillot, D., Jacquin, C., Clark, P.R., and Menoret, A. 2002. Differential acquisition of antigenic peptides by Hsp70 and Hsc70 under oxidative conditions. J. Biol. Chem. 277: 33604–609.

Carrero, P., Okamoto, K., Coumailleau, P., O'Brien, S., Tanaka, H., and Poellinger, L. 2000. Redox-regulated recruitment of the transcriptional coactivators CREB-binding protein and SRC-1 to Hypoxia-inducible factor 1α. Mol. Cell. Biol. 20: 402–415.

Castellino, F., Boucher, P.E., Eichelberg, K., Mayhew, M., Rothman, J.E., Houghton, A.N., and Germain, R.N. 2000. Receptor-mediated uptake of antigen/heat shock protein complexes results in major histocompatability complex class I antigen presentation via two distinct processing pathways. J. Exp. Med. 191: 1957–1964.

Chang, M.K., Binder, C.J., Miller, Y.I., Subbanagounder, G., Silverman, G.J., Berliner, J.A., and Witztum, J.L. 2004. Apoptotic cells with oxidation-specific epitopes are immunogenic and proinflammatory. J. Exp. Med. 200: 1359–370.

Chen, J., Avdonin, V., Ciorba, M.A., Heinemann, S.H., and Hoshi, T. 2000. Acceleration of P/C-type inactivation of voltage-gated K^+ channels by methionine oxidation. Biophys. J. 78: 174–187.

Cherny, V.V., Henderson, L.M., Xu, W., Thomas, L.L., and DeCoursey, T.E. 2001. Activation of NADPH oxidase-related proton and electron currents in human eosinophils by arachidonic acid. J. Physiol. 535.3: 783–794.

Chuang, T.H., Lee, J., Kline, L., Mathison, J.C., and Ulevitcjh, R.J. 2002. Toll-like receptor 9 mediates CpG-DNA signaling. J. Leukocyte Biol. 71: 538–544.

Clejan, L.A. and Cederbaum, A.I. 1992. Role of cytochrome P450 in the oxidation of glycerol by reconstituted systems and microsomes. FASEB J. 6: 765–770.

Coelho, D., Holl, V., Weltin, D., Lacornerie, T., Magnenet, P., Dufour, P., and Bischoff, P. 2000. Caspase-3-like activity determines the type of cell death following ionizing radiation in MOLT-4 human leukaemia cells. Br. J. Cancer 83: 642–649.

Cooney, C.A., Dave, A.A., and Wolff, G.L. 2002. Maternal methyl supplements in mice affect epigenetic variation and DNA methylation of offspring. J. Nutr. 132: 2393S–2400S.

Courageot, M.P., Lépine, S., Hours, M., Giraud, F., and Sulpice, J.C. 2004. Involvement of sodium in early phosphatidylserine exposure and phospholipid scrambling induced by P2X7 purinoceptor activation in thymocytes. J. Biol. Chem. 279: 21815–823.

Crane, J.K. and Vezina, C.M. 2005. Externalization of host cell protein kinase C during enteropathogenic *Escherichia coli* infection. Cell Death Different. 12: 115–127.

Crawley, J.N. 1996. Unusual behavioral phenotypes of inbred mouse strains. Trends Neurosci. 19: 181–182.

Cusato, K., Bosco, A., Rozental, R., Guimaraes, C.A., Reese, B.E., Linden, R., and Spray, D.C. 2003. Gap junctions mediate bystander cell death in developing brain. J. Neurosci. 23: 6413–422.

DeCoursey, T.E. 2002. Voltage-gated proton channels and other proton transfer pathways. Physiol. Rev. 83: 475–579.

Demine, R. and Walden, P. 2005. Testing the role of gp96 as peptide chaperone in antigen processing. J. Biol. Chem. 280: 17573–578.

De Nardo, D., Masendycz, P., Ho, S., Cross, M., Fleetwood, A.J., Reynolds, E.C., Hamilton, J.A., Scholz, G.M. 2005. A central role for the Hsp90.Cdc37 molecular chaperone module in interleukin-1 receptor-associated-kinase-dependent signaling by Toll-like receptors. J. Biol. Chem. 280: 9813–822.

Denny, M.F., Chandaroy, P., Killen, P.D., Caricchio, R., Lewis, E.E., Richardson, B.C., Lee, K.D., Gavalchin, J., and Kaplan, M.J. 2006. Accelerated macrophage apoptosis induces autoantibody formation and organ damage in systemic lupus erythematosus. J. Immunol. 176: 2095–2104.

Dent, P., Yacoub, A., Contessa, J., Caron, R., Amorino, G., Valerie, K., Hagan, M.P., Grant, S., and Schmidt-Ullrich, R. 2003. Stress and radiation-induced activation of multiple intracellular signaling pathways. Radiat. Res. 159: 283–300.

Droge, W. 2002. Free radicals in the physiological control of cell function. Physiol. Rev. 82: 47–95.

Droge, W. 2005. Oxidative aging and insulin receptor signaling. J. Gerontol. 60A: 1378–385.

Esposito, F., Ammendola, R., Faraonio, R., Russo, T., and Cimino, F. 2004. Redox control of signal transduction, gene expression and cellular senescence. Neurochem. Res. 29: 617–628.

Fadeel, B. 2003. Programmed cell clearance. Cell. Mol. Life Sci. 60: 2575–585.

Fadeel, B. and Xue, D. 2006. PS externalization: from corpse clearance to drug delivery. Cell Death Different. 13: 360–362.

Fahrenkrog, B. 2006. The nuclear pore complex, nuclear transport, and apoptosis. Can. J. Physiol. Pharmacol. 84: 279–286.

Fan, J., Frey, R.S., and Malik, A.B. 2003. TLR4 signaling induces TLR2 expression in endothelial cells via neutrophil NADPH oxidase. J. Clin. Invest. 112: 1234–243.

Fathallah-Shaykh, H.M. 2005. Genomic discovery reveals a molecular system for resistance to oxidative and endoplasmic reticulum stress in cultured glioma. Arch. Neurol. 62: 233–236.

Fedoroff, N. 2006. Redox regulatory mechanisms in cellular stress responses. Ann. Bot. 98: 289–300.

Finkel, T. 2003. Oxidant signals and oxidative stress. Curr. Opin. Cell Biol. 15: 247–254.

Finkel, M. and Cohen, H. 2005. Models of acetylation and the regulation of longevity: from yeast to humans. Drug Discov. Today Disease Models 2: 265–271.

Fischer, J.L., Lancia, J.K., Mathur, A., Smith, M.L. 2006. Selenium protection from DNA damage involves a Ref1/p53/Brca1 protein complex. Anticancer Res. 26: 899–904.

Fornace, A.J. Jr. 1992. Mammalian genes induced by radiation: activation of genes associated with growth control. Ann. Rev. Genet. 26: 507–526.

Frankel, S. and Rogina, B. 2006. Sir2, caloric restriction and aging. Pathol. Biol. 54: 55–57.

Frey, R.S., Gao, X., Javaid, K., Siddiqui, S.S., Rahman, A., and Malik, A.B. 2006. Phosphatidylinositol-3-kinase γ signaling through protein kinase Cζ induces NADPH oxidase-mediated oxidant generation and NF-κB activation in endothelial cells. J. Biol. Chem. 281: 16128–138.

Friedman, E.J. 2002. Immune modulation by ionizing radiation and its implications for cancer immunotherapy. Curr. Pharm. Design 8: 1765–780.

Furumoto, K., Inoue, E., Nagao, N., Hiyama, E., and Miwa, N. 1998. Age-dependent telomere shortening is slowed down by enrichment of intracellular vitamin C via suppression of oxidative stress. Life Sci. 63: 935–948.

Geiszt, M. and Leto, T.L. 2004. The Nox family of NAD(P)H oxidases: host defense and beyond. J. Biol. Chem. 279: 51715–718.

Gems, D. and McElwee, J.J. 2005. Broad spectrum detoxification: the major longevity assurance process regulated by insulin/IGF-1 signaling? Mech. Ageing Dev. 126: 381–387.

Gidalevitz, T., Biswas, C., Ding, H., Schneidman-Duhovny, D., Wolfson, H.J., Stevens, F., Radford, S., and Argon, Y. 2004. Identification of the N-terminal peptide binding site of glucose-regulated protein 94. J. Biol. Chem. 279: 16543–552.

Gilmour, P.S., Rahman, I., Donaldson, K., and MacNee, W. 2003. Histone acetylation regulates epithelial IL-8 release mediated by oxidative stress from environmental particles. Am. J. Physiol. Lung Cell. Mol. Physiol. 284: L533–L540.

Gingrich, J.A. and Hen, R. 2000. The broken mouse: the role of development, plasticity and environment in the interpretation of phenotypic changes in knockout mice. Curr. Opini. Neurobiol. 10: 146–152.

Gong, B. and Almasan, A. 2000. Apo2 ligand/Tnf-related apoptosis-inducing ligand and death receptor 5 mediate the apoptotic signaling induced by ionizing radiation in leukemic cells. Cancer Res. 60: 5754–760.

Haendeler, J., Hoffmann, J., Brandes, R.P., Zeiher, A.M., and Dimmeler, S. 2003. Hydrogen peroxide triggers nuclear export of telomerase reverse transcriptase via Src kinase family-dependent phosphorylation of tyrosine 707. Mol. Cell. Biol. 23: 4598–610.

Han, D., Canali, R., Rettori, D., and Kaplowitz, N. 2003. Effect of glutathione depletion on sites and topology of superoxide and hydrogen peroxide production in mitochondria, Mol. Pharmacol. 64: 1136–144.

Harman, D. 1956. Aging: a theory based on free radical and radiation chemistry. J. Gerontol. 11: 298–300.

Hemmi, H., Takeuchi,O., Kawai, T., Kaisho, T., Sato, S., Sanjo, H., Matsumoto, M., Hoshino, K., Wagner, H., Takeda, K., Akira, S. 2000. A Toll-like receptor recognizes bacterial DNA. Nature 408: 740–745.

Herrlich, P., Bender, K., Knebel, A., Bohmer, F.D., Grob, S., Blattner, C., Rahmsdorf, H. J., and Gottlicher, M. 1999. Radiation-induced signal transduction: mechanisms and consequences. Compt. Rend. Acad. Sci. Ser. III, Sci.Vie 322: 121–125.

Ho, G.T., Moodie, F.M., and Satsangi, J. 2003. Multidrug resistance 1 gene (P-glycoprotein 170): an important determinant in gastrointestinal disease? Gut 52: 759–766.

Hopkins, P.A. and Sriskandan, S. 2005. Mammalian Toll-like receptors: to immunity and beyond. Clin. Exp. Immunol. 140: 395–407.

Hoshino, H., Kobayashi, A., Yoshida, M., Kudo, N., Oyake, T., Motohashi, H., Hayashi, N., Yamamoto, M., and Igarashi, K. 2000. Oxidative stress abolishes leptomycin B-sensitive nuclear export of transcription repressor Bach2 that counteracts activation of Maf recognition element. J. Biol. Chem. 275: 15370–376.

Huang, C.L., Huang, N.K., Shyue, S.K., and Chern, Y. 2003. Hydrogen peroxide induces loss of dopamine transporter activity: a calcium-dependent oxidative mechanism. J. Neurochem. 86: 1247–259.

Huber, J., Fürnkranz, A., Bochkov, V.N., Patricia M.K., Lee, H., Hedrick, C.C., Berliner, J.A., Binder, B.R., and Leitinger, N. 2006. Specific monocyte adhesion to endothelial cells induced by oxidized phospholipids involves activation of $cPLA_2$ and lipoxygenase. J. Lipid Res. 47: 1054–1062.

Ishii, K.J. and Akira, S. 2005. Innate immune recognition of nucleic acids: beyond toll-like receptors. Int. J. Cancer 117: 517–523.

Itoh, M., Adachi, M., Yasui, H., Takekawa, M., Tanaka, H., and Imai, K. 2002. Nuclear export of glucocorticoid receptors is enhanced by c-Jun N-terminal kinase-mediated phosphorylation. Mol. Endocrinol. 16: 2382–392.

Jackman, M., Kubota, Y., den Elzen, N., Hagting, A., and Pines, J. 2002. Cyclin A- and cyclin E-Cdk complexes shuttle between the nucleus and the cytoplasm. Mol. Biol. Cell 13: 1030–1045.

Jackson, E.B., Theriot, C.A., Chattopadhyay, Mitra, S., and Izumi, T. 2005. Analysis of nuclear transport signals in the human apurunic/apyrimidinic endonuclease (APE1/Ref1). Nucleic Acids Res. 33: 3303–312.

Jaenisch, R. and Bird, A. 2003. Epigenetic regulation of gene expression: how the genome integrates intrinsic and environmental signals. Nat. Genet. (Suppl.)33: 245–254.

Jans, D.A. 1995. The regulation of protein transport to the nucleus by phosphorylation. Biochem. J. 311: 705–716.

Jazwinski, S.M. 1996. Longevity, genes and aging. Science 273: 54–59.

Kagan, V.E., Borisenko, G.G., Serinkan, B.F., Tyurina, Y.Y., Tyurin, A., Jiang, J., Liu, S.X., Shvedova, A.A., Fabisiak, J.P., Uthaisang, W., and Fadeel, B. 2003. Appetizing rancidity of apoptotic cells for macrophages: oxidation, externalization, and recognition of phosphatidylserine. Am. J. Physiol. Lung Cell. Mol. Physiol. 285: L1–L17.

Kang, H.L., Benzer, S., and Min, K.T. 2002. Life extension in *Drosophila* by feeding a drug. PNAS 99: 838–843.

Kaplan, M.J., Lu, Q., Wu, A., Attwood, J., and Richardson, B. 2005. Demethylation of promoter regulatory elements contributes to performing overexpression in $CD4^+$ lupus cells. J. Immunol. 172: 3652–661.

Kawahara, T., Kuwano, Y., Teshima-Kondo, S., Takeya, R., Sumimoto, H., Kishi, K., Tsunawaki, S., Hirayama, T., and Rokutan, K. 2004. Role of nicotinamide adenine dinucleotide phosphate oxidase 1 in oxidative burst response to Toll-like receptor 5 signaling in large intestinal epithelial cells. J. Immunol. 172: 3051–3058.

Kawahara, T., Kohjima, M., Kuwano, Y., Mino, H., Teshima-Kondo, S., Takeya, R., Tsunawaki, S., Wada, A., Sumimoto, H., and Rokutan, K. 2005. Helicobacter pylori lipopolysaccharide activates Rac1 and transcription of NADPH oxidase Nox1 and its organizer NOXO1 in guinea pig gastric mucosal cells. Am. J. Physiol. Cell Physiol. 288: 450–457.

Kawai, T. and Akira, S. 2006. TLR signaling. Cell Death Different. 13: 816–825.

Kleeberger, S.R., Reddy, S., Zhang, L.Y., and Jedlicka, A.E. 2000. Genetic susceptibility to ozone-induced lung hyperpermeability: Role of Toll-like receptor 4. Am. J. Resp. Cell Mol. Biol. 22: 620–627.

Knoops, L., Haas, R.L., de Kemp, S., Broeks, A., van 't Veer, L.J., and de Jong, D. 2005. Low dose radiation induces a highly effective p53 response and rapid tumor regression in follicular lymphoma. Blood (ASH Annual Meeting Abstracts)106: Abstract 354.

Kodiha, M., Matusiewicz, N., and Stochaj, U. 2004. Multiple mechanisms promote the inhibition of classical nuclear import upon exposure to severe oxidative stress. Cell Death Different. 11: 862–874.

Kourie, J.I. 1998. Interaction of reactive oxygen species with ion transport mechanisms. Am. J. Physiol. Cell Physiol. 275: C1–C24.

Krtolica, A., Parrinello, S., Lockett, S., Desprez, P.Y., and Campisi, J. 2001. Senescent fibroblasts promote epithelial cell growth and tumorigenesis: a link between cancer and aging. PNAS 98: 12072–2077.

Kubota, H., Suzuki, T., Lu, J., Takahashi, S., Sugita, K., Sekiya, S., Suzuki, N. 2005. Increased expression of GRP94 protein is associated with decreased sensitivity to X-rays in cervical cancer cell lines. Int. J. Rad. Biol. 81: 701–709.

Kudo, N., Taoka, H., Toda, T., Yoshida, M., and Horinouchi, S. 1999b. A novel nuclear export signal sensitive to oxidative stress in the fission yeast transcription factor Pap1. J. Biol. Chem. 274: 15151–158.

Kudo, N., Matsumori, N., Taoka, H., Fujiwara, D., Schreiner, E.P., Wolff, B., Yoshida, M., and Horinouchi, S. 1999a. Leptomycin B inactivates CRM1/exportin 1 by covalent modification at a cysteine residue in the central conserved region. PNAS 96: 9112–117.

Kuge, S., Arita, M., Murayama, A., Maeta, K., Izawa, S., Inoue, Y., d Nomoto, A. 2001. Regulation of the yeast Yap1p nuclear export signal is mediated by redox signal-induced reversible disulfide bond formation. Mol. Cell. Biol. 21: 6139–150.

Kumaraguru, U., Pack, C.D., and Rouse, B.T. 2003. Toll-like receptor ligand links innate and adaptive immune responses by the production of heat-shock proteins. J. Leukocyte Biol. 73: 574–583.

Kuo, S.S., Saad, A.H., Koong, A.C., Hahn, G.M., and Giaccia, A.J. 1993. Potassium-channel activation in response to low doses of γ-Irradiation involves reactive oxygen intermediates in nonexcitatory cells. PNAS 90: 908–912.

Kuroda, J., Nakagawa, K., Yamasaki, T., Nakamura, K., Takeya, R., Kuribayashi, F., Imajoh-Ohmi, S., Igarashi, K., Shibata, Y., Sueishi, K., and Sumimoto, H. 2005. The superoxide-producing NAD(P)H oxidase Nox4 in the nucleus of human vascular endothelial cells. Gennes Cells 10: 1139–151.

Lancaster, G.L. and Febbraio, M.A. 2005. Exosome-dependent trafficking of HSP70. J. Biol. Chem. 280: 23349–355.

Laroux, F.S., Romero, X., Wetzler, L., Engel, P., and Terhorst, C. 2005. Cutting Edge: MyD88 controls phagocyte NADPH oxidase function and killing of gram-negative bacteria. J. Immunol. 175: 5596–600.

Leggas, M., Adachi, M., Scheffer, G.L., Sun, S., Wielinga, P., Du, G., Mercer, K.E., Zhuang, Y., Panetta, J.C., Johnston, B., Scheper, R.J., Stewart, C.F., and Schuetz, J.D. 2004. Mrp4 confers resistance to topotecan and protects the brain from chemotherapy. Mol. Cell. Biol. 24: 7612–621.

Leier, I., Jedlitschky, G., Buchholz, U., Center M., Cole, S.P.C., Deeley, R.G., and Keppler, D. 1996. ATP-dependent glutathione disulfide transport mediated by the MRP gene-encoded conjugate export pump. Biochem. J. 314: 433–437.

Lemon, J.A., Boreham, D.R., and Rollo, C.D. 2003. A dietary supplement abolishes age-related cognitive decline in transgenic mice expressing elevated free radical processes, Exp. Biol. Med. 228: 800–810.

Lescot, M., Rombauts, S., Zhang, J., Aubourg, S., Mathé, C., Jansson, S., Rouzé, P., and Boerjan, W. 2004. Annotation of a 95-kb Populus deltoides genomic sequence reveals a disease resistance gene cluster and novel class I and class II transposable elements. TGA Theor. Appl. Genet. 109: 10–22.

Liu, B., Dai, J., and Li, Z. 2005a. Molecular and cellular mechanisms involved in the pathogenesis of autoimmune diseases induced by cell surface Gp96. Immunology 114: 144.

Liu, B., Dai, J., Zheng, H., Stoilova, D., Sun, S., and Li, Z. 2003. Cell surface expression of an endoplasmic reticulum resident heat shock protein gp96 triggers MyD88-dependent systemic autoimmune diseases. PNAS 100: 15824–15829.

Liu, S., Wang, H., Yang, Z., Kon, T., Zhu, J., Cao, Y., Li, F., Kirkpatrick, J., Nicchitta, C.V., and Li, C.L. 2005b. Enhancement of cancer radiation therapy by use of adenovirus-mediated secretable glucose-regulated protein 94/gp96 expression. Cancer Res. 65: 9126–131.

Macara, I.G. 2001. Transport into and out of the nucleus. Microbiol. Mol. Biol. Rev. 65: 570–590.

Martin, G.M., Austad, S.N., and Johnson, T.E. 1996. Genetic analysis of ageing: role of oxidative damage and environmental stresses. Nature Genetics 13: 25–34.

McBride, W.H., Chaing, C.S., Olson, J.L., Wang, C.C., Hong, J.H., Pajonk, F., Dougherty, G.J., Iwamoto, K.S., Pervan, M., and Liao, Y.P. 2004. A sense of danger from radiation. Radiat. Res. 162: 1–19.

McPhail, L.C., Qualliotine-Mann, D., Agwu, D.E., and McCall, C.E. 1993. Phospholipases and activation of the NADPH oxidase. Eur. J. Haematol. 51: 294–300.

Means, T.K., Latz, E., Hayashi, F., Murali, M.R., Golebock, D.T., and Luster, A.D. 2005. Human lupus autoantibody-DNA complexes activate DCs through cooperation of CD32 and TLR9. J. Clin. Invest. 115: 407–417.

Meng, T.C., Lou, Y.W., Chen, Y.Y., Hsu, S.F., and Huang, Y.F. 2006. Cys-oxidation of protein tyrosine phosphatases: its role in regulation of signal transduction and its involvement in human cancers. J. Cancer Mol. 2: 9–16.

Meplan, C., Richar, M.J., and Hainaut, P. 2000. Redox signalling and transition metals in the control of the p53 pathway. Biochem. Pharmacol. 59: 25–33.

Merezak, C., Reichert, M., Van Lint, C., Kerkhofs, P., Portetelle, D., Willems, L., and Kettmann, R. 2002. Inhibition of histone deacytelases induces bovine leukemia virus expression in vitro and in vivo. J. Virol. 76: 5034–5042.

Mevorach, D., Zhou, J.L., Song, X., and Elkon, K.B. 1998. Systemic exposure to irradiated apoptotic cells induces autoantibody production. J. Exp. Med. 188: 387–392.

Milan, M., Campuzano, S., and Garcia-Bellido, A. 1997. Developmental parameters of cell death in the wing disc of Drosophila. PNAS 94: 5691–696.

Miller, D.S., Graeff, C., Droulle, L., Fricker, S., and Fricker, G. 2002. Xenobiotic efflux pumps in isolated fish brain capillaries. Am. J. Physiol. Reg. Integ. Comp. Physiol. 282: 191–198.

Minier, C., Eufemia, N., and Epel, D. 1999. The multi-xenobiotic resistance phenotype as a tool to biomonitor the environment. Biomarkers 4: 442–454.

Mohan, C., Adams, S., Stanik, V., Datta, S.K. 1993. Nucleosome: a major immunogen for pathogenic autoantibody-inducing T cells of lupus. J. Exp. Med. 177: 1367–1381.

Morgan, H.D., Sutherland, H.G.E., Martin, D.I.K., and Whitelaw, E. 1999. Epigenetic inheritance at the agouti locus in the mouse. Nat. Genet. 23: 314–318.

Narayanan, P.K., Goodwin, E.H., and Lehnert, B.E. 1997. Alpha particles initiate biological production of superoxide anions and hydrogen peroxide in human cells. Cancer Res. 57: 3963–971.

Nimchuk, Z., Eulgem, T., Holt III, B.F., and Dangl, J.L. 2003. Recognition and response in the plant immune system. Ann. Rev. Genet. 37: 579–609.

Norbury, C.C., Basta, S., Donohue, K.B., Tscharke, D.C., Princiotta, M.F., Berglund, P., Gibbs, J., Bennink, J.R., and Yewdell, J.W. 2004. CD8[+] T cell cross-priming via transfer of proteasome substrates. Science 304: 1318–321.

Obsil, T., Ghirlando, R., Anderson, D.E., Hickman, A.B., and Dyda, F. 2003. Two 14-3-3 binding motifs are required for stable association of forkhead transcription factor FOXO4 with 14-3-3 proteins and inhibition of DNA binding. Biochemistry 42: 15264–272.

Okamoto, K., Tanaka, H., Ogawa, H., Makino, Y., Eguchi, H., Hayashi, S., Yoshikawa, N., Poellinger, L., Umesono, K., and Makino, I. 1999. Redox-dependent regulation of nuclear import of the glucocorticoid receptor. J. Biol. Chem. 274: 10363–371.

Pardo, L.A., Contreras-Jurado, C., Zientkowska, M., Alves, F., Stuhmer, W. 2005. Role of voltage-gated potassium channels in cancer. J. Membr. Biol. 205: 115–124.

Park, H.S., Jung, H.Y., Park, E.Y., Kim, J., Lee, W.J., and Bae, Y.S. 2004a. Cutting edge: Direct interaction of TLR4 with NAD(P)H oxidase 4 isozyme is essential for lipopolysaccharide-induced production of reactive oxygen species and activation of NF-κB. J. Immunol. 173: 3589–593.

Park, J.Y., Ryang, Y.S., Shim, K.Y., Lee, J.I., and Kim, S.K. 2004b. NF-κB activation and regulation of toll-like receptors expression in human colon cancer cell line stimulated by DNA bis-intercalation agent, echinomycin. Proc. Amer. Assoc. Cancer Res. 45: Abstract #2259.

Pennell, C.A., 2005. Heat shock proteins in immune response in the fall of 2004. Immunology 114: 297–300.

Petosa, C., Schoehn, G., Askjaer, P., Bauer, U., Moulin, M., Steuerwald, U., Soler-Lopez, M., Baudin, F., Mattaj, I.W., and Muller, C.W. 2004. Architecture of CRM1/Exportin1 suggests how cooperativity is achieved during formation of a nuclear export complex. Mol. Cell 16: 761–775.

Pletjushkina, O.Y., Fetisova, E.K., Lyamzaev, K.G., Ivanova, O.Y., Domnina, L.V., Vyssokikh, M.Y., Pustovidko, A.V., Vasiliev, J.M., Murphy, M.P., Chernyak, B.V., and Skulachev, V.P. 2005. Long-distance apoptotic killing of cells is mediated by hydrogen peroxide in a mitochondrial ROS-dependent fashion. Cell Death Different. 12: 1442–444.

Pletjushkina, O.Y., Fetisova, E.K., Lyamzaev, K.G., Ivanova, O.Y., Domnina, L.V., Vyssokikh, M.Y., Pustovidko, A.V., Alexeevski, A.V., Alexeevski, D.A., Vasiliev, J.M., Murphy, M.P., Chernyak, B.V., and Skulachev, V.P. 2006. Hydrogen peroxide inside mitochondria takes part in cell-to-cell transmission of apoptotic signal. Biochemistry (Moscow) 71: 60–67.

Poon, I.K.H. and Jans, D.A. 2005. Regulation of nuclear transport: central role in development and transformation? Traffic 6: 173–186.

Powell, S.N. and Abraham, E.H. 1993. The biology of radioresistance: similarities, differences and interactions with drug resistance. Cytotechnology 12: 325–345.

Prinz, M., Garbe, F., Schmidt, H., Mildner, A., Gutcher, I., Wolter, K., Piesche, M., Schroers, R., Weiss, E., Kirschning, C.J., Rochford, C.D.P., Bruck, W., and Becher, B. 2006. Innate immunity mediated by TLR9 modulates pathogenicity in an animal model of multiple sclerosis. J. Clin. Invest. 116: 456–464.

Proskuryakov, S.Y., Gabai, V.L., and Konoplyannikov, A.G. 2002. Necrosis is an active and controlled form of programmed cell death. Biochem. (Moscow) 67: 387–408.

Qiao, B., Wu, J., Chu, Y.W., Wang, Y., Wang, D.P. Wu, H.S., and Xiong, S.D. 2005. Induction of systemic lupus erythematosus-like syndrome in syngeneic mice by immunization with activated lymphocyte-derived DNA. Rheumatology 44: 1108–114.

Queitsch, C., Sangster, T.A., and Lindquist, S. 2002. Hsp90 as a capacitor of phenotypic variation. Nature 417: 618–624.

Quintana, F.J. and Cohen, I.R. 2005. Heat shock proteins as endogenous adjuvants in sterile and septic inflammation. J. Immunol. 175: 2777–782.

Qureshi, S.T., Zhang, X., Bousette, N., Giaid, A., Shan, P., Medshitov, R.M., and Lee, P.J. 2006. Inducible activation of TLR4 confers resistance to hyperoxia-induced pulmonary apoptosis. J. Immunol. 176: 4950–958.

Radic, M., Marion, T., and Monestier, M. 2004. Nucleosomes are exposed at the cell surface in apoptosis. J. Immunol. 172: 6692–700.

Radsak, M.P., Hilf, N., Singh-Jasuja, H., Braedel, S., Brossart, P., Rammensee, H.G., and Schild, H. 2003. The heat shock protein gp96 binds to human neutrophils and monocytes and stimulates effector functions. Blood 101: 2810–815.

Rapp, U.K. and Kaufmann, S.H.E. 2004. DNA vaccination with gp96-peptide fusion proteins induces protection against an intracellular bacterial pathogen. Int. Immunol. 16: 597–605.

Renes, J., de Vries, E.G.E., Jansen, P.L.M., and Muller, M. 2000. The (patho)physiological functions of the MRP family. Drug Resistance Updates 3: 289–302.

Reznikov, K., Kolesnikova, L., Pramanik, A., Tan-No, K., Gileva, I., Yakovleva, T., Rigler, R., Terenius, L., and Bakalkin, G. 2000. Clustering of apoptotic cells via bystander killing by peroxides. FASEB J. 14: 1754–764.

Richards, T. and Budinger, T.F. 1988. NMR imaging and spectroscopy of the mammalian central nervous system after heavy ion radiation. Radiat. Res. 113: 79–101.

Richardson, J.R., Caudle, W.M., Wang, M., Dean, E.D., Pennell, K.D., and Miller, G.W. 2006. Developmental exposure to the pesticide dieldrin alters the dopamine system and increases neurotoxicity in an animal model of Parkinson's disease. FASEB J. 20: E976–E985.

Robert, J., Menoret, A., and Cohen, N. 1999. Cell surface expression of the endoplasmic reticular heat shock protein gp96 is phylogenetically conserved. J. Immunol. 163: 4133–139.

Rollo, C.D. 2002. Growth negatively impacts the life span of mammals. Evol. Develop. 4: 55–61.

Rollo, C.D. 2006. Radiation and the regulatory landscape of neo^2-Darwinism. Mutat. Res. 597: 18–31.

Rollo, C.D. 2007. Overview of research on giant transgenic mice with emphasis on the brain and aging. In: T. Samaras (Ed.). Human body size and the laws of scaling. Nova Science Publishers, N.Y., pp. 235–260.

Rollo, C.D., Carlson, J., and Sawada, M. 1996. Accelerated aging of giant transgenic growth hormone mice is associated with elevated free radical processes. Can. J. Zool. 74: 606–620.

Ruppersberg, J.P., Stocker, M., Pongs, O., Heinemann, S.H., Frank, R., and Koenen, M. 1991. Regulation of fast inactivation of cloned mammalian $I_k(A)$ channels by cysteine oxidation. Nature 352: 711–714.

Rutherford, S.L. 2003. Between genotype and phenotype: protein chaperones and evolvability. Nat. Rev. Genet. 4: 263–274.

Rutherford, S.L. and Lindquist, S. 1998. Hsp90 as a capacitor of phenotypic variation. Nature 396: 336–342.

Sakakida, Y., Miyamoto, Y., Nagoshi, E., Akashi, M., Nakamura, T.J., Mamine, T., Kasahara, T., Minami, Y., Yoneda, Y., and Takumi, T. 2005. Importin α/β mediates nuclear transport of a mammalian circadian clock component, mCRY2, together with mPER2, through a bipartite nuclear localization signal. J. Biol. Chem. 280: 13272–278.

Sangster, T.A., Qeitsch, C., and Lindquist, S. 2003. Hsp90 and chromatin: where is the link? Cell Cycle 2: 166–168.

Scarlett, D.J.G., Herst, P.M., and Berridge, M.V. 2005. Multiple proteins with single activities or a single protein with multiple activities: The conundrum of cell-surface NADH oxidoreductases. Biochim. Biophys. Acta 1708: 108–119.

Schmidt-Ullrich, R.K., Dent, P., Grant, S., Mikkelsen, R.B., and Valerie, K. 2000. Signal transduction and cellular radiation responses. Radiat. Res. 153: 245–257.

Schuetz, E.G., Umbenhauer, D.R., Yasuda, K., Brimer, C., Nguyen, L., Relling, M.V., Schuetz, J.D., and Schinkel, A.H. 2000. Altered expression of hepatic cytochromes P-450 in mice deficient in one or more mdr1 genes. Mol. Pharmacol. 57: 188–197.

Schulze-Lefert, P. 2004. Plant immunity: the origami of receptor activation. Curr. Biol. 14: R22–R24.

Seo, Y.R., Kelly, M.R., and Smith, M.L. 2002. Selenomethionine regulation of p53 by a ref1-dependent redox mechanism. PNAS 99: 14548–553.

Shi, Y., Evans, J.E., and Rock, K.L. 2003. Molecular identification of a danger signal that alerts the immune system to dying cells. Nature 425: 516–521.

Shin, B.K., Wang, H., Yim, A.M., Le Naour, F., Brichory, F., Jang, J.H., Zhao, R., Puravs, E., Tra, J., Michael, C.W., Misek, D.E., and Hanash, S.M. 2003. Global profiling of the cell surface proteome of cancer cells uncovers an abundance of proteins with chaperone function. J. Biol. Chem. 278: 7607–616.

Sollars, V., Lu, X., Xiao, L., Wang, X., Garfinkel, M.D., and Ruden, D.M. 2003. Evidence for an epigenetic mechanism by which Hsp90 acts as a capacitor for morphological evolution. Nat. Genet. 33: 70–74.

Sommer, T. and Jarosch, E. 2005. Pardon me – no access without ubiquitin. Dev. Cell 8: 4–5.

Sorokin, A. 2004. Cyclooxygenase-2: potential role in regulation of drug efflux and multidrug resistance phenotype. Curr. Pharm. Design 10: 647–657.

Soti, C., Subbarao, A., and Csermely, P. 2003. Apoptosis, necrosis and cellular senescence: chaperone occupancy as a potential switch. Aging Cell 2: 39–45.

Srivastava, P. 2005. Specific immunogenicity of heat shock protein-peptide complexes: new developments. Cancer Immun. 5(Suppl. 1): 11.

Stebbing, J., Bower, M., Nelson, M., Kebba, A.F., Gotch Patterson, S., and Gazzard, B. 2005. The role of HSPs and CD91 in the pathogenesis of HIV infection. Immunology 114: 148.

Stolpen, A.H., Golan, D.E., and Pober, J.S. 1988. Tumor necrosis factor and immune interferon act in concert to slow lateral diffusion of proteins and lipids in human endothelial cell membranes. J. Cell Biol. 107: 781–789.

Stommel, J.M., Marchenko, N.D., Jimenez, G.S., Moll, U.M., Hope, T.J., and Wahl, G.M. 1999. A leucine-rich nuclear export signal in the p53 tetramerization domain: regulation of subcellular localization and p53 activity by NES masking. EMBO J. 18: 1660–672.

Strom, A.C. and Weis, K. 2001. Importin-β-like nuclear transport receptors. Genome Biol. 2: 3008.1–30008.9.

Stuart, J.A. and Brown, M.F. 2006. Energy, quiescence and the cellular basis of animal life spans. Comput. Biochem. Physiol. 143(Part A): 12–23.

Takeshita, F., Leifer, C.A. Gursel, I., Ishii, K.J., Takeshita, S., Gurse, M., and Klinman, D.M. 2001. Cutting edge: Role of Toll-like receptor 9 in CpG DNA-induced activation of human cells. J. Immunol. 167: 3555–558.

Tan, B., Piwnica-Worms, D., and Ratner, L. 2000. Multidrug resistance transporters and modulation. Cancer in AIDS. Curr. Opin. Oncol. 12:450–458.

Tatsuta, T., Naito, M., Oh-hara, T., Sugawara, I., and Tsutuo, T. 1992. Functional involvement of P-glycoprotein in blood-brain barrier. J. Biol. Chem. 267: 20383–20391.

Toone, W.M., Kuge, S., Samuels, M., Morgan, B.A., Toda, T., and Jones, N. 1998. Regulation of the fission yeast transcription factor Pap1 by oxidative stress: requirement for nuclear export factor Crm1 (Exportin) and the stress-activated MAP kinase Sty1/Spc1. Genes Dev. 12: 1453–463.

Tsan, M.F. and Gao, B. 2004a. Endogenous ligands of Toll-like receptors. J. Leukocyte Biol. 76: 514–519.

Tsan, M.F. and Gao, E. 2004b. Cytokine function of heat shock proteins. Am. J. Physiol. Cell Physiol. 286: C739–C744.

Tsunoda, I., Libbey, J.E., Kuang, L.Q., Terry, E.J., and Fujinami, R.S. 2005. Massive apoptosis in lymphoid organs in animal models for primary and secondary progressive multiple sclerosis. Amer. J. Pathol. 167: 1631–646.

Tyurina, Y.Y., Serinkan, F.B., Tyurin, V.A., Kini, V., Yalowich, J.C., Schroit, A.J., Fadeel, B., and Kagan, V.E. 2004. Lipid antioxidant, etoposide, inhibits phosphatidylserine externalization and macrophage clearance of apoptotic cells by preventing phosphatidylserine oxidation. J. Biol. Chem. 279: 6056–6064.

Tyurina, Y.Y., Kini, V., Tyurin, V.A., Vlasova, I.I., Jiang, J., Kapralov, A.A., Belikova, N.A., Yalowich, J.C., Kurnikov, I.V., and Kagan, V.E. 2006. Mechanisms of cardiolipin oxidation by cytochrome c: relevance to pro- and antiapoptotic functions of etoposide. Mol. Pharmacol. 70: 706–717.

Udono, H. and Srivastava, P.K. 1994. Comparison of tumor-specific immunogenicities of stress-induced proteins gp96, hsp90, and hsp70. J. Immunol. 152: 5398–403.

Uhr, M., Steckler, T., Yassouridis, A., and Holsboer, F. 2000. Penetration of amitriptyline, but not of fluoxetine, into brain is enhanced in mice with blood-brain barrier deficiency due to Mdr1a P-glycoprotein gene disruption. Neuropsychopharmacology 22: 380–387.

Underhill, D.M. 2003. Toll-like receptors: networking for success. Eur. J. Immunol. 33: 1767–775.

Velichkova, M. and Hasson, T. 2005. Keap1 regulates the oxidation-sensitive shuttling of Nrf2 into and out of the nucleus via a Crm1-dependent nuclear export mechanism. Mol. Cell. Biol. 25: 4501–513.

Verdel, A., Curtet, S., Brocard, M.P., Rousseaux, S., Lemercier, C., Yoshida, M., and Khochbin, S. 2000. Active maintenance of mHDA2/mHDAC6 histone deacetylase in the cytoplasm. Curr. Biol. 10: 747–749.

Vinken, M., Vanhaecke, T., Papeleu, P., Snykers, S., Henkens, T., and Rogiers, V. 2006. Connexins and their channels in cell growth and cell death. Cell. Signal. 18: 592–600.

Voehringer, D.W., McConkey, D.J., McDonnell, T.J., Brisbay, S., and Meyn, R.E. 1998. Bcl-2 expression causes redistribution of glutathione to the nucleus. PNAS 95: 2956–960.

von Gall, C., Garabette, M.L., Kell, C.A., Frenzel, S., Dehghani, F., Schumm-Draeger, P.M., Weaver, D.R., Kort, H.W., Hastings, M.H., and Stehle, J.H. 2002. Rhythmic gene expression in pituitary depends on heterologous sensitization by the neurohormone melatonin. Nat. Neurosci. 5: 234–238.

Walther-Larsen, H., Brandt, J., Collinge, D.B., and Thordal-Christensen, H. 1993. A pathogen-induced gene of barley encodes a HSP90 homologue showing striking similarity to vertebrate forms resident in the endoplasmic reticulum. Plant Mol. Biol. 21: 1097–1108.

Wang, H.D., Kazemi-Esfarjani, P., and Benzer, S. 2004. Multiple-stress analysis for isolation of Drosophila genes. PNAS 101: 12610–615.

Wang, W., Yang, X., Kawai, T., de Silanes, I.L., Mazan-Mamczarz, K., Chen, P., Chook, Y.M., Quensel, C., Kohler, M., and Gorospe, M. 2004. AMP-activated protein kinase-regulated phosphorylation and acetylation of Importin α1. J. Biol. Chem. 279:48376–388.

Wang, Y. and Tissenbaum, H.A. 2006. Overlapping and distinct functions for a Caenorhabditis elegans SIR2 and DAF-16/FOXO. Mech. Ageing Develop. 127: 48–56.

Warger, T., Hilf, N., Rechtsteiner, G., Haselmayer, P., Carrick, D.M., Jonuleit, H., von Landenberg, P., Rammensee, H.G., Nicchitta, C.V., Radsak, M.P., and Schild, H. 2006. Interaction of TLR2 and TLR4 ligands with the N-terminal domain of Gp96 amplifies innate and adaptive immune responses. J. Biol. Chem. 281: 22545–553.

Wiedlocha, A., Nilsen, T., Wesche, J., Sorensen, V., Malecki, J., Marcinkowska, E., and Olsnes, S. 2005. Phosphorylation-regulated nucleocytoplasmic trafficking of internalized fibroblast growth factor-1. Mol. Biol. Cell 16: 794–810.

Wood, J.G., Rogina, B., Lavu, S., Howitz, K., Helfand, S.L., Tatar, M., and Sinclair, D. 2004. Sirtuin activators mimic caloric restriction and delay aging in metazoans. Nature 430: 686–689.

Xue, L.Y., Butler, N.J., Makrigiorgos, G.M., Adelstein, S.J., and Kassis, A.I. 2002. Bystander effect produced by radiolabeled tumor cells in vivo. PNAS 99: 13765–770.

Yabe, T., Suzuki, N., Furukawa, T., Ishihara, T., and Katsura, I. 2005. Multidrug resistance-associated protein MRP-1 regulates dauer diapause by its export activity in Caenorhabditis elegans. Development 132: 3197–207.

Yamamoto, M., Sato, S., Hemmi, H., Hoshino, K., Kaisho, T., Sanjo, H., Takeuchi, O., Sugiyama, M., Okabe, M., Takeda, K., and Akira, S. 2003. Role of adaptor TRIF in the MyD88-independent Toll-like receptor signaling pathway. Science 301: 640–643.

Yi, A.K., Yoon, H., Park, J.E., Kim, B.S., Kim, H.J., and Martinez-Hernandez, A. 2006. CpG DNA-mediated induction of acute liver injury in D-galactosamine-sensitized mice: The mitochondrial apoptotic pathway-dependent death of hepatocytes. J. Biol. Chem. 281: 15001–5012.

Zaman, G.J.R., Lankelma, J., Van Tellingen, O., Beijnen, J., Dekker, H., Paulusma, C., Oude Elferink, R.P.J., Baas, F., and Borst, P. 1995. Role of glutathione in the export of compounds from cells by the multidrug-resistance-associated protein. PNAS 92: 7690–694.

Zhang, Q., Piston, D.W., and Goodman, R.H. 2002. Regulation of corepressor function by nuclear NADH. Science 295: 1895–897.

Zhang, X., Shan, P., Qureshi, S., Homer, R., Medzhitov, R., Noble, P.W., and Lee, P.J. 2005. Cutting edge: TLR4 deficiency confers susceptibility to lethal oxidant lung injury. J. Immunol. 175: 4834–838.

Zhang, Y. and Xiong, Y. 2001a. A p53 amino-terminal nuclear export signal inhibited by
    DNA damage-induced phosphorylation. Science 292: 1910–915.
Zhang, Y. and Xiong, Y. 2001b. Control of p53 ubiquitination and nuclear export by MDM2
    and ARF. Cell Growth Different. 12: 175–186.
Zhao, X., Gan, L., Pan, H., Kan, D., Majeski, M., Adam, S.A., and Unterman, T.G. 2004.
    Multiple elements regulate nuclear/cytoplasmic shuttling of FOXO1: characterization of
    phosphorylation- and 14-3-3-dependent and –independent mechanisms. Biochem. J. 378:
    839–849.
Zheng, L.M., Zychlinsky, A., Liu, C.C., Ojcius, D.M., and Young, J.D.E. 1991. Extracellular
    ATP as a trigger for apoptosis or programmed cell death. J. Cell Biol. 112: 279–288.
Ziel, K.A., Grishko, V., Campbell, C.C., Breit, J.F., Wilson, G.L., and Gillespie, M.N. 2005.
    Oxidants in signal transduction: impact on DNA integrity and gene expression. FASEB J.
    19: 387–394.

CHAPTER 14

GENETIC ASPECTS OF POLLUTANT ACCUMULATION IN PLANTS

A. KILCHEVSKY[1], L. KOGOTKO[2], A. SHCHUR[3], AND A. KRUK[4]

[1] *Institute of Genetics and Cytology Minsk, Belarus*
e-mail: a.kilchevsky@igc.bas-net.by
[2] *Belarussian State Agricultural Academy, Gorky, Mogilev Region, Belarus*
[3] *Institute of Radiology, Mogilev Department, Mogilev, Belarus*
[4] *University of Gomel, Gomel, Belarus*

Abstract: The object of our research was to study interspecies and intervariety variability of pollutant (nitrates, heavy metals, radionuclides) accumulation in productive parts of vegetable crops, as well as the character of inheritance of nitrate and heavy metal accumulation in tomato fruits in the open ground. We investigated the genetical basis of cadmium, lead and nitrate accumulation using the diallel analysis method. Genotype variation between tomato varieties and hybrids in diallel crosses in cadmium (12.8–15.5-fold) and lead accumulation (7.5–14.1-fold) was established. It gives the possibility to select tomato varieties for growing in the contaminated area, as well as to use them in breeding for development of new varieties and hybrids with minimum accumulation of pollutants in fruits. An independent type of inheritance of cadmium and lead accumulation was revealed. Genotype variation between tomato varieties and hybrids in diallel crosses in nitrate accumulation (3.8–8.3-fold) was revealed. The general type of heavy metal and nitrate accumulation inheritance in tomato fruits – overdominance towards reduction of pollutant value. To make use of heterosis breeding as a method of development of high-yielding hybrids of tomato for open ground with minimal accumulation of pollutant is proposed. Study of uptake of ^{137}Cs and ^{90}Sr in vegetables in Gomel region (Chernobyl zone) revealed the interspecies genetic variation in ^{137}Cs (tomato – 3.1; cabbage –3.3; carrot – 3.0; onion – 0.8-fold) and ^{90}Sr accumulation (tomato – 1.8; cabbage –2.6; carrot – 1.5; onion – 2.3-fold). It gives a possibility to select the genotypes with low content of radionuclides.

Keywords: nitrates; heavy metals; radionuclides; vegetable crops; interspecies genetic variation; hybrids; diallel crosses

C. Mothersill et al. (eds.), Multiple Stressors: A Challenge for the Future, 225–234.
© 2007 *Springer.*

Introduction

At present the quality of farm produce is determined not only by the content
of useful substances (proteins, fats, carbohydrates, vitamins, etc.), but also by
the pollutant accumulation (nitrates, heavy metals, radionuclides, pesticides,
etc.). All high nutrition value of any farm produce can become useless if it
contains some toxic substances exceeding hygienic norms.

The problem of pollution is connected with anthropogenic contamination of agrolandscape (industry, transport, agriculture) and with Chernobyl
catastrophe. In Belarus about 20% of the territory is contaminated with radionuclides. Considerable regions of the Ukraine and Russia are contaminated
as well. Vegetables can actively accumulate some pollutants in the productive
part. Problems of pollutants in vegetable growing in Belarus are complicated
by the fact that more than 80% of vegetables are grown on households including those situated in pollutant contaminated areas.

The process of pollutant accumulation in farm produce depends on three
main factors (Kilchevsky and Khotylyova, 1997): (i) genetical (crop and variety
characters, detering the input, transport, accumulation and detoxication of
the pollutant); (ii) environmental (proximity to the source of pollution and the
intensity of pollution, abiotic and biotic factors of environment, landscape etc);
(iii) agrotechnical (doses and terms of fertilizer and pesticide application, control
of pollutant intake by plants from the soil using agrotechnical methods, etc.).

Andryushchenko (1981), Gamzikova (1992) and Kilchevsky and Khotylyova
(1997) think that the most radical and the cheapest way of lowering pollutant
accumulation in produce is plant breeding. This way is possible in determining
interspecies and intervariety genotype variability in pollutant accumulation,
finding out their genetic determination, donors with minimum pollutant accumulation, the development of breeding strategy in these traits.

So the object of our research is to study interspecies and intervariety variability of pollutant accumulation in productive parts of vegetable crops, as
well as the character of inheritance of nitrate and heavy metal accumulation
in tomato fruits in the open ground.

Materials and Methods

NITRATES

To study nitrate content inheritance in tomato fruits by diallel analysis
method we tested in 1991–1992 in open ground in Gorky Mogilev region
28 hybrid tomato combinations between 8 initial forms using two agricultural
back-grounds of mineral food (normal and higher than normal). Parent material was *Talalihin*(1), *Dohodny*(2), P-7(3), *Beta*(4), *Sub-arctic mini*(5),
Line-7(6), *L pimpinellifolium*(7), *Torosa*(8).The nitrate content in fruits (mg kg^{-1})

was analysed by the potentiometric method. The combining ability of samples was evaluated in method 2 of Griffing (1956), genetic inheritance parameters – in Hayman's method (1954).

HEAVY METALS

In experiment were studied parents *Talalihin*(1), *Dohodny*(2), *Sprint*(3), *Line -7*(4), *Opus*(5), *Povarek*(6), *Radek*(7) and the hybrids between them in the open ground in 1995. During the vegetative period plants were sprayed with the salt solution of cadmium and lead in doses of 0.25 of MPC (maximum permissible concentration) in the soil. Heavy metal content was evaluated in mg kg^{-1} of dry matter using the method of atomic absorption. The combining ability of samples was evaluated in method 2 of Griffing (1956), genetic inheritance parameters – in Hayman's method (1954).

RADIONUCLIDES

Studies were made of the five varieties of four vegetable crops (tomato, cabbage, carrot, onion) in the open ground in Bragin district, Gomel region in 1996–1998 at the density of the soil pollution of 10Ci km^{-2} (^{137}Cs) and 1 Ci km^{-2} (^{90}Sr). ^{137}Cs and ^{90}Sr content was determined in productive organs of vegetables by spectrometrical and radiochemical methods respectively.

Coefficient of accumulation (CA), being equal to the ratio of the specific radionuclides activity per unit plant sample (Bg|kg) to the specific soil activity (Bg|kg), was used as a parameter evaluating radionuclides migration in the soil-plant system.

Results and Discussion

NITRATES

Nitrate content in tomato varieties and hybrids on the experimental plot (control background) ranges from 14.2 to 55.0mg kg^{-1} in 1991, from 10.0 to 37.5mg kg^{-1} in 1992 (Table 1). On the higher background nitrate content in 1991 changed from 13.5 to 13.3mg kg^{-1}, in 1992 – from 10 to 55.3mg kg^{-1} (Table 1). The lowest effect GCA had *L. pimpinellifolium* and Torosa, the highest – Beta, Line - 7 and P-7. The MPC of nitrate content in tomato fruits is 100mg kg^{-1}.

Heterosis effect can be evaluated according to the level of dominance Hp (Table 2). In the majority of cases (39.3–46.4%) the effect of negative superdominance manifested itself that is heterosis towards decreasing nitrate content in fruits. There have been frequent cases of intermediate inheritance (32.1–46.4%). Positive superdominance manifests itself seldom (10.7–21.4%).

TABLE 1. Contents of nitrates in tomato fruits at the high background of mineral nutrition

Years	No.	1	2	3	4	5	6	7	8	Effect GCA	Variance SCA
	Genotypes										
	1	22.9	27.4	35.6	17.4	19.7	23.3	15.5	17.7	−4.2	21.2
	2	24.6		19.3	16.4	31.1	39.0	19.5	14.7	−2.8	50.4
	3			24.9	29.5	25.5	69.4	19.5	33.9	3.8	173.4
1991*	4				113.3	22.0	31.8	16.1	16.3	13.2	751.9
	5					27.4	25.5	18.5	23.3	−2.4	6.6
	6						40.6	17.0	23.2	6.6	155.7
	7							24.2	18.6	−7.1	25.0
	8								13.5	−7.0	11.8
	1	32.7	30.3	35.7	20.3	32.3	26.8	15.3	10.0	0.3	39.9
	2	25.3		18.5	21.0	18.5	55.3	14.5	24.3	0	89.7
	3			29.2	27.8	25.8	50.7	17.7	22.8	2.4	40.5
1992*	4				49.2	27.7	25.0	18.8	20.0	2.6	75.9
	5					31.0	34.7	16.7	26.8	1.1	11.2
	6						37.8	19.8	29.5	8.4	116.1
	7							14.8	13.7	−8.7	2.3
	8								11.3	−6.3	13.0

*Significantly at $p < 0.05$

TABLE 2. Degree of dominance at the character "nitrate content in tomato fruits"

Year	Background	Hybrids	Hp < −1	−1 < Hp < 1	Hp > 1
1991	Control	Quantity	11	10	7
		%	39.3	35.7	25.0
	High	Quantity	13	12	3
		%	46.4	42.9	10.7
1992	Control	Quantity	13	9	6
		%	46.4	32.1	21.4
	High	Quantity	12	13	3
		%	42.9	46.4	10.7

The average level of dominance H/D ranged from 0.95 to 2.74 (Table 3) taking into account testing medium, it also confirms the manifestation of dominance and superdominance. On the higher agricultural backgrounds the frequency of dominant genes increases because the parameters $[\sqrt{(4DH2)}+F]$ / $[\sqrt{(4DH2)}−F]$ overreach 2.17–5.24%.

Correlation between trait meaning and the sum variance and covariance is more often positive which testifies the prevalence of dominance of minimum nitrate content in fruits. Meanings of fully dominant D_{max} and fully recessive R_{max} parent prove it. The studied varieties differ in one group of genes.

Thus nitrate accumulation is more often inherited by heterosis type towards their minimum content in fruits. In this connection we may consider heterosis breeding for producing hybrid F_1, as a method of decreasing nitrate accumulation in the produce.

TABLE 3. Parameters of Hayman at character "nitrate content in tomato fruits"

Background	Year	$\sqrt{H_1/D}$	$\dfrac{H_2}{4H_1}$	$\dfrac{\sqrt{4DH7 + F}}{\sqrt{4DH_2 -} \; F}$	r	D_{max}	R_{max}	$\dfrac{h^2}{H_2}$
Control	1991	0.95	0.27	1.30	0.65	14.96	34.31	0.21
	1992	2.74	0.22	1.45	−0.37	18.53	13.60	0.06
High	1991	1.10	0.12	5.24	0.98	16.87	123.13	0.69
	1992	1.08	0.19	2.17	0.89	12.96	61.02	0.90

HEAVY METALS

The study of cadmium accumulation in tomato fruit revealed the difference between parents by a factor of 5.3 in 1995 and 4.1 in 1996 (Table 4).

All parents and hybrids accumulated cadmium against the contaminated backgrounds more than MPC (0.03 mg/kg). It may be a reflection of poor mechanism of cadmium detoxication in plants and a high potential danger of cadmium under air pollution of agrocenosis. The difference between hybrids in cadmium accumulation was 12.9 in 1995 and 8.6 in 1996; with a whole diallel scheme – 12.8 in 1995 and 15.5 in 1996. Varieties *Talalikhin* 186 and *Dokhodny* may be used as donors of

TABLE 4. Content of cadmium and lead in tomato fruits of diallel crosses (mg/kg)

*No. genotypes	1	2	3	4	5	6	7	Effect QCA	Variance SCA
Content of Cadmium, 1995 (LSD$_{05}$ = 0.22)									
1	0.41	0.65	0.24	0.42	0.64	0.89	0.45	0.05	0.04
2		0.13	0.36	0.13	0.52	0.60	0.32	−0.09	0.04
3			0.53	0.23	0.21	0.22	0.19	−0.12	0.04
4				0.51	1.67	0.28	0.61	0.09	0.21
5					0.69	0.26	0.13	0.14	0.22
6						0.57	0.33	0.01	0.06
7							0.042	−0.08	0.03
Content of Cadmium, 1996 (LSD$_{05}$ = 0.12)									
1	0.36	0.22	0.30	0.29	0.20	0.30	0.30	0.02	0.00
2		0.15	0.17	0.14	0.38	0.25	0.20	−0.05	0.01
3			0.54	0.43	0.22	0.06	0.04	0.02	0.03
4				0.46	0.25	0.24	0.20	0.04	0.01
5					0.55	0.09	0.09	0.02	0.03
6						0.62	0.05	0.01	0.04
7							0.37	−0.06	0.02
Content of Lead, 1995 (LSD$_{05}$ = 0.41)									
1	2.03	0.37	0.39	1.14	0.48	1.13	1.08	0.09	0.28
2		1.25	0.43	1.24	0.88	0.53	0.63	−0.14	0.19
3			2.58	0.43	0.53	0.31	1.16	0.06	0.72
4				2.97	0.21	0.41	0.93	0.27	0.71
5					0.89	0.34	0.46	−0.35	0.20
6						1.74	1.88	0.02	0.45
7							1.08	0.05	0.15
Content of Lead, 1996 (LSD$_{05}$ = 0.26)									
1	2.16	0.44	0.93	0.37	0.59	0.55	1.11	0.12	0.37
2		2.05	0.45	1.31	1.88	0.51	0.49	0.22	0.56
3			0.79	0.31	0.75	0.75	0.92	−0.17	0.05
4				0.68	0.39	0.40	0.57	−0.28	0.10
5					0.94	0.37	0.52	−0.10	0.22
6						1.77	1.80	0.08	0.35

*Talalikhin 186 (1), Dokhodny (2), Sprint (3), Line 7 (4), Opus (5), Povarek (6), Radek (7).

low cadmium accumulation. *Opus* and *Povarek* had the highest content of cadmium.

Parents differed in accumulation of lead against the contaminated background by a factor of 3.3 (1995) and 3.2 (1996), hybrids – 9.0 and 6.1, all genotypes – 14.1 and 7.0 respectively. Variety *Opus* can be used as a donor of minimum accumulation of lead. Varieties *Talalikhin* 186 and *Povarek* accumulated a maximum level of lead in fruits.

Analysis of dominance direction for the studied traits showed that a general type of inheritance of cadmium and lead accumulation in 1995–1996 was an overdominance (negative heterosis) to the decrease of heavy metal accumulation in fruits (Table 5).

Hayman's parameters supported our findings about type of inheritance of heavy metal accumulation in tomato fruits (Table 6).

The average degree of dominance $\sqrt{(H_1/D)}$ was more than one that showed an exhibition of overdominance.

The relationship of dominant and recessive genes was from 1.72 to 1.97 indicating the prevalence of dominant alleles. Correlation coefficient r between trait value and sum of variances and covariances moved to 1. It shows that maximum trait value will be in the genotypes with full recessive

TABLE 5. Degree of dominance Hp at the Cadmium and Lead accumulation in tomato fruits

Type of contamination	Parameters	$H_p > 1$	$-1 \leq H_p \leq 1$	$H_p < -1$
1995				
Cadmium	Quantity genotypes	6	6	9
	%	28.6	28.6	42.8
Lead	Quantity genotypes	1	1	19
	%	4.8	4.8	90.4
1996				
Cadmium	Quantity genotypes	0	5	16
	%	0	23.8	76.2
Lead	Quantity genotypes	1	4	16
	%	4.8	19	76.2

TABLE 6. Hayman parameters at the Cadmium and Lead accumulation in tomato fruits

Trait	Years	$\sqrt{H_1/D}$	$H_2/4H_1$	$\frac{\sqrt{4DH_1}+F}{\sqrt{4DH_1}-F}$	r	D_{max}	R_{max}	$h^2/4H_1$
Cadmium	1995	1.82	0.21	1.72	0.82	0.24	0.98	0.08
accumulation	1996	1.84	0.21	1.97	0.66	0.25	0.72	1.49
Lead	1995	1.34	0.22	1.96	0.98	0.902	3.09	1.85
accumulation	1996	1.15	0.21	1.85	0.96	0.72	2.12	1.96

allele. This information was supported by the value of full dominant D_{max} and full recessive R_{max} parents.

Variation in cadmium and lead accumulation was controlled by one-two loci because the parameter $h^2/4H_1$ changed from 0.08 to 1.96.

Analysis of the adaptive ability and ecological stability of genotypes for traits of heavy metal accumulation was evaluated by parameter b_i (regression coefficient; Eberhart and Russell, 1966) and parameter Sg_i (relative genotype stability; Kilchevsky and Khotylyova, 1997).

It was shown that tomato hybrids on the average accumulate less cadmium and lead in fruits in comparison with parents and differ in the greater stability of accumulation traits (values of b_i and Sg_i were lower).

RADIONUCLIDES

In experiment using four vegetable crops, we established differences between varieties in ^{137}Cs coefficient of accumulation in productive organs of all crops in all years of fruits (Table 7). Maximum differences between varieties were:

TABLE 7. Coefficient of ^{137}Cs accumulation in the varieties of vegetables

Crops	Varieties	Years			Average
		1996	1997	1998	
Tomato	Sprint	0.0047	0.0045	0.0054	0.0049
	Peramoga 165	0.0093	0.0067	0.0071	0.0077
	Kalinka	0.0129	0.0108	0.0078	0.0105
	Dohodny	0.0158	0.0153	0.0142	0.0151
	Rusha	0.0089	0.0078	0.0095	0.0087
	LSD	0.0037	0.0025	0.0030	0.0025
Cabbage	1 Gribovsky 147	0.0222	0.0263	0.0106	0.0197
	Belorusskaya 85	0.0073	0.0045	0.0060	0.0060
	Rusinovka	0.0207	0.0149	0.0127	0.0161
	Amager 611	0.0127	0.0123	0.0078	0.0109
	Turkiz	0.0192	0.0235	0.0151	0.0193
	LSD	0.0047	0.0068	0.0049	0.0068
Carrot	Nantskaya	0.0067	0.0047	0.0065	0.0060
	Losinoostrovskaya	0.0084	0.0134	0.0117	0.0112
	Vitaminnaya	0.0162	0.0101	0.0114	0.0126
	NIIOH 336	0.0192	0.0190	0.0151	0.0178
	Shantene	0.0222	0.0179	0.0142	0.0181
	LSD	0.0035	0.0029	0.0072	0.0051
Onion	Vetraz	0.0246	0.0214	0.0170	0.0210
	Shutgarten rizen	0.0257	0.0194	0.0173	0.0208
	Iantarny	0.0258	0.0235	0.0218	0.0237
	Klivitcky ruzhovy	0.0263	0.0201	0.0158	0.0207
	Skvirsky	0.0231	0.0170	0.0192	0.0198
	LSD	0.0075	0.0039	0.0056	0.0034

tomato – 3.1, cabbage – 3.3, carrot – 3, onion – 0.8 times. Tomato varieties which accumulated least of all of ^{137}Cs were Sprint; carrot – Nantskaya, cabbage – Belarusskaya 85, Amager 611.

Marked differences between varieties in coefficient of ^{90}Sr accumulation in productive organs (Table 8) were found. Maximum difference between varieties was: tomato-1.8; cabbage-2.6; carrot-1.5; onion-2.3 times.

In their ability to accumulate ^{137}Cs vegetables can be placed in the order of priority in the following way: onion, cabbage, carrot, tomato. The same line can be made for vegetables in their ability to accumulate ^{90}Sr: onion, carrot, cabbage, tomato.

Revealed differences in radionuclide accumulation in varieties allow selecting varieties for lower radionuclide intake together with vegetables 2–3 times. The presence of genetic variability makes it possible to carry out breeding aimed at lowering radionuclide accumulation in vegetables.

TABLE 8.   Coefficient of ^{90}Sr accumulation in the varieties of vegetables

| Crops | Varieties | Years | | | |
		1996	1997	1998	Average
Tomato	Sprint	0.0302	0.0191	0.0204	0.0232
	Peramoga 165	0.0596	0.0369	0.0304	0.0423
	Kalinka	0.0498	0.0229	0.0227	0.0318
	Dohodny	0.0445	0.0160	0.0199	0.0268
	Rusha	0.0325	0.0195	0.0176	0.0232
	LSD	0.0179	0.0062	0.0068	0.0094
Cabbage	^{1}1 Gribovsky 147	0.0901	0.0934	0.0756	0.0863
	Belorusskaya 85	0.1942	0.2017	0.1931	0.1963
	Rusinovka	0.1971	0.1896	0.1805	0.1891
	Amager 611	0.1621	0.1776	0.1700	0.1699
	Turkiz	0.2469	0.02200	0.2120	0.2263
	LSD	0.0283	0.0231	0.0284	0.0181
Carrot	Nantskaya	0.1959	0.1612	0.1616	0.1729
	Losinoostrovskaya	0.2171	0.1990	0.2036	0.2066
	Vitaminnaya	0.2093	0.1759	0.1721	0.1858
	NIIOH 336	0.2779	0.2305	0.2267	0.2450
	Shantene	0.2903	0.2580	0.2435	0.2640
	LSD	0.0477	0.0276	0.0234	0.0148
Onion	Vetraz	0.4786	0.4098	0.4062	0.4315
	Shutgarten rizen	0.7925	1.1206	1.0633	0.9921
	Lantarny	0.6359	0.5345	0.5210	0.5638
	Klivitcky ruzhovy	0.5525	0.5506	0.5456	0.5496
	Skvirsky	0.6529	0.5510	0.5420	0.5820
	LSD	0.0953	0.0759	0.0931	0.1876

Conclusion

Analysis of literature and results of carried out experiments show that intervariety variability in pollutant accumulation in vegetables are quite enough to single out genotypes lowering their intake with food 2–5 times. Analysis of inheritance of nitrate and heavy metal accumulation in tomato fruits shows that the main type of inheritance in conditions of produce contamination is superdominance towards decrease of pollutant content. It allows us to admit that heterosis breeding is the method of lowering pollutant accumulation in tomatoes.

In our opinion (Kilchevsky and Khotylyova, 1997) general strategy of breeding towards lowering pollutant accumulation in agricultural produce should include three main stages:

1. Evaluation of the initial material corresponding to a number of useful traits and pollutant accumulation in contaminated areas, selection of initial forms for hybridization which correspond to tasks of breeding.

2. Selection in early generations (F_2–F_5) according to the traits of usefulness as well as to traits of correlation associated with accumulation of pollutants in contaminated areas.

3. Carrying out of competitive and ecological tests on the polluted and clean territory to evaluate selection results.

High price of carrying out analysis of pollutant content will hardly permit us to control their accumulation in many samples in F_2–F_5. Hence it is very important to study correlation of these traits with other morphobiological and physiological traits for indirect selection. Breeding aimed at lowering pollutant accumulation in agricultural produce is the most radical, cheap and economically based means of lowering pollutant intake with food.

References

Andryushchenko, V. K. Methods of optimization of biochemical breeding of vegetables, *Shtiinca*, Kishinev (1981).
Eberhart, S.O., W.F.Russel. Stability parameters for comparing varieties. *Crop Sci.*, 6,1–36–40 (1966).
Gamzikova, O. I. Genetical aspects of wheat responsibility to conditions of mineral nutrition, *Dis. Dr. Sci.*, Novosibirsk (1992).
Griffing, B. Concept of general and specific combining ability in relation to diallel crossing systems, *Austr. J. Biol. Sci.* 9, 463–493 (1956).
Hayman, B. I. The theory and analysis of diallel crosses, *Genetics* 39, 789–809 (1954).
Kilchevsky, A. V. and L.V. Khotylyova, Ecological plant breeding, *Technologia*, Minsk (1997).

CHAPTER 15

RADIATION RISKS IN THE CONTEXT OF MULTIPLE STRESSORS IN THE ENVIRONMENT – ISSUES FOR CONSIDERATION

CARMEL MOTHERSILL AND COLIN SEYMOUR

Medical Physics and Applied Radiation Sciences Department, McMaster University, Hamilton, Ontario, Canada
e-mail: mothers@mcmaster.ca

Abstract: The field of multiple stressors is highly complex. Agents can interact in an additive, antagonist or synergistic manner. The outcome following low dose multiple stressor exposure also is impacted by the context in which the stressors are received or perceived by the organism or tissue. Modern biology has given us very sensitive tools to access change following stressor interaction with biological systems at several levels of organization but the effect-harm-risk relationship remains difficult to resolve. This paper reviews some of the issues, using low dose ionizing radiation as a common stressor and chemicals known to act through similar mechanisms, as examples. Since multiple stressor exposure is the norm in the environment, it is essential to move away from single stressor based protection and to develop tools, including legal instruments, which will enable us to use response-based risk assessment. The problem of radiation protection in the context of multiple stressors includes consideration of humans and non-humans as separate groups requiring separate assessment frameworks. This is because for humans, individual survival and prevention of cancer are paramount but for animals, it is considered sufficient to protect populations and cancer is not of concern. The need to revisit this position is discussed not only from the environmental perspective but because the importance of pollution as a cause of non-cancer disease is increasingly being recognized. Finally a way forward involving experimental assessment of biomarker performance coupled with modeling is discussed.

Keywords: multiple stressors; radiation; low dose exposure; biomarkers

Introduction

Biological systems are highly complex. Modern biology has given us very elegant tools for investigating these systems and understanding mechanisms but where decisions have to be made about the safety of radiation or chemical

C. Mothersill et al. (eds.), Multiple Stressors: A Challenge for the Future, 235–246.

pollutants in the environment, it becomes very difficult to determine the relationship between a detectable effect in a system, and the ultimate consequence for the organism or population, of that insult. This relationship is particularly obscure where the level of exposure to the agent is very low or when multiple agents occur in the system under examination. Much of the uncertainty surrounding the risk of exposure to low doses of single or multiple stressors is due to this inability to determine risk associated with molecular effects. Depending on the perspective of the "stakeholder" molecular effects can be interpreted as highly dangerous or just natural responses to environmental perturbations. Clearly environmental pollutants are not going to disappear so it becomes important to find objective methods for linking effects with risks and also to find regulatory and legal mechanisms for dealing with low doses of multiple pollutants. This paper will address some of the issues which complicate the field and lead to the uncertainty. It will then suggest possible approaches to solving the problems.

Mechanism Issues

It has been known for a long time that low doses of single agents can have fundamentally different biological effects to high doses. As early as 1500 Paracelsus famously said

> *'Alle Ding' sind Gift und nichts ohn' Gift; allein die Dosis macht, das ein Ding kein Gift ist."*

"All things are poison and nothing is without poison, only the dose permits something not to be poisonous."

In environmental protection, it is common to discuss four levels of "dose" of single toxic agents (see Figs. 1a and b), ranging from no effect doses through doses where organisms can accommodate the toxin, through to doses causing reversible damage and finally doses causing irreversible or lethal damage. The grey areas of concern in the multiple stressor field are the boundaries between the categories and how these might be changed if more than one stressor is present.

A less appreciated concern is that the actual mechanisms operating at low doses may not be in a continuum and that mechanistic switches may operate at specific dose thresholds. How these switches might be affected if multiple stressors are present is unknown. Similarly, many low dose non-targeted effects show saturable responses i.e. the dose response relationship is linear initially but then plateaus. The data for the direct dose v bystander effect for radiation is shown in Fig. 2, adapted from Seymour and Mothersill, (2000).

There is a clear saturation of the bystander effect at low doses while the direct effect is not obvious until a dose of 0.5Gy has been delivered.

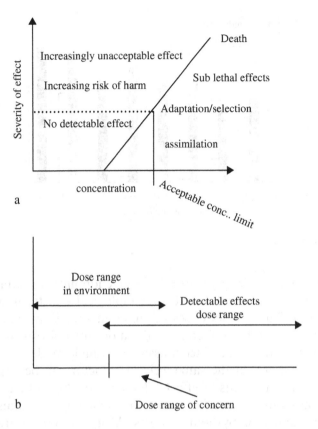

Fig. 1. (a) Levels of dose response categories for mono-stressor regula*tion.* (b) Simple model of mono-stressor protection

Increasing direct radiation then shows a linear dose response relationship. A key question is whether this saturable low dose bystander effect could lead to apparent "protective" effects if two stressors operating by similar mechanisms are present? Stressor 1 might fully use all the receptors or induce the maximum response which the system can produce, meaning that stressor 2 is not seen as inducing any additional effect. This is where the link between effect and harm or risk is critical to understand because "no additional effect" does not mean "no additional harm". The harm could be as a result of the system switching to a new level of response which may not be the one being measured.

Another issue is that of the "critical compound" in the mix. Since some stressors are much more toxic at a given dose than others, it will be necessary to devise "iso-effect" curves linking stressors in terms of some indicator of effect/harm/risk. This was done for single stressors by our group (Mothersill et al., 1998) using the induction of delayed cell death, which is associated

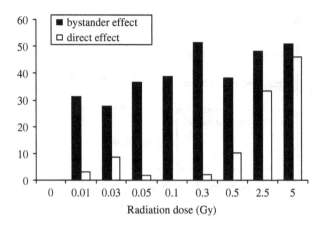

Fig. 2. Direct v bystander effect

with genomic instability (Morgan, 2003), as an endpoint. A useful approach may be to adopt the concept of "relative biological effectiveness" used in radiation biology. This allows doses of different radiation qualities to be added by applying weighting factors to the more toxic forms of radiation. The approach is controversial since it assumes a single mode of damage for all the radiations and that the same weighting factor can be applied for a variety of different endpoints of effect, which is probably not true.

Another mechanistic issue is the role of "enabling mechanisms" such as genomic instability or bystander effects (Morgan, 2003; Lorimore and Wright, 2003; Mothersill and Seymour, 2006). These mechanisms can be induced by one agent and make mutations or apoptosis etc much more likely if another agent is experienced by the system. There are concrete examples of this (Lord, 1999; Hoyes et al., 2000, 2001; Lord and Woolford, 2002; Barber et al., 2006; Barber and Dubrova, 2006) although the stressors were not applied at the same time or even to the same generation, again highlighting the complexity of this area.

Hormesis is a very controversial area but one which must be considered in the multiple stressor field. Hormesis is defined as beneficial effects occurring after low dose exposure to agents which are toxic at high doses. Calabrese has reviewed this field extensively in over 230 papers (e.g. Calabrese and Baldwin, 2001; Calabrese, 2005), and has concluded that beneficial effects at low doses are the norm not the exception. What will happen when multiple stressors are present is unknown. Hormetic mechanisms are thought to include adaptive responses, such as DNA repair induction, which condition cells or organisms making them more able to respond to stress. However, immune system stimulation or metabolic stimulation are also likely mechanisms (Sato et al., 1984; Boonstra et al., 2005; Sakai, 2006). A critical element in the context of multiple stressors must be whether the stressors are all present at the same time or whether one stressor can make the system less or more vulnerable to other agents. There is considerable evidence for reduced effects in the adaptive response field (Broome et al., 1999; Mitchel et al., 1999; Hall et al., 2000) where for example, heat stress can adapt cells to subsequent radiation stress and vive versa. The best evidence for adverse effects occurring is the transgenerational evidence already alluded to where exposure to of parents to ionizing radiation made the progeny more likely to develop cancers in response to a chemical carcinogen.

Apart from the implications of multiple stressor mechanisms for risk assessment, it is important to continue efforts to understand how multiple or single stressor non-targeted low dose effects happen and whether it is possible to modulate them. In the radiation field progress in this area has been reviewed extensively and oxidative stress is now known to be a key cellular effect, which perpetuates non-targeted effects (Morgan, 2003; Little and Morgan, 2003; Prise et al., 2006; Mothersill and Seymour, 2006). Antioxidants work at the cellular level and in tissue models to reduce non-targeted effects of ionizing radiations (Dahle et al., 2005; Prasad, 2005; Seymour et al., 2005; Konopacka and Rzeszowska-Wolny, 2006; Lyng et al., 2006) although there are no data available about their effectiveness in vivo. With chemicals there are problems conclusively demonstrating non-targeted effects because the persistence of the chemical cannot be excluded and therefore it is difficult to distinguish between true delayed or non-targeted effects and effects due to residual chemicals. However, there are convincing data from Glaviano et al. (2006) who looked at Chromium and Vanadium induced delayed chromosomal damage over 30 days in the progeny of cells originally exposed for only 24 h to sub-toxic doses. Other evidence comes from work by our group in collaboration with Salbu's group in Norway (Mothersill et al JNER submitted). Here rainbow trout were exposed to sub-toxic levels of aluminum and cadmium the exposed to low doses of Cobalt 60 gamma rays. About four tissues were examined for production of bystander signals using a reporter system. Results (ms in preparation) show that the metals interact with radiation but the manner of the interaction (synergistic, additive or antagonistic) varies with the tissue and no overall universal pattern is seen. Again this is not surprising from the biological standpoint but it complicates regulatory issues!

Possible Biomarkers

By implication, a move away from dose driven mono stressor regulation, means that biomarkers of exposure or response must be selected and validated. Following multiple stressor exposures, response biomarkers are

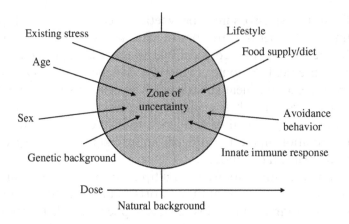

Fig. 3. Factors influencing outcome and uncertainty in multiple stressor scenarios

likely to be more useful and easier to validate because otherwise, causality becomes an issue while when monitoring response, it is possible to define a cell/tissue/organism as, for example, "stressed" without having to identify which of the stressors contributed most to the stress. An important caveat when looking for multiple stressor biomarkers is to accept that a biomarker is probably only good as an indicator of one aspect of the response and results cannot be extrapolated to generalize from for example a biomarker such as elevated ROS to carcinogenic activity or mortality. Validating biomarkers as being relevant and meaningful requires their use in situations where the link between the stressor and the response is already known. Alternatively it is useful just to know a system has been stressed. This could provide first line screening for adverse multiple stressor effects. Generic stress biomarkers at the cellular level could include elevated ROS, P450 up-regulation, calcium influx, mitochondrial membrane depolarization effects and elevated apoptosis. These effects all occur after low doses of many stressors (Mothersill and Mullenders, 2006). They do not however indicate risk, merely that the system has induced a cellular stress response. They are highly dependent on genetic background and on "context" i.e. other environmental factors such as lifestyle, diet and existing health factors (Fig. 3). At the organism level, generic stress responses include elevated cortisol and immune system effects (e.g. Roberts, 2000; Bilbo et al., 2002; Yang et al., 2002). Stress responses at population/species levels are generally behavioral or related to fecundity but discussion of these is outside the scope of this review.

A Possible Way Forward

As discussed earlier, what is needed in the multiple stressor field is a way to determine the risk of low doses of chemicals and radiation for human health and environmental health in its broadest context. The problem is that the new non-targeted effects dissociate dose from risk and there are no simple ways to determine the relationship; i.e.

- Dose is not proportional to effect even at the cellular level
- Effect is not proportional to harm even at the organ level and
- Harm is not proportional to risk at the organism level
- When moving to the environment, none of the above are simply related to survival at the ecosystem level.

In addition to this, the interactions are so complex and the species and stressors involved so diverse that really only a modelling approach can produce testable hypotheses. We propose a tight interaction between experimental biologists and modellers and suggest the approach outlined below as an example using the bystander effect as a test "biomarker of generic stress response at the cellular level.

Modeller–Experimentalist Interactions

To develop models which can predict multiple stressor risk is highly complex and needs to be broken down into a series of sub models. For the purposes of this review we have decided to use the bystander effect induced by radiation as an example and have attempted to define what data are needed in order to allow modelers to develop meaningful models.

Bystander effect example

In the specific case of what the bystander effect might do to the dose – risk equation

1. We need many more phenomenological experiments, repeated in different labs with different or similar systems
2. We need negative results presented and discussed. Much information is lost when negative results are not reported. For example, often enhanced survival following bystander protocols is dismissed as "no effect" or "system not responding or working". Significant effects of any sort are bystander effects.

The working model developed by our group for the bystander effect is shown later (Fig. 4).

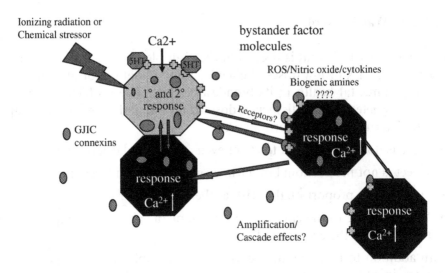

Fig. 4. Model of the bystander effect

This is an experimental model based on all we know from experiments done by our group using medium transfer techniques to test for the presence of the signal. Using this we can develop testable hypotheses and get pointers for the questions which need to be addressed. Some obvious ones are listed below as an example of how this approach might be useful.

Specific examples of data needed in order to model bystander effects:

1. We need quantitative data for several cell lines manifesting different types of bystander effects, specifically we need data about the signal strength in relation to cell number irradiated in defined media volumes. This appears only to have been done once in one cell line (Mothersill and Seymour, 1997). This is needed to determine irradiation volume – likely signal production relationship.

2. We need experimental data concerning dilution effects of signal in media – no data are published concerning this – our own unpublished data suggest even a 1:1 dilution completely loses the effect. This again relates to volume of exposed tissue/blood.

3. We need comparative and recent studies to establish if clastogenic factors and bystander effects are the same phenomenon. This would mean that the old studies of persistent clastogenic effects dating back to 1921 (reviewed in Marozik et al., 2007) could be included in modeling analysis.

4. We need to dissect out the signal production from the response to the signal in a quantitative way. This would allow models to be validated if inhibitors of production/response could be identified and effects quantified in stochiometric ways.

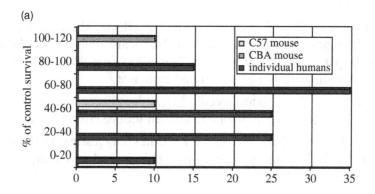

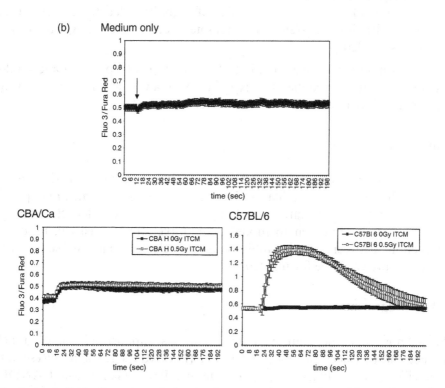

Fig. 5. (a) Individual/genetic variation in the cytotoxic properties of bystander medium (b) Calcium ratios in control and 0.5 Gy TBI CBA/CA and C57BL/6 mice

5. We need to look at bystander effects in animal strains know to be at risk of radiogenic cancer (or known to be resistant). These studies need to be quantitative and done at different doses including low and environmentally relevant doses. CBA strains and C57Bl 6 are obvious starters (see data adapted from Mothersill et al., 2005 Fig. 5a and b) but knockouts would also be good. Mix match experiments with these would also feed into point 4.

6. We need reporting of "positive bystander effects" these tend to get dismissed as negative results. They probably are real effects. Again mix match experiments would be valuable to allow "weight" to be attached to the relative importance of signal and response.

7. We need data about tissue specific bystander effects and whether signals produced by irradiation of one tissue can induce effects in other tissues

8. We need to confirm and extend studies suggesting that bystander effects "drive" genomic instability. This introduces temporal terms into the risk model

9. Signal production may be independent of dose but is not likely to be independent of target cell number or receiving cell number. It is likely that thresholds exist for molecules of signal needed to trigger target cell production of signal and recipient cell response. Quantitative studies needed.

Modelling multiple stressors could follow this type of approach and could help experimentalists define their experiments in a way which would actively support the development of testable models.

Conclusions

This paper discusses several issues relating to the management of risks associated with low doses of radiation. It suggests biomarkers which may prove useful as generic stress markers and proposes, using bystander effects as an example, a new approach to modeling multiple stressor risks using close interactions between experimentalists and modelers so that testable hypotheses can more easily be formulated.

Acknowledgements

We acknowledge the continued and generous support of the National Science and Engineering Research Council (NSERC), The Canada Research Council's Canada Chairs programme, Ontario Power Generation, CANDU Owner's Group and Bruce Power.

References

Barber, R.C., Dubrova, Y.E., 2006, The offspring of irradiated parents, are they stable? *Mutat. Res.* **598**:50–60. Review.

Barber, R.C., Hickenbotham, P., Hatch, T., Kelly, D., Topchiy, N., Almeida, G.M., Jones, G.D., Johnson, G.E., Parry, J.M., Rothkamm, K., Dubrova, Y.E., 2006, Radiation-induced transgenerational alterations in genome stability and DNA damage. *Oncogene*, **25**:7336–7342.

Bilbo, S.D., Dhabhar, F.S., Viswanathan, K., Saul, A., Yellon, S.M., Nelson, R.J., 2002, Short day lengths augment stress-induced leukocyte trafficking and stress-induced enhancement of skin immune function. *Proc. Natl. Acad. Sci. USA.* **99**:4067–4072.

Boonstra, R, Manzon, R.G., Mihok, S., Helson, J.E., 2005, Hormetic effects of gamma radiation on the stress axis of natural populations of meadow voles (Microtus pennsylvanicus). *Environ. Toxicol. Chem.* **24**:334–343.

Broome, E.J., Brown, D.L., Mitchel, R.E., 1999, Adaptation of human fibroblasts to radiation alters biases in DNA repair at the chromosomal level. *Int. J. Radiat. Biol.* **75**:681–690.

Calabrese, E.J. 2005, Historical blunders: how toxicology got the dose-response relationship half right. *Cell Mol. Biol. (Noisy-le-grand)*, **51**:643–654.

Calabrese, E.J., Baldwin, L.A., 2001, Hormesis: U-shaped dose responses and their centrality in toxicology. *Trends Pharmacol. Sci.* **22**:285–291. Review.

Dahle, J., Kvam, E., Stokke, T. 2005, Bystander effects in UV-induced genomic instability: antioxidants inhibit delayed mutagenesis induced by ultraviolet A and B radiation. *J. Carcinog.* **4**:11–17.

Glaviano, A., Nayak, V., Cabuy, E., Baird, D.M., Yin, Z., Newson, R., Ladon, D., Rubio, M.A., Slijepcevic, P., Lyng, F., Mothersill, C., Case, C.P., 2006, Effects of hTERT on metal ion-induced genomic instability. *Oncogene* **25**:3424–435.

Hall, D.M., Xu, L., Drake, V.J., Oberley, L.W., Oberley, T.D., Moseley, P.L., Kregel, K.C., 2000, Aging reduces adaptive capacity and stress protein expression in the liver after heat stress. *J. Appl. Physiol.* **89**:749–759.

Hoyes, K.P, Hendry, J.H., Lord B.I., 2000, Modification of murine adult haemopoiesis and response to methyl nitrosourea following exposure to radiation at different developmental stages. *Int. J. Radiat. Biol.* **76**:77–85.

Hoyes, K.P., Lord, B.I., McCann, C., Hendry, J.H., Morris, I.D., 2001, Transgenerational effects of preconception paternal contamination with (55)Fe. *Radiat. Res.* **156**:488–494.

Konopacka, M., Rzeszowska-Wolny, J., 2006, The bystander effect-induced formation of micronucleated cells is inhibited by antioxidants, but the parallel induction of apoptosis and loss of viability are not affected. *Mutat. Res.* **593**:32–38.

Little, J.B., Morgan, W.F. (eds), 2003, Special issue of on Genomic Instability, *Oncogene* **13**, 6977.

Lord, B., 1999, Transgenerational susceptibility to leukaemia induction resulting from preconception, paternal irradiation. *Int. J. Radiat. Biol.* **75**:801–810. Review.

Lord, B.I., Woolford, L.B., 2002, Induction of stem cell cycling in mice increases their sensitivity to a chemical leukaemogen: implications for inherited genomic instability and the bystander effect. *Mutat. Res.* **501**:13–17.

Lorimore, S.A., and Wright, E.G., 2003, Radiation Induced genomic instability by stander effects; (see final page).

Lyng, F.M., Maguiree, P., Kilmurray, N., Mothersill C., Shac Folkard M., Prise, K.M., 2006, Apoptosis is initiated in.

Marozik, P., Mothersill, C., Seymour, C.B., Mosse, I., Melnov, S., 2007, Bystander effects induced by serum from survivors of the Chernobyl accident *Exp. Haematol.*, **35**:55–63.

Mitchel, R.E, Jackson, J.S., McCann, R.A., Boreham, D.R. 1999, The adaptive response modifies latency for radiation-induced myeloid leukemia in CBA/H mice. *Radiat. Res.* **152**:273–279.

Morgan, W.F. 2003, Non-targeted and delayed effects of exposure to ionizing radiation: I. Radiation-induced genomic instability and bystander effects in vitro. *Radiat. Res.* **159**: 567–580 Review.

Mothersill, C., Mullenders, L., 2006, Eds, Special Issue of Mutation Research, **597**:1–2.

Mothersill, C., Seymour, C.B., 2006, Actions of radiation on living cells in the "post-bystander" era. *EXS,* **96**:159–177. Review.

Mothersill, C., Lyng, F., Seymour, C., Maguire, P., Lorimore, S., Wright, E., 2005, Genetic factors influencing bystander signaling in murine bladder epithelium after low-dose irradiation in vivo. *Radiat Res.* **163**:391–399.

Mothersill, C., Crean, M., Lyons, M., McSweeney, J., Mooney, R., O'Reilly, J., Seymour, C.B., 1998, Expression of delayed toxicity and lethal mutations in the progeny of human cells surviving exposure to radiation and other environmental mutagens. *Int J Radiat Biol.* **74**:673–680.

Prasad, K.N., 2005, Rationale for using multiple antioxidants in protecting humans against low doses of ionizing radiation. *Br. J. Radiol.* **78**:485–492. Review.

Prise, K.M., Folkard, M., Kuosaite, V., Tartier, L., Zyuzikov, N., Shao, C. 2006. What role for DNA damage and repair in the bystander response? *Mutat. Res.* **597**:1–4.

Roberts, J.E., 2000, Light and immunomodulation. *Ann. N.Y. Acad. Sci.* **917**:435–445. Review.

Sakai, K., 2006, Biological responses to low dose radiation-hormesis and adaptive responses. *Yakugaku Zasshi.* **126**:827–831.

Sato, K., Flood, J.F., Makinodan, T., 1984, Influence of conditioned psychological stress on immunological recovery in mice exposed to low-dose X irradiation. *Radiat. Res.* **98**:381–388.

Seymour, C.B., Mothersill, C. 2000, Relative contribution of bystander and targeted cell killing to the low-dose region of the radiation dose-response curve. *Radiat. Res.,* **153**: 508–511.

Yang, E.V., Bane, C.M., MacCallum, R.C., Kiecolt-Glaser, J.K., Malarkey, W.B., Glaser R. 2002, Stress-related modulation of matrix metalloproteinase expression. *Neuroimmunology* **133**:144–150.

CHAPTER 16

PROTECTION BY CHEMICALS AGAINST RADIATION-INDUCED BYSTANDER EFFECTS

PAVEL MAROZIK[*,1], IRMA MOSSE[1],
CARMEL MOTHERSILL[2], AND COLIN SEYMOUR[2]

[1]Institute of Genetics and Cytology National Academy of Sciences of Belarus, Akademicheskaya 27, 220072, Minsk, Belarus
[2]Medical Physics and Applied Radiation Science Unit, McMaster University, Nuclear Research Building, 1280 Main Street West, Hamilton, Ontario, Canada L8S 4K1

Abstract: The purpose of this work was to study possible mechanisms of radiation-induced bystander effect proceeding from its modification using different radioprotective substances: melanin, melatonin and α-tocopherol. All substances were able to statistically significant decrease the damaging effect of bystander factor from irradiated cells on non-irradiated. The protective effect against bystander irradiation was much less than against direct irradiation. Melatonin showed the best protective effect against both direct and bystander irradiations, and vitamin E – the least. According to the results, bystander factor may have physical component and oxidative nature.

Keywords: radiation-induced bystander effect; melanin; melatonin; α-tocopherol (vitamin E); HPV-G cells; radioprotection

Introduction

In April, 2006, 20 years have been passed since the explosion of the fourth block of Chernobyl nuclear power plant. This accident affected millions of people, and large territories were contaminated by radionuclides. As a result, background radiation levels increased, and people from contaminated territories are living constantly in low dose radiation conditions. The situation became more complicated after the discovery of non-direct radiation effects, which were not taken into account in procedures of radiation dose evaluation. Bystander effect is one of such indirect radiation effects.

The radiation-induced bystander effect is a phenomenon whereby the cellular damage is expressed in unirradiated neighboring or bystander cells,

247

C. Mothersill et al. (eds.), Multiple Stressors: A Challenge for the Future, 247–262.
© 2007 Springer.

connected or not to an irradiated cell or cells. The mechanism of this transfer of damage from irradiated cells to non-irradiated neighbours is still unknown. There is evidence that the bystander effect may have at least two separate pathways: through gap junctions (Nagasawa and Little, 1992; Deshpande et al., 1996; Lorimore et al., 1998) or by cell-culture mediated factors (Mothersill and Seymour, 1997). The nature of these factors remains unknown, but it was hypothesised that it may be a protein. A series of studies (Lehnert and Goodwin, 1997; Narayanan et al., 1999) suggest a mechanism in which the irradiated cells secrete cytokines or other factors that act to increase intracellular levels of reactive oxygen species in unirradiated cells. Barcellos-Hoff and Brooks (2001) have also hypothesised that TGFβ1, an extracellular sensor of damage, may also be involved in the bystander effect. Another possible mediator of the bystander effect is the apoptosis inducing factor, secreted by mitochondria in response to oxidative stress (Kroemer, 1997).

Both high LET alpha-particles and low LET γ-irradiation have been shown to induce this effect; however, it remains unclear whether the same signal is involved for both types of radiation. Also recently, bystander effect was induced using UV- (Dahle et al., 2005) and laser (Mosse et al., 2006) irradiation.

This effect has been studied extensively since 1992 by many researchers, but to date there have been no any published reports on the use of radio-protectors in modifying bystander responses. All studies are focused on understanding possible mechanisms of bystander effects and the nature of bystander factor. Meanwhile, modification of this phenomenon may help to understand its possible mechanisms, proceeding from the properties of modifying substances.

We used three different radioprotective substances to modify radiation-induced bystander effect – melanin, melatonin and α-tocopherol. All these substances were shown to have antioxidant properties. Melanin is a photo-protective pigment, which is also able to take-up and retain for a long period many xenobiotics and to convert all types of physical energy into heat. Melatonin is a neurohormone, which is involved in reproductive physiology, control of circadial rhythms, immune function and cancer growth. α-Tocopherol (vitamin E) participates in stabilisation of biological membranes and prevents many deceases. Its main protective effect is connected with reparation processes. All these substances are non-toxic, widespread, effective in low concentrations and able to reduce genetic effects of radiation.

The aim of this study was to assess the direct and bystander effect of low level γ-radiation on human keratinocytes, immortalised with HPV-virus *in vitro* and the possibility to modify these effects using radioprotective substances.

Materials and Methods

CELL CULTURE

The HPV-G cell line is a human keratinocyte line, which has been immortalised by transfection with the HPV virus, rendering the cells p53 null. They grow in culture to form a monolayer, display contact inhibition and gap junction intracellular communication.

HPV-G cells were cultured in Dulbecco's MEM: F12 (1:1) medium supplemented with 10% Foetal bovine serum, 1% penicillin-streptomycin, 1% L-glutamine and 1 μg/mL hydrocortisone. The cells were maintained in an incubator at 37°C, with 95% humidity and 5% carbon dioxide and routinely subcultured every 8–10 days. When 80–100% confluent, the medium was poured from the flask and replaced with 1:1 solution of versene (1nM solution) to trypsin (0.25% in Hank's Balanced Salt Solution) (Gibco, Irvine, UK) after washing with sterile PBS. The flask was placed in the incubator at 37 degrees Celsius for about 11 min until the cells started to detach. The flask was then shaken to ensure that all cells had been removed from the base of the flask. The cell suspension was added to an equal volume of DMEM F12 medium to neutralise the trypsin. From this solution new flasks could be seeded at the required cell quantity.

Radioprotective Substances

MELANIN

Melanin was isolated from animal hair by Belarus Pharmaceutical Association (Minsk). By analysis, it was determined to be eumelanin. Both ortochinoid and indolic fragments were present. Melanin was added to the cell medium at 10 mg/L concentration 30 min–1 h before irradiation for directly irradiated cells and 1 h after irradiation to the irradiated cell-culture medium (ICCM) before filtration for bystander recipient cells.

MELATONIN

Melatonin (N-Acetyl-5-methoxytryptamine) was received from Sigma (Germany) as white powder, synthetic. Melatonin was added to the cell medium at 10 mg/L 30 min–1 h before irradiation for directly irradiated cells and 1 h after irradiation to the ICCM before filtration for bystander recipient cells.

α-TOCOPHEROL

α-Tocopherol water-soluble analogue – Trolox was received from Sigma (Germany) as a brown liquid, synthetic. Tocopherol was added to the cell medium at $2 \mu g/mL$ concentration 30 min–1 h before irradiation for directly irradiated cells and 1 h after irradiation to the ICCM before filtration for bystander recipient cells.

Y-IRRADIATION

HPV-G cells were treated 12–24 h after plating in culture flasks, size $25 \, cm^2$. By this time they had attached to the bottom of the flask. The dose was delivered at room temperature using a ^{60}Cobalt teletherapy source, delivering approximately 1.9 Gy/min at a source-to-cell distance of 80 cm. The control cultures were removed from the incubator and brought to the ^{60}Co teletherapy unit with the irradiated cultures but were not irradiated. All cells in the flasks received the same dose of 0.5 Gy. Once irradiated, the cells were immediately replaced in the CO_2 incubator und left undisturbed before analysis.

CLONOGENIC ASSAY

The cell suspension after dilution was counted using a Coulter counter (Coulter Z1). Appropriate cell numbers were plated according to the Puck and Marcus (1956) technique in 5 mL medium in $25 \, cm^2$ NUNC flasks. There were three types of flasks: direct irradiation, bystander donor and bystander recipient. Bystander donor flasks were very heavily seeded with cells (0.5×10^6 cells per flask) in order to produce the bystander factor into the medium after irradiation. Bystander recipient flasks were set up with the ordinary cloning number (300 cells per flask) and received no treatment except the bystander medium (ICCM) from the bystander donor flasks. The direct irradiation flasks were ordinary survival measurement (seeded with 300 cells per flask), after irradiation they received no further treatment. Each of three types of flasks had four sets in triplicate: control, melanin, irradiated cells and irradiated cells with melanin added.

After seeding cells, the flasks were left at 37°C in the incubator to attach for 12 h. Then bystander donor and directly irradiated flasks were irradiated and replaced back in the incubator at 37°C for 1 h. The medium from bystander donor flasks was removed and the radioprotective substance added in appropriate concentration for 30–60 min. Further, the medium was filtered through NALGENE 0.22 μm sterile syringe filters (to ensure that no cells were present in the medium) and used to replace the medium from bystander recipient flasks. Then all flasks were returned to the incubator and left untouched for 9–10 days (until colonies were visible) and then stained with carbol fuchsin and colonies were counted and surviving fraction calculated.

The data are presented as mean ± standard error in all cases. Significance was determined using the t-test.

For the clonogenic assay experiments, plating efficiency is the proportion of in vitro plated cells that form colonies and it is calculated as a percentage of the final number of colonies counted over the initial number of cells plated.

Surviving fraction of cells is calculated from the plating efficiency of the irradiated cells divided by the plating efficiency of the control cells (it is expressed as a percentage of the control plating efficiency).

MICRONUCLEUS ASSAY

The cell suspension after dilution was counted using a Coulter counter (Coulter Z1). For direct irradiation and bystander recipient experiments, about 6,000 cells were plated on glass coverslips (diameter 23 mm) in Petri dishes (diameter 60 mm) in 1 mL of medium for 6 h to attach. Then another 5 mL of cell-culture medium were added to the dishes. Bystander donor cells were plated in 5 mL of medium in 25 cm^2 NUNC flasks and were very heavily seeded with cells (0.5×10^6 cells/flask) in order to produce the bystander factor into the medium after irradiation. Each of three types of flasks had 4 sets in triplicate: control, melanin, irradiated cells and irradiated cells with melanin added.

After seeding, cells were left at 37°C in the CO$_2$ incubator to attach for 12 h. Then bystander donor and direct flasks were irradiated and replaced back in the incubator at 37°C for 1 h. Later, the medium from bystander donor flasks was removed and the radioprotective substance was added in appropriate concentration for 30 min. Then ICCM was filtered through NALGENE 0.22 μm sterile syringe filter to ensure that no cells were present in the medium and no melanin added to recipient flasks thus showing the usual protective effect. This sterile filtered ICCM medium was used to replace the medium from bystander recipient cells. The direct irradiation cells were exposed to the radioprotective substance 30–60 min before irradiation and then left untouched as a direct irradiation test sample.

Then cells were moved back to incubator. After 1–1.5 h cytochalasin B was added at 7 μg/mL concentration, and the cells were incubated for 24 h. After this the cell-culture medium was removed, the cells were washed with PBS and fixed with chilled Karnua solution (1 part of glacial acetic acid and 3 parts of methanol, 10–15 mL three times for 10–20 min). Later coverslips were dried and stained by 10% Giemsa solution. Using mounting medium (Sigma), coverslips were attached to the microscope slides.

The micronuclei count was carried out under inverted microscope (×400). Micronuclei were counted only in binucleated cells (1,000 binucleated cells per flask).

All the data is calculated as the micronuclei number recorded per 1,000 binucleated cells (micronuclei were analyzed only in binucleated cells).

SPECTROMETRY

The absorbance of intact filtered culture medium, culture medium with melanin added and filtered culture medium with melanin added has been analyzed using a Perkin Elmer Lamda 900 UV/VIS/NIR Spectrometer. This spectrometer is a double-beam, double monochromator ratio recording system with pre-aligned tungsten-halogen and deuterium lamps as sources. The wavelength is from 175 to 3300 nm with an accuracy of 0.08 nm in the UV-visible region and 0.3 nm in the NIR region guaranteed.

Results and Discussions

CLONOGENIC EXPERIMENTS

Figure 1 presents the results obtained for clonogenic assay for four direct irradiation and bystander recipient sets of flasks: control, control with melanin, cells irradiated at 0.5 Gy and cells irradiated at 0.5 Gy with melanin added.

Figure 1 shows a significant decrease in the survival of directly irradiated or bystander recipient HPV-G cells irradiated at 0.5 Gy (108 ± 3.5 and 119 ± 0.6) and irradiated at 0.5 Gy with melanin added (123 ± 1.5 and 130 ± 1.2) compared to controls (146 ± 3.2 and 147 ± 2.2). Addition of melanin to the medium of irradiated cells shows a significant increase in the number of colonies compared with cells irradiated without melanin in the medium ($t = 3.86$, $P < 0.01$). And addition of melanin to the ICCM before transfer to recipient

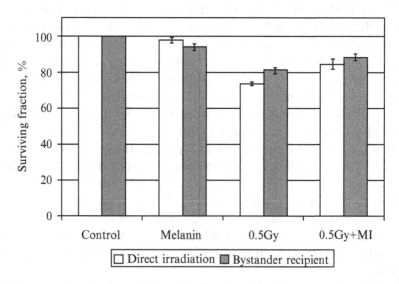

Fig. 1. The surviving fraction of direct irradiated and bystander recipient HPV-G cells (melanin added before irradiation).

cells shows not as significant an increase in the number of colonies compared with cells treated with ICCM without melanin ($t = 3.30$; $P < 0.01$). It can be clearly seen in Fig. 1, where the difference in survival between directly irradiated cells with and without melanin and bystander recipient cells treated with and without melanin is especially evident. Melanin treatment alone was not found to alter the survival of HPV-G cells (143 ± 1.2 and 138 ± 1.2; $P > 0.05$).

These results suggest that melanin is capable of decreasing low-dose radiation effects both direct and bystander cells, but it is more effective after direct irradiation.

Earlier it was shown (Mothersill and Seymour, 1997) that medium alone, irradiated in the absence of cells had no effect on survival of unirradiated cultures. This would seem to exclude the possibility that hydrolysis of medium to give radicals is involved. The time over which the effect persists also excluded any possibility that short-lived species are causing the cell death.

In these experiments, melanin was added to bystander donor cells before irradiation and filtration. It was important to insure that melanin was not present in the medium after filtration so that it could not influence bystander recipient cells, thus showing the usual radioprotective effect. The absorbance of intact culture medium, medium with melanin added and filtered medium with melanin added was compared. Figure 2 shows that filtered medium is identical to culture medium, indicating that the melanin is no longer present in the medium after filtration.

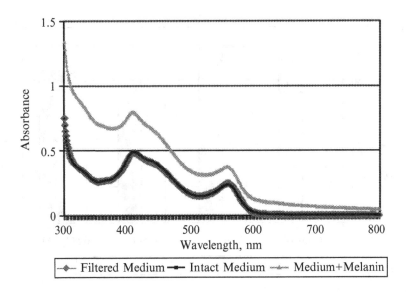

Fig. 2. Absorbance spectra of HPV-G cell-culture intact medium, medium containing melanin and filtered medium containing melanin.

   Also it was supposed that addition of melanin before irradiation to
bystander donor cells decreases the effect of direct irradiation and that is
why a decreased bystander effect was observed. In order to further define
the effect, melanin was added after irradiation. To show that melanin could
decrease bystander effect in recipient cells, melanin was added to the ICCM
1 h after irradiation (when the bystander factor was produced) before filtration and transfer to recipient cells. The results are presented in Fig. 3.

   Figure 3 shows that direct irradiation and irradiated bystander medium
clearly reduces the survival of HPV-G cells compared with controls. No
cytotoxic effect of melanin on HPV-G cells was observed. Addition of melanin
to directly irradiated cells clearly reduces the radiation effects, significantly
increasing survival of HPV-G cells ($t = 7.55$; $P < 0.01$). And addition of
melanin to the ICCM after irradiation increased the survival of HPV-G cells
compared to cells treated with ICCM without melanin, but the effect was not
as significant ($t = 2.54$; $P < 0.05$).

   In previous cases when melanin was added before irradiation $P < 0.01$,
but when melanin was added after irradiation to ICCM, $P < 0.05$. This
shows that addition of melanin before irradiation may influence bystander
donor cells and protect them from direct irradiation, producing not as much
of the damaging bystander factor. When added to ICCM, bystander factor
was formed and melanin was not in contact with bystander donor cells.

   The data show that melanin clearly decreases direct irradiation effects
and not as clearly bystander effects. In bystander cells, the effect of melanin

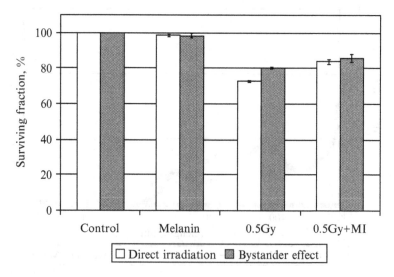

Fig. 3. The surviving fraction of direct irradiated and bystander recipient HPV-G cells
(melanin added after irradiation).

depends of the time of its addition to the medium – if it was added before irradiation, it decreases the bystander effect significantly, but if it was added after irradiation before filtration, the decrease is less significant. The radio-protective action of melanin was not as effective for bystander cells as for directly irradiated cells.

As melanin can absorb all types of physical energy, the result indicates that the bystander signal may have a physical nature (component). At the same time, melanin has very effective antioxidant properties. We decided to compare the effect of melanin with another effective antioxidant – melatonin.

Figure 4 presents the results of the effect of melatonin on survival of directly irradiated and ICCM exposed HPV-G cells. Melatonin was added before irradiation to directly irradiated cells and after irradiation to bystander donor cells.

The results from Fig. 4 show that there is no significant influence of melatonin on HPV-G ($P > 0.05$). Thus, no cytotoxic or proliferation induc-ing effect of melatonin was observed – the survival of control cells with melatonin added is very close to controls.

The data shows a significant decrease in survival of HPV-G cells irradi-ated at 0.5 Gy (26% compared to control, $P < 0.01$) and after bystander donor medium transfer (20%, $P < 0.01$). Addition of melatonin to the medium of directly irradiated cells results in a significant increase in sur-vival on 12% compared to cells irradiated without melatonin present in the medium ($t = 4.84$; $P < 0.01$), although it is still less than in control. And for

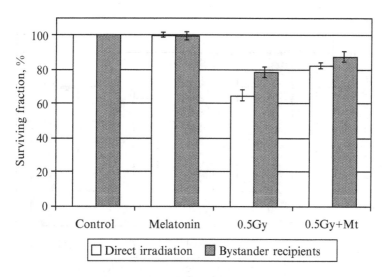

Fig. 4. The surviving fraction of direct irradiated and bystander recipient HPV-G cells (melatonin added after irradiation).

bystander experiments, addition of melatonin also increased the survival of bystander recipient HPV-G cells, but not as significant as in direct irradiation experiments – on 9% ($t = 2.16$; $P < 0.05$).

Thus, melatonin can reduce the damaging effect of the bystander factor after its production, but again not as effective as the protection against the damaging effect of direct irradiation. Both, melanin and melatonin were able to decrease bystander effect, possibly generally because of their antioxidant effect.

In our further study we used vitamin E, which is also very effective antioxidant substance, but it acts in a different pathway – by influencing reparation processes. Also, melanin and melatonin have very close chemical structure, so it was important to use different substance. The results obtained for four sets of flasks: control, control with tocopherol, cells irradiated at 0.5 Gy and cells irradiated at 0.5 Gy with tocopherol added are presented in Fig. 5. Vitamin E was added to the medium of bystander donor cells after irradiation.

Data presented in Fig. 5 show the effect of vitamin E on the survival of HPV-G cells after irradiation. Cells irradiated at 0.5 Gy have clearly lower survival compared with control cells ($P < 0.01$). Addition of tocopherol to the medium of non-irradiated cells does not have any cytotoxic effect.

The survival of irradiated cells with tocopherol added to the medium is lower, than controls, but higher, than the survival of cells irradiated without tocopherol ($t = 4.92$; $P < 0.01$). And addition of vitamin E to the ICCM increased the survival of bystander recipient HPV-G cells, but not as significantly as in direct irradiation experiments ($t = 2.12$; $P < 0.05$).

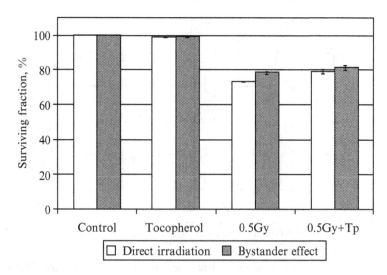

Fig. 5. The surviving fraction of direct irradiated and bystander recipient HPV-G cells treated with α-tocopherol (tocopherol added before irradiation).

Here again we observe the same picture as for melanin and melatonin: α-tocopherol is able to protect HPV-G cells against direct radiation and bystander factors, but its protection against bystander factors is less significant and the least effective between all examined substances. The best protection against bystander factor for both direct and bystander clonogenic experiments showed melatonin.

MICRONUCLEUS EXPERIMENTS

The micronucleus assay currently is widely used in genetic monitoring of populations and for evaluation of the mutagenic effects in vitro. This assay is also used to evaluate individual sensitivity to physical and chemical mutagens. The purpose of the micronucleus test used in the present research was to detect if all substances could modify chromosome structure and segregation in such a way as to lead to induction of micronuclei in interphase cells. HPV-G cells are very sensitive to micronucleus assay.

The results of melanin influence on micronuclei frequency using cytochalasine B block in HPV-G cells irradiated at 0.5 Gy are presented in Fig. 6. The micronuclei frequency in the controls is comparatively low and indicates the level of spontaneous mutagenesis. The number of the cells with 2 and 3 micronuclei is much lower than the number of the cells with 1 micronucleus.

As it can be seen from Fig. 6, the average micronuclei frequency increased in irradiated cells and after bystander donor medium transfer (230.00 ± 10.87

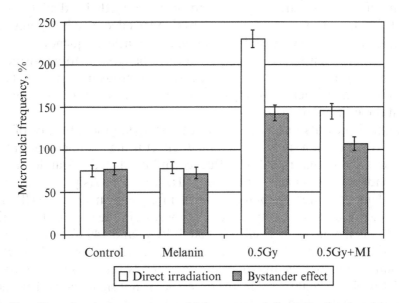

Fig. 6. The effect of melanin on micronuclei frequency of direct irradiated and bystander recipient HPV-G cells (melanin added after irradiation).

and 149.67 ± 9.21, correspondingly) compared with control (75.33 ± 6.81 and 76.67 ± 6.87). Melanin was not found to induce micronuclei – after addition of melanin to the non-irradiated cells the micronuclei frequency was very close to the control (78.00 ± 6.92 and 72.00 ± 6.67, correspondingly; the difference is not significant). This proves that melanin does not have any toxic effect on the cells and that it has no micronuclei inducing or suppressing ability.

After addition of melanin to the medium of the directly irradiated cells the micronuclei frequency (144.00 ± 9.07) was higher, than in the controls, but much less than in irradiated cells ($t = 6.08$; $P < 0.01$). And after addition of melanin to ICCM after irradiation before filtration and transfer to bystander recipient cells the micronuclei frequency (114.01 ± 8.21) was again higher, than in the controls, but less than in cells treated with ICCM without melanin ($t = 2.89$; $P < 0.01$).

The data shows highly significant total micronuclei frequency ($P < 0.01$) for both direct and bystander effects of radiation. If we look at the total number of cells with micronuclei, for directly irradiated cells the addition of melanin to the medium again significantly reduces the number of cells with micronuclei compared to cells irradiated without melanin ($t = 5.35$; $P < 0.01$). But the number of micronuclei cells in bystander recipient cells treated with ICCM + melanin added is not as significant if compared with cells treated with ICCM without melanin ($t = 2.44$; $P < 0.05$). The same situation occurs for the number of the cells with 1 micronucleus – for direct irradiation the difference is highly significant ($t = 4.72$; $P < 0.01$), but it is slightly significant for bystander cells ($t = 2.12$; $P < 0.05$).

This suggests that melanin is able to protect directly irradiated cells and partially to modify radiation-induced bystander effect (by both preventing the decrease in cell survival and increase in micronuclei frequency), suppressing the signals, transferring damage from irradiated cells to non-irradiated cells. And again, as for the clonogenic assay, the protection of HPV-G cells against bystander factors is slightly significant, in contrast to protection against direct irradiation.

Figure 7 presents results obtained after studying the influence of melatonin on the radiation-induced bystander effect using the micronuclei test.

Table 1 shows that radiation influence and irradiated bystander medium clearly induces micronuclei formation in HPV-G cells. As previously shown with melanin, the average total micronuclei frequency in control cells (75.33 ± 6.81 and 76.67 ± 6.87, respectively) is lower than the average total micronuclei frequency in treated cells (230.00 ± 10.87 and 149.67 ± 9.21, respectively). Compared to controls, no micronuclei inducing activity of melatonin was found – the average micronuclei frequency is very close to the control (74.67 ± 6.79 and 76.00 ± 6.84 – the difference is not significant). This means that melatonin has no cytotoxic effect on HPV-G cells.

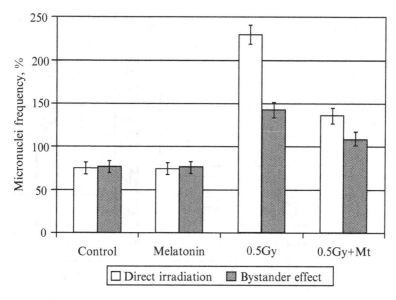

Fig. 7. The effect of melatonin on micronuclei frequency of direct irradiated and bystander recipient HPV-G cells (melatonin added after irradiation).

The results indicate that the average micronuclei frequency of the irradiated cells with melatonin added and after bystander donor medium transfer to recipient cells with melatonin added (136.00 ± 8.85 and 106.00 ± 7.95) is higher than in the control, but lower, than in cells irradiated or treated with ICCM without melanin added ($t = 6.7$ and $t = 3.59$; $P < 0.01$ in both cases). As the protective effect of melatonin is highly significant in both cases, it is again more effective for directly irradiated cells.

Figure 8 presents data obtained in study of the effects of α-tocopherol on micronuclei frequency of directly irradiated and bystander recipient HPV-G cells.

The data from Fig. 8 show that the number of micronuclei in control cells (spontaneous micronuclei frequency) is comparatively low. The micronuclei frequency in control cells with tocopherol added is very close to control (the difference is not significant). Thus, α-tocopherol doesn't have cytotoxic effect on HPV-G cells.

Irradiation at 0.5 Gy and ICCM transfer has a very high micronuclei inducing ability: the number of micronuclei in treated cells is four times higher than in controls. Addition of α-tocopherol to the medium before irradiation significantly reduces micronuclei frequency compared with cells irradiated without α-tocopherol ($t = 5.05$; $P < 0.01$). And micronuclei frequency in bystander recipient cells treated with ICCM with vitamin E

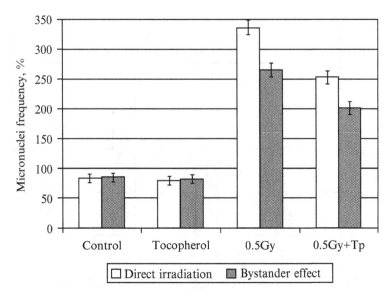

Fig. 8. The effect of vitamin E on micronuclei frequency of direct irradiated and bystander recipient HPV-G cells (vitamin E added after irradiation).

added is higher, than in controls, but lower, than in recipient cells treated with ICCM without vitamin E added ($t = 2.13$, $P < 0.05$). Again Vitamin E shows less effectiveness of protection against bystander factors compared with directly irradiated cells.

The results suggest that α-tocopherol is able to reduce the effects of direct γ-irradiation in human keratinocyte cells, but is less effective against bystander factors. In previous studies, the same effect was observed with melanin and melatonin.

[60]Co irradiation source emits rays with an energy of 1.25 MeV. Thus, most of the interactions of these rays with cells are Compton interactions. These interactions generate many free radicals, which play a very important role in cell damage following irradiation.

All three radioprotective substances used in the present research were shown to have antioxidant effects and are able to neutralize free radicals. This is possibly the main mechanism of their radioprotective action.

For both, direct irradiation and the bystander effect, α-tocopherol is the less-effective protector against radiation damage. And melatonin appears to provide most effective protection for HPV-G cells against radiation. At the same time, all the substances have more effective protection against direct radiation damage compared with bystander damage. The same was observed for micronuclei assay.

A possible explanation of such different protective effect after direct and bystander irradiations is that in direct irradiation experiments the radioprotective substances were directly in contact with cells since irradiation and were able to protect them against all kinds of damaging factors, starting from energy input (some of these substances may convert all types of energy into heat) and concluding with emerging free radicals. In bystander experiments, the substances were not in contact with cells until added to the ICCM 1 h after irradiation when the damaging factor was already produced. So melanin/melatonin/tocopherol could neutralise only some of the damaging factors (e.g. long-living free radicals) and are unable to protect against the main damaging factor (bystander factor), the nature of which is still unknown.

It is not known the mechanism by which they reduce bystander effect – if they neutralise free radicals or it is possibly another mechanism or factor. But it is definitely clear that these substances are not toxic or stimulating to the cells, increasing their survival or viability or decreasing micronuclei frequency.

Conclusions

The analysis of the effects of radioprotective substances on survival and micronuclei frequency of HPV-G cells after direct and bystander low dose radiation influence allows us to conclude that:

- All substances does not have any cytotoxic effect on HPV-G cells;
- The absorption spectrum of the filtered medium, which contained melanin is identical to one of the intact culture medium. This means that melanin is not present in the filtered medium, so it cannot directly protect recipient cells except by reducing the bystander factor;
- Substances are able to significantly increase the survival of HPV-G cells in vitro after direct low dose radiation influence;
- Added to ICCM before filtration, substances perform protective effect against low doses of radiation (decreasing bystander effect), but not as significant as the effect against direct irradiation;
- The results of micronuclei test show that all substances are able to decrease the level of micronuclei frequency after direct low dose radiation influence and ICCM transfer. They have more effective protective effect in directly irradiated cells than in bystander recipient cells;
- The observed effect of decreased bystander effect may be due to the ability to absorb all types of physical energy, to take-up and retain xenobiotics (as sieves) or because of the antioxidant effect of substances.

References

Barcellos-Hoff, M.H. and Brooks, A.L., 2001, Extracellular signalling through the microenvironment: a hypothesis relating carcinogenesis, bystander effects, and genomic instability, *Radiat Res* **156**(5 Pt 2):618–627.

Dahle, J., Kvam, E., and Stokke, T., 2005, Bystander effects in UV-induced genomic instability: antioxidants inhibit delayed mutagenesis induced by ultraviolet A and B radiation, *Carcinogenesis* **4**:11

Deshpande, A., Goodwin, E.H., Bailey, S.M., Marrone, B.L., and Lehnert, B.E., 1996, Alpha-particle induced sister chromatid exchange in normal human lung fibroblasts - evidence for an extranuclear target, *Radiat Res* **145**:260–267.

Kroemer, G., 1997, The proto-oncogene bcl-2 and its role in regulating apoptosis, *Nat Med* **3**(6):614–620.

Lehnert B.E. and Goodwin E.H., 1997, A new mechanism for DNA alterations induced by alpha particles such as those emitted by radon and radon progeny, *Environ Health Perspect* **105**:1095–1101.

Lorimore, S.A., Kadhim, M.A., Pocock, D.A., Papworth, D., Stevens, D.L., Goodhead, D.T., and Wright E.G., 1998, Chromosomal instability in the descendants of unirradiated surviving cells after alpha-particle irradiation, *Proc Natl Acad Sci USA* **95**(10):5730–5733.

Mosse, I.B., Marozik, P.M., and Ksenzova, T., 2006, Bystander effect induced by laser rays, *ERRS Meeting 2006*, Kiev, Ukraine, 23–26 August 2006.

Mothersill, C. and Seymour, C., 1997, Medium from irradiated human epithelial cells but not human fibroblasts reduces the clonogenic survival of unirradiated cells, *Int J Radiat Biol* **71**:421–427.

Nagasawa, H. and Little, J.B., 1992, Induction of sister chromatid exchanges by extremely low doses of alpha-particles, *Cancer Res* **52**:6394–6396.

Narayanan, P.K., LaRue, K.E.A, Goodwin, E.H., and Lehnert, B.E., 1999, Alpha particles induce the production of interleukin-8 by human cells, *Radiat Res* **152**(1):57–63.

Puck, T.T. and Marcus, P.I., 1956, Action of x-rays on mammalian cells, *J Exp Med.* **103**(5): 653–666.

CHAPTER 17

CONSIDERATIONS FOR PROTEOMIC BIOMARKERS IN RAINBOW TROUT ECOTOXICOLOGY

RICHARD W. SMITH[1], IURGI SALABERRIA[2], PHIL CASH[3], AND PETER PÄRT[4]

[1]*Department of Biology, McMaster University, Hamilton, Canada. e-mail: rsmith@mcmaster.ca*
[2]*Institute of Environment and Sustainability, European Commission Joint Research Centre, Ispra, Italy.*
[3]*Department of Medical Microbiology, University of Aberdeen, United Kingdom.*
[4]*DG JRC, European Commission Joint Research Centre, Ispra, Italy.*

Abstract: The rainbow trout (*Oncorhynchus mykiss*) is one of the most extensively researched and characterised species of fish. In addition its low tolerance to poor water quality has established it as one of the most useful sentinel species for aquatic toxicology. The subject of proteomics offers a potentially powerful approach to ecotoxicology, particularly with respect to providing biomarkers of environmental contamination. Therefore there is a valid rationale for combining this species with this experimental approach. However evidence exists that the rainbow trout liver proteome can be influenced by dietary composition. Furthermore individuals, within a trout population, are known to exhibit widely differing food consumption rates and therefore growth. As a result there may be fundamental issues to consider before this combination can be fully exploited. Using the subject of endocrine disruption preliminary data are presented which demonstrate that individual growth rates moderate the proteomic responses to the injection of a single dose of β-estradiol. The injection dose is less than required to induce vitellogenin synthesis, the recognised endpoint of endocrine disruption, which suggests these proteomic changes may in fact be more sensitive biomarkers. This study therefore provides evidence and suggested guidelines for proteomic toxicology in rainbow trout. The study also provides evidence that if these considerations are met proteomic changes in the trout liver could be a valuable addition to existing biological markers of aquatic contamination.

C. Mothersill et al. (eds.), Multiple Stressors: A Challenge for the Future, 263–269.

Keywords: rainbow trout hierarchy; growth rate; β-estradiol; endocrine disruption; proteomic biomarkers

Introduction

When considering experimental tools for investigating ecotoxicology, particularly in terms of establishing biomarkers of environmental contamination, two definitions emphasise the potential proteomics has to offer: (i) "Proteomics collectively analyses the proteins that are regulated, expressed or modified in the cell under different conditions" (Liebler, 2002) and (ii); biomarkers have been defined as "any biochemical, histological and/or physiological alterations or manifestations of stress" (Holdway et al., 1995). In proteomics we have an experimental approach which allows the detection of multiple protein changes, any of which could be implicated in any of the biomarker categories listed. However, whilst the value of proteomics in understanding human cellular mechanistic responses has been recognised (Möller et al., 2001), this technique has yet to be extensively applied to aquatic toxicology.

The central technique used to study proteomics is the two-dimensional (2D) electrophoresis gel, whereby a protein mixture (cell or tissue homogenate) is separated according to both the iso-electric point and molecular size. Individual proteins are thus resolved to a unique coordinate. 2D gel image analysis by powerful software then determines differences in protein induction, deletion or up- or down-regulation. Selected proteins can then be removed and identified using mass-spectrometry of peptide fragments and comparison with information in sequence databases. Several reviews of the techniques associated with proteomics exist; e.g. Pandet and Mann (2000). Similarly Sanchez et al. (2001) review the SWISS-2D database, one of the more complete and most commonly used.

In terms of being a sentinel species the rainbow trout offers a number of important advantages: (i) its universal application allows comparison between research groups, (ii) its demand for high water quality means it is sensitive to changes in the aquatic environment, (iii) its genome is one of the most well characterised of all fish species and (iv) it can act as a surrogate for other commercially important salmonids (e.g. Pacific and Atlantic salmon; other *Oncorhynchus* species and *Salmo salar*, respectively). Nevertheless, within the context of proteomics, there is evidence that certain considerations may be necessary if the full potential of this species is to be properly exploited.

The liver plays a central role in toxicology (e.g. Holdway et al., 1995). As such the relevance and importance of hepatic proteome analysis is evident. However, feeding rainbow trout a diet enriched with plant proteins has been shown to alter the expression of approximately 4% of the liver proteome

(Martin et al., 2003). Therefore this finding provides evidence to suggest proteomic responses to waterborne contaminants may be moderated by dietary composition which, in reality, is likely to vary to some degree in trout in naturally occurring water bodies.

Continuing the theme of dietary influence on the trout liver proteome, a more fundamental question remains unanswered. Trout are known to form social hierarchies based on individual food consumption rates (McCarthy et al., 1992). However, despite total nutritional intake varying considerably within a group, to date no evaluations of the effect of social position/ differential feeding have been incorporated into toxicological studies involving rainbow trout proteomics.

Clearly, if proteomic responses in rainbow trout liver are to be used in ecotoxicology, particularly as biomarkers, it would be advantageous to establish any further limitations and/or considerations which are applicable in this model species.

Differential Growth and Hepatic Proteomic Changes Following β-Estradiol Injection

Aside from pollutant detoxification one of the major roles of the trout liver is the production of the egg protein vitellogenin which is then exported, via the blood, to eggs developing in the ovary. Since vitellogenin should therefore only be synthesised in large quantities by sexually mature female fish its presence in male or sexually immature fish has been established as a biomarker of the class of aquatic contaminants which mimic the action of estrogen; i.e. Endocrine Disrupting Compounds (EDCs) (e.g. for reviews refer to Nicolas, 1999; Vos et al., 2000). Consequently it is particularly pertinent to investigate rainbow trout hepatic proteomic changes associated with endocrine disruption by estrogen. Therefore, using estrogen treatment as a model stressor, the aim of this preliminary investigation was to determine the influence of hierarchical dietary variation on rainbow trout liver proteomics and how this might relate to using proteomic changes as biological markers.

Fish, husbandry and estradiol injection. Groups of 10 juvenile rainbow trout (overall mean mass = 35.7 ± 9.5 g) were each fitted with an internal Passive Integrated Transponder (PIT) tag (Biomark, Idaho, USA) and then placed in 100 L aquaria. These aquaria were supplied with sand- (100 μm interstitial pore size) and activated carbon-filtered water delivered at a flow rate of 2.0 L min⁻¹. Water chemistry: Ca = 375, Mg = 133, Na = 116, K = 3.9 (all μM) and DOC = 1.61 mg L⁻¹). The fish were fed 1.5% total body mass day⁻¹, with a commercial diet. Weekly measurements of length and mass were recorded and the food ration adjusted accordingly.

After 3 weeks the fish in each group were given a single intra-peritoneal injection (1% volume/body mass) of 0.005, 0.5 or 500 μg kg^{-1} β-estradiol (E2). Another group were injected with an equal volume of corn oil (the E2 solvent) and another group remained non-injected. The fish were then returned to their aquaria. About 2 weeks after injection the fish were killed, blood samples were taken for plasma vitellogenin analysis and the liver was collected, wrapped in aluminium foil and immediately frozen in liquid nitrogen, and stored at −80°C. The liver proteome was then investigated using 2D electrophoresis as described by Smith et al. (2005). Plasma vitellogenin concentrations of E2-treated fish were also determined semi-quantitatively using a fast protein liquid chromatography system (FPLC) (Pharmacia) (Silversand et al., 1993), the aim being to compare this recognised endpoint with proteomic changes.

Vitellogenin induction and changes to the proteome arising from differential growth and estradiol injection. Specific Growth Rates (SGR) (Ricker, 1979), calculated from the initial and final body mass of each fish, varied from −0.5 to 1.1% day^{-1}. No significant difference in overall mean growth was found between the treatments (data not shown). However, within each treatment group, individuals could be assigned as "high" or "low" SGR depending on whether they fell outside of the upper or lower values of the range defined by the mean ± one standard deviation, respectively.

Vitellogenin was not detected in the plasma from any fish irrespective of growth rate or estradiol injection (data not shown). Although we did not excise and actually identify any of the proteins resolved on the 2D gels, this study clearly revealed differences between the proteomes of fish exhibiting a "high" and "low" SGR. Examples of the major proteomic changes are illustrated by Fig. 1. The only consistent proteome changes were in low SGR fish; i.e. proteome changes which occurred in 100% of the liver samples. In high SGR fish there were proteome changes, relative to the control treatments, but these were not consistent and occurred in a few individuals only. Essentially four types of qualitative proteome change were identified: (A) proteins expressed in all low SGR fish (but not in any high SGR fish) irrespective of E2 treatment; (B) proteins expressed by all E2 treatments in low SGR fish, only; (C) proteins expressed by the highest E2 dose (500 μg kg^{-1}) in low SGR fish only and (D) proteins expressed in non-injected (and corn oil solvent control injected) low SGR fish, not expressed following injection with 0.005 and 0.5 μg kg^{-1} E2 in low SGR fish, but expressed following injection with 500 μg kg^{-1} E2 in low SGR. In all of the earlier examples the descriptions apply to 100% of the fish in the categories described. There were also numerous quantitative and semi-quantitative changes; i.e. increases or decreases in the normalised spot volumes of some proteins expressed in all fish, in all categories,

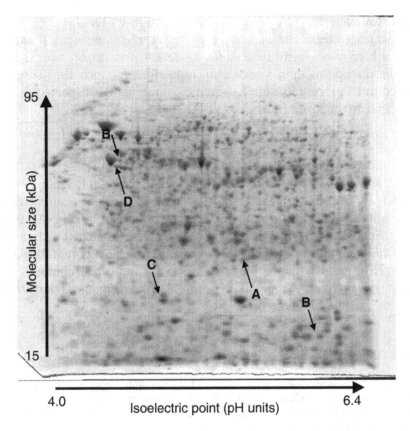

Fig. 1. A typical 2D gel derived from the rainbow trout liver of a low SGR fish, illustrating the most consistent proteome changes which occurred following a single injection with 0.005, 0.5 and 500 μg kg^{-1} β-estradiol (E2). Four categories of expression were identified: (A) proteins expressed in all low SGR fish (but not in any high SGR fish) irrespective of E2 treatment; (B) proteins expressed by all E2 treatments in low SGR fish only; (C) proteins expressed by the highest E2 dose (500 μg kg^{-1}) in low SGR fish only and (D) proteins expressed in non-injected (and corn oil solvent control injected) low SGR fish, not expressed following injection with 0.005 and 0.5 μg kg^{-1} E2, but expressed following injection with 500 μg kg^{-1} E2.

and also similar changes in some proteins expressed in some individuals within an experimental category.

Even without protein identification this study has revealed a significant amount of modulation of proteomic responses to E2 injection, within the rainbow trout liver, by growth (and therefore feeding) rate and social position. The importance of this is increased if one considers that, to induce vitellogenin (the accepted EDC endpoint) a single injection of 5.0 mg kg^{-1} (Korte et al., 2000) or two injections of 10 mg kg^{-1} (Folmar et al., 2000) have been previously employed in fathead minnow (*Pimephales promelas*) and summer flounder (*Paralichthys dentatus*), respectively. Thus we propose the proteomic responses to the single, lower E2 concentration singular injections described here may constitute a much more sensitive set of biomarkers to endocrine disruption than vitellogenin induction. As such they should be evaluated in terms of exposure to aquatic contaminants suspected of being EDC which are of lower potency than E2.

Conclusions

In conclusion it is proposed to effectively utilise proteomic changes in the rainbow trout liver as biological indicators the challenge for ecotoxicologists is to be able to distinguish between the proteomic biomarkers of multiple stressors. Specifically; responses to the inherent stress associated with rainbow trout social biology has to be distinguished from any potential external stressor. However if this distinction can be accomplished changes to the proteome may be considerably more sensitive to environmental perturbation than some existing biomarkers.

References

Folmar, L.C., Gardner, G.R., Schreibman, M.P., Magliulo-Cepriano, L., Mills, L.J., Zaroogian,G., Gutjahr-Gobell, R., Haebler, R., Horowritz, D.B., and Denslow, N.D., 2001, Vitellogenin-induced pathology in male summer flounder (*Paralichthys dentatus*). *Aquat. Toxicol.* **51**: 421–441.
Holdway, D.A., Brennab, S.E., and Ahokas, J.T., 1995, Short review of selected fish biomarkers of xenobiotic exposure with an example using fish hepatic mixed-function oxidase. *Aust. J. Ecol.* **20**: 34–44.
Korte, J.J., Hakl, M.D., Jensen, K.M., Mumtaz, S.P., Parks, L.G., LeBlanc, G.A., and Ankley, G.T., 2000, Fathead minnow vitellogenin: complementary DNA sequence and messenger RNA and protein expression after 17 β-estradiol treatment. *Env. Toxicol. Chem.* **19**: 972–981.
Liebler, D.C., 2002, *Introduction to proteomics: tools for the new biology*. Humana Press, Inc, Totowa, NJ.
Martin, S.A.M., Vilhelmsson, O., Médale, F., Watt, P., Kaushik, S., and Houlihan, D.F., 2003, Proteomic sensitivity to dietary manipulations in rainbow trout. *Biochim. Biophys. Acta* **1651**: 17–29.
McCarthy, I.D., Carter, C.G., and Houlihan, D.F., 1992, The effects of feeding hierarchy on individual variability in daily feeding of rainbow trout, *Oncorhynchus mykiss* (Walbaum). *J. Fish Biol.* **41**: 257–263.
Möller, A., Soldan, M., Völker, U., and Maser, E., 2001, Two-dimensional gel electrophoresis: a powerful method to elucidate cellular repsonses to toxic compounds. *Toxicology* **160**: 129–138.
Nicolas, J-M., 1999, Review: Vitellogenesis in fish and the effects of polycyclic aromatic hydrocarbon contaminants. *Aquat. Toxicol.* **45**: 77–90.
Pandet, A. and Mann, M., 2000, Proteomics to study genes and genomics. Insight review articles. *Nature* **405**: 837–845.
Ricker, W.E., 1979, Growth rates and models. In: Hoare, W.S., and Randall, D.J., and Brett, J.R, *Fish Physiology*. Vol 8, pp 677–743. Academic Press, New York.

Sanchez, J-C., Chiappe, D., Converset, V., Hoogland, C., Binz, P-A., Paesano, S., Appel, R.D., Wang, S., Sennitt, M., Nolan, A., Cawthorne, M.A., and Hochstraser D.F., 2001, The mouse SWISS-2D PAGE database: a tool for proteomics study of diabetes and obesity. *Proteomics* **1**: 136–163.

Silversand, C., Hyllner, S. J., and Haux, C., 1993, Isolation, immunochemical detection, and observations of the instability of vitellogenin from four teleosts. *J. Exp. Zool.* **267**: 587–597.

Smith, R.W., Wood, C.M., Cash, P., Diao, L., and Pärt, P., 2005, Apolipoprotein AI could be a significant determinant of epithelial integrity in rainbow trout gill cell cultures: a study in functional proteomics. *Biochim. Biophys. Acta* **1749**: 81–93.

Vos, J.G., Dybling, E., Greim, H.A., Ladefoged, O., Lambré, C., Tarazona, J.V., Brandt, I., and Vethaak, A.D., 2000, Health effects of endocrine-disrupting chemicals on wildlife, with special reference to the European situation. *Crit. Rev. Toxicol.* **30**: 71–133.

CHAPTER 18

GENETIC EFFECTS OF COMBINED ACTION OF SOME CHEMICALS AND IONIZING RADIATION IN ANIMALS AND HUMAN CELLS

IRMA MOSSE, L.N. KOSTROVA, AND V.P. MOLOPHEI

Institute of Genetics and Cytology, National Academy of Sciences, Minsk, Belarus
e-mail: i.mosse@igc.bas-net.by

Abstract: Environmental contamination by radionuclides and different chemical substances results in exposure of human and other beings to a complex of physical and chemical factors. Results of combined influence of diverse agents can be unpredictable, because observed effects can differ from the sum of effects of each one taken separately. We studied cytogenetic effects of sodium nitrite and nitrate in drosophila and mice. It was found that these substances didn't possess mutagenic activity and sensibilized significantly (2–4 times) genetic effect of ionizing radiation. Genetic effects of herbicide zenkor were found to be completely different. Zenkor has mutagenic activity and irradiation of zenkor-treated mice results in decreasing aberration levels. Chronic influence of zenkor and gamma-rays leads to more strict "antagonistic" effect than acute one. Such effect can be explained by increased death of cells treated with both mutagenic factors – zenkor and irradiation. In the same time many food stuffs contain radioprotectors or antimutagens. We shown that melanin is very effective radioprotector not only against acute irradiation, but even against chronic one. Influence of different substances (melanin, tocopherol or zenkor) on fractionated irradiation is also different. So, phenomena of sensibilization, antagonism, protection or inhibition were observed. Influence of combined action of chemical and physical factors on hereditary structures are worthy of great attention and must be considered during taking environmental protection measures.

Keywords: acute and chronic irradiation; melanin; tocopherol; zenkor; sodium nitrite and nitrate; mutation; mice germ and somatic cells; human cells; *Drosophila*

C. Mothersill et al. (eds.), Multiple Stressors: A Challenge for the Future, 271–286.
© 2007 *Springer.*

Introduction

Environmental contamination by radionuclides and different chemical substances results in exposure of human and other beings to a complex of physical and chemical factors. Not only ionizing radiation, but 80% of chemical pollutants also have mutagenic and carcinogenic activity. Results of combined influence of diverse agents can be unpredictable, because observed effects can differ from the sum of effects of each one taken separately.

There are different types of factor interaction:

A + B = AB – additive effect
A + B > AB – sensibilization if one factor is not mutagenic
A + B > AB – synergism if both factors are mutagenic
A + B < AB – protection if one factor is not mutagenic
A + B < AB – antagonism if both factors are mutagenic

Genetic monitoring – control of genetic disturbances – is especially actual and important in this situation for both human and other organisms. Information about genetic effects of combined factors is necessary in order to make a decision about measures for decreasing or preventing dangerous consequences of complex factor influence. After the Chernobyl disaster great areas of agricultural land in Belarus, the Ukraine and Russia were radiocontaminated and it is known that various chemical substances are actively used in agriculture. There are very few data about genetic consequences of the joint effect of radiation and some chemicals. Many factors which are not under control can change significantly biological effects of radiation. For instance, some of substances are present in our food – residual amounts of fertilizers or herbicides can be mutagenic or influence mutagenic action of radiation. Others, such as pigment melanin or antioxidant tocopherol, can be antimutagenic on the contrary.

So, we investigated genetic effects of combined action of some chemicals and ionizing radiation in mice and human cells.

Genetic Effects of Sodium Nitrite and Nitrate Separately and in Combination with Irradiation

Nitrites and nitrates penetrate into human organism with food and drinking water. Nitrosocompounds exhibiting mutagenic and carcinogenic effect can be potential products of nitrite - and nitrate biotransformation.

When studying genetic activity of 89 different samples of foodstuffs, treated with nitrite at pH = 4.2, it was revealed that fish and other sea products were mutagenic without metabolic activation (Mosse et al., 1990). Mutagenity of vegetables, coffee, tea and drinks (wine, juice, beer) was also increased.

Alkaline fraction of beer exhibited the highest mutagenic activity after nitrate treatment. Two tetrahydro-β-carboline derivatives were identified in this fraction. Formation of direct mutagens after nitrate treatment of soy-bean sauce, widely used in Japan as a food flavouring, is shown (Mosse et al., 1990).

We investigated genetic effects of sodium nitrite and nitrate separately and in combination with irradiation in germ cells of mice and *Drosophila*. Males Af line at the age of 2.0–2.5 months were exposed to treatment. Mice were given sodium nitrite and nitrate solutions with drinking water in concentration of 10 mg/L for 2.5 months.

Acute irradiations with neutrons and gamma-rays were used at dose 0.5 Gy.

The frequency of reciprocal translocations in germ cells was analysed at the stages of spermatocytes and spermatogonia after E.P. Evans method (Evans et al., 1970). Statistical data processing was conducted by Student *t*-test (the number of tests was taken as "*n*").

It was shown that these fertilizers do not possess a mutagenic activity but sensibilize significantly (3–4 times) genetic action of ionizing radiation (gamma-rays and neutrons) (Fig. 1).

Laboratory drosophila population taken from eight normal isogenic lines was used. In drosophila experiments sodium nitrite and nitrate were administered into a nutrient medium at 100 mg/l. Flies grown on the normal medium and the medium with the substances were simultaneously irradiated with X-ray at the dose of 15 Gy.

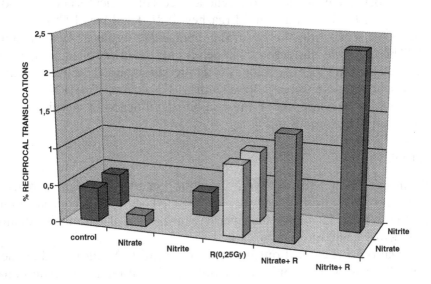

Fig. 1. Influence of sodium nitrite and sodium nitrate on mutagenic action of radiation in mice germ cells

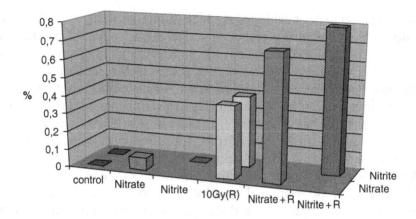

Fig. 2. NaNO$_2$ and NaNO$_3$ influence on radiation induced recessive lethal mutation level

The frequency of recessive sex-linked lethal mutations was determined after Meller method.

The obtained results were very similar to those observed in mice experiments – sodium nitrite and nitrates had no significant mutagenic effect and strongly decreased the recessive mutation frequency (Fig. 2)

Combined Action of Herbicide Zenkor and Ionizing Radiation

In some papers there are data about genetic effects of pesticides and herbicides (Stroev, 1968, 1970; Sultanov and Ergashev, 1981; Khalikov, 1990)

Herbicide zenkor was shown to be a mutagenic substance (Azatyan et al., 1984). This herbicide is widely used in Belarus for treatment of tomato, potato and cereals planting. Zenkor belongs to sim-triazin group, its active substance (entity) is 4-amino-6-tetrabutyl-3(methylthya)-1,2,4-triazin-5(4H)OH.

There were no data about the combined influence of herbicides and radiation on living organisms.

METHODS

Male mice (CBA×C57Bl) F$_1$, age 2.5 months and 20–25 g weight were used. Zenkor solution in the distilled water was injected perorally in concentrations from 0.1 mg/kg to 150 mg/kg once per day in the case of chronic irradiation or 24 h before acute irradiation.

Irradiation was carried out in Obninsk Scientific Medical Radiological Centre (Russia). For chronic irradiation special installation "Panorama" was used with dose-rate of 1.17×10^{-4} Gy/min, and acute irradiation was conducted with Co^{60} (γ-cell installation, Canada), 1 Gy/min dose-rate.

The levels of reciprocal translocations in metaphase of spermatocytes were analysed cytologically by the modified method described by Evans et al. (1970) Reciprocal translocation can be inherited, thus we studied real genetic effect mutations, which emerge in germ cells and which can pass from generation to generation.

RESULTS AND DISCUSSION

The data about zenkor influence on reciprocal translocation rate in mice germ cells are presented in Fig. 1. There is no strict correlation between zenkor concentration and activity. So, zenkor in the concentrations of 0.1 mg/kg and 1 mg/kg did not increase the control mutation level, but this herbicide in the doses of 0.25 mg/kg; 10 mg/kg and 150 mg/kg was shown to be a week mutagenic substance, mutagenic activity did not differ significantly in the range of the above concentrations. A possible cause of such a phenomenon is the peroral method of zenkor injection – a concentration of herbicide in the animal blood can vary depending on the quantity and quality of food in mice stomach because experiments in which zenkor was used in diverse concentrations were conducted in different days. Nevertheless the data, presented in Fig. 3, allow a decision to be made concerning mutagenic capacity of zenkor, especially in high concentrations.

Results of the study on the combined action of zenkor and acute irradiation are presented Fig. 4. Irradiation 3 Gy induced 1.64% of reciprocal translocations in mice germ cells. The influence of zenkor together with radiation resulted in the mutation rate values less than expected sums of both factors effects, i.e. antagonistic phenomena have been observed in all experiments, in which different concentrations of zenkor were used.

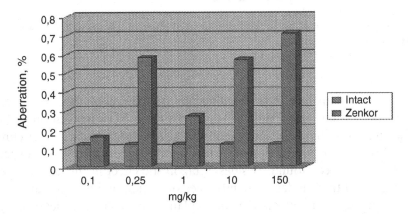

Fig. 3. Genetic effects of herbicide zenkor in mice germ cells

I. MOSSE ET AL.

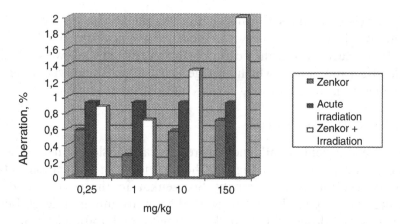

Fig. 4. Combined action of zenkor and acute irradiation on reciprocal translocation levels in mice germ cells

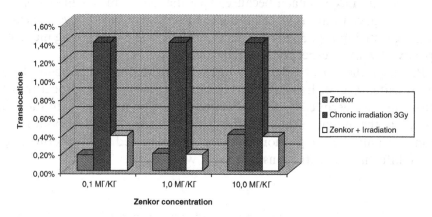

Fig. 5. Combined action of zenkor and chronic irradiation on reciprocal translocation levels in mice germ cells

Analogous data have been obtained in studying combined influence of zenkor and chronic irradiation (Fig. 5). Chronic 3 Gy irradiation induced 1.48% mutations, that is in agreement with the literature data. Under chronic action of both factors reduction in mutation levels as compared to the expected total values has been found. Besides antagonistic effect in this case was even more expressed than under acute influence of zenkor and gamma rays.

Such antagonistic effects can be explained by increased death of cells treated with both damaging factors, sensitive cells with mutations being eliminated first. These effects are corroborated by the highest reduction in the individual number of the *Drosophila* populations exposed to the combined effect of both factors, these findings being obtained in our laboratory (Mosse and Makeeva, 1996).

So, phenomena of sensibilization or antagonistic influence of combined chemical and physical factors on hereditary structures are worthy of great attention, because some chemicals having no mutagenic or week mutagenic activity can increase significantly genetic effects of radiation.

Genetic Effects of Combined Action of Pigment Melanin and Radiation

Pigment melanin is wide-spread in the living world. It has many different functions, only one of which is radioprotection (Hill, 1992). It has been found that melanin under some conditions increases survival of fungi after a lethal dose of ionizing radiation (Baraboi, 1984). Dark pigmented fungi live in areas of high radiation background due to radionuclide contamination. Some of these organisms have survived even after irradiation of the soil with 6400 Gy (Shilds and Durell, 1961). It has been shown that intensively pigmented plants can exist in areas contaminated with ^{90}Sr (Krivolutzky et al., 1972). Wallace and King (1958) reported that primarily melanin-containing fungi survived on the Bikini atoll after the atomic bomb explosion.

Hollaender and Steplton (1953) found an analogous effect in frogs – melanin formation in frog melanophores increased due to hypophysis irradiation with γ-rays. These authors noted that the increased oxidation of tyrosine took place in the tissue of irradiated animals in the first hours after irradiation (Hollaender and Steplton 1953).

Some attempts have been made to use melanin to increase biological radioresistance by Berdishev (1964), and Malama (1965) who found increased survival and increased life expectancy of irradiated white mouse due to intraperitoneal injection of melanin before irradiation with 8 Gy.

All investigations of melanin radioprotective ability were concerned its influence on lethal radiation effects. We were the first to study the antimutagenic activity of this pigment. We found that melanin significantly decreased frequencies of different types of mutations induced by radiation in germ cells of animals (drosophila, mouse; Mosse, 1990; Mosse et al., 1996). Melanin's ability to reduce the amount of genetic lesions inherited from generation to generation that accumulate in populations as "genetic load" is especially valuable (Mosse and Lyach, 1994).

In this paper we present results of melanin influence on radiation mutagenic effects in mice and human cells.

MATERIALS AND METHODS

Melanin

Melanin pigments of different types were used – synthetic or isolated from human or animal hair. Synthetic pigments were produced via DOPA-oxidation. Melanins from human and animal hair were produced by Belarus Farmaceutic Association (Minsk). Analysis determined to be an melanin. Both ortochinoid and indolic fragments were present. Pyrochromatograhic profiles were similar to synthetic melanin from L-DOPA. The sample in a dry state possessed the ability to produce free radicals in amounts of 2.2×10^{19} spins/g.

Human lymphocyte culture

Blood was taken from healthy donors of 25–45 years old at the blood transfusion station. Human lymphocytes were collected and cultured according to the method of Moorhead et al. 1960, for 52 h at 37°C. Colchicine was added 2 h before the cells were fixated with ethyl alcohol and glacial acetic acid mixture.

Cytological tests included dicentric- ring- and fragment analysis.

Melanin suspension in distilled water was sterilized and added to culture media at the onset of culturing (G_0/G_1 stage) or at 47-th h (G_2 stage) in the following concentrations: 0.1; 0.3; 1.0; 3.0; 10.0 and 30.0 mg/L.

Human lymphocytes were irradiated 40 min after melanin addition. The laboratory ^{137}Cs-machine for microbiological and biochemical investigation (LMB – γ – 1M) was used. The dose rate of acute irradiation was 40 cGy/min. Radiation doses of 0.5; 1.0 and 2.0 Gy were used.

Mouse germ and bone marrow cells

Mice Af of 2–2.5 month old (weight 18–20 g) were used. Mice were exposed in agreement with the «Principles of laboratory animal care» (NIH publication no. 85–23, revised 1985) and the requirements of National ethics committee. The starch gel or its melanin suspension was injected into stomach every day with a special needle. This was shown in our previous experiments to be an effective and preferable for practice in comparison with intraperitoneal injections. Melanin was supplied in concentrations from 0.3 to 30 mg/kg 2 hours before acute irradiation and once a day during chronic one. Mice were exposed to 1–3 Gy of γ-rays of Cs^{137} at the dose rate of 0.007 Gy/h (chronic irradiation) and 420 Gy/h (acute one). Animals were killed 24 hour after whole body acute irradiation and 2.5–3.0 months after the chronic exposure was stopped. This interval was necessary for repairing irradiated spermatogonia. The levels of reciprocal translocations in metaphase of spermatocytes were analysed cytologically after Iven's method (Evans et al., 1970). Chromosome aberration frequencies in metaphases of bone marrow cells were analysed cytologically by the standard method (Preston et al., 1947). Aberrations of chromosome and chromatid types were estimated.

Statistic data processing was carried out with the computer IBM PC/AT. The significance of the experimental data was confirmed by Student-Fisher T-test criterion, where "*n*" was equal to the number of analysed mouse or to the number of analysed metaphases in human cells. Standard deviation from the mean was estimated.

RESULTS AND DISCUSSION

Melanin influence on mutagenic action of chronic irradiation in mice germ cells

Investigations of melanin influence on the spontaneous mutation level have demonstrated that melanin itself does not possess a mutagenic activity in all concentrations used, even being supplied for 30 days.

Melanin in all concentrations was shown to reduce effectively mutagenic action of acute γ-irradiation (Mosse et al., 2006). The melanin influence on genetic effect of chronic irradiation was even more effective. The data presented in Fig. 6 show that the pigment in all used concentration greatly reduced the percentage of induced mutations at different doses of chronic irradiation.

It is very difficult to compare antimutagenic activity of melanin under acute and chronic irradiation because in the first case only one injection of melanin has been used, but in the second case melanin has been injected many times (once a day for 10–20 days). Nevertheless, it is possible to draw a conclusion that melanin is no less and even more effective under chronic irradiation than under acute one.

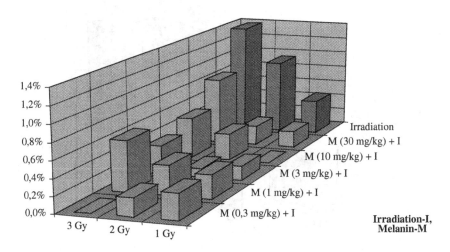

Fig. 6. The melanin influence on chronic irradiation induced mutation frequency in mouse germ cells

It was revealed that melanin activity does not depend on concentration used. There is evidence that only small amount of melanin can penetrate into cells and melanin quantity inside cells does not increase with rise in outside melanin concentration (Grossi et al., 1998). This fact can explain the absence of correlation between melanin dose and activity.

Radioprotective action of this pigment is associated with its ability to accept and to release electrons and with antiradical activity. It is clear that when low-dose irradiation is used, the possibility for melanin to catch free radicals or electrons is better.

So, by means of cytogenetic analysis of mouse germ cells, we have demonstrated the possibility to decrease genetic effects of chronic irradiation using pigment melanin.

Radioprotective effect of melanin in human cells

The aberration frequency in intact human cells (control) ranged from 0.42 ± 0.19 to $1.0 \pm 0.7\%$. These values agree with literature data.

Melanin had no mutagenic ability – it did not increase the control aberration level.

Melanins from human and animal hair at concentrations from 0.1 to 30 mg/L were very effective in reducing the aberration level induced by radiation (Fig. 7). The results were similar to those in mice – pigment activity did not depend on the concentration.

Melanin action at different cell phases was examined. In previous experiments, cells were irradiated at the G_0/G_1 stage. Table 1 presents the

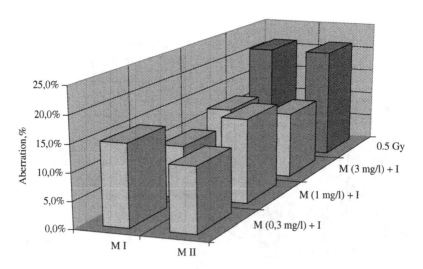

Fig. 7. Melanin (M) influence on aberration level induced by irradiation (I) in human M I – melanin from human hair, M II – melanin from animal hair, I – irradiation

TABLE 1. Influence of melanin (0.3 mg/l) from human hair on aberration frequency induced by 0.5 Gy irradiation of human lymphocytes at G1 and G2 stages

Treatment	Metaphase number		Aberration number				Aberration frequency %
	Total	With aberrations	Chromo-some	Chro-matid	Total	Gap	
Intact	2384	12	5	7	12	1	0.50 ± 0.14
G₁ stage							
Melanin	600	2	0	2	2	0	0.33 ± 0.23
Irradiation	1616	235	316	16	332	1	20.54 ± 1.00
Melanin + Irradiation	455	36	44	5	49	0	10.76 ± 1.45**
G₂ stage							
Melanin	200	0	0	0	0	0	0
Irradiation	200	36	20	18	38	0	19.00 ± 2.77
Melanin + Irradiation	200	25	16	11	27	0	13.50 ± 2.42*

$*p < 0.1$; $**p < 0.001$.

results of an experiment with irradiation at G_2 stage (46 h after the onset of culturing and 40 min after melanin addition). The antimutagenic effect of melanin was almost the same as at the G_0/G_1 stage.

These results suggest that melanin action does not depend (or depends little) on the biochemical repair system. The same conclusion was drawn earlier, when melanin had been investigated in drosophila and mice (Mosse, 1990) and agree with the data of Hopwood et al. (1985).

Combined Action of Some Chemicals and Fractionated Irradiation

The phenomenon of radioadaptive response is currently being investigated widely. It is known that one of the mechanisms of adaptive response is the cell repair system stimulation, so some repair inhibitors are able to stop or to decrease adaptive reaction. We decided to try another modification – to use an effective radioprotector, which is capable of removing the small conditioning radiation dose. Earlier we found that the lower the radiation dose, the higher the melanin protection (Mosse et al., 2000), and that the mechanism of melanin protection is not related with repair system. In addition, study of the adaptive response provides an excellent means to investigate very low doses effects.

We found an adaptive response in mouse bone marrow cells in vivo. Mice Af were exposed to X-rays with dose rate 5.6 cGy/min. One animal group was subjected to single 1.7 Gy dose, the other to 1.7 Gy divided into priming (conditioning) 0.2 Gy dose followed by the second (challenge) 1.5 Gy dose with 4-h interval. Chemicals (melanin, zenkor or tocopherol) were injected 2 h before irradiation with single or divided dose or before challenge dose. Animals were killed 24 h after irradiation of 1.7 or 1.5 Gy. Chromosome aberrations in bone marrow cells were analysed cytologically.

Irradiation of mice with a fractionated dose (0.2 Gy + 1.5 Gy with 4 h interval) led to a significant decrease in the chromosome aberration level in comparison with single 1.7 Gy dose effect (Table 2).

Melanin injection (3 mg/kg) for 2 h before irradiation with 1.7 Gy significantly decreased its clastogenic effect.

The adaptive response alone reduces aberration frequency by about 50%. Likewise melanin alone, absence of a conditioning dose, reduces the aberration frequency by about 50%. Melanin and the conditioning dose together have the same effect as either alone. Thus the two factors are not additive when melanin is given before the conditioning dose. However, they are additive when melanin is given between the conditioning dose and the highly dose.

Melanin has a very effective radical scavenging capacity and it possesses a high ability to accept and to give back electrons. Melanin has polymeric structure, which ensures the capturing, stabilization and inactivation of emerging free radicals. The complicated netted structure of melanin molecules makes them ideal "molecular sieves," in which active molecule fragments, radicals and other products of irradiation are trapped (Hill, 1992). Thus melanin exerts its protective action at the initial stage of irradiation, preventing DNA damage and not affecting biochemical repair system. That's why melanin can exert its protection after the conditioning effect has occurred. In this case the chromosome aberration level was 4-fold lower than the clastogenic effect of single 1.7 Gy dose (Table 2). Thus, both adaptive response and melanin protection has been observed.

The same results were obtained in mouse germ cells (Fig. 8).

We studied the same phenomenon with antimutagen tocopherol use. If this substance was applied before conditioning dose adaptive response was not observed as well as under melanin influence (Fig. 9).

However tocopherol being used between the first and the second radiation doses did not also change their mutagenic effect in contrary to melanin. It is known that mechanism of tocopherol action is based on repair system stimulation. It means that this antimutagen influences the same system as a priming radiation dose. It's impossible to stimulate one the same repair system twice. That is why it is unknown or the antimutagenic action of tocopherol only occurs or priming dose effect is observed.

TABLE 2. Melanin influence on fractionated irradiation in mouse born marrow cells

	Number		Aberration type				Aberration		
	Mice	Metaphases	Single fragments	Double fragments	Exchanges	Multi-fragments	Number of aberrant cells	Number of aberrations	$\% \pm S_x$
The first experiment									
Intact	4	300	3	1	0	0	4	4	1.33 ± 0.66
1.7Gy	4	300	39	31	5	0	53	75	25.00 ± 2.50
0.2Gy + 1.5Gy	4	300	16	14	3	0	29	33	$11.00 \pm 1.81**$
M + 1.7Gy	4	300	15	17	5	0	32	37	$12.33 \pm 1.90**$
M + 0.2Gy + 1.5Gy	4	300	20	15	2	0	33	37	12.33 ± 1.90
0.2Gy + M + 1.5Gy	4	300	9	7	1	0	14	17	$5.7 \pm 1.34*$
The second experiment									
Intact	4	300	4	2	0	0	6	6	2.00 ± 0.81
1.7Gy	4	300	48	31	1	0	57	80	26.67 ± 2.55
0.2Gy + 1.5Gy	4	300	25	17	1	0	33	43	$14.33 \pm 2.02**$
M + 1.7Gy	4	300	30	15	2	0	30	47	$15.67 \pm 2.10**$
M + 0.2Gy + 1.5Gy	4	300	25	18	2	1	35	45	15.00 ± 2.06
0.2 + M + 1.5	4	300	9	9	0	0	17	18	$6.00 \pm 1.37**$

Note: M refers to melanin
$*p < 0.05$, $**p < 0.01$.

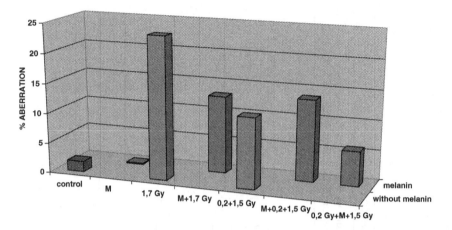

Fig. 8. Melanin influence on radioadaptive response in mice bone marrow cells.

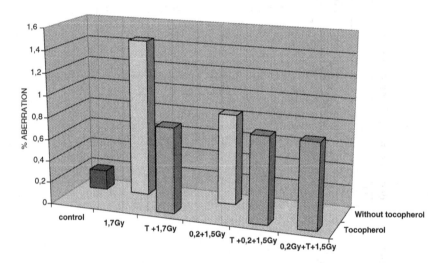

Fig. 9. Influence of tocopherol (vitamin E) on radioadaptive response in mice bone marrow cells

On the other hand we found that some mutagens such as herbicide zenkor prevent adaptive response by inhibiting repair processes. Thus different chemicals can modify radioadaptive phenomenon in different ways (Fig. 10).

Thus, many factors which are not under control can change significantly biological effects of radiation and by this can be responsible for serious mistakes of biodosimetry. Some of substances are present in our food-residual amounts of fertilizers or herbicides can be mutagenic or influence mutagenic

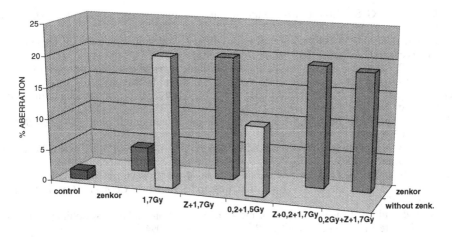

Fig. 10. Influence of herbicide zenkor on radioadaptive response in mice bone marrow cells

action of radiation. In the last case phenomena of sensibilization, synergism, antagonism or inhibition can be observed. In the same time many food stuffs contain radioprotectors or antimutagens. So, tea, coffee, cocoa, chocolate, mushrooms and others have melanin, which is very effective radioprotector not only against acute and fractionated irradiation, but even against chronic one. Many vitamins, such as tocopherol, are in our food and people use them also as a special medical additive substances. All of them can influence genetic effect of ionizing radiation.

So, some chemicals can change results of irradiation very strongly. Some drugs, stress, virus diseases and so on can influence biological effects of radiation too. Thus combined action of chemical and physical factors are worthy of great attention and must be considered during taking environmental protection measures.

References

Azatyan, R. A., V. A. Avakyan, and G. I. Mirzoyan, Cytogenetic effect of pesticides zenkor, bazagrane and difenamide, *Cytol. Genet.* 18(6), 460–462 (1984).

Baraboi, V. A. *Plants phenols and human health* (Nauka, Moscow, 1984).

Berdishev, G. D. About protective action of melanin in irradiated mouse, *Radiobiologia*, 4, 644–645 (1964).

Evans, E. P., C. E. Ford, A. G. Searle, and B. J. West, Studies on the induction of translocations in mouse spermatogonia. III. Effect of X-irradiation, *Mutat. Res.*, 9(5), 501–506 (1970).

Grossi, G. F., M. Durante, G. Gialanella, M. Pugliese, and I. Mosse, Effects of melanin on high- and low- linear energy of human epithelial cells, *Radiat. Environ. Biophys.* 37, 63–67 (1998).

Hill, H. Z. The function of melanin or six blind people examine an elephant, *Bioassays* 14, 49–56 (1992).

Hollaender and G. Steplton, Ionizing radiation and cell metabolism, *Physiol. Rev.* 33, 77–81 (1953).

Hopwood, L. E., H. M. Swartz, and S. Pajak, Effect of melanin on radiation response of CHO cells, *Int. J. Radiat. Biol.* 47, 531–537 (1985).

Khalikov, P. Kh. Chromosomal mutation level in bone marrow cells of wild mice from the area of intensive pesticide use. *Cytol. Genet.* 24(5), 10–13 (1990).

Krivolutzky, L. A., A. B. Smirnov, and M. A. Snetkov, Influence of soil radiocontamination with ⁹⁰Str on some organisms variability, *General Biol.*, 33, 581–591 (1972).

Malama, A. and P. A. Bulanov, Melanin influence on Erlich tumour, *Rep. Acad Sci. BSSR*, 9, 627–629 (1965).

Mosse, I. B. *Radiation and Heredity. Genetic aspect of antiradiation protection* (University Press, Minsk, 1990).

Mosse, I. B. and E. N. Makeeva, Modifying action of some chemicals on radiation genetic effects, 10th Int. Congress of Radiation Research, *Vurzburg, Germany, Congr. Proc.* 1, 1–38 (1996).

Mosse, I. B. and I. P. Lyach, Influence of melanin on mutation load in Drosophila population after long-term irradiation, *Radiat. Res.* 139, 356–358 (1994).

Mosse, I. B., S. I. Plotnikova, and I. P. Lyakh, Genetic changes induced by ionizing radiation in complex with sodium nitrite and nitrate in animals (BELNIITI, Minsk, 1990).

Mosse, I. B., B. V. Dubovic, S. I. Plotnikova, L. N. Kostrova, and S. T. Subbot, Melanin decreases remote consequences of long-term irradiation, Int Congr. on Radiation Protection, Austria, *Congr. Proc., part 4:*Vienna, pp. 168 – 170 (1996).

Moorhead, P. S., P. C. Navell, W. J. Meliman, D. H. Battips, and D. A. Hungerford, Chromosome preparation of leukocytes cultured from human peripheral blood, *Exp. Cell Res.* 20, 613–616 (1960).

Mosse, I. B., P. Marozik, C. Seymour, and C. Mothersill, Melanin influence on bystander effect in human keratinocytes, *Mutat. Res.* 597(1–2), 133–137 (2006).

Mosse, I. B., L. N. Kostrova, S. T. Subbot, I. Maksimenya, and V. P. Molophei, Melanin decreases clastogenic effects of ionizing radiation in human and mouse somatic cells and modifies the radioadaptive response, *Radiat. Environ. Biophys.* 39(1) 47–52 (2000).

Preston, R. I., B. J. Dean, Sh. Galloway, Mammalian in vivo cytogenetic assays analysis of chromosome aberration in bone marrow cells, *Mutat. Res.* 189, 157–165 (1987).

Shilds, L. M. and L. W. Durell, Preliminary observations on radiosensitivity of Algae and fungi from soils of the Nevada test site, *Ecology*, 42, 440–441 (1961).

Stroev, V. S. Cytogenetic activity of herbicides simazine and gidrasine, *Genetika*, 4(12), 130–134 (1968).

Stroev, V. S. Cytogenetic activity of herbicides atrazine and parakvate, *Genetika*, 6(3), 31–37 (1970).

Sultanov, S. and A. K. Ergashev, Comparative study of mutagenic effects of herbicides cotorane and toluene in the cells of some plants, *Genetika*, 27(11), 2057–2059 (1981).

Wallace, B. and J. C. King, A genetic analysis of the adaptive values of populations, *Proc. Natl. Acad. Sci. USA*, 38, 706–713 (1958).

CHAPTER 19

CYTOGENETIC BIOMARKERS FOR EXPOSURE
TO MULTIPLE STRESSORS

MARCO DURANTE

Department of Physics, University Federico II, Naples, Italy
e-mail: marco.durante@unina.it

Abstract: Risk associated to exposure to a single genotoxic agent can be assessed from the dose or concentration of the pollutant. For instance, radiation risk is estimated from physical dose (energy per mass unit), using radiation and tissue weighting factors and risk coefficients derived from epidemiology. However, for exposure to multiple stressor this approach is complicated, and deviations from simple additive models are common. Chromosomal aberrations in peripheral blood lymphocytes have long been considered a biomarker of cancer risk. In fact, cytogenetic damage in blood cells reflect similar events occurring in target organs, and take into account the interaction of different clastogenic agents, individual sensitivity, genetic background, physiological status, synergisms, etc. New molecular techniques allow a careful and unequivocal identification of several chromosomal rearrangements and can be used to provide biologically motivated risk estimates, to be compared with the estimates coming from field measurements of dose, concentration, etc. One example of exposure to multiple stressors where cytogenetic biomarkers have been widely used is manned space exploration. During space flight, astronauts are exposed to cosmic radiation while sustaining several other stresses, primarily weightlessness. It is well known that spaceflight environment has large effects on the immune system and produce several physiological alterations. The interaction of these stresses, especially microgravity, with radiation exposure has been long studied in vitro, with contradictory results. However, measurements of cytogenetic biomarkers in astronauts have demonstrated that the radiation risk is not higher than expected from ground-based studies. The use of biomarkers is therefore particularly useful for the exposure to multiple stressors.

Keywords: chromosomal aberrations; fish; densely ionizing radiation

C. Mothersill et al. (eds.), Multiple Stressors: A Challenge for the Future, 287–293.
© 2007 *Springer.*

Introduction

Estimating the risk of exposure to multiple pollutants is challenging. Any given morbidity (e.g. cancer) is a complicated, usually long-term effect of the exposure. Predicting the risk from the dose of physical or chemical mutagenic–clastogenic–teratogenic–tumorigenic agents can be misleading. In fact, the final outcome will be strongly influenced by the interaction of the different agents, and by other factors, such as individual susceptibility, genetic background, diet, immune status, age, gender, and so on.

It is then advisable to resort to *biomarkers* of risk, i.e. to biological parameters that can be measured in a given subject and are in some way related to the late effect. Brooks (1999) identified three different classes: biomarkers of exposure, sensitivity, and disease. Biomarkers of exposure are biological parameters for which a dose-response relationship can be established, and can be broadly indicated as *biodosimeters*. Biomarkers of sensitivity are genetic markers associated with an increase in the individual susceptibility to mutagenic agents. For instance, *ataxia-telangectasia* mutated gene (ATM) heterozygotes may be hypersensitive to radiation as compared to the normal population (Worgul et al., 2002). ATM heterozygosity would then represent a biomarker of sensitivity to ionizing radiation. Finally, biomarkers of disease are those biological events that can be used to anticipate the clinical diagnosis of a specific illness. In the case of carcinogenesis, these biomarkers have also been called *intermediate endpoints of cancer* (IEC), i.e. a detectable lesion, or a cellular or molecular parameter with some of the histological or biological features of preneoplasia or neoplasia (Bonassi and Au, 2002). There is no need for a causal relationship between the intermediate endpoint and the disease, although biomarkers that are part of the multistep process leading from initiation to the occurrence of invasive cancer are obviously perceived as more relevant.

For exposure to multiple pollutants, biomarkers can be divided in markers of *dose* and markers of *risk* (Durante, 2005). Biomarkers of dose are of no use for combined exposure, because the concept of dose is not defined in the presence of different agents. On the other hand, biomarkers of risk are extremely useful, if the correlation between the (early) biomarker and the (late) disease is proven (*validated* biomarker). In fact, it has been already argued that an individual's response using a validated biomarker should be part of their medical record (Albertini, 2001).

Cytogenetic Biomarkers

Several biodosimeters are available for a number of genotoxic and cytotoxic agents. However, only a few can be used to predict risk, and not all of them are validated. Chromosomal aberrations (CA) probably represent at the moment the only validated biomarker of risk. It has long been hypothesized that the frequency of genetic damage in peripheral blood lymphocytes reflects equivalent yields of damage in precursor cells of the carcinogenic processes in target tissues. The link between CA (in blood cells) and cancer (in any organ) is further strengthened by the evidence that CA are an indicator of genomic instability, which plays a key role in cancer development (Sieber et al., 2003). Epidemiological evidence that CA in lymphocytes are positively correlated with cancer risk has been recently obtained in a large cohort study, performed by the European Study Group on Cytogenetic Biomarkers and Health (ESCH), where a group of almost 22,000 healthy subjects from different European countries were screened for CA over a period of three decades (Norppa et al., 2006). Subjects were divided into 3 categories (low, medium, or high) based on percentiles of CA frequency and followed-up for cancer incidence or mortality (Fig. 1). A significant increase of both outcomes was found for the "high-frequency" group (66–100 percentile), where the occurrence of cancer was more than double that of subjects classified as "low-frequency" group (Bonassi et al., 2000). The correlation was independent from the exposure to the clastogenic agents, suggesting a causal relationship and not a simple association (Durante et al., 2001). The ESCH results have been supported by epidemiological investigations in the Czech Republic and Taiwan (reviewed in Bonassi et al., 2004). This epidemiological evidence appears strong enough to conclude that CA are biomarkers of risk. However, it is still unclear which aberration(s) should be scored, i.e. which rearrangement is more closely related to the risk, and the possible influence of the time factor, i.e. the time between exposure to the clastogen and the CA test.

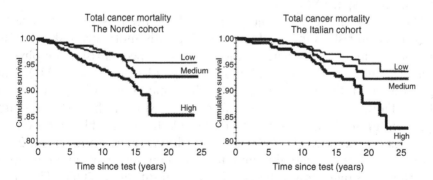

Fig. 1. Survival curves for total cancer incidence (time from chromosomal aberration test to the first diagnosis of cancer) or cancer mortality (time from chromosomal aberration test to death) in the ESCH cohort study. The three curves correspond to subjects classified as low chromosomal aberrations (1–33 percentile), medium chromosomal aberrations (33–66 percentile), and high chromosomal aberrations (66–100 percentile). Reproduced from Bonassi et al. (2000) with permission

Space Environment

Given their strong correlation to risk, measurements of CA in subjects exposed to pollutants should be recommended to decide about medical surveillance. This test is indeed applied to astronauts involved in space missions. The main goal here is to evaluate the risk associated to the exposure to galactic cosmic radiation, which could lead to an increased cancer incidence following long-term missions (Cucinotta and Durante, 2006). Radiation risk in space is carefully estimated using physical dosimetry coupled to risk models derived from epidemiology (NCRP, 2000). However, this estimate is affected by large uncertainties. One of the factors increasing uncertainty is that astronauts are exposed to multiple stressors during the spaceflights. Apart from cosmic radiation, the space environment is characterized by (ESA, 2003): psychological stress, isolation, chemical pollution (mostly contamination inside the spacecraft by acetaldehyde, dichloromethane, benzene, and formaldehyde), non-ionizing radiation, null magnetic field, and especially microgravity, which has several strong physiological effects. Is there interaction between these stressors? The answer is unclear, despite several studies in the field, especially concerning the interaction between microgravity and ionizing radiation (Kiefer and Pross, 1999). Clearly, biomarkers can play a decisive role to estimate the risk in such a complicated environment. Biomarkers take automatically into account all the multiple stressor, and the individual sensitivity and predisposition to the disease. Biological dosimetry in astronauts started several years ago using CA in peripheral blood lymphocytes (reviewed in Durante, 2005), and it is routinely performed at the NASA Johnson Space Center on crewmembers of the International Space Station.

The results of the space biodosimetry have been useful to reduce the uncertainty on risk estimates. In fact, biologically-motivated risk estimates from biodosimetry are generally fairly close to expected risk from physical dosimetry and epidemiological models (e.g. George et al., 2001), thus suggesting that the uncertainty is smaller than the upper estimates of 1500% (Durante et al., 2001). However, the result of the test is strongly dependent upon the time between the spaceflight and the blood draw: both dicentrics (Durante et al., 2003) and translocations (George et al., 2005) decline after the spaceflight, and for cosmonauts involved in multiple spaceflight it is much lower than expected (Durante et al., 2003). Therefore, a lack of correlation of correlation is observed between time in space and CA (Fig. 2). This could be due to adaptation or interactions with other factors, for instance affecting the immune system (Esposito et al., 2001) and the survival of the blood cells.

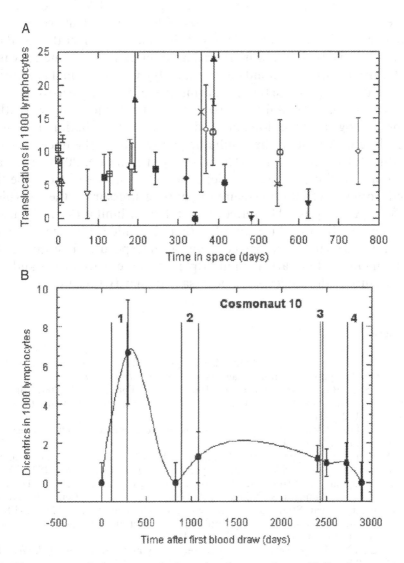

Fig. 2. Time-course of chromosomal aberration in a study on 22 Russian cosmonauts. A. Relationship between translocation frequency (WGE) and total duration of space sojourns for cosmonauts involved in multiple space flights. Each symbol represents a different cosmonaut. B. Kinetics of dicentrics in one cosmonaut involved in 4 spaceflights. The dicentric frequency (per 1,000 lymphocytes) is plotted versus the time from the first blood draw used for the cytogenetic test. Vertical bars indicate the time of launch and landing for the mission, and the number between the bars the flight number for the cosmonaut. The line connecting the datapoints is a guide for the eye. Plots from reference Durante et al. (2003), reproduced with permission from Karger Publishers, Basel

Conclusions

Biomarkers are necessary for estimating risk following chronic or acute exposure to multiple pollutants. Chromosomal aberrations are acknowledged as the only validated biomarker currently available, although it is likely that new markers based on molecular genetics will be available in the future (Amundson et al., 2001). Chromosomal aberrations have been used to estimate the risk during spaceflight, when astronauts are exposed to a complex multi-stressor environment, including weightlessness, psychological stress, ionizing and non-ionizing radiation, mutagenic chemicals, etc. Although the results helped to reduce the uncertainty on risk estimates in space travel (Cucinotta and Durante, 2006), they also point to a time factor – i.e. to a dependence of the result from the time elapsed from spaceflight (exposure to multiple stressor) to the test (blood draw for CA analysis). This effect may reflect a limitation in the use of biomarkers, or be an indication of a time-dependence of the risk itself, which not necessarily remains constant after exposure (Durante, 2005). Measurements of biomarkers in multiple stressor environments and careful follow-up of the subjects are necessary to clarify this issue.

References

Albertini, R.J., 2001, Validated biomarker responses influence medical surveillance of individuals exposed to genotoxic agents. *Radiat. Prot. Dosim.* **97**: 47–54.

Amundson, S.A., Bittner, M., Meltzer, P., Trent, J., and Fornace, A.J., 2001, Biological indicators for the identification of ionizing radiation exposure in humans. *Expert Rev. Mol. Diag.* **1**: 211–219.

Bonassi, S., and Au, W.W., 2002, Biomarkers in molecular epidemiology studies for health risk prediction. *Mutat. Res.* **511**: 73–86.

Bonassi, S., Znaor, A., Norppa, H., and Hagmar, L., 2004, Chromosomal aberrations and risk of cancer in humans: an epidemiologic perspective. *Cytogenet. Genome Res.* **104**: 376–382.

Bonassi, S., Hagmar, L., Strömberg, U., Huici Montagud, A., Tinnerberg, H., Forni, A., Heikkilä, P., Wanders, S., Wilhardt, P., Hansteen, I.-L., Knudsen, L.E., Norppa, H., for the European Study Group on Cytogenetic Biomarkers and Health (ESCH), 2000, Chromosomal aberrations in lymphocytes predict human cancer independently of exposure to carcinogens. *Cancer Res.* **60**: 1619–625.

Brooks, A.L., 1999, Biomarkers of exposure, sensitivity and disease. *Int. J. Radiat. Biol.* **75**: 1481–503.

Cucinotta, F.A., and Durante, M., 2006, Cancer risk from exposure to galactic cosmic rays: implications for space exploration by human beings. *Lancet Oncol.* **7**: 431–435.

Durante, M., 2005, Biomarkers of space radiation risk. *Radiat. Res.* **164**: 467–473.

Durante, M., Bonassi, S., George K., and Cucinotta, F.A., 2001, Risk estimation based on chromosomal aberrations induced by radiation. *Radiat. Res.* **156**: 662–667.

Durante, M., Snigiryova, G., Akaeva, E., Bogomazova, A., Druzhinin, S., Fedorenko, B.S., Greco, O., Novitskaya, N., Rubanovich, A., Shevchenko, V., von Recklinghausen, U., and Obe, G., 2003, Chromosome aberration dosimetry in cosmonauts after single or multiple space flights. *Cytogenet. Genome Res.* **103**: 40–46.

ESA, 2003, *HUMEX: Study on Survivability and Adaptation of Humans to Long-Duration Exploratory Missions*. ESA SP-1264, ESTEC, Nordwijk.

Esposito, R.D., Durante, M., Gialanella, G., Grossi, G., Pugliese, M., Scampoli, P., and Jones, T.D., 2001, On the radiosensitivity of man in space. *Adv. Space Res.* **27**: 345–354.

George, K., Willingham, V., and Cucinotta, F.A., 2005, Stability of chromosome aberrations in blood lymphocytes of astronauts, measured after space flight by FISH-chromosome painting. *Radiat. Res.* **164**: 474–480.

George, K., Durante, M., Wu, H., Willingham, V., Badhwar, G., and Cucinotta, F. A., 2001, Chromosome aberrations in the blood lymphocytes of astronauts after space flight. *Radiat. Res.* **156**: 731–738.

Kiefer, J., and Pross, H.D., 1999, Space radiation effects and microgravity. *Mutat. Res.* **430**: 299–305.

NCRP, 2000, *Radiation Protection Guidance for Activities in Low-Earth Orbit*. Report no. 132. NCRP, Bethesda, MD.

Norppa, H., Bonassi, S., Hansteen, I.L., Hagmar, L., Stromberg, U., Rossner, P., Boffetta, P., Lindholm, C., Gundy, S., Lazutka, J., Cebulska-Wasilewska, A., Fabianova, E., Sram, R.J., Knudsen, L.E., Barale, R., and Fucic, A., 2006, Chromosomal aberrations and SCEs as biomarkers of cancer risk. *Mutat. Res.* **600**: 37–45.

Sieber, O.M., Heinimann, K., and Tomlinson, I.P., 2003, Genomic instability - the engine of tumorigenesis? *Nat. Rev. Cancer* **3**: 701–708.

Worgul, B.V., Smilenov, L., Brenner, D.J., Junk, A., Zhou, W., and Hall, E.J., 2002, ATM heterozygous mice are more sensitive to radiation-induced cataracts than are their wild-type counterparts. *Proc. Natl. Acad. Sci. USA* **99**: 9836–839.

CHAPTER 20

REDOX PROTEOMICS – A ROUTE TO THE IDENTIFICATION OF DAMAGED PROTEINS

DAVID SHEEHAN*, RAYMOND TYTHER, VERA DOWLING, AND BRIAN MCDONAGH

*To whom correspondence should be addressed. Department of Biochemistry, University College Cork
Lee Maltings, Prospect Row, Mardyke, Cork, Ireland.
e-mail: D.Sheehan@ucc.ie

Abstract: The "oxygen paradox" is that molecular oxygen is both essential for aerobic life but also can be toxic to cells largely because of the effects of oxygen-derived species collectively called "reactive oxygen species" (ROS) such as the hydroxyl radical. Cells have evolved elaborate defences against ROS but if these defences are decreased (as in ageing) or if the ROS challenge becomes too great (as in toxicity), a state of oxidative stress (OS) ensues. Proteins are the principal targets of ROS and redox proteomics uses proteomics tools to study redox-based effects on the cell's protein complement. We have long used bivalve molluscs as sentinel organisms for study of pollution effects in estuaries, in particular looking at effects on stress-response proteins such as antioxidative enzymes, detoxification enzymes and heat shock proteins. Stress-response proteins are often affected by more than one stressor so these targets are likely to be of interest in other stress contexts. We are now applying redox proteomics approaches to study stress effects in bivalves. We detect carbonylation, glutathionylation, ubiquitination, effects on disulphide bridge patterns and changes in protein expression signatures in a range of electrophoresis formats. The effects are tissue- and treatment-specific. We find that many proteins targeted by OS are associated with either actin or protein disulphide isomerase. Many of the tools we use are species-independent and are appropriate for other stress scenarios.

Keywords: bivalve; proteome; oxidative stress; redox; protein oxidation; 2D SDS–PAGE; hydroxylation; glutathione; carbonyl; racemization, ubiquitin

Introduction

The "oxygen paradox" is that molecular oxygen is both essential for life but can also cause toxicity via formation of reactive oxygen species

C. Mothersill et al. (eds.), Multiple Stressors: A Challenge for the Future, 295–308.
© 2007 Springer.

(ROS; Holland, 1994; Halliwell, 1996). This may be mediated by a variety of xenobiotics (e.g. metals, polyaromatic hydrocarbons, quinones; Valavanidis et al., 2006). ROS can interact rapidly and quantitatively with lipids, DNA and proteins with rate constants approaching diffusion-controlled values ($\sim 10^8 - 10^{10}$ M^{-1} s^{-1}). Some ROS are free radicals which are produced by mitochondrial electron transport and in inflammation. Cells evolved complex defences against ROS at an early stage of aerobic evolution. These defences (Fig. 1) include antioxidant enzymes, thioredoxin-like proteins and natural antioxidants such as glutathione (GSH) and vitamin E (Masella et al., 2005). The reduced state of the cell interior is ultimately maintained by a reduced:oxidized glutathione ratio (GSH:GSSG) maintained in the range 30–100 at the expense of NADPH oxidation; a low ratio can trigger apoptosis (Mates and Sanchez-Jimenez, 2000). Much research has focused on oxidative stress, especially from the perspective of cell signalling (Bigelow and Squier, 2005), ageing (Levine and Stadtman, 2001), and the etiology of human disease (Halliwell and Gutteridge, 2007).

Proteomics studies the total complement of cell proteins or of defined protein subsets; sub-proteomes (He and Chiu, 2003). It uses high-throughput techniques to detect change in level/status of specific

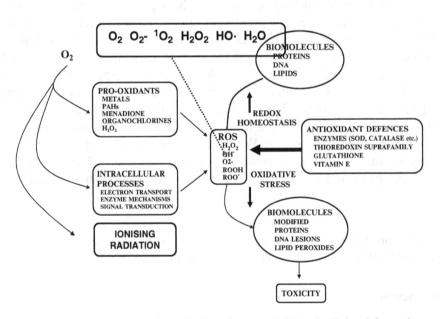

Fig. 1. Oxidative stress arises when the balance between ROS and cellular defences is upset. This can happen if the levels of ROS increase or if defences decrease as occurs in ageing

proteins. Two dimensional sodium dodecyl sulphate polyacrylamide gel electrophoresis (2D SDS–PAGE) is especially popular for these studies. Proteins are first separated in a pH gradient based on pI differences followed by a second dimension based on differing Mr (O'Farrell, 1975; Görg, 1991). Oxidative stress can cause change in levels of specific proteins detectable by protein staining and image analysis (Patton, 2002). Similarly, redox-based processes altering the pI or Mr of proteins (e.g. charge isomerization, protein backbone cleavage, crosslinking) are detected as altered 2D SDS–PAGE spots. In practice, a surprisingly small number of changes to the absolute amounts of individual proteins is usually observed. Notwithstanding this, redox-based modification of proteins in certain proteins is often quite extensive. This article describes how 2D SDS–PAGE separations can be probed to detect redox-based protein modification and thus to identify proteins targeted by oxidative stress.

Biological Targets of Reactive Oxygen Species

Molecular oxygen may be reduced to water by successively being converted into ROS such as H_2O_2 and the hydroxyl radical (HO·) in a strongly thermodynamically favoured process ($\Delta G \sim -110\,kCal/Mol$). Individual ROS may react at differing rates with a range of biological targets (Davies, 2005). For example, the most important ROS in Biology, HO·, reacts with cysteine 1,000 times faster than with glycine. Conversely, a given biological target such as methionine can react at quite different rates with individual ROS. Fast reacting ROS are therefore thought to react with proteins rather non-specifically while slow reacting ones have the potential to react more specifically. Proteins absorb approximately 70% of ROS in cells while nucleic acids and lipids provide other important targets. Reaction with ROS can change the structure of proteins in a number of ways including backbone fragmentation, hydroxylation and carbonylation of amino acid residue side-chains, glutathionylation and effects on disulphide bridges. Thus, oxidation introduces considerable structural diversity into the proteome. Redox proteomics takes advantage of these structural changes to identify specific effects on individual proteins even where the absolute amount of the protein may be unchanged. The following section outlines some experimental strategies that have proven especially useful for exploring the complex interactions of ROS with proteins. These approaches offer potential to radiation biologists since it is known that exposure to radionuclides causes oxidative stress.

Probing the Proteome for Redox-Based Modification

Several workers have pioneered an approach to environmental proteomics involving making comparisons between matched control and test samples to discover up- or down-regulated proteins followed by spot identification by mass spectrometry methods (Shepard et al., 2000; Bradley et al., 2002; Romero-Ruiz et al., 2006). The protein expression signature acts as a "fingerprint" of the proteome in the given set of experimental circumstances. In practice, a very small number of proteins are found to change in most field or laboratory exposure studies. A limitation of this classical proteomics approach is that it can be difficult to identify proteins from organisms that are not well represented in sequence databases. This is especially true of sentinel species widely used in ecotoxicology such as mussels and clams. A possible strategy for overcoming this limitation in the context of redox proteomics is to use molecular probes capable of identifying biologically widespread chemical structures. Commercially-available antibodies provide a very convenient means of detecting features such as glutathionylation, ubiquitination and carbonylation in proteins separated by 2D SDS–PAGE. They are also useful in selecting sub-proteomes by immunoprecipitation. Alternatively, some stress-response proteins are sufficiently structurally conserved that they can be detected with antibodies raised to their mammalian orthologues. Good examples are the heat shock proteins, GSH transferases and ubiquitin. Redox-based lesions which we have found especially useful include carbonylation, glutathionylation, effects on disulphide bridges and ubiquitination.

CARBONYLATION

Many amino acid side-chains are readily converted to aldehyde or ketone groups on exposure to ROS, a process called carbonylation (Levine and Stadtman, 2001). This is an irreversible modification leading usually to inactivation, crosslinking and turnover of the damaged protein. Its irreversibility makes carbonylation an especially useful measure of redox-based damage because the lesion withstands extraction and proteomic separation (Fig. 2). Protein carbonyls react quantitatively with hydrazines to form hydrazones (Levine et al., 1990). After 2D SDS–PAGE, proteins can be transferred to nitrocellulose membranes and carbonylated proteins labelled with 2, 4-dinitrophenyl hydrazine (hydrazine-DNP) followed by detection with anti-DNP (England and Cotter, 2004). As mentioned earlier, proteins derivatized with hydrazine-DNP can be immunoprecipitated with anti-DNP prior to 2D SDS–PAGE (McDonagh et al., 2006). Carbonylation is mainly

a)

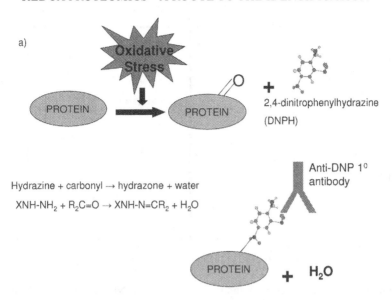

2,4-dinitrophenylhydrazine
(DNPH)

Hydrazine + carbonyl ⟶ hydrazone + water

$XNH\text{-}NH_2 + R_2C{=}O \rightarrow XNH\text{-}N{=}CR_2 + H_2O$

Anti-DNP 1^0
antibody

PROTEIN   + H_2O

b)

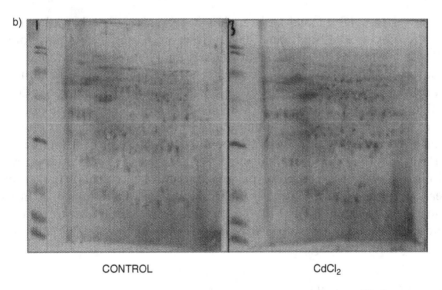

CONTROL $CdCl_2$

Fig. 2. Protein carbonyls. (a) Proteins can be carbonylated by oxidative stress. 2, 4-Dinitrophyl hydrazine (hydrazine-DNP) reacts with carbonyl groups to form hydrazone-DNP, which is detectable with anti-DNP. (b) 2D SDS–PAGE separation of *Mytilus edulis* extracts of gills dissected from animals exposed to $CdCl_2$ and stained with fast stain.

(continued)

c)

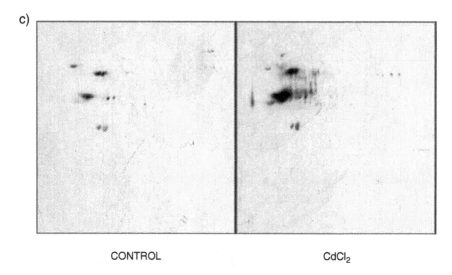

CONTROL   CdCl₂

Fig. 2. (continued) (c) The same separation probed with anti-DNP. Note that only some proteins are carbonylated

caused by the HO· radical and this has been suggested to be rather non-specific (Levine and Stadtman, 2001). However, recent studies have found selective protein carbonylation. In mussels and clams we have found that exposure to pro-oxidants causes extensive carbonylation of specific proteins against a largely unchanged proteomic background (Dowling et al., 2006; McDonagh and Sheehan, 2006; McDonagh et al., 2006). Intriguingly, the effect does not depend on protein quantity in that some abundant proteins are not carbonylated at all while some low-abundance proteins are heavily carbonylated. We also find that the effect is surprisingly rapid (appearing within 2 hours of exposure to pro-oxidants) and shows both tissue- and treatment-specificity. Quantitative information can be obtained by measuring the total staining intensity of spots and comparing tests to controls. Alternatively, image analysis software such as PDQuest facilitates spot counting and matching.

GLUTATHIONYLATION

GSH can form mixed disulphides with sulphydryl groups of proteins that are key targets of oxidation (Dalle-Donne et al., 2001). Unlike most other lesions, this is a reversible modification. Using 1D SDS–PAGE followed by immunoblotting with anti-GSH, we have identified actin as a key target for glutathionylation (McDonagh et al., 2005). Normally, the second dimension of 2D SDS–PAGE is run under reducing conditions which would reduce mixed disulphides thus losing the GSH from its target protein. This can be avoided by running the second dimension under reducing conditions or else by immunoprecipitating with anti-GSH before 2D SDS–PAGE and silver staining (McDonagh et al., 2006).

UBIQUITINATION

Ubiquitin is a highly conserved protein with an approximate mass of 8.5 kDa. It plays a key role in targeted turnover of short-lived proteins in cell (Fig. 3). Oxidatively damaged proteins have been reported as being cleared via ubiquitinylation followed by digestion in the 20S core of the 26S proteasome in the cytosol and nucleus (Davies, 2001; Marques et al., 2004; Petroski and Deshaies, 2005; Friguet, 2006) and by the Lon protease in the mitochondrial matrix (Bota et al., 2002). Commercially-available antibodies to ubiquitin and polyubiquitin facilitate identification of ubiquitinated proteins in blots of 2D SDS–PAGE separations (Fig. 3) (McDonagh and Sheehan, 2006). In mammalian systems, carbonylated proteins have been

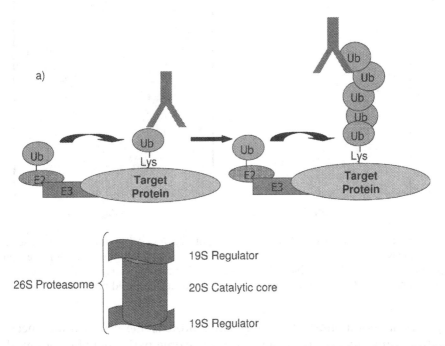

Fig. 3. Protein ubiquitinylation. (a) Proteins can be labelled with ubiquitin via the ubiquitin proteolytic pathway leading to polyubiquitination. Anti-ubiquitin reacts specifically with ubiquitinated proteins

(continued)

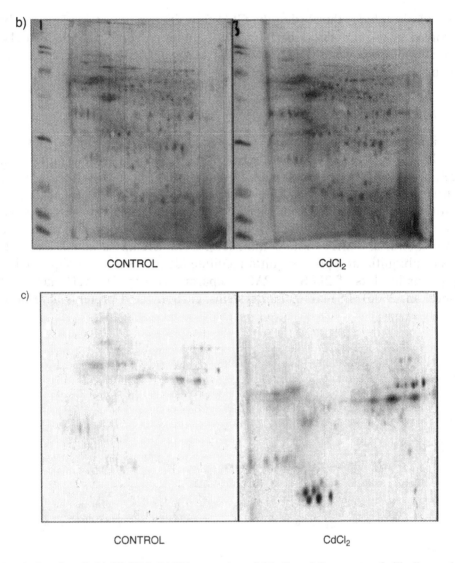

Fig. 3. (continued) (b) 2D SDS–PAGE separation of *Mytilus edulis* extracts of gills dissected from animals exposed to $CdCl_2$ and stained for total protein with fast stain. (c) The same separation probed with anti-ubiquitin. Note that ubiquitination increases with exposure to pro-oxidant but that the pattern is distinct from that for carbonylated proteins (Fig. 2)

reported as good substrates for the 20S proteasome core without necessarily passing through the ubiquitin proteasome pathway (Marques et al., 2004). Our recent work with oxidatively stressed molluscs suggests that there may not always be a correlation between carbonylation and ubiquitination (McDonagh and Sheehan, 2006).

PROTEIN THIOLS

Cysteines are key targets for ROS and it is now being increasingly realized that this is important in signal transduction in cells. Protein thiols do not react with oxidants at their biological concentrations (Eaton, 2006). However, thiol pKa values can be lowered by their surrounding environment which makes individual cysteines especially redox-sensitive. Thiol modifications include direct oxidation (to sulphenic, sulphinic and cysteic acids), formation of mixed disulphides (e.g. with GSH, cysteine and homocysteine) and formation of intra/intermolecular protein disulphides (Davies, 2005; Eaton, 2006). Oxidized variants of –SH generally do not react with thiol-specific reagents such as maleimides, iodoacetic acid, iodoacetamide and thiosulphates. Thus it is possible to compare samples using labels covalently attached to these chemical functionalities. Examples include biotin (selectable by binding to avidin) (Eaton et al., 2002), polyethylene glycol (increases Mr) (Marques et al., 2004), fluorescein (immunodetectable with anti-fluorescein) (Baty et al., 2002) and radionuclides (Brooker and Slayman, 1983). A rare reversible redox modification is represented by nitrosothiol (SNO) resulting from reaction of specific thiols with nitric oxide. This can be detected with the "biotin switch" assay which depends on the fact that ascorbate specifically reduces SNO but not disulphides (Jaffrey et al., 2001). Reaction with N-ethyl maleimide-biotin allows specific labelling of new –SH groups with subsequent selection on avidin columns and 2D SDS–PAGE (Fig. 4). Thiol-specific chemical functionalities can also be used to detect available –SH groups in proteins. Iodoacetamide-fluorescein labels available –SH (i.e. those not involved in disulphides) and these proteins can be identified by immunoblotting with anti-fluorescein. Since –SH groups are readily oxidized to species which do not react with iodoacetamide, comparisons of oxidized with control separations reveal proteins containing susceptible cysteines. A modification of this approach involves first treating the proteins with NEM followed by reduction with DTT and treatment with iodoacetamide-fluorescein which selects for cysteines involved in disulphide bridges. Another way of probing effects of oxidative stress on disulphide bridge patterns of proteins is to use diagonal gels (McDonagh and Sheehan, 2006). A 1D SDS–PAGE separation is performed first under non-reducing conditions (Fig. 5). The track of proteins is excised, reduced and run orthogonally on a reducing gel. Proteins with no disulphide bridges run along a diagonal. Proteins with intrachain disulphides run above the diagonal (since their structure is now more "open") while proteins with interchain disulphides run below the diagonal. Thus, a characteristic pattern of disulphide-bridged proteins is generated. An elaboration of this technique is to probe the separation for oxidative lesions such as protein carbonyls.

a)

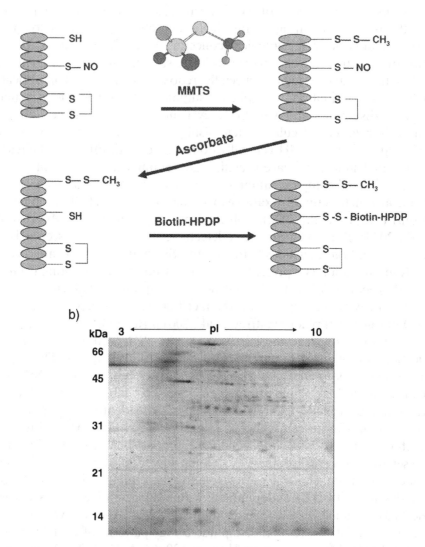

Fig. 4. The Biotin switch assay. (a) Ascorbate specifically reduces -SNO to SH without reducing disulphides. (b) Silver-stained 2D SDS–PAGE gel of streptavidin-agarose purified proteins from rat kidney medulla, following biotin switch assay

Future Perspectives

Several bottlenecks prevent full exploitation of redox proteomics in environmental science (Dowling and Sheehan, 2006). Chief amongst these is lack of sequence information to facilitate protein identification. As this

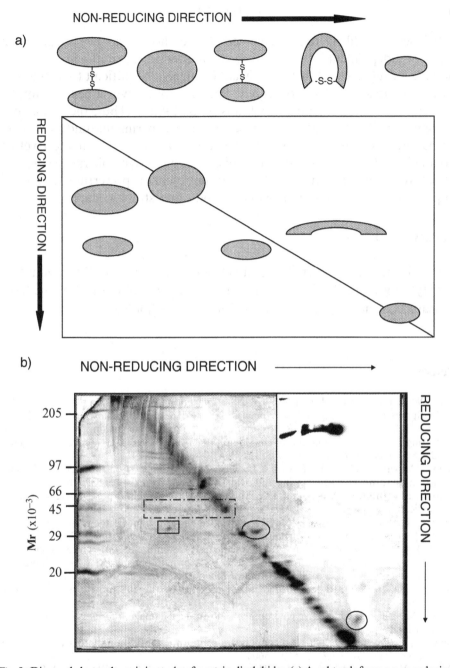

Fig. 5. Diagonal electrophoresis in study of protein disulphides. (a) A gel track from a non-reducing gel is excised, exposed to reducing buffer and laid orthogonally on a reducing SDS–PAGE gel. Proteins lacking disulphides migrate to a diagonal position as they electrophorese identically in reducing and non-reducing conditions. Proteins with interchain disulphides migrate as separate polypeptides below the diagonal while those with intrachain disulphides migrate more slowly under reducing conditions and thus appear above the diagonal. (b) Separation of extract of gill from *Mytilus edulis* stained with silver. Inset shows an immunoblot with anti-actin corresponding to dashed box

deficiency is unlikely to be remedied in the short term, we suggest that "species-independent" chemical-based approaches as outlined here may provide a route around this bottleneck. It can also be difficult to distinguish specific causes of effects observed. For example, environmental samples notoriously contain complex cocktails of pollutants. Thus, there is a continuing need for well-designed dose-response experiments under controlled laboratory conditions. Redox proteomics allows study at the level of organelle, cell, tissue and whole organism and has a dynamic range covering three orders of magnitude. It therefore represents a powerful set of techniques for the exploration of oxidative and other stress scenarios.

Acknowledgements

Our laboratory is supported by grants from the Irish Higher Education Authority Programme for Research in Third Level Institutions and by the Irish Research Council for Science Engineering and Technology.

References

Baty, JW, Hampton, MB, Winterburn, CC 2002. Detection of oxidant sensitive thiol proteins by fluorescence labelling and two dimensional electrophoresis. *Proteomics* 2: 1261–1266.

Bigelow, DJ, Squier, TC 2005. Redox modulation of cellular signalling and metabolism through reversible oxidation of methionine sensors in calcium regulatory proteins. *Biochimica et Biophysica Acta* 1703: 121–134.

Bota, DA, Van Remmen, H, Davies, KJA 2002. Modulation of Lon protease activity and aconitase turnover during ageing and oxidative stress. *FEBS Letters* 532: 103–106.

Bradley, BP, Shrader, EA, Kimmel, DG, Meiller, JC 2002. Protein expression signatures: an application of proteomics. *Marine Environmental Research* 54: 373–377.

Brooker, RJ, Slayman, CW 1983. [14C] N-ethylmaleimide labelling of the plasma membrane [H+]-ATP-ase of *Neurospora crassa*. *Journal of Biological Chemistry* 258: 222–226.

Dalle-Donne, I, Rossi, R, Milzani, A, Di Simplicio, P, Colombo, R 2001. The actin cytoskeleton response to oxidants: from small heat shock protein phosphorylation to changes in the redox state of actin itself. *Free Radical Biology and Medicine* 31: 1624–1632.

Davies, KJA 2001. Degradation of oxidized proteins by the 20S proteasome. *Biochimie* 83: 301–310.

Davies, MJ 2005. The oxidative environment and protein damage. *Biochimica et Biophysica Acta* 1703: 93–109.

Dowling, V, Sheehan, D 2006. Proteomics as a route to identification of toxicity targets in ecotoxicology *Proteomics* 6: 5597–5604.

Dowling, V, Hoarau, P, Romeo, M, O'Halloran, J, van Pelt, FNAM, O'Brien, NM, Sheehan, D 2006. Protein carbonylation and heat shock response in *Ruditapes decussatus* following p,p'-dichlorodiphenyldichloroethylene (DDE) exposure: a proteomic approach reveals DDE causes oxidative stress. *Aquatic Toxicology* 77: 11–18.

Eaton, P 2006. Protein thiol oxidation in health and disease: techniques for measuring disulfides and related modifications in complex protein mixtures. *Free Radical Biology and Medicine* 40: 1889–1899.

Eaton, P, Byers, HL, Leeds, N, Ward, MA, Shattock, MJ 2002. Detection, quantitation, purification, and identification of cardiac proteins S-thiolated during ischemia and reperfusion. *Journal of Biological Chemistry* 277: 9806–9811.

England, K, Cotter, T 2004. Identification of carbonylated proteins by MALDI-TOF mass spectroscopy reveals susceptibility of ER. *Biochemical and Biophysical Research Communications* 320: 123–130.

Friguet B 2006. Oxidized protein degradation and repair in ageing and oxidative stress. *FEBS Letters* 580: 2910–2916.

Görg, A 1991.Two-dimensional electrophoresis. *Nature* 349: 545–546.

Guo, ZY, Chang, CCY, Lu, XH, Chen, J, Chang, TY 2005. The disulfide linkage and the free sulfhydryl accessibility of acyl-coenzyme A: cholesterol acyltransferase 1 as studied by using mPEG5000-maleimide. *Biochemistry* 44: 6537–6546.

Halliwell, B 1996. Oxidative stress, nutrition and health: experimental strategies for optimization of nutritional antioxidant intake in humans. *Free Radical Research* 25: 57–74.

Halliwell, B, Gutteridge, JMC 2007. *Free Radicals in Biology and Medicine* Fourth Edition Oxford University Press, Oxford.

He, QY, Chiu, JF 2003. Proteomics in biomarker discovery and drug development. *Journal of cellular biochemistry* 89: 868–886.

Holland, HD 1994. Early proterozoic atmospheric change. In: *Early Life on Earth*, Nobel Symposium No. 84: 237–244.

Jaffrey, SR, Erdjument-Bromage, H, Ferris, CD, Tempst, P, Snyder, SH 2001. Protein S-nitrosylation: a physiological signal for neuronal nitric oxide. *Nature Cell Biology* 3: 193–197.

Levine, RL, Stadtman, ER 2001. Oxidative modification of proteins during aging. *Experimental Gerontology* 36: 1495–1502.

Levine, RL, Garland, D, Oliver, CN, Amici, A, Climent, I, Lenz, A, Ahn, BW, Shaltiel, S, Stadtman, ER 1990. Determination of carbonyl content in oxidatively modified proteins. *Methods in Enzymology* 186: 464–478.

Marques, C, Pereira, P, Taylor, A, Liang, JN, Reddy, VN, Szweda, LI, Shang, F 2004. Ubiquitin-dependent lysosomal degradation of the HNE-modified proteins in lens epithelial cells. *FASEB Journal* 18: 1424–1426.

Masella, R, Di Benedetto, R, Vari, R, Filesi, C, Giovannini, C 2005. Novel mechanisms of natural antioxidant compounds in biological systems: involvement of glutathione and glutathione-related enzymes. *Journal of Nutritional Biochemistry* 16: 577–586.

Mates, JM, Sanchez-Jimenez, FM 2000. Role of reactive oxygen species in apoptosis: implications for cancer therapy. *International Journal of Biochemistry and Cell Biology* 32: 157–170.

McDonagh, B, Sheehan, D 2006. Redox proteomics in the blue mussel *Mytilus edulis*: carbonylation is not a pre-requisite for ubiquitination in acute free radical-mediated oxidative stress. *Aquatic Toxicology* 79: 325–333.

McDonagh, B, Tyther, R, Sheehan, D 2006. Redox proteomics in the mussel *Mytilus edulis*. *Marine Environmental Research* 62: S101–104.

McDonagh, B, Tyther, R, Sheehan, D 2005. Carbonylation and glutathionylation of proteins in the blue mussel *Mytilus edulis* detected by proteomic analysis and Western blotting: actin as a target for oxidative stress. *Aquatic Toxicology* 73: 315–326.

O'Farrell, PH 1975. High-resolution two-dimensional electrophoresis of proteins. *Journal of Biological Chemistry* 250: 4007–4021.

Patton, WF 2002. Detection technologies in proteome analysis. *Journal of Chromatography. B – Analytical Technologies in the Biomedical and Life Sciences* 771: 3–31.

Petroski, MD, Deshaies, RJ 2005. Mechanism of lysine-48-linked ubiquitin chain synthesis by the cullin-RING ubiquitin-ligase complex SCF-Cdc34. *Cell* 123: 1107–1120.

Romero-Ruiz, A, Carrascal, M, Alhama, J, Gomez-Ariza, JL, Albian, J, Lopez-Barea, J 2006. Utility of proteomics to assess pollutant response of clams from the Donana bank of Guadalquivir estuary (SW Spain). *Proteomics* 6: S245–S255.

Shepard, JL, Olsson, M, Tedengren, M, Bradley, BP 2000. Protein expression signatures identified in *Mytilus edulis* exposed to PCBs, copper and salinity stress. *Marine Environmental Research* 50: 337–340.

Valavanidis, A, Vlahogianni, T, Dassenakis, M, Scoullos, M 2006. Molecular biomarkers of oxidative stress in aquatic organisms in relation to toxic environmental pollutants. *Ecotoxicology and Environmental Safety* 64: 178–189.

CHAPTER 21

EXPOSURE ASSESSMENT TO RADIONUCLIDES TRANSFER IN FOOD CHAIN

MARIA DE LURDES DINIS AND ANTÓNIO FIÚZA

Geo-Environment and Resources Research Centre (CIGAR) Engineering Faculty, Porto University, Rua Dr. Roberto Frias, 4200–465, Porto, Portugal.
e-mail: mldinis@fe.up.pt; afiuza@fe.up.pt

Abstract: Generally sites with radioactive contamination are also simultaneously polluted with many other different toxics, especially heavy metals. Besides the radioactivity, these wastes may also hold different amounts of chemicals, toxic pollutants and precipitates. The radionuclides released into the environment can give rise to human exposure by the transport through the atmosphere, aquatic systems or through soil sub-compartments. The exposure may result from direct inhalation of contaminated air or ingestion of contaminated water, or from a less direct pathway, the ingestion of contaminated food products. Contamination of the trophic chain by radionuclides released into the environment will be a component of human exposure to ionizing radiations by transferring the radionuclides into animal products that are components of the human diet. This can occur by first ingestion of contaminated pasture by animals and then by ingestion of animal products contaminated. The relevant incorporation of the radionuclides into cow's milk is usually due to the ingestion of contaminated pasture. This transfer process is often called the *pasture-cow-milk* exposure route. A compartment dynamic model is presented to describe mathematically the radium behaviour in the *pasture-cow-milk* exposure route and predict the activity concentration in each compartment. The dynamic model is defined by a system of linear differential equations with constant coefficients based in a mass balance concept. For each compartment a transient mass balance equation defines the relations between the inner transformations and the input and output fluxes. The concentration within each compartment is then transcribed to doses values based on a simplified exposure pathway and a pre-defined critical group.

Keywords: exposure assessment; radium, dynamic model; differential equations

C. Mothersill et al. (eds.), Multiple Stressors: A Challenge for the Future, 309–323.

Introduction

The environmental effects originated by uranium mining activities result mainly from the large volume of residues produced by the ore processing. Besides the radioactivity these wastes may also hold different amounts of chemicals, toxic pollutants and precipitates originated by pH or Eh alterations. The radionuclides released from these wastes can give rise to human exposure by transport through the atmosphere, aquatic systems or through soil sub-compartments. The exposure may result from direct inhalation of contaminated air or ingestion of contaminated water, or from a less direct pathway – the ingestion of contaminated food products. Nevertheless this pathway can be quite significant as a result of biological concentration in the foodstuff.

The exposure resulting from airborne particulates containing ^{230}Th, ^{226}Ra, and ^{210}Pb as well as uranium, is primary by the inhalation of particles and or through the food chain. The predominant target effective dose from these radionuclides is to the bones. Non-radioactive metals and other chemical reagents may also induce chronic or acute health effects. The harmful effects of radionuclides do not come from their chemistry within tissue, but from the radiation associated with radioactive decay which increases the risk of cancer.

Radionuclides deposition can be a significant pathway to human exposure by first ingestion of contaminated pasture by animals and then by the ingestion of animal products contaminated (dairy or meat). Plants in general tend to accumulate radionuclides in a scale dependent on many factors and within animals and humans, certain tissues tend to accumulate selected radionuclides. The relevant incorporation of the radionuclides in the milk is usually due to the ingestion of contaminated pasture. This transfer process is often called the *pasture-cow-milk* exposure route. We developed a compartment dynamic model to describe mathematically the radionuclide behaviour in the *pasture-cow-milk* exposure route and predict the activity concentration in each compartment following an initial radionuclide deposition.

Methods and Results

DESCRIPTION OF THE MODEL

The dynamic model consists of a system of linear differential equations with constant coefficients describing the mass balance in different compartments taking in account the fluxes in and out of the compartment and the radionuclides decay. For each compartment, a transient mass balance equation defines the relations between the inner transformations and the input and output fluxes. The fluxes between the compartments are estimated with a transfer rate proportional to the amount of the radionuclide in the compartment. The model also considers possible transformations within the compartment.

The first model considered for the propagation through the food chain is relatively simple and classic and considers as initial state a contaminated pasture that is consumed by a cow that produces a certain quantity of milk. The transfer coefficients for soil and pasture compartments are expressed as function of soil characteristics and ecological parameters. A more sophisticated model is also described taking into account the spread of radium within the cow by including the sub-compartments involved: the gastrointestinal system (GIT), the plasma and the bones. The transfer coefficients for the sub-compartments within the cow are combined with biological half-lives which is the time taken for the radionuclide activity concentration in tissues or milk be reduced by one half of its initial value.

The processes involved in the radionuclide transfer to cow milk resulting from consumption of contaminated pasture are: (i) pasture deposition; (ii) deposition on the soil; (iii) retention of radionuclides by pasture over a certain period of time; (iv) root uptake; (v) consumption of contaminated pasture by the cow and (vi) the secretion of radionuclides into the milk. A scheme of the conceptual model is given in Fig. 1.

For the compartment 1 the input fluxes results from the fraction of the radionuclides deposition (d, $Bq \cdot m^{-2} \cdot d^{-1}$) that is not intercepted by pasture (1–F) and goes directly to the soil, the radionuclides weathering from pasture surface (k_{21}, d^{-1}), after the material has been deposited onto the pasture surface it will

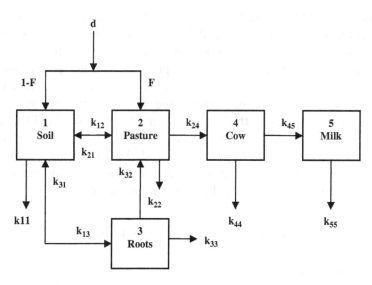

Fig. 1. Conceptual scheme of the model for the pasture-cow-milk exposure route

begin to weather off the surface, and the transfer from pasture to soil (k_{31}, d^{-1}), considering that root uptake is a reversible process described with first order kinetic equations for the root uptake balance (IAEA, 2002; Teale et al., 2003). The outputs fluxes results from losses in the compartment 1 due to radioactive decay and environmental processes related to radionuclide migration in soil (k_{11}, d^{-1}), resuspension of radioactive particles from soil and subsequent deposition onto pasture surface (k_{12}, d^{-1}) and root uptake (k_{13}, d^{-1}). The transient mass balance equation for compartment 1 is defined by the following equation:

$$\frac{dA_1}{dt} = d \cdot (1-F) \cdot A_s - k_{11}A_1 - k_{13}A_1 - k_{12}A_1 + k_{21}A_2 + k_{31}A_3 \tag{1}$$

For the compartment 2 the inputs fluxes results from the fraction of radio-nuclides deposition intercepted by pasture surface (F), the resuspension of radioactive particles from soil and subsequent deposition onto pasture surface (k_{12}, d^{-1}) and the translocation from pasture roots to the interior of the grass or to any other edible parts (k_{32}, d^{-1}). The output fluxes results from the losses in this compartment due to radioactive decay (k_{22}, d^{-1}), the radionuclides weathering from pasture surface (k_{21}, d^{-1}) and by the ingestion of contaminated pasture (k_{24}, d^{-1}). The transient mass balance equation for compartment 2 is defined by the following equation:

$$\frac{dA_2}{dt} = d \cdot F \cdot A_P + k_{12}A_1 - k_{22}A_2 - k_{21}A_2 - k_{24}A_2 + k_{32}A_3 \tag{2}$$

The input flux in compartment 3 results from the radionuclides transfer from soil to pasture by root uptake (k_{13}, d^{-1}). The outputs fluxes in this compart-ment results from the losses by radioactive decay (k_{33}, d^{-1}), the translocation from pasture roots to the interior of the grass or to other edible parts (k_{32}, d^{-1}) and the transfer from pasture roots to soil (k_{31}, d^{-1}), considering that root uptake is a reversible process. In the simpler approach root uptake transfer (k_{13}) is estimated with the soil to grass transfer factor, TF (Bq·kg^{-1}, fresh mass/Bq·kg^{-1}, dry mass), pasture roots biomass, M_p (kg·m^{-2}), the soil mass in soil compartment, M_s (kg·m^{-2}), and the equilibrium root uptake transfer, k_{31} (d^{-1}) (Teale et al., 2003). The transient mass balance equation for compartment 3 is defined by the following equation:

$$\frac{dA_3}{dt} = k_{13}A_1 - k_{33}A_3 - k_{32}A_3 - k_{31}A_3 \tag{3}$$

and k_{13} is defined by

$$k_{13} = TF \cdot \frac{M_P}{M_S} \cdot k_{31} \qquad (4)$$

For the compartment 4, the input flux results from the cow intake of contaminated pasture (k_{24}, d^{-1}) and the outputs results from the losses in this compartment due to radioactive decay (k_{44}, d^{-1}) and the secretion into cow's milk (k_{45}, d^{-1}). The transient mass balance equation for compartment 4 is defined by the following equation:

$$\frac{dA_4}{dt} = k_{24}A_2 - k_{44}A_4 - k_{45}A_4 \qquad (5)$$

Finally, for the compartment 5, the input flux results from radionuclides secretion into cow's milk (k_{45}, d^{-1}) and the output is the loss due to radioactive decay (k_{55}, d^{-1}). The transient mass balance equation for compartment 5 is defined by the following equation:

$$\frac{dA_5}{dt} = k_{45}A_4 - k_{55}A_5 \qquad (6)$$

In the previous equations, d represents the total deposition ($Bq \cdot m^{-2} \cdot d^{-1}$), A_s the soil compartment area (m^2), A_p the pasture compartment area (m^2), F is the interception factor (dimensionless) defined as the fraction of the activity deposited on the ground (soil and pasture) which is intercepted by vegetation during the time of deposition, k_{ii} are the losses from compartment i (d^{-1}) and k_{ij} are the kinetic transfer from compartment i to compartment j (d^{-1}). The equations system can be represented in the following matrix form and solved numerically in appropriate software Fig. 2.

In the complete model, the radium distribution within the cow is modelled by including three more sub-compartments: the gastrointestinal system (GIT), the plasma and the bones. The scheme for the conceptual model describing the radionuclide transfer within the cow adapted from the

$$
\begin{bmatrix} \frac{dA_1}{dt} \\ \frac{dA_2}{dt} \\ \frac{dA_3}{dt} \\ \frac{dA_4}{dt} \\ \frac{dA_5}{dt} \end{bmatrix} =
\begin{bmatrix}
-k_{11}-k_{13}-k_{12} & +k_{21} & +k_{31} & 0 & 0 \\
+k_{12} & -k_{22}-k_{21}-k_{24} & +k_{32} & 0 & 0 \\
+k_{13} & 0 & -k_{33}-k_{32}-k_{31} & 0 & 0 \\
0 & +k_{24} & -k_{44}-k_{45} & 0 & 0 \\
0 & 0 & 0 & +k_{45} & -k_{55}
\end{bmatrix}
\cdot
\begin{bmatrix} A_1 \\ A_2 \\ A_3 \\ A_4 \\ A_5 \end{bmatrix}
+
\begin{bmatrix} d \cdot (1-F) \cdot A_s \\ d \cdot F \cdot A_p \\ 0 \\ 0 \\ 0 \end{bmatrix}
$$

Fig. 2. Matrix form of the balance equations from the conceptual model (simple)

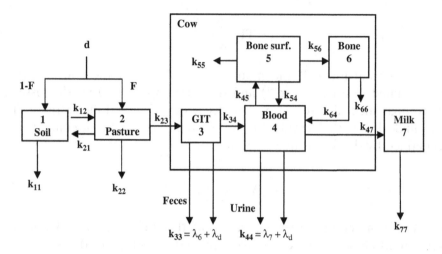

Fig. 3. Conceptual model for radium kinetics in the cow (IAEA, 2004; Leggett et al., 2003)

International Commission on Radiological Protection (ICRP) biokinetic models (Leggett et al., 2003) is represented in Fig. 3.

Within animals and humans certain tissues tend to accumulate selected radionuclides. From what is known about radium retention and distribution in the organism after oral exposure, radium is quickly distributed to body tissues followed by a rapid decrease of its content in blood. It later appears in the urine and faeces. Retention in tissues decreases with time following maximal uptake, not long after the intake to blood. Bones become the principal radium repository in the body and this is due to its chemical similarity with calcium. Two compartments are considered for bones: a bone surface compartment (5) in which radium is retained for short periods and a bone volume compartment (6) in which radium is retained for long periods (Leggett et al., 2003). The system of linear equations for this conceptual model is obtained as previous by defining a transient mass balance equation for each compartment. The resulting matrix with constant coefficients is represented below Fig. 4.

EXAMPLE OF MODEL SIMULATION

The first model considered for the radionuclides propagation through the food chain is relatively simple and classic. It considers as initial state a contaminated pasture that is consumed by a cow that produces a certain quantity of milk. In this simpler model the processes involved in radionuclides transfer to pasture are deposition, resuspension and root uptake. Deposition is the only

$$
\begin{bmatrix}
\dfrac{dA_1}{dt} \\[4pt]
\dfrac{dA_2}{dt} \\[4pt]
\dfrac{dA_3}{dt} \\[4pt]
\dfrac{dA_4}{dt} \\[4pt]
\dfrac{dA_5}{dt} \\[4pt]
\dfrac{dA_6}{dt} \\[4pt]
\dfrac{dA_7}{dt}
\end{bmatrix}
=
\begin{bmatrix}
-k_{11}-k_{12} & k_{21} & 0 & 0 & 0 & 0 & 0 \\
k_{12} & -k_{22}-k_{21}-k_{23} & 0 & 0 & 0 & 0 & 0 \\
0 & k_{23} & -k_{34}-k_{33} & 0 & 0 & 0 & 0 \\
0 & 0 & k_{34} & -k_{44}-k_{45}-k_{47} & k_{54} & k_{64} & 0 \\
0 & 0 & 0 & k_{45} & -k_{56}-k_{55}-k_{54} & 0 & 0 \\
0 & 0 & 0 & 0 & k_{56} & -k_{66}-k_{64} & 0 \\
0 & 0 & 0 & k_{47} & 0 & 0 & -k_{77}
\end{bmatrix}
\cdot
\begin{bmatrix}
A_1 \\ A_2 \\ A_3 \\ A_4 \\ A_5 \\ A_6 \\ A_7
\end{bmatrix}
+
\begin{bmatrix}
d\cdot(1-F)\cdot A_s \\
d\cdot F\cdot A_p \\
0 \\
0 \\
0 \\
0 \\
0
\end{bmatrix}
$$

Fig. 4. Matrix form of the balance equations from the conceptual model (complete)

contamination source and there is no pasture irrigation considering that rainfall water is enough for pasture growing.

A simulation was done for a constant radionuclide input of 1 Bq·m^{-2}·d^{-1} in a hypothetical case study. The necessary parameters were adopted from different sources: some parameters were adopted from measurements referring to a particular contaminated site (Falcão et al., 2005) and others were adopted from published data on radionuclide behaviour in animals, such as distribution or retention in different organs and tissues and subsequent excretion routes (IAEA, 2004). The unknown parameters were estimated from available data.

The rate at which the radionuclides on pasture surface are weathered off onto the soil has been estimated using a weathering half-life. A default value of 14 days has been adopted which is consistent with literature values and the value for the kinetic constant representing this transfer process (k_{21}) is 0.495 d^{-1} (Teale, 2003).

The radionuclide migration down the column soil and out of reach of the roots can be modelled using a loss from the soil compartment based on a soil migration half-life. The default value adopted from literature is 2.5×10^4 days (Hoffman et al., 1984) and the value for the kinetic constant, k_{11}, which represents the radium decay losses and the losses due to its migration in soil, is 2.82×10^{-5} d^{-1}.

Available data on root uptake is in the form of a transfer factor defined as the ratio of activity concentration in the edible part of the plant (Bq·kg^{-1}, fresh mass) to that in the soil (Bq·kg^{-1}, dry mass), once equilibrium has been reached. This transfer pathway was included in the conceptual model although it is not relevant when the contamination is due to deposition. To model this process, a pair of transfer rates between the soil and the roots compartments was used. The transfer factor implies equilibrium conditions which should be reached quickly to legitimate its use (Teale, 2003). To achieve this, the equilibrium rate was set to one day and the resulting value for k_{31} is

MARIA DE L. DINIS AND A. FIÚZA

$0.693\,d^{-1}$. The soil to plant transfer factor for radium in pasture, k_{13}, suggested in literature is 0.08 (IAEA, 1994).

The initial conditions were defined for the two first compartments (soil and pasture). For the other compartments the initial activity was considered to be inexistent. Pasture and soil compartments have an area of $1,000\,m^2$ each and the grass biomass is $1\,kg\cdot m^{-2}$. The same value was considered for the grass roots biomass. Pasture consumption is $50\,kg\cdot d^{-1}$ and the daily milk production is $12\,kg$.

For the exploration of the model several radionuclides were defined as relevant but, for the present, only radium was considered in the calculations, due to the availability of data. The endpoints are radium concentrations in soil, pasture, cow (whole body) and milk, in the simpler model. The results of the model calculations for radium activity can be seen in the Figs. 5–6.

To transcribe the results to doses values it is necessary to estimate the concentration in each compartment. This can be done by computing the ratio between the resulting activity and the volume of the respective compartment. The results of the model calculations for concentration in each compartment with units of $Bq\cdot kg^{-1}$ are represented in the Figs. 7–12

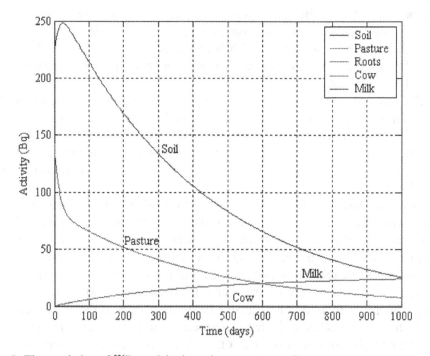

Fig. 5. Time variation of ^{226}Ra activity in each compartment, Bq

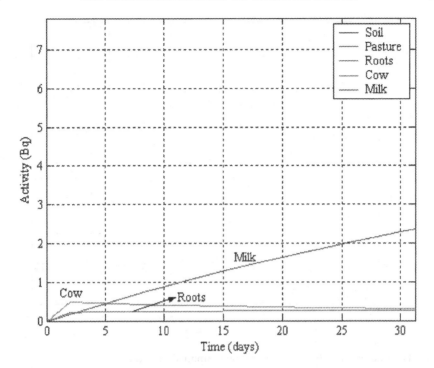

Fig. 6. Time variation of ^{226}Ra activity in each compartment (amplified), Bq

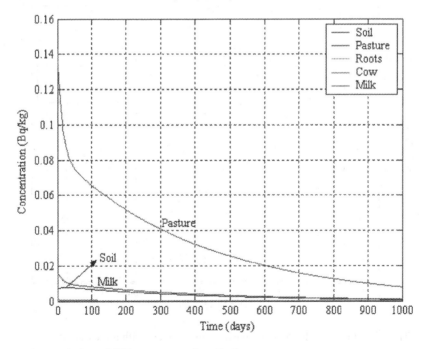

Fig. 7. Time variation of ^{226}Ra concentration, Bq/kg

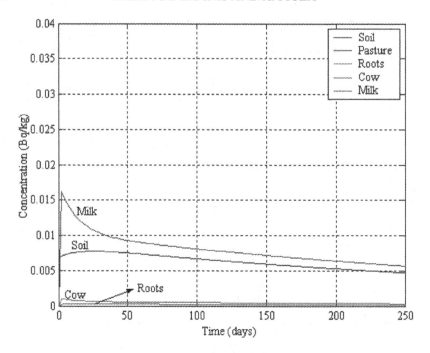

Fig. 8. Time variation of ^{226}Ra concentration (amplified), Bq/kg

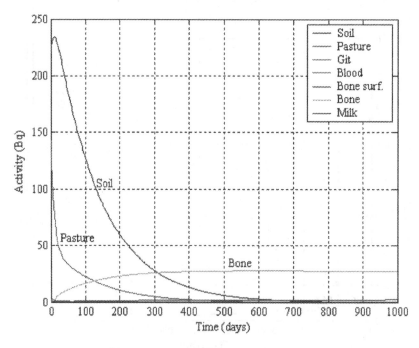

Fig. 9. Time variation of ^{226}Ra activity in each compartment within the cow, Bq

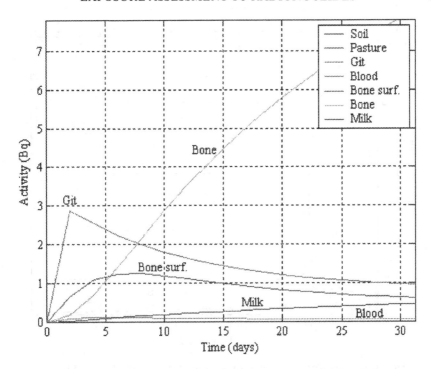

Fig. 10. Time variation of ^{226}Ra activity in each compartment within the cow (amplified), Bq

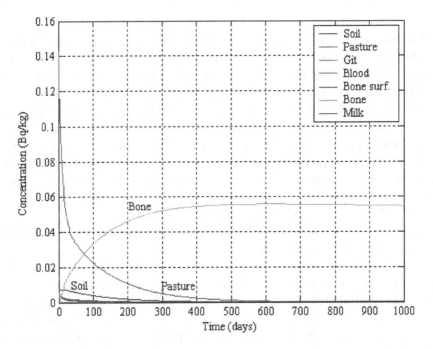

Fig. 11. Time variation of ^{226}Ra concentration within the cow, Bq/kg

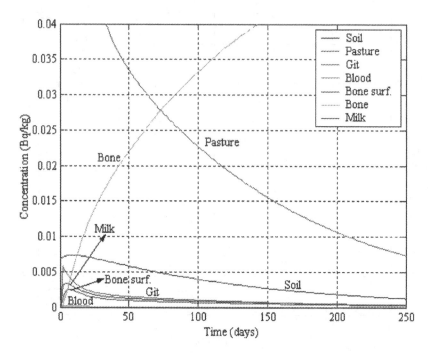

Fig. 12. Time variation of ^{226}Ra concentration within the cow (amplified), Bq/kg

Considering the radionuclide transfer within the cow for modelling the distribution, retention and elimination in each compartment, the endpoints are GIT, plasma, bone and milk. For the exploration of this model the same conditions from the previous exploration were considered. Root uptake was negligee for model simplicity and considering the fact this transfer pathway is not relevant when the pasture contamination is primary due to deposition.

Radium transfer rates for excretion (λ_6), git-blood (included in k_{34}), urinary excretion (λ_7), milk excretion (included in k_{47}), blood-bone surface (included in k_{45}), bone surface-blood (include in k_{54}), bone surface-bone (included in k_{56}) and bone-blood (included in k_{64}) were selected from specialized literature (Leggett et al., 1987; 2003; IAEA, 2004).

The results of the model calculations for radium activity and concentration are represented in the next figures:

Radium concentration of pasture grass decreases with time after deposition. The concentration in soil also decreases with time although, initially it has a slight increasing probably due to the activity removed from pasture by the environmental processes and transferred to soil. The radium concentration in soil is quite lower than concentration in pasture which can be explained by the fact that grass leaves intercept almost the deposition and beside that, soil surface is covered with the pasture limiting the deposition onto the soil. Roots concentration is very low which confirms that this pathway is negligible in case pasture contamination is due to deposition.

Radium concentration in cow (whole organism) reaches its maximum quickly and then it decreases with time. Radionuclides present in cow's diet are probably absorbed from the gastrointestinal tract to blood. Radium is eliminated from the body, mainly from the faeces with smaller amounts being excreted in the urine (Leggett et al., 2003). Some organs and tissues, notably bones, have the capacity to concentrate radium from blood and although some of the radium is excreted through the faeces and urine over a long time, a portion will remain in the bones throughout the organism's lifetime. ICRP considers that radium removal from bone volume to blood occurs very slowly during organism life having a biological half-time of about 1,200 days (Leggett et al., 1987). Less notably but also important, radium can be concentrated in mammary glands and be present in the milk of lactating animals. For this reason, the radium transfer from cow diet to milk has received particular attention.

Conclusions and Discussion

A compartment model has been developed to predict the activity concentration in *pasture-cow-milk* exposure route. A more complete model is also described for radium distribution, retention and elimination within the cow by including the sub-compartments affected by these processes: GIT, plasma and bones.

The important routes of contamination have been identified as bases for determining the modeling approach and pasture intake was considered to be a major source of radium excreted in milk (Kirchmann et al., 1972).

There are some research topics that should be addressed in order to obtain a more realistic quantification of radium transport in food chain than is currently possible, in the absence of site-specific data.

The parameters independent of the radionuclide, measures environmental and ecological characteristics which vary from site to site and also with the spatial scale considered. These parameters related to vegetation, soil and fauna characteristics, can be directly obtained from site investigations or estimated with simple models.

The parameters dependent of the radionuclide are related with processes responsible for differences in radionuclides distribution in the compartments, their retention and elimination from the system. Most of these parameters present a substantial data gaps. Some of these gaps could be filled with data obtained from on-going site investigation. However, we should be aware that these experimental data will always and only represent a limited set of environmental conditions. For example, plant uptake of radium has been proven to occur on a number of radium contaminated sites although the extent to which it occurs is variable. For a plant to take up radium, a soluble or exchangeable form of this radionuclide must be present (Baker and Toque, 2005). Without these conditions any direct contamination of the pasture surface will dominate the activity concentration in the pasture. Processes such as root uptake and direct soil contamination will only be important when there is no direct deposition onto pasture surface (Watson et al., 1983).

The transfer model to animals requires further testing, which could be done by performing measurements of radionuclide concentrations in animals and in their feed. Literature documenting trophic transfer of radium from feed to animal products is not much and although there are numerous publications presenting radium content of milk and meat products from market basket surveys (Watson *et al.*, 1983) most of literature does not contain the necessary information of radium concentration in the animal's diet from which the products were obtained.

Concentrations in the specific organs could be estimated from empirical measures ratios between concentrations in organs and in the whole body or with the help of more detailed kinetic models (Avila and Facilia, 2006). The quantities that can be monitored during organism life include whole-body radium content, blood concentration, urinary and fecal excretion rate and milk secretion. Until now direct observation *in vivo* of radium retention in bone compartment has not been possible and what has been learned about it has been inferred from *postmortem* observations and with modeling studies (BEIR, 1988).

Acknowledgement

This work has been carried out with the financial support of Foundation for Science and Technology (FCT-MCES).

References

Avila, R., Facilia, A. B., Model of long-term transfer of radionuclides in forests. Technical report TR-06-08. Swedish Nuclear Fuel and Waste Management Co. Sweden 2006. Journal of Radiological Protection 25, 127–140, 2005.
Baker, A. C., Toque C., A review of the potential for radium luminising activities to migrate in the environment. Journal of Radiological Protection 25, 127–140, 2005.
BEIR, Health Effects of Radon and Other Internally Deposited Alpha-Emitters, National Academy Press, Washington D.C., BEIR IV, Committee on the Biological Effects of Ionizing Radiations, 1988.
Falcão, J. M., Carvalho, F. P., Leite, M. R., Alarcão, M., Cordeiro, E., Ribeiro, J., MinUrar, Minas de Urânio e seus Resíduos: Efeitos na Saúde da População. Relatório científico I, 2005.

Hoffman, F. O. Bergstrom, U. et al., Comparison of predictions from internationally recognizes assessment models fort the transfer of selected radionuclides through terrestrial food chains. Nuclear Safety 25, 533–564, 1984.

IAEA (International Atomic Energy Agency), Handbook of parameters values for the prediction of radionuclide transfer in temperate environments – produced in collaboration with the International Union of Radioecologists, Vienna, Technical Report 364, IAEA 1994.

IAEA, Biomass 1, Modelling the migration and accumulation of radionuclides in forest ecosystems. Report of the Forest Working Group of the Biosphere Modelling and Assessment (Biomass) Programme, Theme 3, 2002.

IAEA, Biomass 7, Testing of environmental transfer models using data from the remediation of a radium extraction site. Report of the Remediation Assessment Working Group of Biomass, Theme 2, 2004.

Kirchmann, R., LaFontaine, A., Van den Hoek, J. and Koch, G., Comparison of the rate of transfer to cow milk of ^{226}Ra from drinking water and ^{226}Ra incorporated in hay. Comptes Rendus des Seances de la Societe de Biologie et de ses Filiales 166(11): 1557–1562, 1972.

Leggett, R. W. and Eckerman, K. F., Dosimetric Significance of the ICRP's Updated Guidance an Models, 1989–2003, and Implications for U. S. Guidance. Oak Ridge National Laboratory, 2003. ORNL/TM-2003/207.

Leggett, R. W., Eckerman, K. F. et al., Age-specific Models for Evaluating Dose and Risk from Internal Exposures to Radionuclides. Oak Ridge National Laboratory, 1987. ORNL/TM-10080.

Teale, J. B., Modelling Approach for the Transfer of Actinides to Fruit Species of Importance in the UK. NPRB-W46, 2003.

Watson, A. P., Etnier, E. L. and McDowell-Boyer, L. M., Radium-226 in Drinking Water and Terrestrial Food Chains: A Review of Parameters and an Estimate of Potential Exposure and Dose. Oak Ridge National Laboratory, 1983. ORNL/TM-8597.

CHAPTER 22

RADIATION, OXIDATIVE STRESS AND SENESCENCE; THE VASCULAR ENDOTHELIAL CELL AS A COMMON TARGET

PAUL N. SCHOFIELD AND JOSE GARCIA-BERNARDO

Department of Physiology, Development and Neuroscience, University of Cambridge, United Kingdom
e-mail: ps@mole.bio.cam.ac.uk

Abstract: Much evidence exists to implicate the endothelial cell as an important target for radiation in vivo. The nature and delayed manifestation of radiation damage to the endothelium is consistent with the induction of stress associated premature senescence (SIPS) a phenomenon known to occur in response to many types of oxidative stress. We propose the hypothesis that, particularly at low dose levels, induction of SIPS is the most important form of cellular damage sustained by the endothelium, and that this may be exacerbated by additional inducers of oxidative stress such as heavy metals and other pollutants. This paradigm raises new possibilities for therapies designed to slow or reverse endothelial cell senescence as part of the long term treatment for radiation exposure.

Keywords: endothelial cell; senescence; radiation; oxidative stress

Introduction

Until approximately a decade ago the accepted paradigm for estimation of the risk of cellular and organismal damage due to low dose radiation was based on the Linear Non-Threshold (LNT) hypothesis, using extrapolation from higher dose epidemiological data and the assumption that the energy deposited in cells by radiation had its deleterious effects through stochastic damage to DNA, either directly or through secondary damage caused by radicals, resulting in mutations and chromosome lesions. Progress in establishing the real nature of risk at low doses of ionising radiation and testing of the LNT assumption has been hampered by lack of relevant data in this dose region from human beings and experimental systems using a 'low dose' range – variously taken as below 1 Gy and below 100 cGy according to

C. Mothersill et al. (eds.), Multiple Stressors: A Challenge for the Future, 325–334.
© 2007 *Springer.*

different opinions. The doses used in most experimental systems are greatly above the estimated average human environmental exposure of 2.4 mSv per year (UNSCEAR, 2000) and exceed the currently accepted threshold safety dose of 50 mSv. In most experiments the dose is delivered acutely or at best in relatively large fractions.

With the exception of medical radiation delivered in effectively acute doses either diagnostically (3–25 mSV) or in radiotherapy (RT), the risk from most environmental sources, either natural or man made, will generally be the result of chronic exposure or combined chronic exposure with other toxic agents (multi-stressors). These levels are likely to be of the order of 20–100 mSv assuming a range consistent with exposure of workers in the nuclear industry to therapeutic accidents (Brenner et al., 2003).

When considering environmental exposure it is obviously then vitally important to assess the degree of interaction of other environmental stressors with low dose radiation and to do this we need to have a clear view of the predominant mechanisms of action of different classes of agent at the concentrations or doses which are relevant to the population at large. Most important and as yet unknown is whether the effects of multi-stressors, such as heavy metals and pesticides are additive or synergistic, whether they all ultimately converge on one class of cellular damage or if they cause a characteristic and differential spectrum of effects (Preston, 2005). The LNT hypothesis itself suffers from the weakness that there are no human data at low doses to support the extrapolation from high, and its valid application depends on the premise that the sole important effect of radiation is through damage to DNA and subsequent generation of cancer. The assumption that cancer is the most important and the predominant endpoint of radiation exposure is one which is increasingly being challenged at low doses and a further question we need to ask is if the focus on neoplasia, caused through DNA damage, is genuinely the most important contributor to morbidity and mortality and is any more a sustainable paradigm at low acute or protracted dose exposure?

Non-neoplastic sequelae of low dose radiation are now emerging as common effects in a variety of clinical procedures and accidental exposure events. There is emerging evidence of excess risk of non- neoplastic late health effects in the lifespan study (LSS) of atomic bomb survivors; in particular, excess radiation-associated mortality and morbidity due to circulatory, digestive and respiratory diseases (Wong et al., 1993; Preston et al., 2003). Whilst the atomic bomb survivor analyses are often been considered as high-dose studies, the mean dose in the exposed group in the LSS cohort is only 200 mSv, with 50% of the exposed individuals in the cohort having doses of 50 mSv. Preston and co-workers (Preston et al., 2003) have showed that the numbers of radiation induced non-cancer-attributable deaths may be of the same order as neoplasm associated. Whilst this conclusion is not shared by Berrington and co-workers (2001) in their investigations of occupational exposure of UK radiologists, an excess of mortality due to circulatory diseases was reported by Cardis and Carpenter in a survey of radiation workers (Cardis et al., 1995) and more recently an excess of cardiovascular mortality in a cohort of Swedish women treated for breast cancer (Darby et al., 2003). Similar data has recently been reported by Nilsson and co-workers (Nilsson et al., 2005) for cerebral infarction following RT for breast cancer. Interestingly in both studies there was a long period of ten or more years between RT and death and in general vascular radiation pathology is a delayed injury (Fajardo, 2005). Classically it includes capillary telangiectasia, rupture and reduction in the microvascular network resulting in ischaemia. In larger vessels intimal fibrosis, occlusive thrombosis and foam cell containing plaques are seen.

Results to date suggest that there is no radiation-related excess non-cancer mortality in Russian Chernobyl liquidators, although radiation-related excess cardiovascular disease mortality has been reported and increasingly there is evidence of both increased incidence of cardiovascular disease and related symptoms in exposed individuals (Kamarli and Abdulina, 1996; Ivanov et al., 2001; Bandazhevskaya et al., 2004; Becker, 2005). Interestingly recent data from a Swedish cohort of patients treated with low doses of radiation during childhood indicate a long term effect on cognitive function (Hall et al., 2004) and given the known effects of radiation on brain microvasculature it will be interesting to see to what extent the reported psychiatric sequelae reported in the Chernobyl liquidators might be related to such pathology.

The evidence that the vasculature is a major target for radiation, particularly low dose radiation, is now established (Basavaraju and Easterly, 2002; Trivedi and Hannan, 2004; Becker, 2005) and in fact has been known for some time (Gassmann, 1899) In many cases the manifestation of the pathology occurs many years after the initial exposure and this may be a clue as to the radiation dependent processes leading to cardiovascular disease.

Low Dose Radiation and Oxidative Stress

Simple calculations (Feinendegen, 2005) may be made of the consequence of energy deposition within cells. 1 mGy of low LET radiation, such as X-rays will generate within an average cell one energy deposition event; 150 reactive oxygen species (ROS); 2 DNA alterations of any kind; 10^{-2} DNA double-strand breaks, 10^{-4} chromosome aberrations and a probability of oncogenic transformation of 10^{-13}–10^{-14}. Whilst doses above 100 mSV are accepted as likely to cause neoplasia, those in the region of normal exposure are very difficult to assess. The problems are discussed exhaustively in (Brenner et al., 2003) who conclude that from epidemiological data for cancer incidence as an endpoint, the level at which there is a serious risk of neoplasia is at an exposure of 50 mSV (effectively 50 mGy) and argue that linear extrapolation below that level is justifiable on the basis of biophysics. They do address scenarios where such an extrapolation is modified by factors such as the bystander effect and other non-cell autonomous processes (Mothersill and Seymour, 2003) and the effects of genetic background, but conclude that in the absence of experimental data in the low dose region, the existence of upwardly and downwardly curving dose/response curves below 50 mSv cannot be determined. A striking observation however is that at these low doses the generation of ROS remains a major effect of radiation whereas direct effects on DNA are predicted to be rather small by comparison (Pollycove and Feinendegen, 2003).

This leads us to the hypothesis that at low dose the biological effects of irradiation may be mediated predominantly through the generation of oxidative stress rather than direct DNA damage. There is considerable evidence that low dose radiation generates significant amounts of ROS (Lyng et al., 2001; Little et al., 2002; de Toledo et al., 2006; Haghdoost et al., 2006; Little, 2006) and that the pattern of ROS generation in response to radiation is likely to cause oxidative stress (Feinendegen, 2003; 2005). ROS will affect DNA itself, but more importantly will activate protective mechanisms designed to limit oxidative damage; amongst these being antioxidant activation and the HIF-1 response (Moeller et al., 2004). Recently the bystander response has itself also been linked to oxidative stress (Azzam et al., 2003) and persistent stress linked to chromosomal instability (Clutton et al., 1996). Other targets within the cell may be intermediary metabolites such as folate and the nucleotide pool (Haghdoost et al., 2006).

Endothelial Cells, Oxidative Stress and Senescence

Under normal circumstances endothelial cells rarely divide, with a turnover rate of about every 3 years. They may be induced to divide by damage and may be repaired by bone marrow derived endothelial progenitor cells (Hunting et al., 2005). However their ability to proliferate is limited and in culture they finally enter a state of proliferative quiescence; senescence. Senescent endothelial cells are characterized by alterations in morphology, gene expression, and function. For example they express senescence associated β-galactosidase (Kurz et al., 2000), believed to reflect an increase in lysosomal content of senescent cells; increased ROS content (Haendeler et al., 2004); increased superoxide production (probably through increased expression of NADPH oxidase) (Oudot et al., 2006); downregulation of cdk2 (Freedman and Folkman, 2005); accumulation of p16, p21 and p27; changes in cell surface properties such as alterations in ICAM –1 activation and clustering (Zhou et al., 2006); reduction of NO production (Sato et al., 1993); and lower levels of NO synthase (Matsushita et al., 2001). These in vitro phenomena broadly reflect the phenotype of senescent endothelial cells in vivo (Erusalimsky and Kurz, 2005). Interestingly endothelial cells, unlike any others mainly growth arrest with a supradiploid DNA content of 4N and a subcomponent of 8N. (Wagner et al., 2001).

Wagner et al showed that senescent cells were associated with a high level of apoptosis and whilst Kalashnik and co-workers (Kalashnik et al., 2000) showed no change in apoptosis, Hampel and co-workers (Hampel et al., 2006) demonstrated increased sensitivity to apoptosis induction by ceramide with senescence, probably as a consequence of increased ganglioside synthesis, which can trigger mitochondrial leakiness. The conflict is probably explained by the availability of survival factors in culture medium, as senescent endothelial cells show exquisite sensitive to the availability of VEGF for example.

The phenomena seen in vitro are also seen in vivo, and senescent cells are found in porcine and human atherosclerotic plaques (Minamino et al., 2002) Uniquely senescent endothelial cells show the formation of giant multinucleate post-mitotic cells which are also seen occasionally in the human vasculature and atherosclerotic lesions (Tokunaga et al., 1989; Satoh et al., 1998).

The current paradigm for cellular senescence involves the attrition of telomeres and gradual loss of proliferative capacity with each cell division, however some cellular stressors may rapidly elicit growth arrest and senescence; referred to as stress-induced premature senescence (SIPS)(Chen, 2000; Foreman and Tang, 2003; Unterluggauer et al., 2003; Chen and Goligorsky, 2006). Stressors giving rise to premature senescence include activated oncogenes, DNA damage, and oxidative stress. Oxidative stress has been known to generate premature senescence in fibroblasts for some while, (Honda and Matsuo, 1983; von Zglinicki et al., 1995) and it is now clear that even mild oxidative stress induced by reducing endogenous glutathione levels (Unterluggauer et al., 2003) will induce the phenotype of premature senescence and ROS scavengers will inhibit ROS induced SIPS (Haendeler et al., 2004).

Preliminary data from our laboratory indicate that X-irradiation of early passage HUVEC cells at doses above 0.1 Gy will generate SIPS (PNS, Wang and Garcia-Bernardo unpublished). This observation might go some way to explaining phenomena seen in irradiated endothelial cells in vivo and in vitro, such as the impairment of NO mediated vascular relaxation in irradiated arteries (Sugihara et al., 1999). Interestingly endothelial senescence is inhibited by NO (Hayashi et al., 2006) and loss of NO synthase is characteristic of endothelial cell senescence. Irradiated endothelial cells in vitro fail to organize into capillary-like structures (Ahmad et al., 2007) and show upregulation of vascular cell adhesion molecule, ICAM-1, and P-selectin amongst others, leading to a proinflammatory, procoagulent and prothrombotic – broadly atherogenic phenotype. Alterations in the properties of the vascular wall are also seen in vivo, with loss of smooth muscle tone and fibrosis. TGF β has here been implicated in cross talk (Milliat et al., 2006) between the damaged endothelium and the vessel wall. See (Barcellos-Hoff, 2005) for discussion of the role of TGFβ in the oxidative stress response.

Conclusions

Pulling together the threads, it is a small leap to conjecture that the endothelial cell dysfunction seen in vivo following irradiation, particularly at low doses, might not be due simply to DNA damage and its manifestation when cells divide, but to SIPS induced by oxidative stress.

Several predictions follow from this hypothesis:

Firstly, other agents which cause oxidative stress might act additively or even synergistically with low dose radiation. There is already extensive evidence for oxidative stress induced change in endothelial cell function as a consequence of heavy metal exposure (Jeong et al., 2004; Ni et al., 2004; Dickhout et al., 2005; Bernhard et al., 2006; Wiseman et al., 2006; Wolf and Baynes, 2007).

Secondly, we might expect quite wide variation in the response due to the heterogeneity in telomere length seen in the human population. This has already been suggested to be responsible for predisposition to cardiovascular amongst other degenerative diseases (discussed in Erusalimsky and Kurz, (2005)).

Thirdly, oxidative stress has been frequently linked to the mechanism of the bystander effect and it will be intriguing to revisit the literature of abscopal effects (discussed in Mothersill et al., (2005)) in the light of the possibility of intra-vascular signaling through radiation induced oxidative damage of endothelial cells.

Focus on the senescence of the endothelial cell as a target for therapy provides a new approach to the design of agents to protect against acute or chronic radiation damage, such as statins and free-radical scavengers, the latter having already seen some success in the use of the peroxynitrite scavenger and antioxidant ebselen (Brodsky et al., 2004). Our increased understanding of endothelial stem cells may also offer a way to mobilize stem cells to effect repair of senescent endothelium.

In summary a new paradigm for the dominant biological effects of low dose radiation which involves the induction of premature senescence as the predominant effect of low dose radiation seems attractive because it explains many of the phenomena seen in vivo whist being consistent with existing knowledge concerning the normal ageing of the endothelium. Should oxidative damage be the dominant factor at these radiation doses then factors which add to the oxidative burden may be expected to contribute significantly to the effects of radiation. Given that these factors may involve heavy metals and other extrinsic oxidative stressors as well an intrinsic atherogenic agents such as oxidized LDL we will expect to see major contribution of genetic background, diet, smoking and environment to the cardiovascular outcomes of irradiation making an epidemiological approach to low dose risk all the more difficult.

References

Ahmad, M., Khurana, N.R. and Jaberi, J.E., 2007, Ionizing radiation decreases capillary-like structure formation by endothelial cells in vitro. *Microvasc Res.* **73**: 14–19.

Azzam, E.I., de Toledo, S.M. and Little, J.B., 2003, Oxidative metabolism, gap junctions and the ionizing radiation-induced bystander effect. *Oncogene*, **22**: 7050–7057.

Bandazhevskaya, G.S., Nesterenko, V.B., Babenko, V.I., Yerkovich, T.V. and Bandazhevsky, Y.I., 2004, Relationship between caesium (^{137}Cs) load, cardiovascular symptoms, and source of food in 'Chernobyl' children – preliminary observations after intake of oral apple pectin. *Swiss Med Wkly*, **134**: 725–729.

Barcellos-Hoff and M.H., 2005, How tissues respond to damage at the cellular level: orchestration by transforming growth factor-{beta} (TGF-{beta}). *Br J Radiol*, Supplement **27**: 123–127.

Basavaraju, S.R. and Easterly, C.E., 2002, Pathophysiological effects of radiation on atherosclerosis development and progression, and the incidence of cardiovascular complications. *Med Phys*, **29**: 2391–403.

Becker, R.C., 2005, Radiation exposure and coronary atherothrombosis. *J Thromb Thrombolysis*, **20**: 191–192.

Bernhard, D., Rossmann, A., Henderson, B., Kind, M., Seubert, A. and Wick, G., 2006, Increased serum cadmium and strontium levels in young smokers: effects on arterial endothelial cell gene transcription. *Arterioscler Thromb Vasc Biol*, **26**: 833–838.

Berrington, A., Darby, S.C., Weiss, H.A. and Doll, R., 2001, 100 years of observation on British radiologists: mortality from cancer and other causes 1897–1997. *Br J Radiol*, **74**: 507–519.

Brenner, D.J., Doll, R., Goodhead, D.T., Hall, E.J., Land, C.E., Little, J.B., Lubin, J.H., Preston, D.L., Preston, R.J., Puskin, J.S., Ron, E., Sachs, R.K., Samet, J.M., Setlow, R.B. and Zaider, M., 2003, Cancer risks attributable to low doses of ionizing radiation: Assessing what we really know. *PNAS*, **100**: 13761–766.

Brodsky, S.V., Gealekman, O., Chen, J., Zhang, F., Togashi, N., Crabtree, M., Gross, S.S., Nasjletti, A. and Goligorsky, M.S., 2004, Prevention and reversal of premature endothelial cell senescence and vasculopathy in obesity-induced diabetes by ebselen. *Circ Res*, **94**: 377–384.

Cardis, E., Gilbert, E.S., Carpenter, L., Howe, G., Kato, I., Armstrong, B.K., Beral, V., Cowper, G., Douglas, A., Fix, J. and et al., 1995, Effects of low doses and low dose rates of external ionizing radiation: cancer mortality among nuclear industry workers in three countries. *Radiat Res*, **142**: 117–132.

Chen, J. and Goligorsky, M.S., 2006, Premature senescence of endothelial cells: Methusaleh's dilemma. *Am J Physiol Heart Circ Physiol*, **290**: H1729–739.

Chen, Q.M., 2000, Replicative senescence and oxidant-induced premature senescence. Beyond the control of cell cycle checkpoints. *Ann NY Acad Sci*, **908**: 111–125.

Clutton, S.M., Townsend, K.M., Walker, C., Ansell, J.D. and Wright, E.G., 1996, Radiation-induced genomic instability and persisting oxidative stress in primary bone marrow cultures. *Carcinogenesis*, **17**: 1633–639.

Darby, S., McGale, P., Peto, R., Granath, F., Hall, P. and Ekbom, A., 2003, Mortality from cardiovascular disease more than 10 years after radiotherapy for breast cancer: nationwide cohort study of 90 000 Swedish women. *BMJ*, **326**: 256–257.

de Toledo, S.M., Asaad, N., Venkatachalam, P., Li, L., Howell, R.W., Spitz, D.R. and Azzam, E.I., 2006, Adaptive responses to low-dose/low-dose-rate gamma rays in normal human fibroblasts: the role of growth architecture and oxidative metabolism. *Radiat Res*, **166**: 849–857.

Dickhout, J.G., Hossain, G.S., Pozza, L.M., Zhou, J., Lhotak, S. and Austin, R.C., 2005, Peroxynitrite causes endoplasmic reticulum stress and apoptosis in human vascular endothelium: implications in atherogenesis. *Arterioscler Thromb Vasc Biol*, **25**: 2623–629.

Erusalimsky, J.D. and Kurz, D.J., 2005, Cellular senescence in vivo: its relevance in ageing and cardiovascular disease. *Exp Gerontol*, **40**: 634–642.

Fajardo, L.F., 2005, The pathology of ionizing radiation as defined by morphologic patterns. *Acta Oncol*, **44**: 13–22.

Feinendegen, L.E., 2003, Relative implications of protective responses versus damage induction at low dose and low-dose-rate exposures, using the microdose approach. *Radiat Prot Dosim*, **104**: 337–346.

Feinendegen, L.E., 2005, Significance of basic and clinical research in radiation medicine: challenges for the future. *Br J Radiol* (Suppl) **27**: 185–195.

Foreman, K.E. and Tang, J., 2003, Molecular mechanisms of replicative senescence in endothelial cells. *Exp Gerontol*, **38**: 1251–257.

Freedman, D.A. and Folkman, J., 2005, CDK2 translational down-regulation during endothelial senescence. *Exp Cell Res*, **307**: 118–130.

Gassmann, A., 1899, Zur histologie der roentgenulcer. *Fortschr Geb Roentgenstr*, **2**: 199–211.

Haendeler, J., Hoffmann, J., Diehl, J.F., Vasa, M., Spyridopoulos, I., Zeiher, A.M. and Dimmeler, S., 2004, Antioxidants inhibit nuclear export of telomerase reverse transcriptase and delay replicative senescence of endothelial cells. *Circ Res*, **94**: 768–775.

Haghdoost, S., Sjolander, L., Czene, S. and Harms-Ringdahl, M., 2006, The nucleotide pool is a significant target for oxidative stress. *Free Rad Biol Med*, **41**: 620–626.

Hall, P., Adami, H.-O., Trichopoulos, D., Pedersen, N.L., Lagiou, P., Ekbom, A., Ingvar, M., Lundell, M. and Granath, F., 2004, Effect of low doses of ionising radiation in infancy on cognitive function in adulthood: Swedish population based cohort study. *BMJ*, **328**: 19.

Hampel, B., Fortschegger, K., Ressler, S., Chang, M.W., Unterluggauer, H., Breitwieser, A., Sommergruber, W., Fitzky, B., Lepperdinger, G., Jansen-Durr, P., Voglauer, R. and Grillari, J., 2006, Increased expression of extracellular proteins as a hallmark of human endothelial cell in vitro senescence. *Exp Gerontol*, **41**: 474–481.

Hayashi, T., Matsui-Hirai, H., Miyazaki-Akita, A., Fukatsu, A., Funami, J., Ding, Q.F., Kamalanathan, S., Hattori, Y., Ignarro, L.J. and Iguchi, A., 2006, Endothelial cellular senescence is inhibited by nitric oxide: implications in atherosclerosis associated with menopause and diabetes. *Proc Natl Acad Sci USA*, **103**: 17018–7023.

Honda, S. and Matsuo, M., 1983, Shortening of the in vitro lifespan of human diploid fibroblasts exposed to hyperbaric oxygen. *Exp Gerontol*, **18**: 339–345.

Hunting, C.B., Noort, W.A. and Zwaginga, J.J., 2005, Circulating endothelial (progenitor) cells reflect the state of the endothelium: vascular injury, repair and neovascularization. *Vox Sang*, **88**: 1–9.

Ivanov, V.K., Gorski, A.I., Maksioutov, M.A., Tsyb, A.F. and Souchkevitch, G.N., 2001, Mortality among the Chernobyl emergency workers: estimation of radiation risks (preliminary analysis). *Health Phys*, **81**: 514–21.

Jeong, E.M., Moon, C.H., Kim, C.S., Lee, S.H., Baik, E.J., Moon, C.K. and Jung, Y.S., 2004, Cadmium stimulates the expression of ICAM-1 via NF-kappaB activation in cerebrovascular endothelial cells. *Biochem Biophys Res Commun*, **320**: 887–892.

Kalashnik, L., Bridgeman, C.J., King, A.R., Francis, S.E., Mikhalovsky, S., Wallis, C., Denyer, S.P., Crossman, D. and Faragher, R.G., 2000, A cell kinetic analysis of human umbilical vein endothelial cells. *Mech Ageing Dev*, **120**: 23–32.

Kamarli, Z. and Abdulina, A., 1996, Health conditions among workers who participated in the cleanup of the Chernobyl accident. *World Health Stat Q*, **49**: 29–31.

Kurz, D.J., Decary, S., Hong, Y. and Erusalimsky, J.D., 2000, Senescence-associated (beta)-galactosidase reflects an increase in lysosomal mass during replicative ageing of human endothelial cells. *J Cell Sci*, **113**(20): 3613–622.

Little, J.B., 2006, Lauriston S. Taylor lecture: nontargeted effects of radiation: implications for low-dose exposures. *Health Phys*, **91**: 416–426.

Little, J.B., Azzam, E.I., de Toledo, S.M. and Nagasawa, H., 2002, Bystander effects: intercellular transmission of radiation damage signals. *Radiat Prot Dosimetry*, **99**: 159–162.

Lyng, F.M., Seymour, C.B. and Mothersill, C., 2001, Oxidative stress in cells exposed to low levels of ionizing radiation. *Biochem Soc Trans*, **29**: 350–353.

Matsushita, H., Chang, E., Glassford, A.J., Cooke, J.P., Chiu, C.P. and Tsao, P.S., 2001, eNOS activity is reduced in senescent human endothelial cells: Preservation by hTERT immortalization. *Circ Res*, **89**: 793–798.

Milliat, F., Francois, A., Isoir, M., Deutsch, E., Tamarat, R., Tarlet, G., Atfi, A., Validire, P., Bourhis, J., Sabourin, J.C. and Benderitter, M., 2006, Influence of endothelial cells on vascular smooth muscle cells phenotype after irradiation: implication in radiation-induced vascular damages. *Am J Pathol*, **169**: 1484–495.

Minamino, T., Miyauchi, H., Yoshida, T., Ishida, Y., Yoshida, H. and Komuro, I., 2002, Endothelial cell senescence in human atherosclerosis: role of telomere in endothelial dysfunction. *Circulation*, **105**: 1541–544.

Moeller, B.J., Cao, Y., Li, C.Y. and Dewhirst, M.W., 2004, Radiation activates HIF-1 to regulate vascular radiosensitivity in tumors: role of reoxygenation, free radicals, and stress granules. *Cancer Cell*, **5**: 429–441.

Mothersill, C. and Seymour, C., 2003, Radiation-induced bystander effects, carcinogenesis and models. *Oncogene*, **22**: 7028–7033.

Mothersill, C., Moriarty, M.J. and Seymour, C.B., 2005, Bystander and other delayed effects and multi-organ involvement and failure following high dose exposure to ionising radiation. *BJR Suppl*, **27**: 128–131.

Ni, Z., Hou, S., Barton, C.H. and Vaziri, N.D., 2004, Lead exposure raises superoxide and hydrogen peroxide in human endothelial and vascular smooth muscle cells. *Kidney Int*, **66**: 2329–336.

Nilsson, G., Holmberg, L., Garmo, H., Terent, A. and Blomqvist, C., 2005, Increased incidence of stroke in women with breast cancer. *Eur J Cancer*, **41**: 423–429.

Oudot, A., Martin, C., Busseuil, D., Vergely, C., Demaison, L. and Rochette, L., 2006, NADPH oxidases are in part responsible for increased cardiovascular superoxide production during aging. *Free Rad Biol Med*, **40**: 2214–222.

Pollycove, M. and Feinendegen, L.E., 2003, Radiation-induced versus endogenous DNA damage: possible effect of inducible protective responses in mitigating endogenous damage. *Hum Exp Toxicol*, **22**: 290–306.

Preston, R.J., 2005, Bystander effects, genomic instability, adaptive response, and cancer risk assessment for radiation and chemical exposures. *Toxicol Appl Pharmacol*, **207**: 550–556.

Preston, D.L., Shimizu, Y., Pierce, D.A., Suyama, A. and Mabuchi, K., 2003, Studies of mortality of atomic bomb survivors. Report 13: Solid cancer and noncancer disease mortality: 1950–1997. *Radiat Res*, **160:** 381–407.

Sato, I., Morita, I., Kaji, K., Ikeda, M., Nagao, M. and Murota, S., 1993, Reduction of nitric oxide producing activity associated with in vitro aging in cultured human umbilical vein endothelial cell. *Biochem Biophys Res Commun*, **195:** 1070–1076.

Satoh, T., Sasatomi, E., Yamasaki, F., Ishida, H., Wu, L. and Tokunaga, O., 1998, Multinucleated variant endothelial cells (MVECs) of human aorta: expression of tumor suppressor gene p53 and relationship to atherosclerosis and aging. *Endothelium*, **6:** 123–132.

Sugihara, T., Hattori, Y., Yamamoto, Y., Qi, F., Ichikawa, R., Sato, A., Liu, M.Y., Abe, K. and Kanno, M., 1999, Preferential impairment of nitric oxide-mediated endothelium-dependent relaxation in human cervical arteries after irradiation. *Circulation*, **100:** 635–641.

Tokunaga, O., Fan, J.L. and Watanabe, T., 1989, Atherosclerosis- and age-related multinucleated variant endothelial cells in primary culture from human aorta. *Am J Pathol*, **135:** 967–976.

Trivedi, A. and Hannan, M.A., 2004, Radiation and cardiovascular diseases. *J Environ Pathol Toxicol Oncol*, **23:** 99–106.

UNSCEAR, U.N.S.C.o.t.E.o.A.R. 2000. UNSCEAR 2000 REPORT Volume 1: Sources.

Unterluggauer, H., Hampel, B., Zwerschke, W. and Jansen-Durr, P., 2003, Senescence-associated cell death of human endothelial cells: the role of oxidative stress. *Exp Gerontol*, **38:** 1149–1160.

von Zglinicki, T., Saretzki, G., Docke, W. and Lotze, C., 1995, Mild hyperoxia shortens telomeres and inhibits proliferation of fibroblasts: a model for senescence? *Exp Cell Res*, **220:** 186–193.

Wagner, M., Hampel, B., Bernhard, D., Hala, M., Zwerschke, W. and Jansen-Durr, P., 2001, Replicative senescence of human endothelial cells in vitro involves G1 arrest, polyploidization and senescence-associated apoptosis. *Exp Gerontol*, **36:** 1327–347.

Wiseman, D.A., Wells, S.M., Wilham, J., Hubbard, M., Welker, J.E. and Black, S.M., 2006, Endothelial response to stress from exogenous Zn^{2+} resembles that of NO-mediated nitrosative stress, and is protected by MT-1 overexpression. *Am J Physiol Cell Physiol*, **291:** C555–C568.

Wolf, M.B. and Baynes, J.W., 2007, Cadmium and mercury cause an oxidative stress-induced endothelial dysfunction. *Biometals*. **20:** 73–81.

Wong, F.L., Yamada, M., Sasaki, H., Kodama, K., Akiba, S., Shimaoka, K. and Hosoda, Y., 1993, Noncancer disease incidence in the atomic bomb survivors: 1958–1986. *Radiat Res*, **135:** 418–430.

Zhou, X., Perez, F., Han, K. and Jurivich, D.A., 2006, Clonal senescence alters endothelial ICAM-1 function. *Mech Ageing Dev*, **127:** 779–785.

CHAPTER 23

SENSITIVITY OF IRRADIATED ANIMALS TO INFECTION

V.S. NESTERENKO[1], I.S. MESHCHERJAKOVA[2],
V.A. SOKOLOV[1], R.S. BOUDAGOV[1], AND A.F. TSYB[1]

[1]*Medical Radiological Research Center, RAMS, Obninsk, Russia*
[2]*Research Institute of Experimental Microbiology, RAMS, Moscow, Russia*

Abstract: The study was carried out to investigate the combined action of γ-radiation and infection on mice. Death rate of animals was estimated in dependence of radiation and vaccination doses.

Keywords: γ-radiation; *Francisella tularensis*; mice

Introduction

One of the main causes of death due to radiation-induced damage is development of endogenous infection at the early period of clinical manifestation of acute radiation disease. On the other hand, the decrease of immunological reactivity of the body following the action of ionizing radiation may aggravate deleterious effects of such kind of weapon of mass destruction as biological weapon (World, 2004; Casadeval and Pirofski, 2004). Easiness of production and spreading, delayed beginning of the disease, diagnostic difficulties make rather real to consider microorganisms as a powerful biological weapon. The use of live vaccine after radiation impact may lead to dissemination of infection in the vaccinated organism.

Exposure to relatively low radiation doses (<1 Gy to a human) causes suppression of immunoreactivity and antibacterial resistance. Risk of infectious disease caused by extremely dangerous pathogens is known to be higher in those exposed to radiation compared to unexposed subjects (Brook et al., 2004, 2005; Elliot et al., 2002).

The most dangerous pathogens are *Variola major, Bacillus anthracis, Yersinia pestis, Clostridium botulinum, Francisella tularensis*. A vast amount of information about extremely dangerous infections, improvement of countermeasures and medical preparedness for and response to pathogenic

335

C. Mothersill et al. (eds.), Multiple Stressors: A Challenge for the Future, 335–340.

infections has been published for the recent years. However, whether the low resistance to vaccine cultures associates with specific immunoreactivity and what effect will be produced by vaccine culture on a exposed human should be the subjects of further studies.

Experimental results which allow quantification of relationship between dose and sensitivity of irradiated animals to extremely dangerous pathogens may be of significant importance for medical proposes and preparedness.

This study was aimed at investigation of the influence of total γ-irradiation in non-lethal doses on the course of vaccination process in mice.

Material and Methods

White non-linear mice served as biological objects. Certificated strain No 15 *F. tularensis* (Research Institute of Experimental Genetics, Moscow) with high level of "residual virulence" for these animals was used in the experiments. 30 mice were exposed to 1 Gy, 30 mice to 4 Gy, and 30 mice served as a control group (non-irradiated animals). Following 5 days after irradiation the mice were vaccinated subcutaneously with immunization doses 3.2×10^1, 3.2×10^2, and 3.2×10^3 of tularemia cells (per 10 animals for each dose in all three groups). Animals were exposed to γ-radiation at a dose rate 10 cGy/min. In other experiments animals were vaccinated with immunization dose 1.5×10^4 in 26 days after irradiation with dose 4 Gy (%).

Results

Surveillance over the animals during 20 days after vaccination demonstrated an aggravation of vaccination process in the exposed animals by criteria of mortality and life-shortening. The highest mortality in control groups (vaccinated with 3.2×10^3 *F. tularensis.* cells, non-irradiated) and in the group of irradiated (1 Gy) and vaccinated mice was observed on the 8th day following vaccination. The highest mortality in the group of mice irradiated with dose of 4 Gy was observed on the 7th day after vaccination (Fig. 1). Increase in dose of vaccine caused death of irradiated mice on the 6th day and the highest mortality in the control group was observed on the 8th day after vaccination (Fig. 2). Dose of 3.2×10^3 *F. tularensis.* cells caused the highest death of mice in control and irradiated groups on the 6th day following vaccination, i.e. the interval between exposure to agents (radiation and vaccine) and death of mice reduced (Fig. 3).

The results point that the toxicity effect of vaccination took place in the first week after irradiation (Table 1).

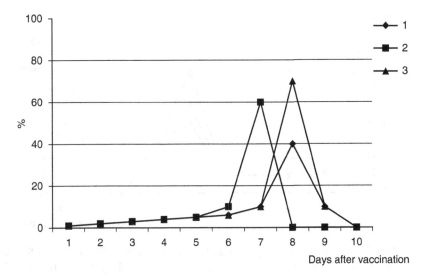

Fig. 1. Mortality in mice (control group and irradiated with different doses) following inoculation of 3.2×10 *F. tularensis* cells (%). $1-3.2 \times 10$; $2-4\,Gy + 3.2 \times 10$; $3-1\,Gy + 3.2 \times 10$

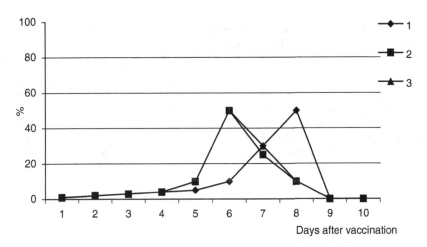

Fig. 2. Mortality in mice (control group and irradiated with different doses) following inoculation of 3.2×10^2 *F. tularensis* cells (%). $1-3.2 \times 10^2$; $2-4\,Gy + 3.2 \times 10^2$; $3-1\,Gy + 3.2 \times 10^2$

In the next row of experiments the influence of vaccination (its virulence) at the survival of mice injected with tularemia cells after 26 days following whole-body exposure to 4Gy was studied in period reconvalescens (Table 2). Experiments were repeated three times. Vaccinated in the same doses non-irradiated mice served as a control (Figs. 4–6).

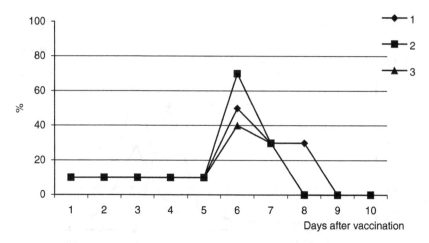

Fig. 3. Mortality in mice (control group and irradiated with different doses) following inoculation of 3.2×10^3 *F. tularensis* cells (%). $1-3.2 \times 10^3$; $2-4\,\mathrm{Gy} + 3.2 \times 10^3$; $3-1\,\mathrm{Gy} + 3.2 \times 10^3$

TABLE 1. Lifespan of mice in control and irradiated groups following vaccination with *F. tularensis* cells in 5 days after irradiation (days, $M \pm \sigma$)

	Vaccination doses		
Groups mice	3.2×10	3.2×10^2	3.2×10^3
Vaccination without irradiation	9.0 ± 0.6	7.0 ± 0.3	7.1 ± 0.7
Vaccination and irradiation in dose 1 Gy	8.6 ± 0.7	7.4 ± 0.5	6.9 ± 0.4
Vaccination and irradiation in dose 4 Gy	$7.1 \pm 0.9^*$	6.8 ± 0.6	6.3 ± 0.5

*$p < 0.05$

TABLE 2. Lifetime between inoculation of different doses of *F. tularensis* cells in 26 days following irradiation (4 Gy) and death of mice (days, $M \pm \sigma$).

Vaccination dose	Control (without irradiation)	Vaccination after irradiation in dose 4 Gy
1.5×10^4	6.9 ± 0.7	6.2 ± 0.6
1.5×10^3	6.9 ± 0.6	6.8 ± 0.7
1.5×10^2	7.8 ± 0.7	8.2 ± 0.3
1.5×10	8.6 ± 1.1	9.1 ± 1.4

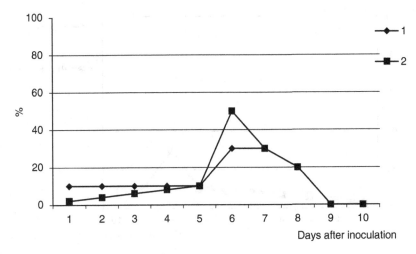

Fig. 4. Mortality of control and irradiated (4 Gy) mice after vaccination of 1.5×10^4 *F. tularensis* cells (%). $1-1.5 \times 10^4$; $2-4$ Gy $+ 1.5 \times 10^4$

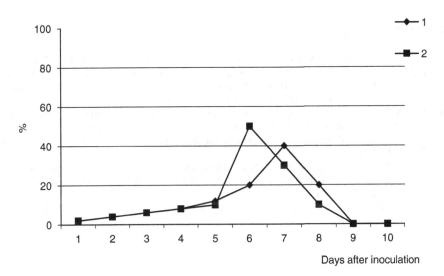

Fig. 5. Mortality of control and irradiated (4 Gy) mice after vaccination of 1.5×10^4 *F. tularensis* cells (%). $1-1.5 \times 10^4$; $2-4$ Gy $+ 1.5 \times 10^4$

An absence of significant difference in the level of "residual virulence" (LD50) and time of death between irradiated and non-irradiated cells was registered. The data demonstrated an aggravation of vaccination process of *F. tularensis* at the initial stage of the development of radiation damage and insignificant difference between irradiated and non-irradiated mice during the recovery period of immunological resistance of the body.

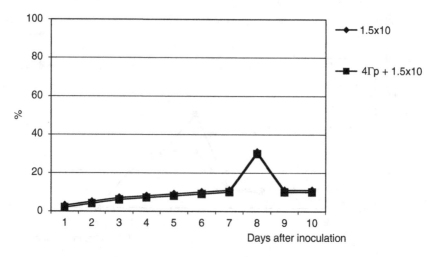

Fig. 6. Mortality of control and irradiated (4 Gy) mice after vaccination of 1.5×10 *F. tularensis* cells (%). $1-1.5 \times 10$; $2-4 \text{ Gy} + 1.5 \times 10$

References

Brook, I., D. E. Girardo, A. Germana, D. P. Nicolau, W. E. Jackson, T. B. Elliott, J. H. Thakar, M. O. Shoemaker, G. D. Ledney, Comparison of clarithromycin and ciprofloxacin therapy for *Bacillus anthracis Sterne* infection in mice with or without ^{60}Co gamma-photon irradiation, *J. Med. Microbiol.* 54(12), 1157–962 (2005).

Brook, I., T. B. Elliot, J. D. Ledney, M. O. Knudsen, Management of postirradiation infection: lessons learned from animal models, *Mil. Med.* 169(3), 194–197 (2004).

Casadeval, A., L. A. Pirofski, The weapon potential of a microbe, *Trends Microbiol.* 12(6), 259–262 (2004).

Elliot, T. B., I. Brook, R. A. Haring, S. S. Bouhaouola, S. J. Peacock, G. B. Knudsen, *Bacillus anthracis* infection in irradiated mice: susceptibility, protection, and therapy, *Mil. Med.* 167(Suppl. 2), 103–104 (2002).

World, M. J. Bioterrorism: the need to be prepared, *Clin. Med.* 4(2), 161–164 (2004).

SECTION 5

MULTIPLE STRESSORS: APPLIED ASPECTS

CHAPTER 24

FEATURES OF SOMATIC GENE MUTAGENESIS IN DIFFERENT AGE GROUPS OF PERSONS EXPOSED TO LOW DOSE RADIATION

ALEXANDER S. SAENKO* AND
IRINA A. ZAMULAEVA

Medical Radiological Research Center of RAMS, Korolev str.-4, Obninsk, 249036, Russian Federation

Abstract: The aim of this work was to study the level of somatic gene mutagenesis in persons exposed to ionizing radiation at doses up to 200 mSv and determine features of this process in different age groups. Frequency of lymphocytes bearing mutations at T-cell receptor (TCR) locus was assessed by flow cytometry in 1386 persons, including 215 unexposed control donors, in 1995–2005. Exposed group consisted of employees of Nuclear Power Engineering, cleanup workers of the Chernobyl accident and residents of radiation contaminated territories of Bryansk, Kaluga, Orel, Tula oblasts of the Russian Federation. Results of group analysis demonstrated an increase in frequency of TCR-mutant cells after low dose irradiation compared to that in age-matched groups ($p < 0.05$, Mann-Whitney test). Radiation effect depended on developmental stage at the moment of beginning of exposure. The most pronounced elevation of the TCR-mutant cell frequency was found in the individuals irradiated in utero. Only a proportion (12–18%) of persons exposed in postnatal period had the TCR–mutant cell frequencies exceeding the 95% confidence interval in the control groups. This regularity was observed in all examined categories: in the residents of radiocontaminated regions, the Chernobyl cleanup workers and employees of Nuclear Power Engineering. The proportion of individuals with elevated mutant cell frequencies was inversely proportional to age at exposure. The relative number of irradiated individuals with elevated TCR-mutant frequencies was higher in group with benign neoplasms (in thyroid, mammary glands or uterus) than in group without this pathology.

*To whom correspondence should be addressed; e-mail: asaenko@mail.ru

C. Mothersill et al. (eds.), Multiple Stressors: A Challenge for the Future, 343–349.

Keywords: somatic mutation; T-cell receptor; ionizing radiation; low doses; flow cytometry

Introduction

The study of somatic mutagenesis after low dose radiation exposure is one of the urgent problems of radiobiology, radioecology and radiation medicine. This problem is of great interest, both theoretical and applied, because radiation-induced somatic cell mutations are likely to be the principal cause of cancer risk elevation after radiation exposure. The overwhelming majority of somatic mutagenesis studies in human cells in vivo were performed by cytogenetic analysis of structural mutations (Noppa et al., 2006). Gene mutations after low dose irradiation were studied much worse. During the last 10 years we were focusing on investigations of the frequencies of somatic cells harboring gene mutations in T-cell receptor (TCR) locus in various groups of individuals exposed to low doses of radiation. These are residents of the territories of the Russian Federation contaminated with radionuclides after the Chernobyl accident and those exposed for occupation conditions such as the Chernobyl cleanup workers and nuclear industry workers. Our measurements were performed 9–40 years after the beginning of exposure.

The aim of this study was to compare the level of somatic gene mutagenesis in individuals exposed to ionizing radiation at doses up to 200 mSv with that in unexposed individuals and determine features of this process in different age groups.

Materials and Methods

The blood samples from 215 unexposed persons, 343 employees of Nuclear Power Engineering Institute (NPEI, Obninsk), 184 cleanup workers of the Chernobyl accident and 644 residents of radiation contaminated territories of Bryansk, Kaluga, Orel, Tula oblasts of the Russian Federation were investigated. Individual doses were retrieved from the Russian National Radiation Registry database affiliated with our Center. Mean cumulative doses (±SE) were 100.8 ± 6.6 mSv in the employees of NPEI, 100.0 ± 9.1 mGy in the Chernobyl cleanup workers, 5.8 ± 0.7 mSv in residents whose exposure began in utero ($N = 119$), 5.1 ± 0.4 mSv in residents who were at childhood or adolescent age at the moment of the accident ($N = 233$) and 13.8 ± 0.4 mSv in adult residents at the moment of the Chernobyl accident ($N = 292$).

The determination of the TCR-mutant cell frequencies is done by means of flow cytometry technique. Mutant cells are identified among peripheral blood T-cells with fluorochrome-labeled monoclonal antibodies to CD3 and CD4 surface antigens. The mutant cells display CD4 + CD3-phenotype.

As shown by the inventors of this method – Doctors Mitoshi Akiyama and Seishi Kyoizumi of Japan, about 85% of T-lymphocytes with such phenotype possess mutations in the TCR locus (Kyoizumi et al., 1990).

The TCR assay was performed as described earlier in details (Saenko et al., 1998). Mutant cell frequency in an individual was rather stable. According to our data, repetitive measurements during several years in the same control persons yielded an intraindividual variation not exceeding 18%.

Results and Discussion

Numbers of mutant somatic cells in humans and animals are known to increase with age (Cole, Skopek, 1994). In group of 215 control persons we found that the increase in the TCR-mutant cell scores is about 3% per year, and this fact is necessary to be taken into account with regard to the control group.

Our results indicated that mutant cell frequencies were significantly elevated in all groups of exposed individuals compared to the respective controls (Table 1).

TABLE 1. Comparison between frequency of the TCR-mutant cells in control donors and that in persons exposed to low dose radiation

Study groups	N	Mean age[a], years	TCR-mutant cell frequency, $\times 10^{-4}$		P[b]
			Median	Range	
Control 1	67	55	3.8	1.4–10.0	0.03
Employees of Nuclear Power Engineering	343	56	4.2	0.9–63.0	
Control 2	133	45	3.6	1.1–10.0	<0.01
Cleanup workers of the Chernobyl accident	184	46	4.1	0.6–20.0	
Control 3	100	21	2.8	1.0–7.9	0.001
Residents of radiation contaminated territories (0–17 years old at the moment of the Chernobyl accident)	233	20	3.5	1.0–18.6	
Control 4	38	16	2.4	0.9–5.6	<0.001
Residents of radiation contaminated territories (in utero at the moment of the Chernobyl accident)	119	16	3.5	1.0–10.0	

[a]Mean age at the moment of the investigation
[b]Comparison with control group by Mann-Whitney test

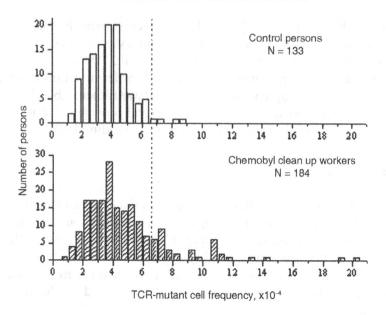

Fig. 1. Comparison between frequencies of the TCR-mutant cells in the group of Chernobyl cleanup workers and the control group. Vertical dashed line is an upper limit of the 95% confidence interval of the mutant frequencies in the group of control donors

Distributions of control persons and the Chernobyl cleanup workers according to mutant cell frequencies are shown in Fig. 1. The mutant cell frequency in the control group was normally distributed (according to Kolmogorov-Smirnov test). Upper limit of 95% confidence interval in the control group was calculated as mean + 2SD. In the group of exposed persons there is an appreciable proportion of individuals with elevated mutant cell frequencies (exceeding the 95% confidence interval in control). If such persons are excluded from the analysis, mutant cell frequency in the remaining group does not differ from that in the age-matched control group. These findings imply that statistically significant increase of mutant cell frequencies observed in the irradiated group is attributable namely to these subset of individuals with high levels of mutant cells. Similar results were found in groups of the NPEI employees and the residents of radiation contaminated territories who were at their childhood or adolescent age at the moment of the accident. The proportion of persons with elevated mutant cell frequencies is not too large (12–18% in various groups).

Acquisition of a great number of individual mutant cell frequency measurements in persons irradiated with low doses made it possible to assess age-associated changes. From the data described before it became clear that

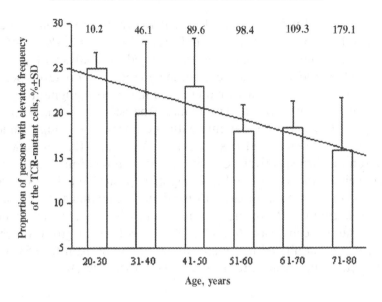

Fig. 2. Proportion of employees of NPEI with the elevated TCR-mutant frequency in different age groups. Mean cumulative doses (mSv) in the groups are shown

the analysis should be done by comparing the proportion of individuals with high mutant scores in the respective age groups but not only by evaluating the mean mutant frequency. Figure 2 shows a proportion of individuals with elevated mutant frequencies among employees of NPEI examined 5–40 years after the beginning of occupational exposure. It can be seen that such proportion is greater in the persons of younger age at exposure in spite of lower cumulative doses. Similar results were obtained in liquidators examined 9–17 years after irradiation.

With regard to the individuals whose irradiation started in prenatal period, their mutant frequency pattern was distinct from those of any other examined groups. After the exclusion of a subset of persons with high mutant scores, the mutant frequency in the remaining subgroup appeared to be significantly different from that of age-matched control group. Median values were 3.3×10^{-4} versus 2.4×10^{-4} accordingly ($p < 0.05$, Mann-Whitney test). The results imply that nearly all individuals irradiated in utero display a reaction seen here as an increased level of TCR-mutant cells even long term (16–19 years) after beginning of exposure.

In contrast to cleanup workers and NPEI employees, the response of irradiated residents (children, adolescents and adults together) displayed to some extent a correlation with radiation dose. Among those who live in the regions with Cesium-137 contamination exceeding $100 \, \text{kBq/m}^2$, the prevalence of individuals with elevated mutant scores was higher than that in residents of less contaminated areas: 14.5% vs 8.8% ($p = 0.03$).

Using case-control approach, we performed an analysis of mutant frequencies in the residents of radiocontaminated regions who developed benign tumors of the thyroid, breast or uterus. Similarly to malignant neoplasms, such tumors mostly associate with clonal expansion and thus with somatic mutagenesis. Furthermore, patients with these types of benign tumors are known to have an increased risk of cancer. We found that in patients with benign tumors, irrespectively of the age at examination – and therefore irrespectively of the age at the beginning of exposure, the TCR-mutant frequencies were elevated compared to irradiated individuals of the same age but without tumors. Proportion of residents with elevated frequencies of the TCR-mutant cells was 16.2% (62/381) in combined group of children, adolescents and adults with benign tumors in comparison to 9.5% (27/283) in combined group without tumors ($p = 0.01$, Fisher test).

Conclusion

Results of group analysis demonstrate an increased frequency of mutant cells after low dose irradiation compared to that in age-matched groups. Radiation effect correlates with the developmental stage at the moment of beginning of exposure and it is more pronounced in the individuals irradiated in utero. If radiation exposure takes place in postnatal period, only a proportion (12–18%) of all exposed display elevated TCR-mutant frequencies. It is worth noting that this regularity is observed in all examined categories: in the residents of radiocontaminated regions, the Chernobyl cleanup workers and employees of Nuclear Power Engineering.

The relative number of individuals "reacting" to radiation is inversely proportional to age at exposure. The proportion of residents of radiocontaminated regions "reacting" to radiation is significantly higher among those living in regions with Cesium-137 contamination exceeding $100\,kBq/m^2$, compared to less contaminated areas. The proportion of irradiated individuals with high TCR-mutant frequencies is increased in those diagnosed for some types of benign neoplasms compared to others.

Acknowledgments

This work was supported by the Federal agency on science and innovations in the framework of the Federal Program "Researches and elaborations in priority directions of science and technique development" for 2002–2006 (N 02.434.11.3003).

References

Cole, J. and Scopek, T.R., 1994, Somatic mutant frequency, mutation rates and mutational spectra in the human population in vivo, *Mutat. Res.* **304:** 33–105.

Kyoizumi, S., Akiyama, M., Hirai, Y. et al., 1990, Spontaneous loss and alteration of antigen receptor expression in mature CD4 + T cells, *J. Exp. Med.* **171:** 1981–999.

Noppa, N., Bonassi, S., Hansten, I.-L. et al., 2006, Chromosomal aberrations and SCE as biomarkers of cancer risk, *Mutat. Res.* **600:** 37–45.

Saenko, A.S., Zamulaeva, I.A., Smirnova, S.G. et al., 1998, Determination of somatic mutant frequencies at glycophorin A and T-cell receptor loci for biodosimetry of prolonged irradiation, *Int. J. Radiat. Biol.* **73:** 613–618.

CHAPTER 25

STATE OF ECOSYSTEMS AT LONG-TERM CONTAMINATION WITH TRANSURANIUM RADIONUCLIDES

V. KUDRJASHOV AND E. KONOPLYA

Institute of Radiobiology, National Academy of Sciences of Belarus, Minsk, Belarus

Abstract: The levels of radioactive contamination with transuranium elements on territory of Belarus as a result of nuclear weapon tests and Chernobyl NPP accident have been assessed by 17 actinides.

The study of the atmosphere contamination with TUE in Republic of Belarus is being held since 1980 to now. A gradual decreasing of TUE content in the surface air to $3.2\,nBq/m^3$ in April 1986 as to $^{239+240}$ Pu. In the period of Chernobyl NPP accident (end of April - beginning of May, 1986) the $^{239+240}$ Pu content reached value of $120\,\mu Bq/m^3$ in North part of Belarus. The period of half-removal of plutonium from nuclear weapon test and Chernobyl origin from the atmosphere according the observation period is the same and constitutes 14 ± 2 month. The mechanism of radioactive air pollution from April, 1986 is determined by dust transfer from radioactive contaminated regions. The value of this transfer is influenced considerably by agricultural activities on contaminated territory, forest fires and other anthropogenic factors.

A characteristic peculiarity of the Chernobyl NPP accident is the injection of small dispersed fuel particles containing TUE into biosphere.

The transfer coefficients in the soil- plant system vary from 7.1×10^{-6} to 4.0×10^{-3} for $^{239+240}$ Pu and 1.2×10^{-5} to 1.4×10^{-2} for ^{241} Am and have plant species dependence. The behavior of TUE in environment is discussed.

Keywords: transuranic elements; resuspension, plutonium; Chernobyl NPP accident; radioactive contamination; radionuclides

C. Mothersill et al. (eds.), Multiple Stressors: A Challenge for the Future, 351–357.

Introduction

The nuclear weapon tests and different accidents at nuclear power plants caused the appearance of transuranium elements (TUE) in the environment. Their content is constantly increasing in separate components of biosphere. TUE have long half-life and include themselves into the circulation of substances by trophic chains. They will be radiologically dangerous for human during thousands of years. The alpha particles with energy higher than 5 Mev are present in the radiation spectrum of the majority of actinides. Therefore, by penetration of actinides into the organism, the alpha-radiation obtains the leading role by inducing of biological effects. High energy of alpha particles creates high ionization density in microvolumes of cells and tissues. Therefore the reparation processes are practically absent at the alpha-radiation action. As a result, the injuries caused by TUE sum up in the course of time. All this testifies the great danger of TUE incorporation into human organism and the necessity of thorough study of regularities of their behavior in the environment and biological effects.

Levels of Surface Soil Contamination and Isotopic Composition of Actinides Deposition

The territory of Republic of Belarus underwent the anthropogenic pollution with transuranium elements from two sources: firstly, the global fall-out after nuclear weapon tests, and, secondly, the fall-out as a result of Chernobyl NPP accident.

As a result of nuclear weapon tests, the principal levels of pollution of soil surface on Belarus territory were formed in the middle of 1970s and the level of pollution with $^{239+240}$Pu constituted 53 ± 17 Bq/m^2. Beside of $^{239+240}$Pu, other actinides precipitated also on the soil surface. Their relative composition is adduced on Table 1.

The Chernobyl NPP accident led to very uneven pollution of soil surface in Belarus with transuranium elements. The content of $^{239+240}$Pu of "Chernobyl" origin varied from 1.1×10^5 Bq/m^2 on territories adjacent to Chernobyl NPP to average global levels in the north of Republic. At the active stage of Chernobyl NPP accident, the pollution of main components of biosphere was determined by short-living actinides ^{239}Np and ^{242}Cm. Their contents were higher than the $^{239+240}$Pu activity 5.6×10^5 and 40 times respectively. In the global fall-out, these transuranium elements were practically absent. The constants growth of ^{241}Am contents in all components of Belarus ecosystem is observed as a result of radioactive decay of ^{241}Pu. The maximum level of pollution with ^{241}Am will be reached at 2060 and will exceed that of $^{239+240}$Pu 2.7 times. In our estimation, the areas with density of surface pollution of

TABLE 1. Composition of actinides on Belarus territory in the radioactive releases from nuclear weapon tests and Chernobyl NPP accident

| Radionuclide | Half-life | Relative content by the activity (at the Chernobyl NPP accident moment–26.04.1986) | |
		"Chernobyl" fall-out	Global fall-out from the nuclear weapon test
^{235}U	$7.13 \times 10^8 y$	$(1.9 \pm 0.4) \times 10^{-4}$	–
^{236}U	$2.39 \times 10^7 y$	$(1.0 \pm 0.7) \times 10^{-3}$	–
^{238}U	4.56×10^9 y	$(3.1 \pm 0.6) \times 10^{-3}$	–
^{237}Np	2.4×10^6 y	$(2.9 \pm 0.7) \times 10^{-4}$	–
^{239}Np	$2.35 d$	$(5.2 \pm 1.1) \times 10^4$	–
^{236}Pu	$2.85 y$	$(1.1 \pm 0.3) \times 10^{-4}$	–
^{238}Pu	$87.1 y$	0.99 ± 0.02	0.054 ± 0.010
^{239}Pu	2.41×10^4 y	1	1
^{240}Pu	$6540 y$	1.44 ± 0.04	0.58 ± 0.03
^{241}Pu	$14.4 y$	210 ± 10	6.4 ± 0.3
^{242}Pu	3.76×10^5 y	$(2.1 \pm 0.2) \times 10^{-3}$	$(2.4 \pm 0.1) \times 10^{-3}$
^{241}Am	$452 y$	0.17 ± 0.03	0.57 ± 0.15
^{242m}Am	$152 y$	$(2.9 \pm 0.7) \times 10^{-3}$	–
^{243}Am	$7380 y$	$(2.2 \pm 0.6) \times 10^{-3}$	–
^{242}Cm	$163 d$	28 ± 5	–
^{243}Cm	$28.5 y$	$(4.5 \pm 0.2) \times 10^{-3}$	–
^{244}Cm	$18.1 y$	0.15 ± 0.03	–

soil with $^{238+239+240}Pu + {}^{241}Am$ up to $3.7 \, kBq/m^2$ will expand out of the limits of alienation zone in west and north-west direction by 20–30 km.

The Air Contamination in Belarus with Transuranium Elements

The pollution of air in the Republic of Belarus with transuranium elements before the Chernobyl NPP accident took place as a result of nuclear weapon tests in the atmosphere. After the intensive nuclear weapon tests in 1961–1962, the maximum annual contents of $^{239+240}Pu$ in the air of Belarus reached the value of $0.21 \, mBq/m^3$ in 1963.

The study of radioactive pollution of the atmosphere in the Belarus with transuranium elements was carried out from 1980 to the present (Fig. 1). In the dynamics of radioactive pollution of the near ground air with $^{239+240}Pu$ before the Chernobyl NPP accident, the tendency is registered to the decrease of $^{239+240}Pu$ content from $57 \, nBq/m^3$ in 1980 to $3.2 \, nBq/m^3$ in

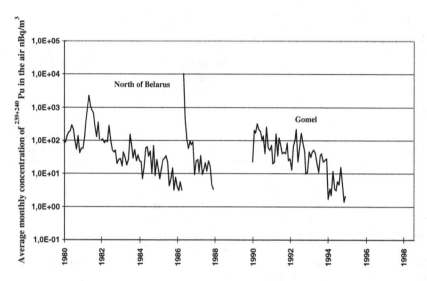

Fig. 1. The dynamic of the plutonium content in near ground air of Belarus.

April 1986. The increase of $^{239+240}$Pu in near ground air up to $3.5\,\mu$Bq/m^3 in May, 1981 was the result of stratospheric fall-out from 26th Chinese nuclear test in atmosphere in October 1980.

There are three stages in the dynamics of radioactive pollution in the near ground air with $^{239+240}$Pu after Chernobyl NPP accident:

I. Sufficient increasing of radionuclides content in surface air during transfer of the radioactive cloud. At this period (end of April–beginning of May 1986) the $^{239+240}$Pu content has increased approximately by a factor of 5×10^4 in the north of Belarus and has reached the value of $120\,\mu$Bq/m^3. The ratio of activities of ^{238}Pu/$^{239+240}$Pu has increased from 0.031 to 0.47.

II. Fast decreasing of radionuclides content in the air. This phase has lasted approximately till the mid of July 1986. At this period the levels of soil radioactive contamination have been arranged. $^{239+240}$Pu content in the air has decreased to 300 nBq/m^3 in the north of Belarus.

III. Slow decreasing of radionuclides content in the surface air. This stage has lasted and is being lasted up-to the present and is attributed to resuspension processes mainly.

The primary pollution of ground surface occurred in approximately two first weeks from the moment of accident. The overwhelming majority of radionuclides was released from the active zone of reactor into the atmosphere in that period. The dry precipitation and washout of radioactive aerosols from atmosphere formed the primary pollution of littering surface.

The systematic control of radioactive air pollution on sites within the alienation zone and adjacent zone is carried out. The observations of radioactive air pollution were carried out on the territories with different pollution density of soil (from 0.04 to 14 kBq/m^2 of $^{239+240}$Pu). The samplers were stood in the regional centers of Belarus (Gomel, Mogilev, Minsk, Mozyr, Brest), in the alienation zone of Chernobyl NPP and in the places where the agricultural and other activities are being held and the people are living.

The ratio of activities of ^{238}Pu/$^{239+240}$Pu in near ground air on all territory of Belarus correspond to "Chernobyl" one and constitutes 0.47 $\pm$ 0.05. This means that the pollution of air with transuranium elements after Chernobyl NPP accident is determined by the processes of transfer from radiocontaminated areas adjacent to Chernobyl NPP. As a result, the $^{239+240}$Pu content in near ground air in cities situated near alienation zone (Gomel, Mozyr) is 2–3 times as high as in other cities.

It can highlight the next components in the dynamics of average monthly levels of radioactive contamination of atmosphere: the determined trend of mid level, the cyclic component conditioned by seasonal variations and the occasional component which is the result of influence of multiple factors of anthropogenic and natural origin.

In the dynamics of average monthly levels of $^{239+240}$Pu content, there is seasonal cyclic component which conditions the spring growth of $^{239+240}$Pu content in near ground air as 4–5 times as compared with average annual level.

The forest fires on radioactively polluted territory in 1992 led to the increase of transuranium elements contents in air on all territories in the summer. Whereas the average annual content of $^{239+240}$Pu in Mozyr and Gomel in 1992 exceeded the level of 1991 and constituted 80 nBq/m^3.

As a result of study of mechanism of formation of radioactive pollution with transuranium elements, it was found that the levels of radioactive pollution of air with transuranium elements in the alienation zone and adjacent areas were determined both by the content of dust in the air and its specific activity. In the alienation zone, due to the cessation of economic activity, the concentration of dust in the air is approximately equal in all sites and does not exceed 10 μg/m^3. Therefore the transuranium elements content in the air of this locality depends linearly on the density of surface pollution of soil. In the areas adjacent to the alienation zone, the intensive agricultural activity in spring period leads to significant increase of dust content in the air (up to 1 mg/m^3). As a result, the transuranium elements content in the air of these area reaches and, in separate time periods, exceeds their content in the mostly polluted sites of alienation zone. The analysis of dynamics of average annual concentrations of radionuclides in the air of Belarus cities testifies to the tendency of slow decrease of atmosphere pollution with transuranium elements. This process may be described quantitatively with the period of semi-cleaning of atmosphere.

The half-life in the atmosphere of the plutonium of "Chernobyl" origin and from nuclear weapon tests by the data for the period of 1980–1999 was equal for all observation sites and constituted 14 ± 2 months.

It was revealed by the methods of α-autoradiography and neutron-fragment radiography that there is inversely proportional square dependence of the specific activity of transuranium elements containing aerosol particles on their diameter. The existence of such dependence indicates that transuranium elements containing aerosol particles are not fuel ones at all, but they are non-radioactive aerosol particles, on the surface of which the smallest fuel particles are distributed. The dimension of elementary fuel particle on the average constitutes $0.1\,\mu m$, and the specific $^{239+240}Pu$ activity in such a particle is $\sim 10^8$ Bq/g.

The Accumulation of Transuranium Elements by Plants and Animals

The range of varying the coefficients of plutonium and americium accumulation in wild species of plants on forest and meadow sites constituted by our data from 3.0×10^{-5} to 4.3×10^{-4} for $^{239+240}Pu$ and from 1.2×10^{-4} to 1.8×10^{-3} for ^{241}Am in forest phytocoenosis and 1.9–6.7×10^{-4} for $^{239+240}Pu$ and 4.6×10^{-4}–2.7×10^{-3} for ^{241}Am in meadow phytocoenosis.

The radiochemical analysis of transuranium elements content in above-ground phytomass of sowed fodder grasses in the alienation zone of Chernobyl NPP has shown in field experiment that biological mobility of transuranium elements in condition of cultivated soil increases as compared with non-cultivated soil of natural complexes: the values of transfer coefficients in cultivated species of grasses are higher by 10–100 than these in wild species.

It has been found the dependence of level of transuranium elements accumulation on the species of both wild and agricultural plant: the bush dominants – Cytisus ruthenicus, Vaccinium myrtillus, Calamagrostis epigeios – are accumulators in forest phytocoenoses, and among meadow grasses (in agrocoenoses and natural communities) - sedges and legumes.

The value of transfer coefficients of "Chernobyl" transuranium elements from soil to plants is within the limits determinated for "bomb" transuranium elements. The transfer coefficients from soil to plant of ^{241}Am are higher that those for plutonium because the americium is more soluble. It is necessary to note that the transuranium elements presently are transported to plant via the root system.

The specific activity in water of mostly radioactively polluted Belarus lakes reaches $0.98\,Bq/m^3$ for $^{239+240}Pu$ and $11\,Bq/m^3$ for ^{241}Am. The main forms of plutonium in water are colloidal particles with adsorbed Pu(IV) and complex compounds of Pu(V) and Pu(VI).

The concentrations of $^{239+240}Pu$ and ^{241}Am in wild animal body are $3.1 \pm 1.4\,mBq/kg$ and 1.2 ± 0.4 respectively, for global fall out, and reaches values $120\,mBq/kg$ for $^{239+240}Pu$ and $1300\,mBq/kg$ for ^{241}Am for some "hot" regions in Belarus after Chernobyl NPP accident.

CHAPTER 26

TECHNOLOGICALLY ENHANCED NATURALLY OCCURRING RADIOACTIVE MATERIALS (TENORM) IN NON-NUCLEAR INDUSTRY AND THEIR IMPACT INTO ENVIRONMENT AND OCCUPATIONAL RADIATION RISK

BOGUSLAW MICHALIK*

Central Mining Institute, Laboratory of Radiometry, Plac Gwarków 1, 40-166 Katowice, Poland.

Abstract: Despite the history of artificial radionuclides (^{137}Cs, ^{90}Sr, etc.) influence on biota is limited to several decades the overwhelming majority of present-day radioecological investigations is connected to study of redistribution in environment and biological action of this group of radionuclides. Far less attention was paid to radiation risk to people and environment caused by exposition to enhanced ionizing radiation originating from naturally occurring radioactive materials. But after the occurrence of natural radioactivity had been considered thoroughly it turned out that such phenomena are present very frequently in our environment. Enhanced natural radioactivity (TENORM) touches a lot of aspects of our common life, starting from occupational risk at work places, trough some "contaminated" goods or even a visit in spa, and ending on huge amount of bulk waste materials very often dumped in our vicinity. Each particular way of occurring of natural radioactivity determinates some unique scenario of exposition usually differing from those ones caused by artificial radionuclides. Moreover, consequences of natural radioactivity's occurrence can be assessed from different points of view. Sometimes the public comprehension of this phenomenon is a plentiful source of very serious effects far more detrimental and painful than the direct exposition to radiation. Risk caused by naturally occurring radioactivity is a case where enforcing of ALARA rule became very complex and multidimensional.

The way of creation the risks by enhanced natural radioactivity sizeable vary with the situation related to the presence of artificial radioactive sources. Usually the amount of TENORM-type waste materials can amount to hundreds of thousands of cubic metres, they had been created or dumped directly into environment and associated with other pollutants as heavy metals or hydrocarbons. Therefore the application of routine rules used for

---

*To whom correspondence is addressed. e-mail: b.michalik@gig.katowice.pl

C. Mothersill et al. (eds.), Multiple Stressors: A Challenge for the Future, 359–372.

assessment of risk caused by artificial radioactivity and work practices can lead to the completely misunderstandings.

To prepare a coherent system rules and recommendation designated for the monitoring and control of risk related to the presence of natural radioactivity is necessary at very beginning to precise causes of concern and define the clear terms for describing risk scenarios. Then it is possible to assess potential detrimental effects starting from radionuclides' inventory and migration and ending on genotoxicity.

Keywords: TENORM; radiation; radiation protection; natural radioactivity

Introduction

Radioactivity is primordial property of matter and the human environment. Since life emerged on the Earth every living organism has been exposed to ionizing radiation. According to the state of art radiation protection some specific amount of radiation, called "natural background" is considered not to be harmful to human beings and the natural environment. But in some cases such harmless level of radiation can be enhanced due to the people's activity. Such alteration of natural state resulted in an increment of radiation risk to the people as well as to the environment and non-human biota. The main reason for this is that in nature exist at least 400 minerals containing elevated concentration of natural radionuclides belonging to uranium and thorium decay series. Some of them are processed by industry in order to get a useful material or goods. The most famous industrial process where problem of radiation risk caused by natural radioactivity exist is uranium ore extraction and fissile fuel production. But it is a part of the nuclear industry so that always was subject to radiation protection. Besides it there are a lot of different branches of industry where radiation risk is significant but they usually have been out of concern of radiation protection.

The importance of radiation risk caused by natural radioactivity firstly have been underlined in the European Council Directive 96/29EURATON laying down basic safety standards for the protection of the health of workers and the general public against the danger arising from ionizing radiation. Namely, in the paragraph 40 of this Decree, was clearly stated that common rules of radiation protection must be applied in following cases:

- work activities involving operations with, and storage of, materials, not usually regarded as radioactive but which contain naturally occurring radionuclides, causing a significant increase in the exposure of workers, and, where appropriate, members of the public.

• work activities which lead to the production of residues not usually regarded as radioactive but which contain naturally occurring radionuclides, causing a significant increase in the exposure of members of the public, and, where appropriate workers.

Since this, a lot of research works dealing with this matter have been done but up to now any uniform and coherent approach to assessment of the risk did not been enforced, especially in case of environmental risk assessment. It results in big gap in international and national regulation.

Usually the problem of natural radioactivity is not included in regulations dealing with radioactive waste and totally excluded from decrees regulating environmental protection. Finally such hazard is rarely taken into consideration when environmental risk assessment (ERA) is carried out. But on the other hand, natural radioactivity, from point of view ERA, should be treated in a special way because usually is associated with other harmful agents, as heavy metals or hydro carbons, typically occurring in industrial waste.

To shed light on problem related to natural radioactivity presence and then prepare a coherent system rules and recommendation designated for the monitoring and control of risk related to the its presence, it is necessary at very beginning to precise causes of concern and define the clear terms for describing risk scenarios.

System of Natural Radioactivity Occurrence Classification

The first time the NORM (naturally occurring radioactive material) and technologically enhanced NORM and associated risk were distinguished by Gesell and Pritchard in 1975. They use the abbreviation TENR (*technologically enhanced natural radioactivity*). After the EC Directive 96/29 had been published problem of risk caused by natural radioactivity was discussed widely. Some authors have followed terminology applied by Gesell and Pritchard (i.e. Kathren, 1998; Righi et al., 2000) but a lot of different abbreviations have emerged to describe these phenomena. For example Baxter (1996) applied TERM (*technologically enhanced radioactive material*), Vandenhove (2000) used "materials containing natural radionuclides in enhanced NORs." In Canada two names: NORM-*contaminated* and *Naturally Occurring Nuclear Substances* have been applied to distinguish non-nuclear industry from uranium ore extraction and processing (Ministry of Health, Canada, 2000). Martin et al. (1997) in the first report completely dealing with European industry coping with natural radioactivity used simple descriptive name "materials containing natural radionuclides in enhanced concentrations." IAEA (2003) report has distinguished two situations when natural radioactivity in non-nuclear industry can cause significant increment of radiation risk: NORM (*naturally occurring radioactive materials*) and *technologically enhanced* NORM. Such approach was followed in many articles presented during periodic conference NORM IV held in Poland (IAEA, 2005). All mentioned above abbreviations are used by authors interchangeably to describe situation when presence of natural radioactivity causes not negligible radiation risk. But after even rough analysis of branches of industry processing minerals it is clear that such enhanced risk can be created in two different way and acronyms NORM and technologically enhanced NORM (TENORM) reflected them the best.

The term NORM (naturally occurring radioactive materials) should be used, accordingly to the definition, only for cases, when radiation hazard is due to the presence of materials with elevated concentration of natural radionuclides, significantly above average level of radioactivity, albeit not related to or caused by any type of human's activity. It have to be pointed out, that NORMs are taken into account in the radiation hazard assessment scenarios only in cases, when appear in the natural or work environment due to industrial activity, otherwise are treated as sources of the natural background and not taken into considerations as enhanced radiation risk.

The acronym TENORM(s) means technologically enhanced naturally occurring radioactive materials. This term is used for the description of any raw material, product or waste, in which concentrations of the natural radionuclides have been altered (enhanced) as a result of technological processes to the levels, causing significant increase of the radiation hazard above the natural background. It does not matter, if the enhancement is intentional or not. It can be seen, that in some cases NORMs are used as a substrate(s) for the process(es) where TENORMs are created as products or by-products. On the other hand, it's possible to create TENORMs in processes, where no NORMs have been used as raw materials. An example of such processes is coal combustion for power generation. Hard coal is well known as a material with relatively low concentration of natural radio-nuclides, and can not be treated as NORM at all. The combustion leads to the very big reduction of the initial fuel mass. Owing mainly to the elimination of organic component of hard coal, during combustion process there is approximately one order of magnitude enhancement of the radioactivity concentration from fuel to ash.

One should remember, that usual exemption from TENORM classification concerns raw materials and substances, used in the nuclear industry either for civil or for military purposes. But radiation protection in both types of mentioned above activities is their immanent part (due to current terminology applied in radiation protection such activities are classified as "practices" on contrary to these dealing with NORM or TENORM that are named "work activities").

The third kind of processes leading to enhanced radiation risk related to the occurrence of natural radioactive nuclides, especially important from risk to living organisms' point of view, but completely not appreciated enough are processes going in living matter. There are a lot of data dealing with the behaviour of natural radionuclides in biota after intake. Frequently the processes of metabolism lead to concentration of some long-lived natural radionuclides in particular tissues of plants as well as animals (McDonald et al., 1996). The committed effective dose resulted from their presence in such a situation many times exceeds total doses caused by associated artificial radionuclides (Aarkrog et al., 1997). By the analogy with technological processes one can use in these cases the term BENORM – biologically enhanced naturally occurring radioactive materials.

Radiation Risk Scenarios

Accordingly to definitions mentioned earlier, processes leading to the increase of the radiation hazard due to natural radioactivity, can be distinguished and enclosed into two groups.

The first one encloses exploitation, transfer or disposal to the natural environment raw materials or waste without any changes of their properties. Such processes are mainly performed in the industrial branches, relating to exploitation of mineral resources with elevated concentrations of natural radionuclides – phosphates or thin, titanium, niobium and rare earth metal ores. The spoils produced in the exploitation processes are in form of natural materials, with the same chemical composition and physical properties as the raw material. The enhancement of radiation hazard is usually a result of the direct exposure to its radiation and may be due to the increase of the amount of waste in the installation or its release to the natural environment. Also different exploitation technologies, applied in underground mining or drilling, may lead to unintentional release of waters or gases with elevated levels of natural radioactivity. The presence or use of NORM is in this case crucial to create significant radiation risk. Following waste materials can be stated as mentioned above substances:

- Waste rocks from mining exploitation and raw materials enrichment, together with possible contamination resulting from leaching of natural radionuclides;
- Radium-bearing waters and radioactive deposits in underground galleries and settling ponds, scales in the dewatering systems in mining industry;
- Brines, released from oil and gas rigs and scales in the dewatering systems in this industry;

- Radon and its progeny in caves, underground mines, in tunneling and in natural gas;
- Dust from systems of air cleaning.

Inside this class of radiation risk scenarios it is necessary to take into consideration other situations, not exactly related to the industrial activity, by rather influenced by changes of style life or habits. Baxter was correct pointing out that after those human beings had moved into a cave they were exposed to radon and radon progeny much more than on open air areas.

The second group consists of processes leading to significant mass reduction and/or with changes of chemical composition or state of aggregation which might influence further behaviour in the environment. Such processes are characteristic for industrial branches, where processing of raw materials takes place. Mass reduction in these processes may cause concentration of all impurities in produced waste materials, together with natural radioactivity. Spoils with concentration of natural radionuclides several tens higher as in the raw materials are often created as the results of such processes. Besides fossil fuels combustion, also all parts of the processes of metal ores treatment (metallurgy) as well as particular stages of certain chemical technologies can be included to this category. In this case to create radiation risk it is not necessary to process NORM materials and these raw materials in practice are often contain less natural radioactivity than usually taken as average for Earth's crust. As the fact of the matter, these technologies of concern are not aimed toward production of natural radionuclides, therefore usually radioactive isotopes are accumulated in waste or by-products of these processes.

In general, properties of the waste materials from the second group of technologies are related to the level of accumulation of the natural radionuclides due to the mass reduction within the process. Following waste materials can be taken into account in this category:

- Spoils from fossil fuel combustion (Coal Combustion Products);
- Waste materials from chemical technologies (mainly production of phosphoric acid and titanium dioxide pigment);
- Spoils from metallurgical industry (iron ore processing and steel production);
- Waste products from non-ferrous ore processing – mainly tin, copper and niobium;
- Spoils from rare earth metals industry.

Besides the occupational risk and possible risk to the members of the public inhabiting neighbouring areas NORM/TENORM type waste impacts upon the environment where they have been dumped. As the result of the lack of appropriate regulation it was common practice to dump such kind of waste into a heap or settling pond, relatively to the state of aggregation, without any means of protection. Usually the total amount of waste collected in one dump reach hundred thousands cubic meters or tonnes. NORM/TENORM placed into environment set some additional processes in motion, leading to the selective transfer and accumulation of particular radionuclides and disequilibrium in chains of natural radioisotopes. Finally, the possible detrimental effect to environment and non-human biota is not so easy to assess, especially if one take into account that in such waste usually occur other pollutants that had been concentrated in the same way as natural radionuclides.

Problems with NORM/TENORMs are not related only to waste materials or by-products. Sometimes natural radionuclides are transferred during production into the final products of particular technologies in both groups of mentioned above processes. Therefore among final products and goods, available on the market, specific category of products can be recognized. For such goods, concentration of the natural radionuclides cannot be stated as negligible from the point of view of radiological protection. On the other hand, often cleaning or removal of natural radioactivity from these products is very difficult and too costly, from the economical point of view or not required by radiological protection regulations. Following products can be included to this category:

- Refractory materials
- Fertilizers;
- Glass and ceramic;
- Abrasive materials;
- Flints for lighters.

Very peculiar group of TENORMs are products, into which natural radionuclides have been added intentionally for the improvement of their properties, not connected directly with the used of radioactivity. Sometimes these products are made of natural radionuclides entirely. In this category following products can be recognized:

- TIG welding rods;
- Mantles for gas lamps;
- Dentistry materials;
- Products made of depleted uranium (ballast or counterbalance)
- Optical glassware;
- Fluorescence dyes, (in the past).

TENORM at Areas Affected by Mining Activity

The most common process leading to NORM release and TENORM creation is energy generation from fossil fuel. If one considers this process as a whole, it means starting from fossil fuel extraction and ending on combustion products utilization, it is easy to distinguish few several stages of the process when different NORM or TENORM are created (Fig. 1). Coal combustion products (CCP) are very well known as waste with slightly enhanced concentration of natural radioactivity. But by far less known is TENORM-type waste created by the way of underground water cleaning.

The Upper Silesian Coal Basin (USCB) is located in the southern part of Poland and there had been working up to 65 underground coal mines. Total outflow of waste waters from these mines reached 900,000 m³/day. Due to their very high salinity (sometimes higher than 200 g/L) they have caused severe damages to the natural environment. Additionally, these waters have often elevated concentrations of radium isotopes ^{226}Ra, ^{228}Ra as well as barium and other metals. This phenomenon has been recognized since the 1960s. Such waters have been found also in German coal mines and it is very well known that the oil and gas extraction industry produces saline waters with enhanced radium concentration, too.

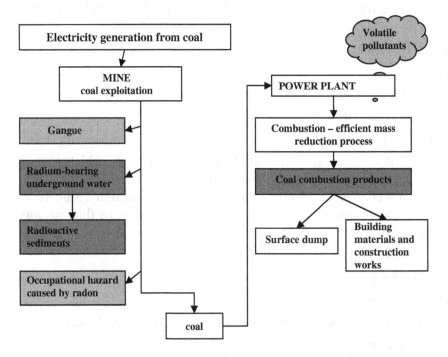

Fig. 1. NORM, TENORM waste and pathways of enhanced radiation risk to people and the environment

In USCB waters with high radium concentration occur mainly in the southern and central part of the coal basin, where coal seams are overlaid by a thick layer of impermeable deposits. First investigation showed that concentration of radium in formation water is correlated with its salinity and it has been found that the behaviour of radium during transportation of radium-bearing brines in the gutters of underground galleries, settling tanks and ponds, pipelines and rivers depends mainly on chemical composition of the brines. Hence two types of radium-bearing waters have been distinguished. Waters type A contains not only radium ions but barium ions as well. Concentration of barium in these waters is at least six orders of magnitude higher than that of radium and reaches 2 g/L. High barium concentrations enable co-precipitation of radium with barium sulphate when radium-bearing water type A is mixed with water containing sulphate ions, which are very common in nature. In case of radium-bearing waters type B, where barium ions are not present, concentration of radium ions is too low to enable precipitation of radium sulphate because concentration of $[Ra^{2+}]$ * $[SO_4^{2-}]$ does not exceed the solubility product. Due to differences in their chemical proprieties, the effect of release of radium-bearing waters type A and type B into the natural environment is completely different. Radium from radium-bearing waters type A is precipitated out in underground mine workings and in settling ponds, pipelines and little rivers. Concentration of radium in such precipitates is usually high reaching 400 kBq/kg in underground galleries and 270 kBq/kg on the surface, but the precipitation and sedimentation takes place rather close to the point where radium-bearing waters type A mix with waters containing sulphates, so that in the distance over few or several kilometres downstream from the discharge point of mine water the river water is free of radium. In opposite, from radium-bearing waters type B radium is not precipitated but transported with water to large rivers. Although, concentrations of radium in bottom sediments are in this case not very high, contamination of river waters and bottom sediments is observed over a large distance even up to hundred kilometres from the discharge point (Lebecka et al., 1996).

In hard coal exploitation process a settling ponds were applied to allow mechanical suspensions, carried out by underground brines, to settle. For this purpose some natural lakes or fishing ponds were adapted. The sediments with concentration of both radium isotopes exceeding 200 Bq/kg were found inside 25 settling ponds. The total capacity of all these settling ponds reaches 5 millions cubic meters (Michalik 2004).

Some of settling ponds have been used for over 25 years and during this period about 200 million m³ of waste waters had been discharged into. It can resulted in 100,000 m³ (about 150,000 tonnes) of total amount of the suspension, deposited in the pond. Significant increase of dose rates can be observed up to $42\,\mu Gy/h$ near the point of inflow of waters into a settling pond (Chałupnik et al., 2001). Besides enhancement of natural radioactivity precipitation and sedimentation processes concentrate heavy metals as well as chlorides and sulphates.

The occurrence of enhanced natural radioactivity in Polish coal mines gives rise to radiation hazard for mining crews. In the underground mining industry in Poland, monitoring of the radioactivity of mine waters and precipitates, as well as gamma doses and radon progeny in air, has been obligatory since 1989 in frame of occupational risk control system. On the contrary, the monitoring of environment, in light of lack formal regulation has been denied.

Recommended Limits of Activity Concentration and Dose Constraints

After the review of literature reports the general question has been emerged: when NORM or TENORM must be treating as source of enhanced, above natural background, radiation risk? The EC Directive 96/29/EURATOM does not precise the answer for this question so that it became a real problem. Leaving the risk caused by NORM and TENORM free for the each EC member justification leads to the situation where a lot of different approaches to decide whether NORM/TENORM is a source of radiation risk or not exist. This difficulty is especially emphasized when one would like to carry out an ERA.

All recommendations concerning radiation protection, issued by international bodies base on dose constraints and the derived activity concentration limits. Exemption and clearance levels approaches a priori assume that some, certain level of radioactivity should be released from any control (IAEA, 2004). If such levels were low enough it would work properly even in case of application to ERA. The levels established for artificial radioactivity are really extremely low. On the contrary such levels recommended for natural radioactivity sometimes are three orders of magnitude higher. The best example is the last ICRP recommendations as in table below (Table 1).

TABLE 1. Recommended exemption levels (according to the project of the ICRP 2005 recommendation)

Nuclides	Exclusion Activity Concentration Bq q^{-1}
Artificial	0.01
artificial β/γ emitters	0.1
Head of chain activity level [238] U, [232]TH (including all progeny)	1.0
[40]K	10

For the persons who are at least a little knowing physics such differentiation between natural an artificial radiation seems a bit strange. Why ionizing radiation coming from natural sources is deemed to be less detrimental than this one coming from the artificial sources? In uranium and thorium decay series there are at least 20 α-emitters! Why one have to accept their occurrence in our vicinity in concentrations 100 times higher than the artificial ones? But on the other hand, why the artificial ones seem to be so dangerous?

The main reason that causes such mysterious, underestimating approach to the assessment of potential risk related to natural radioactivity is an application of routine rules used in risk assessment caused by artificial radioactivity. Artificial radioactivity it is something odd in environment and it is the field when the radiation protection was started and developed. Hence majority of radiation protection experts have origin in this field. They are used to meet the radiation risk caused by relatively small sources with high activity concentration. This risk was limited in space and very often in time. It would be working in case when, i.e. radium, the most important radionuclide among natural ones, was used as radioactive source but application of such radiation risk scenario to the case of NORM/TENORM occurrence leads to the completely misunderstandings. If in case of occupational risk assessment (ORA) it can explain natural radioactivity away then in case of environment there is no matter what is the origin of radiation.

The second reason of natural radioactivity depreciation is inappropriate way in which such phenomenon is comprehended. The main source of wrong comprehension lays in the adjective "natural". Such a word before a noun describing a substance usually is concerned to be equalled "not dangerous." Hence it is not so far to the interpretation that natural radioactivity is less detrimental than artificial one.

One can find in reports dealing with the radiation protection problems different limits of activity concentration deemed to be exempted from radiation risk monitoring (Table 2)The numbers in this table have different origin but exactly concern the same radionuclides. If it is necessary which limit should be applied? Which one should be applied to ERA?

The fact of the matter is that the origin of the proposed by ICRP (ICRP, 2005) exemption levels can not be justified by any rational reason. Such value as 1 Bq/g seems to be a magic one because it is round but it does not comply with general dose limit level established for the natural radioactivity at the level 1 mSv per year. Assuming the big amount of TENORM with 1 Bq/g of natural radionuclides homogeneously distributed (what is very common situation) one can calculate the dose rate above the ground basing on recommendation of ICRU (ICRU, 1994). Such calculation can give the dose about 12 mSv per year caused only by external radiation. Values of activity concentration of natural radionuclides, calculated on the base the

TABLE 2 Recommended exemption levels by the specific mass activity concentration

	Canadian NORM Guideline	RP 112[a] [RP 1999]	ICRP	IAEA BSS [IAEA 2003]		RP 122 [RP 2001]	
				Activity Exemption Level	Derived from $10\,\mu Sv/y$ Dose Constraint[b]	Activity Exemption Level	Derived from $0.3\,mSv$ Dose Constraint[b]
	Bq/kg						
^{238}U	0.3	0.05	1	1	0.001	0.5	0.025
^{232}Th	0.3	0.033	1	1	0.001	0.5	0.025
^{40}K	17 (*)	600	10	100	–	5	0.2

(*)natural abundance of ^{40}K in potassium chloride
[a]Calculated on base of criteria: $f_1 < 0.5$
[b]Air kerma rate 1 m above ground, basing on recommendation of ICRU 53 for the radionuclides distributed homogeneously in the ground.

same equation as above, and derived from the dose constraint assumed for the risk caused by artificial radionuclides can be 1000 times lower than recommended in the same report values. Is there any reason, therefore, why the materials with activity concentration of about 1000 Bq/kg should be excluded from any radiation hazard control?

Conclusion

In the light of other European regulations that very often contain a lot of details concerning subjects that in comparison with radiation risks appear not to be so serious and plenty with negative consequences, the EC Directive 96/29/EURATOM seems as if enclosing not so many details as necessary. It results in that the problems of NORM/TENORM are mostly out of any regulation and non-nuclear industry of concern was till now not aware of the problems connected with natural radioactivity or would expect negative consequences in case of implementing radiation protection measures. Therefore many companies do not provide radioactivity data as long as no precise regulation exists. Finally, in spite of the number of reports issued recently and describing effects of TENORM and NORM occurrence in industry and environment this kind of risk is not appreciated enough.

For the keeping environment in good condition it not enough to limit the releases of potentially detrimental agents. Approach basing on limitation of concentration in many cases is effective but on the other hand not efficient from economical point of view. The potential detriment to human beings and environment depends not only on concentration. Chemical compounds, total amount or at least, certain site of occurrence significantly impact on it. That's why it is worth applying for the assessment quality of the environment more complex method giving precise result in each particular risk scenario. Especially in case of NORM/TENORM occurrence application of a multi-stage ERA seems promising. It should consist of following stages:

- The scale of the pollution (contaminants' inventory)
- Radionuclides' migration and availability
- Biota exposition to external alpha and gamma radiation
- Radionuclides' transfer factors into biota
- Radiation effects on biota (at molecular level)
- Associated pollutants' impact (antagonistic/synergistic effects assessment)

Such approach can deliver enough, good quality information about mutual relationship among different parameters describing state of the environment. The last stage of proposed ERA gives the most important information. Interaction of contaminants with biota firstly takes place at the cellular level making cellular responses. It is not only the first manifestation of harmful effects, but also can be a suitable tool for an early detection of pollution. Therefore, just genetic test-systems seem to be most effective and efficient for an early and reliable displaying of the alterations in the ecosystems resulting from the human industrial activity. Knowing direct effects on biota one can bind those and draw a conclusion about acceptable dose or concentration.

If one did such integral assessment of the man-caused impact, taking into account the whole set of harmful factors (as well as their ssynergistic and antagonistic effects) including the radioactive contamination and chemical pollution for non-nuclear industry, the European Waste Catalogue would be completed with new, reasonable clearance levels for natural radioactivity.

References

Aarkrog A., Baxter M. S., Bettencourt A. O., Bojanowski R., Bologa A., Charmasson S., Cunha I., Delfanti R., Duran E., Holm E., A comparison of doses from ^{137}Cs and ^{210}Po in marine food: A major international study, Journal of Environmental Radioactivity, 34(2), 217–218, 1997.

Baxter M. S., Technologically enhanced radioactivity: An Overview, Journal of Environmental Radioactivity, 32(1–2), 3–17, 1996.

Canadian Guideline for the Management of Naturally Occurring Radioactive materials, Ministry of Health, Canada 2000.

Chałupnik S., Michalik B., Wysocka M., Skubacz K., Mielnikow A., Contamination of settling ponds and rivers as a result of discharge of radium - bearing waters from Polish coal mines, Journal of Environmental Radioactivity, 54 (2001) 85–98.

B. MICHALIK

Gesell T. F. and Pritchard N. M. The technologically enhanced natural radiation environment. Health Physics 28, 361, 1975.

IAEA (2003), Technical Report Series No. 419, Extent of Environmental Contamination by Naturally Occurring Radioactive Material (NORM) and Technological Options for Mitigation, Vienna

IAEA (2003) Safety Series No. 115/CD, International Basic Safety Standards for Protection against Ionizing Radiation and for the Safety of Radiation Sources (CD-ROM Edition, 2003)

IAEA (2004) Safety Standards Series No. RS-G-1.7, Application of the Concepts of Exclusion, Exemption and Clearance Safety Guide.

IAEA (2005) TEC-DOC-1472, Naturally Occurring Radioactive Materials (NORM IV), Proceedings of an International Conference held in Szczyrk, Poland, May 2004, issued October 2005.

ICRP (2005), 2005 Recommendation of International Committee on Radiation Protection – draft for consultation.

ICRU (1994), Gamma ray Spectrometry in the Environment, ICRU Report No 53, 1994.

Kathren, R. L., NORM Sources and Their Origins, Applied Radiation and Isotopes (Incorporating Nuclear Geophysics) 49(3), 149–168, (March 1998).

Lebecka, J., Mielnikow, A., Chalupnik, S., Wysocka, M., Skubacz, K., Michalik, B., Radium in mine waters in Poland: occurrence and impact on river waters - Proceedings of the Natural Radiation Environment VI, Montreal, in: Environment International 1996.

Martin, A., Mead, S., Wade, B. O., Materials containing natural radionuclides in enhanced concentrations, Directorate-General Environment, Nuclear Safety and Civil Protection, contract No B4-3070/95/00387/MAR/C3 Final report No EUR 17625 EN, 1997.

McDonald P., Baxter M. S., Scott E. M., Technological enhancement of natural radionuclides in the marine environment, Journal of Environmental Radioactivity, 32(1–2), 67–90, July-August 1996.

Michalik, B. "Environmental pollution caused by natural radioactivity occurring in mining industry – the scale of the problem" Sustainable Post-Mining Land Management, Edited by: Euromines, CBPM CUPRUM Wrocław and Mineral and Energy Economy Research Institute Polish Academy of Science, Kraków, ISBN 83-906885-9-Y, Wrocław 2004, pp 145–154.

Radiation Protection 112 (1999), Radiation Protection Principles concerning Natural Radioactivity in Building Materials, European Commission Directorate General Environment, Nuclear Safety and Civil Protection.

Radiation Protection 122, (2001) Practical use of the concepts of clearance and exemption levels, part II: Application of the concepts of clearance and exemption to natural sources of radiation, European Commission Directorate General Environment, Nuclear Safety and Civil Protection.

Righi, S., Betti, M., Bruzzi, L., Mazzotti, G., Monitoring of natural radioactivity in working places, Microchemical Journal, 67, 119–126, 2000.

Vandenhove, H., European sites contaminated by residues from the ore-extracting and –processing industries, Proceedings of the 5th International Conference on High Levels of Natural Radiation and Radon Areas, Munich, Germany, September 2000, Elsevier International Congress Series No. 1225, 2002, ISBN: 0-444-50863-5 (ICHLNRR, 2000).

CHAPTER 27

STEPPE SOILS BUFFER CAPACITY AND THE MULTIPOLLUTION IMPACT OF INDUSTRIAL ENTERPRISES IN UKRAINE

MYKOLA M. KHARYTONOV[1], ANN A. KROIK[2], AND LARISA V. SHUPRANOVA[2]

[1]*The State Agrarian University, Voroshilov st. 25, Dnipropetrovsk, 49600, Ukraine; e-mail: mykola_kh@yahoo.com*
[2]*The State Agrarian University, Dnipropetrovsk National University, Gagarina av. 44, Dnepropetrovsk, 49600, Ukraine. e-mail: envteam@ukr.net; larchyk@rambler.ru*

Abstract: It has been shown that the variability of chemical composition of soils depending on the different polluted sites lead to active reorganization in cell protein system. By the soil sorption degree, heavy metals can be arranged in line in the order of decrease: $Pb > Cu > Zn > Cd > Co > Ni > Mn$. The comparison of the obtained results shows that some minerals introduction provides an increase for the soil buffer capacity in case of industrial inorganic pollutants.

Keywords: buffer capacity; field trials; heavy metals; protein spectrum; minerals

Introduction

The heavy metals (HM) get into the soil in various forms and compounds. After the expiry of a certain period of time they turn from the combined state into free, mobile form in the soil solution and migrate by food chain. Detoxication of the heavy metals is determined to a considerable extent by the ability of the soil itself to combine them into compounds. Recently some industrial regions of the country encountered the processes, connected with the soil degradation due to its aerotechnogenic pollution. Technogenic soil contamination with heavy metals has been found in the mining–metallurgical regions at the southeast part of Ukraine (National report on environment state in Ukraine, 1995 (1997); National report on the environmental conditions in Ukraine for 1998; Kharytonov et al., 2002).

The amount and the type of emissions in different places of the Dnipropetrovsk region depend on the type of industrial enterprises involved.

373

C. Mothersill et al. (eds.), Multiple Stressors: A Challenge for the Future, 373–380.

Major stationary emissions sources in the Dnipropetrovsk city are centralized heat and power production, metallurgical industries, chemical and petroleum industries, and so on.

One of the means of improving the soil fertility, buffer capacity, and physiological state is the usage of organic fertilizers and minerals. The purpose of our work was the studying of peculiarities of the heavy metals sorption improvement. The prospects to increase the buffer ability of the soils with aid of some amendments in the model condition were estimated.

Materials and Methods

The multipollution exposure assessment was made for the several Dnipropetrovsk city sites. Soil samples were taken on third distances (50, 100, 150 m) near several enterprises and one recreation zone as following: "Varnishes & Paints Enterprise" (VPE), "Tire Enterprise" (TE), Metallurgical Enterprise (ME), Dnipropetrovsk Heat Power Station (DHPS), Botany Orchard (BO). The multifactorial soil pollution influence on functional state of test radish seedlings was studied. The 4-day radish seedlings were treated in the water-soluble soil extracts in the Petri dishes. Protein spectra in the 4-day radish roots (variety "French breakfast") were determined with SDS electrophoresis.

The amendments of artificial fertilizers, minerals into contaminated soils were used as rehabilitation methods for protection from heavy metals pollution.

The investigations were carried out under laboratory and field conditions on the soils with different technogenic load. The objects under research were sand–clay mix and ordinary medium-loamy chernozem.

Field model experiment with nitrogen fertilizers on the heavy metals background: Several field trials consisted of control; heavy metals (HM); phosphates; and two nitrogen forms (carbamide or N_c and NH_4NO_3 or N_{NN}) were introduced into the soil separately and with HM.

Heavy metals salts ($ZnSO_4$, $CoSO_4$ $7H_2O$, $NiCl_2$, $Pb(NO_3)_2$, $CdSO_4$) were introduced in experimental plots, respectively, to 1 MPC. Plot square: 1 m²; repetition: 4 times.

Soil samples were prepared for heavy metal analyses by extraction with 1 N HCl.

The content of heavy metals in the samples was determined by flame atomic absorption spectrophotometer. Agrochemical analyses of the soil samples were made by the standard methods. The preliminary agrochemical analyses performed for the samples of ordinary medium-loamy black soil have shown that soil was unsalted. The humus content was equal to 4%.

The predominant metals by the muriatic extract data are iron and manganese. Their quantities in the soil are 1,150 and 230 mg/kg correspondingly.

The soil sorption properties and mobility of the heavy metals were determined at the static and dynamic conditions. In the laboratory experiments on studying of the physical–chemical features of heavy metals' sorption (Pb, Zn, Cu, Ni, Co, Mn), the soil-to-solution ratio was equal to 1:100. The metals' concentrations were changed in experimental conditions from 5 to 1000 mg/L. In the course of experiment, the corresponding aliquot of the heavy metal under study was added to the soil sample.

LABORATORY EXPERIMENTS WITH BARLEY AND SOYBEAN

The experiments were realized under artificial lighting. The exposure duration was equal to 20 days. The vessel capacity: 0.5 kg; repetition: 4 times.

The first two experiments were performed in succession one after another with the usage of quartz sand as the substrate and addition of Pryanishnikov nutrient mixture (control variant: 0). Montmorillonite clay was tested as the sorbent in proportion 1:4 with sand.

Heavy metals salts were introduced in the laboratory vessels in the level of 1 MPC (maximum permissible concentration of the heavy metals regarding to the sanitary methodological approaches in the NIS countries): $CuSO_4$–100 mg/kg, $ZnSO_4$–140 mg/kg, $MnSO_4$–1,500 mg/kg, $Pb(C_2H_3O_2)_2$ –30 mg/kg, $CdCl_2 \cdot 2,5H_2O$–3 mg/kg.

Laboratory experiment with soybean: the third experiment was made on the ordinary medium-loamy black soil (chernozem). The salts of zinc, copper, lead, cadmium, nickel, and cobalt were brought in doses of 1 and 5 MPC. The minerals were tested – zeolite and glauconite. In particular, the heavy metals salts were introduced in the laboratory vessels as following: $CuSO_4$–100 mg/kg, $ZnSO_4$–140 mg/kg, $Pb(NO_3)_2$–30 mg/kg, $CdSO_4$–3 mg/kg, $CoSO_4 \cdot 7H_2O$–25 mg/kg in the level of 1 MPC.

The fourth experiment was based on black soil, contaminated by lead, cadmium, and arsenic in dose of 5 MPC. The minerals tested (zeolite, glauconite, palygorskite, saponite, and bentonite) were introduced in doses of 3.3 g/kg (10 t/ha).

Results and Discussion

The multipollution exposure influence on functional state of 4-day test radish seedlings was studied for the several Dnipropetrovsk city sites (Fig. 1).

Comparative assessment for 2-year microfield experiments does not allow fixing negative heavy metals impact for the barley growth. It became

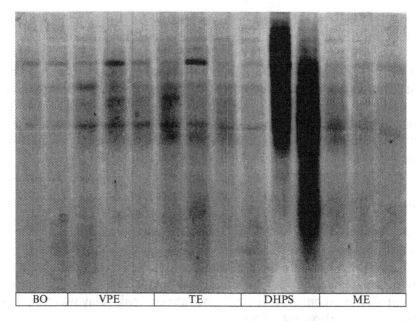

Fig. 1. The components composition of the easily soluble proteins of the radish roots grown in the water soil solutions of the samples taken in the different Dnipropetrovsk city sites: Botany Orchard (BO), "Varnishes & Paints Enterprise" (VPE), "Tire Enterprise" (TE), Dnipropetrovsk Heat Power Station (DHPS), Metallurgical Enterprise (ME). It has been shown that the variability of different level of the soil chemical pollution in the different technogenic zones leads to active reorganization in protein system. Taking it in account, we have conducted the microfield experiments to model multielement pollution in the two nitrogen fertilizers background. The polyelements spectrum was determined because of their priority existence in the industrial dust (Kharytonov et al., 2002; Babiy et al., 2003). The data obtained in the microfield experiments with barley growing in the different trials with heavy metals pollution are presented in Table 1.

TABLE 1. Barley yield the model experiments with multielements pollution in the two nitrogen fertilizers background, t/ha

Trial	Grain yield	
	First year	Second year
Heavy metals (1 MPC)	3.50 ± 0.08	1.30 ± 0.14
$(N_{NN} P)60$	5.15 ± 0.1	2.07 ± 0.14
(NcP)60	5.30 ± 0.1	2.26 ± 0.20
$(N_{NN} P)60 + BM$	5.35 ± 0.1	1.88 ± 0.14
(NcP)60 + BM	5.4 ± 0,04	2.09 ± 0.13
Control	3.4 ± 0,06	1.23 ± 0,07
LSD	0.24	0.38

possible due to some mitigation reasons including black soil high-level buffer capacity, barley firmness to the heavy metals, and so on. Probably the rest of nonabsorbed metals took part as essential trace elements fertilizers in the trials with NP background (Kabata-Pendias and Pendias, 1987).

The data obtained in the microfield experiment were the reason for ordinary chernozem buffer capacity assessment regarding to the heavy metals pollution. As the example, the values of maximum sorption capacity are given in the Table 2 for the ordinary chernozem in respect of Pb, Cu, and Zn.

The studying of physical–chemical peculiarities of the heavy metals sorption by black soil permits to determine that the toxic elements sorption largely depends on pH of the starting soil solution. The presence of carbonates in soil leads to the alkalization of the solution and, consequently, increases the chernozem ability to absorb metals. It was also noted that the sorption of lead and copper cations by soil is accompanied by certain pH reduction in the solution. The sharpest pH change was observed in case of doses with metal concentration being not great. The studying of the sorption dependence on the metal concentration in the solution has shown that under the change of initial element concentration from 5 to 50 mg/L the absorption is directly proportional to the metal concentration in the solution. For determination of the maximum saturation of chernozem with heavy metals, their concentrations were increased up to 1000 mg/L. By the data of the sorption isotherms, we calculated the values of maximum sorption capacity. Similar investigations were made also for the other metals. By the sorption degree, they can be arranged in line in the order of decrease: Pb > Cu > Zn > Cd > Co > Ni > Mn.

Thus, the problem detoxication of technogenically contaminated soils looks complicated enough in its perspective solution. One of the prospects on detoxication of technogenically contaminated soils is the usage of minerals – sorbents.

In Table 3 the registration data are presented for two model experiments, obtained after successive growing of barley and soybean on the sand substrate with different level of heavy metals contamination.

TABLE 2. Maximum sorption capacity of the ordinary chernozem

Sample	N_{max}, mg/g			Carbonates' Content, mg/100 g		Exchange Cations, mg/100 g	
	Pb	Cu	Zn	$CaCO_3$	$MgCO_3$	Ca	Mg
Ordinary chernozem	37.5	19.8	16.5	2.0	1.05	29.5	4.0

TABLE 3. Influence of claying on the plant toxicity reduction of heavy metals

| | Bioproductivity (g/vessel) | | | |
| | Barley | | Soybean | |
Trial	Sand	Sand/Clay	Sand	Sand/Clay
0	1.0	1.2	5.9	5.8
1 MPC HM	0.2	1.0	3.8	3.9
5 MPC HM	0.0	0.3	0.0	1.9
LSD	0.1	0.12	0.9	1.3

The positive effect of claying was mostly revealed in the variant with heavy metals introduction into the sand substrate in dose of 5 MPC. The copper, zinc, and lead concentration in biomass of barley and soya in the variant of 1 MPC was 2–5 times higher than the control variant, and for manganese – by the order of magnitude. The introduction the montmorillo-nite clay in the polluted sand substrate to an even greater degree promoted the reduction of the heavy metals entering into vegetative mass of the barley. The tested montmorillonite clay belongs to one of the rock types opened under development of minerals in Dnipropetrovsk region. In the following model experiment, the efficiency of heavy metals detoxication with the sorbents was studied. The result of the experiment on soybean growing in the ordinary chernozem is presented in Table 4.

The total action of heavy metals led to the reduction of soybean bio-productivity in the variant of 1 MPC HM by 25%, in the variant of 5 MPC HM by 49%, in comparison with the control. The zeolite acted better than glauconite for the increasing of the plants bioproductivity. The introduction of heavy metals in dose of 5 MPC in the soil led to the zinc and cobalt concentration increasing 4 times, cadmium – 3 times, copper – by 60%, nickel and lead correspondingly by 50% and 30%, in comparison with the control value. Zeolite has shown itself as the sorbent better for detoxication.

TABLE 4. Heavy metals content and bioproductivity of soybean in the experiment with the minerals

Trial	Biomass, g/vessel	Cu	Zn	Co	Ni	Pb	Cd
Control	4.1	5.4	14.0	4.0	6.0	8.0	0.4
1 MPC HM	3.1	6.4	26.0	4.0	8.0	8.0	0.6
5 MPC HM	2.1	8.6	54.0	18.0	9.0	10.0	1.2
5 MPC HM + zeolite	3.7	6.0	44.0	14.0	9.0	9.0	0.6
5 MPC HM + glauconite	3.4	8.0	56.0	15.0	9.0	9.4	0.8
LSD	1.0						

TABLE 5. Heavy metals (Pb, Cd, As) phytotoxicity
for the barley growing in black soil

Trials	Roots Biomass, g/vessel
Control	0.52
HM	0.37
Zeolite + HM	0.62
Saponite + HM	0.73
Bentonite + HM	0.58
Palygorskite + HM	0.67
Glauconite + HM	0.89
LSD	0.185

The registration data of the biomass change in barley roots in the fourth
vegetation experiment are presented in Table 5.

The comparison of the obtained results shows that the introduction of
the minerals provides increasing in the root mass at the level of 20–61% (in
respect of the control value).

Concluding Remarks

1. The soil pollution influence on functional state of radish seedlings was
   studied. It has been shown that the variability of chemical composition
   of soils from different polluted sites leads to active reorganization in cell
   protein system.

2. Comparative assessment for 2-year microfield experiments does not cause
   the selected heavy metals dose (1 MPC) negative impact for the barley
   growth. It became possible due to some reasons including black soil high-
   level buffer capacity, barley firmness to the heavy metals, and so on. Rest
   nonabsorbed metals took part as essential trace elements fertilizers in the
   trials with NP background.

3. By the soil sorption degree, heavy metals can be arranged in line in the
   order of decrease: $Pb > Cu > Zn > Cd > Co > Ni > Mn$.

4. The comparison of the obtained results shows that the introduction of
   the minerals provides an increasing for the soil buffer capacity in case of
   industrial inorganic pollutants. At the same time, these means allow to
   decrease the heavy metals accumulation by plants. Long-term application
   for some amendments can transform the pollutants into difficulty
   soluble forms in the soils. Thus model experiments showed distinct
   detoxication effectiveness of soil depending on dose, soil buffering, kinds
   of amendments, and kind of cultivated plant.

References

Babiy, A.P., Kharytonov, M.M., and Gritsan, N.P., Connection between emissions and concentrations of atmospheric pollutants. In: Melas D., and Syrakov D. (eds.), Air Pollution Processes in Regional Scale, *NATO Science Series, IV: Earth and Environmental Sciences.* Kluwer Academic Publishers, Printed in the Netherlands, 2003, pp. 11–19.

Kabata-Pendias, A., and Pendias, H. *Trace Elements in Soils and Plants*, 2nd ed. Levis Publ., Boca Raton, FL, 1987, 365 p.

Kharytonov, M., Gritsan, N., and Anisimova, L. Environmental problems connected with air pollution in the industrial regions of Ukraine. In: Barnes, I. (ed.), Global Atmospheric Change and its Impact on Regional Air Quality. *NATO Science Series, IV: Earth and Environmental Sciences.* Kluwer Academic Publishers, Vol. 16, 2002, pp. 215–222.

National report on the environmental conditions in Ukraine for 1998 prepared by the Ministry for Environment Protection and Nuclear Safety of Ukraine (CD-ROM publication, Web pages created by the Ministry for Environmental Protection and Nuclear Safety of Ukraine, www.grida.no).

National report on environment state in Ukraine, 1995 (1997). Raevsky Press, Kiev, 96 p. (Ukrainian).

CHAPTER 28

CANCER RISK ASSESSMENT IN DRINKING WATER OF IZMIR, TURKEY

SUKRU ASLAN[1] AND AYSEN TURKMAN[2]

[1]*Cumhuriyet University, Department of Environmental Engineering, 58140, Sivas/Turkey;*
e-mail: saslan@cumhuriyet.edu.tr
[2]*Dokuz Eylul University, Department of Environmental Engineering, 35160, Buca, Izmir/Turkey.*
e-mail: aysen.turkman@deu.edu.tr

Abstract: In this study, the occurrence of trihalomethanes (THMs) of the tap water in Izmir City was investigated and the lifetime cancer risk of THMs through oral ingestion, dermal absorption, and inhalation exposure were estimated. The total THMs in samples taken from the Tahtali and Balçova Water Treatment Plants (TWTP, BWTP), which are the major water sources of the Izmir City were about 72 and 88 µg/L, respectively. Chloroform existed at the highest concentrations in samples. Although the cancer risk evaluation of $CHBr_3$ through oral route for both sexes was below the EPA level, the highest lifetime cancer risk was originating due to $CHCl_2Br$; 5.2×10^{-5} and 4.3×10^{-5} for males and 4.76×10^{-5} and 5.8×10^{-5} for females, for the samples from BWTP and TWTP, respectively. While overall the average lifetime cancer risks through oral route, dermal absorption, and inhalation exposure for THMs were higher than the EPA acceptable risk of 10^{-6} by about 87, 340, and 5.7 times in the samples from TWTP and 99, 390, and 7.9 times in the samples from BWTP, respectively. The average lifetime cancer risk for THMs in both sources was in decreasing order, $CHCl_2Br$, $CHCl_2Br$, $CHCL_3$, and $CHBr_3$ for both sexes.

Keywords: chlorination; THMs; risk assessment; drinking water

Introduction

The chlorination has been reported and well-documented to cause the formation trihalomethanes (THMs) (Rook, 1974), which are only one class of halogenated disinfection by-products (DBPs) produced in water

C. Mothersill et al. (eds.), Multiple Stressors: A Challenge for the Future, 381–389.

chlorination. Until now, hundreds of halogenated DBPs have been identified but they account for less than 50% of the total organic halide concentration in drinking water. A number of these DBPs have adverse health effects (Vahala et al., 1999; Yoon et al., 2003). Nevertheless, the chlorination process leads to the formation of not only THMs but also numerous other volatile and nonvolatile organic halogen compounds (Peters et al., 1991; Sansebastione et al., 1995).

Current Regulations in the United States require drinking water utilities to maintain the total concentration of THMs below 80 µg/L. The United States Environmental Protection Agency (USEPA, 1986) proposed a two-stage DBP rule; in stage I, the maximum contaminant level (MCL) of 60 µg/L will be set for the total of five haloacetic acids (HAAs); in stage II, the MCLs are expected to be further decreased to 40 and 30 µg/L for THMs and HAAs, respectively. The European Community (EC) and Turkish drinking water quality standards for total THMs (TTHMs) are 100 and 150 µg/L, respectively (EECD, 1997; TMH, 2005). In this study, the experimental results were compared with the USEPA, EC, and Turkish standards.

The THMs concentration in drinking water changes seasonally due to the organic content of raw water and also the higher doses of chlorine used in warm season than in winter (Golfinopoulos, 2000; Shafy and Gruenwald, 2000; Lee et al., 2001; Golfinopoulos and Arhonditsis, 2002; Duonga et al., 2003; Tokmak et al., 2004). Although THMs concentration have been shown to increase within the distribution system in some studies (Golfinopoulos, 2000; Tokmak et al., 2004); Nissinen et al. (2002) and Duonga et al. (2003) found that levels of THMs decreased within the distribution systems since the residual chlorine was consumed by the deposits in the water pipes and a biotic or abiotic degradation of THMs under low oxygen conditions.

In this study, the occurrence of THMs level in the effluent water of Tahtali and Balcova Water Treatment Plants (TWTP, BWTP), the major drinking water sources of Izmir City, was investigated. In addition, a cancer risk evaluation through oral ingestion, dermal adsorption, and inhalation exposure was estimated.

Materials and Methods

STUDY AREA

Izmir is the third biggest city of Turkey with a population of 3.5 million. In Izmir, two water treatment plants, Tahtali and Balcova serving more than 50% of the total population. The TWTP is the biggest water source with capacity of $520{,}000\,m^3$/day and $1{,}733\,L/s$. BWTP has a capacity of $70{,}000\,m^3$/day. The treatment applied in the TWTP is aeration, prechlorination, coagulation, flocculation, sedimentation, and rapid sand filtration and postchlorination. Because of high-quality water in Balcova Dam, BWTP has no chemical treatment unit and water is aerated by cascade aeration system and then sent to rapid sand filtration. Treated water is disinfected with chlorine gas adjusted considering the raw water quality at the water treatment plant. NaOCl is injected several times at the intermediate chlorine pumping station in the distribution system for supplying additional residual chlorine.

WATER SAMPLE COLLECTION

In order to evaluate the quality of drinking water with respect to THMs in Izmir, water samples from the major surface water sources, TWTP and BWTP were taken from the influent and effluent for analysis in warm season (Fig. 1).

ANALYTICAL METHODS

Total organic carbon (TOC) was determined by using TOC analyzer (Dohrmann DC-190). Some of the data for TOC were not detected, since TOC analyzer (Dohrmann 190) used did not have the desired sensitivity to measure low organics concentrations.

THM ANALYSIS

The samples were analyzed using a gas chromatograph (GC) (Agilent 6890N) equipped with a mass selective detector (Agilent 5973 inert MSD) and a purge and trap (OI Analytical, Eclipse 4660). Analytes were purged from $5\,mL$ of aqueous solution with a helium flow rate of $40\,mL/min$ for 11 min maintaining the trap temperature at 30°C. Then, the volatile organic compounds were desorbed by opening the vents at 190°C for 0.5 min. Once the analytes have been desorbed, a bake step is programmed at 220°C for 10 min. The purge-and-trap system was directly connected to the gas chromatograph by means of a heated transfer line (at 200°C) in order to avoid analyte condensation during analysis. The analytical column connected to the system was a HP5-ms capillary column $30\,m \times 0.25\,mm$, 0.25-μm film thickness. Helium was used as the carrier gas at a flow rate of $1.0\,mL/min$ and $36\,cm/s$ linear velocity in the constant flow mode. The initial oven temperature was 40°C held for 3 min, then the temperature was programmed from 40°C to 120°C at 5°C/min for 1 min. Injection was in split mode (split ratio 1:40). The injector temperature was 240°C. The MSD was operated in electron impact

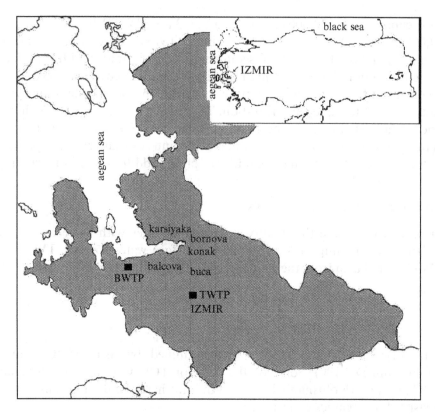

Fig. 1. Map of the Study area in Izmir City

(EI) mode with 70 eV electron energy. The ion source and quadropole temperatures were 230°C and 150°C, respectively. The MSD was run in selected ion monitoring. Compounds were identified based on their retention times (within ±0.05 min of the retention time of calibration standard), target, and qualifier ions. Identification was confirmed by the abundance of the qualifier ion relative to the target ion (relative intensity of qualifier ion). If the relative intensity in the sample spectrum was within ±20% of the relative intensity in the standard spectrum, identification was confirmed. Identified compounds were quantified using the external standard calibration procedure. A stock standard solution containing 54 VOCs at 2 mg/mL (Chem Service Inc., West Chester, PA, USA) was used to prepare the calibration standards. The analytical system (Purge and Trap-GC-MS) was calibrated using five levels (1.0–100 μg/L) of aqueous standard solutions prepared by spiking 1 μL diluted standard solutions in methanol into 5 mL deionized and prepurged water. In all cases, linear fit was good with $r^2 > 0.99$.

CANCER RISK ASSESSMENT OF THMS IN IZMIR CITY

The risk assessment of THMs in drinking water was carried out for the major sources, TWTP and BWTP in Izmir City. In this study, the approach used for human health cancer risk assessment used was based on the USEPA (1999, 2002) and adopted by Lee et al. (2004). The cancer risks of exposure to THMs through ingestion, dermal adsorption, and inhalation exposure were considered. In these estimations, body weight was taken as 72 and 65 kg and the average life span for males and females as 71 and 72 years, respectively (Tokmak et al., 2004). The average water ingestion rate considered for oral cancer risk calculations was 2.0 L/day, as assumed for adults by the USEPA. In inhalation risk calculations, the daily dose was calculated by assuming $20 m^3$ aspirated air per day (Lee et al., 2004). The risk cancer risk assessment for ingestion route, dermal absorption, and inhalation are described elsewhere (Lee et al., 2004). The chloroform concentration in air used for the estimation of risk through inhalation was calculated using a volatilization factor of 0.5 as suggested by USEPA (1991).

Results and Discussion

THMS IN CHLORINATED WATER IN IZMIR

Tap water samples were collected and analyzed from the TWTP and BWTP effluents. The resulting data were compared with the European Community (EC) drinking water quality standard of 100 µg/L for THMs, USEPA (1986), and 150 µg/L for TMH (2005). Although the THMs values measured were all below the EC and TMH guideline value, the total concentrations of THMs in tap water samples from the TWTP and BWTP were close to USEPA guideline values, and the total concentrations of THMs in tap water samples from TWTP (72 µg/L) and BWTP (88 µg/L) were close to the maximum total THM level of 100 µg/L set in many countries (Lee et al., 2004). THM concentrations, $CHCl_3$, $CHCl_2Br$, $CHClBr_2$, and $CHBr_3$ in chlorinated water were 27.6, 24.94, 16.65, and 3.01 µg/L in TWTP effluents and 37.8, 30.2, 17.2, 2.76 µg/L in BWTP effluents, respectively. Chloroform was the major species of THMs in the samples. The bromo-THMs were present in lower concentrations than $CHCl_3$.

EVALUATION OF THE CANCER RISK FOR THMS

The results of cancer risk evaluation of the THMs through oral route, dermal absorption, and inhalation exposure are shown in Figs. 2 and 3 for males and females in Izmir City, respectively. The calculation of cancer risk of

THMs through inhalation was carried out for chloroform (Lee et al., 2004). The cancer risk through oral ingestion for $CHCl_3$, $CHCl_2Br$, and $CHClBr_2$ for males and females were higher than 10^{-6} the USEPA's risk level. Among the THMs, the highest lifetime cancer risk was originating due to $CHCl_2Br$; 5.2×10^{-5} and 4.3×10^{-5} for males and 4.76×10^{-5} and 5.8×10^{-5} for females, for the samples from BWTP and TWTP, respectively. The cancer risk for the BWTP exceeded the EPA level by a factor of over 40 times for both sexes. While overall the average lifetime cancer risks through oral route, dermal absorption, and inhalation exposure for THMs in TWTP were 8.7×10^{-5}, 3.4×10^{-4}, and 5.7×10^{-6} higher than 10^{-6} by about 87, 340, and 5.7, respectively, in the samples from BWTP were 9.9×10^{-5}, 3.9×10^{-4}, and 7.9×10^{-6}, which were higher than the EPA acceptable risk by about 99, 390, and 7.9 times, respectively. The cancer risk evaluation of $CHBr_3$ through oral route for both sexes was below the EPA level by a factor 1.5 and 1.64 in the samples from the TWTP and BWTP, respectively.

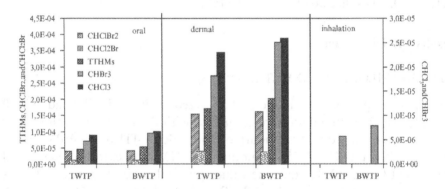

Fig. 2. Cancer risk for males from THMs, dermal contact, inhalation, and oral route in tap waters

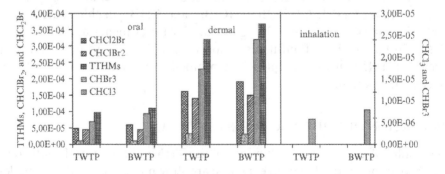

Fig. 3. Cancer risk for females from THMs, dermal contact, inhalation, and oral route in tap waters

The percentage contribution of average cancer risks of each THMs through oral route, dermal absorption, and inhalation exposure for male and female are shown in Figs. 4 and 5. As can be seen from the figures, while the lifetime cancer risks for $CHCl_2Br$ made the highest percentage contribution, $CHBr_3$ is below the EPA acceptable level of 10^{-6} for oral route for both sexes. The average lifetime cancer risk for THMs in both sources was in decreasing order for $CHCl_2Br$, $CHClBr_2$, $CHCL_3$, and $CHBr_3$ for both sexes.

Conclusions and Recommendations

Although THMs concentration have been expected to increase within the distribution systems, the level of THMs decreased in Izmir. This may be explained by the adsorption of some part of THMs in calcium and iron salts deposits in the distribution system as well as physicochemical decay.

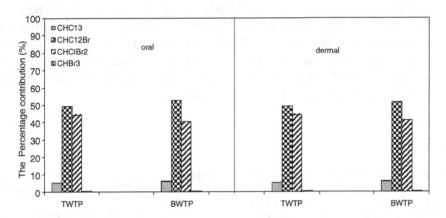

Fig. 4. The percentage contribution of average cancer risks of the each THMs for male

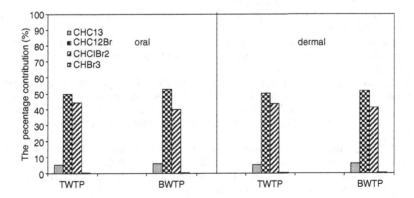

Fig. 5. The percentage contribution of average cancer risks of the each THMs for female

Even though the THM values measured were all below the EC and TMH guideline values, the total concentrations of THMs in tap water samples from the TWTP and BWTP close to USEPA guideline level. $CHCl_3$ was the major species of THMs in the samples, and followed by $CHCl_2$, $CHClBr_2$, and $CHBr_3$.

The cancer risks through oral ingestion by $CHCl_2Br$ in TWTP and BWTP for both sexes were higher than USEPA's risk level. The average lifetime cancer risk for THMs in both sources was in decreasing order, $CHCl_2Br$, $CHClBr_2$, $CHCl_3$, and $CHBr_3$ for both sexes. The total cancer risk analysis indicates that Izmir residents could get cancer from the daily intake of water.

In this study, cancer risk evaluation was performed using THMs concentrations in the water treatment plant effluents, but THMs level in the water changes seasonally and along the pipeline, so further study should be performed considering pipeline deposition which is the problem in Izmir City and the seasonal variations of the THMs level in the distribution system.

In order to reduce the cancer risk, ozonation instead of prechlorination is recommended in Tahtali Water Treatment Plant. In fact, Izmir Water and Sewerage Administration is considering this alternate. Activated carbon adsorption would decrease THMs also, but due to its additional cost, this alternate is not being considered at the moment.

References

Duonga, H. A., Berg, M., Hoanga, M. H., Phama, H. V., Gallardb, H., Guntenb, U. V. (2003) Trihalomethane formation by chlorination of ammonium- and bromide-containing groundwater in water supplies of Hanoi, Vietnam, Wat. Res. 37, 3242–3252.

EECD (European Economic Community Directive). (1997) Amended proposal for a Council Directive concerning the quality of water intended for human consumption-common position. In: Proceedings of the Council of the European Union, Directive 80/778/EEC, Com (97) 228 final 95/0010 SYN, Brussels.

Golfinopoulos, S. K. (2000) The occurrence of trihalomethanes in the drinking water in Greece, Chemosphere 41, 1761–1767.

Golfinopoulos, S. K. and Arhonditsis, G. B. (2002) Multiple regression models: A methodology for evaluating trihalomethane concentrations in drinking water from raw water characteristics, Chemosphere 47, 1007–1018.

Lee, S. C., Guo, H., Lam, S. M. J., and Lau, S. L. A. (2004) Multipathway risk assessment on disinfection by-products of drinking water in Hong Kong, Environ. Res. 94, 47–56.

Lee, K. J., Kim, B. H., Hong, J. E., Pyo, H. S., Park. S., and Lee, D. W. (2001) Study on the distribution of chlorination by-products (CBPs) in treated water in Korea, Wat. Res. 35(12), 2861–2872.

Nissinen, T. K., Miettinen, I, T., Martikainen, P. J., and Vartianien, T. (2002) Disinfection by-products in Finnish Drinking Waters, Chemosphere 48, 9–20.

Peters, R. J. B., Erkelens, C., De Leer, E. W. B., and De Galon, L. (1991) The analysis of halogenated acetic acids in Dutch drinking water, Wat. Res. 25, 473–477.

Rook, J. J. (1974) Formation of haloforms during chlorination of natural waters, J. Water Treat. Exam. 23, 234–243.

Sansebastione, G., Rebizzi, V., Bellelli, E., Sciarrone, F., Reverberi, Saglia, S. Braccaioli, A., and Camerlengo, P. (1995) Total organic halogenated organic (TOX) measurements in ground waters treated with hypochlorite and with chlorine dioxide. Initial results of an experiment carried out in Emilia-Romagna (Italy), Wat. Res. 29(4), 1207–1209.

Shafy, A. and Gruenwald, A. (2000) THM formation in water supply in South Bohemia, Czech Republic, Wat. Res. 34(13), 3453–3459.

TMH (Turkish Ministry of Health). (2005) Official Paper, Number 25730 (in Turkish).

Tokmak, B., Capar, G., Dilek, F. B., and Yetis, Ü. (2004) Trihalomethanes and associated potential cancer risks in the water supply in Ankara, Turkey, Environ. Res. 96, 345–352.

USEPA. (1986) Guidelines for Carcinogen Risk Assessment. US Environmental Protection Agency, Washington, D.C. EPA/600/8-87/045.

USEPA. (1991) Risk Assessment Guidance for Superfund – Vol. I, Part B. EPA/540/R-92/003, US Environmental Protection Agency, Washington, D.C.

USEPA. (1999) Guidelines for Carcinogen Risk Assessment. NCEA-F- 0644 (Revised draft) Risk Assessment Forum, US Environmental Protection Agency, Washington, D.C.

USEPA. (2002) Integrated Risk Information System (electronic data base). US Environmental Protection Agency, Washington, D.C. Available online: http://www.epa.gov/iris

Vahala, R., Langvik, V. A., and Laukkanen, R. (1999) Controlling adsorbable organic halogens (AOX) and trihalomethanes (THM) formation by ozonation and two-step granule activated carbon (GAC), filtration, Wat. Sci. Tech. 40(9), 249–256.

Yoon, J., Choi, Y., Cho, S., and Lee, D. (2003) Low trihalomethane formation in Korean Drinking Water, Sci. Total Environ. 20, 302(1–3):157–166.

CHAPTER 29

ENHANCED ADSORPTION OF ATRAZINE IN DIFFERENT SOILS IN THE PRESENCE OF FUNGAL LACCASE

NATALIA A. KULIKOVA[1], VALENTINA
N. DAVIDCHIK[2], ELENA V. STEPANOVA[2], AND
OLGA V. KOROLEVA[2]*

[1]Department of Soil Science, Lomonosov Moscow State
University, Leninskie Gory, GSP-2, 119992 Moscow, Russia
[2]Bach Institute of Biochemistry of the Russian Academy of
Sciences, Leninsky prospect 33, 119071 Moscow, Russia

Abstract: Adsorption–desorption behavior of atrazine was studied in three soils belonging to different soil geographical zones. Experimental investigations focused on the effect of laccase addition on adsorption and desorption of target chemical when present in solution. Addition of laccase resulted in a dramatic increase in adsorption of the atrazine. Desorption was little or negligible. Hysteresis, represented by hysteresis indices, was significantly enhanced upon laccase addition. Increases in Freundlich K_F values upon laccase addition were attributed to the covalent bonding of atrazine to soil organic matter by oxidative coupling mechanism.

Keywords: atrazine; adsorption; desorption; soil; laccase; oxidative coupling

Introduction

Agricultural chemicals, while often benefiting agricultural productivity, can have detrimental environmental effects when applied improperly. Herbicides, such as chlorotriazines in general or atrazine (2-chloro-4-ethylamino-6-isopropylamino-1,3,5-triazine) in particular, are most frequently detected in both terrestrial and aquatic ecosystems due to their relatively high mobility and persistence in the environment (Huber, 1993; Solomon et al., 1996; Graymore et al., 2001). Atrazine is one of the most widely used pesticides for the control of both grasses and broadleaf weeds in many crops and in nonagricultural

---

*To whom correspondence should be addressed. e-mail: koroleva@inbi.ras.ru

C. Mothersill et al. (eds.), Multiple Stressors: A Challenge for the Future, 391–403.

situations such as on railways, highways, and industrial sites. The concentration of atrazine can exceed the general quality standard for surface water, with maximum reported values of 9–25 mg/L in Europe (Croll, 1991; Legrand et al., 1991), 87–100 mg/L in North America (Steinheimer, 1993; Graymore et al., 2001), and up to 150 mg/L in Australia (Graymore et al., 2001).

Atrazine has several different fates in the environments including adsorption–desorption, migration, decomposition, and so on. (Gao et al., 1998; Mersie et al., 1998). However, adsorption–desorption processes are often supposed to control the other fates including both translocation into plants, movement in soil profiles or aqueous systems, abiotic or biological decay, and so on.

To describe the overall sorption characteristics of a particular soil at equilibrium with a range of contaminant concentrations adsorption isotherms, such as the Freundlich isotherm, are often used. The Freundlich model for atrazine adsorption can be described by the equation:

$$\text{Atrazine}_{\text{sorbed}} = K_\text{F}[\text{Atrazine}]^{n_\text{F}},$$

(1)

where $\text{Atrazine}_{\text{sorbed}}$ and [Atrazine] represent the solid- and aqueous-phase equilibrium concentrations of the target chemical, respectively; the Freundlich constant K_F is a measure of the sorption capacity of the sorbent; and the Freundlich n_F measures sorption linearity, which is related to the heterogeneity of sorption site energy.

There are several physical–chemical characteristics of the soil affecting adsorption and desorption of atrazine such as organic carbon (OC) contents, acidity, surface area, electric potential of the clay surface, and others (Lee et al., 1990), but organic matter are known to be the leading one among them (Seta and Karathanasis, 1997). This was numerously shown in the literature as positive correlation between atrazine Freundlich constant K_F or distribution coefficient Kd (equal to K_F when $n_\text{F}=1$) and content of OC (Mersie and Seybold, 1996; Moreau and Mouvet, 1997; Moorman et al., 2001; Ben-Hur et al., 2003). The Kd coefficient can be normalized to the fractional OC content of soil to give the relatively invariable partitioning coefficient, K_{OC} (Karickhoff, 1984), widely used for predictions of pesticides retention by soils in environmental and agricultural practice. In the database compiled by Wauchope et al. (2002), an average K_{OC} of 100 L/kg and a range of K_{OC} values from 38 to 174 L/kg are listed for atrazine; the upper limit for this value found in literature is 650 L/kg (Ben-Hur et al., 2003).

While adsorption is an important mechanism to examine, desorption is equally significant since it is directly related to herbicide runoff and leaching that leads to surface water and aquifer contamination, respectively. Barriuso et al. (1994) found that atrazine adsorption was generally reversible for smectites, suggesting weak van der Waals forces or hydrogen bonds between the atrazine molecules and the clay surfaces. For soil samples, particularly those high in OC, however, adsorption is typically less reversible. Many researchers have found that atrazine exhibits hysteresis effects, that is, less herbicide desorbs than is predicted by adsorption isotherms. Hysteresis reflects the portion of contaminant that is very strongly or irreversibly bound to the soil or OC, and has been recently described by the hysteresis index H (Celis et al., 1997):

$$H = n_{Fa} / n_{Fd},$$
(2)

where n_{Fa} and n_{Fd} are indexes of nonlinearity of the adsorption and desorption isotherms, respectively.

Although adsorption–desorption processes between atrazine and soil components have been well studied (Lesan and Bhandari, 2000; Ben-Hur et al., 2003; Magezan et al., 2003), chemical reactions between these pollutants and soil/sediment matrices have not been investigated in great detail. There is increasing evidence, however, that chemical interactions between organic pollutants and soil components, specifically reactions catalyzed by transition metal oxides or extracellular soil enzymes (so-called "oxidative coupling" or "oxidative polymerization reactions"), can significantly affect the fate of contaminants in soils and sediments, and potentially alter the associated health risks from the chemicals. The latter was numerously demonstrated for organic pollutants of phenol and amine structure (Voudrais and Reinhard, 1986; Wang et al., 1986; Nannipieri and Bollag, 1991; Bollag, 1992; Gianfreda and Bollag, 1994). It was elucidated, for example, that in the case of phenols the contaminants could be fixed or "trapped" within soil matrices as a result of enzymatic processes that imitate humus formation (ibid.).

Soils contain a large background concentration of extracellular enzymes that catalyze degradation and biosynthesis reactions in the soil environment. These organic catalysts are often protected against natural degradation by their attachment to soil constituents. Several soil enzymes including peroxidases, laccases, and polyphenol oxidases are capable of catalyzing chemical reactions that result in the polymerization of hydroxylated aromatic compounds (Sjobald et al., 1976; Suflita and Bollag, 1980; Sarkar et al., 1988). However, only single study on incorporation of atrazine by oxidative polymerization reactions into soils was found (Barriuso and Koskinen, 1996) though atrazine is known to belong to herbicides readily forming nonextractable bound residues, which are closely related to OC content in soil (Calderbank, 1989; Loiseau and Barriuso, 2002; Munier-Lamy et al., 2002).

Among enzymes catalyzing oxidative coupling, laccase seems to be the most important both from designing engineered remediation systems and controlling bound residues formation in nature points of view. This supposition is governed by the fact that this enzyme is only keeping relatively constant activity in soils all over the year (Criquet et al., 2000; Mayer and Staples, 2002). Besides, laccases possess wide substrate specificity catalyzing the oxidation of a variety of aromatic hydrogen donors with the concomitant reduction of oxygen to water. Moreover, laccases have been reported to oxidize many recalcitrant substances, such as chlorophenols, PAHs, lignin-related structures, organophosphorous compounds, nonphenolic lignin model compounds, phenols, and aromatic dyes in the presence of appropriate redox mediator (Mayer and Staples, 2002). The main feature of laccase action is its ability to produce a free radical from a suitable substrate. The mechanism of xenobiotics bioremediation using laccase is based on this reaction.

The aim of the study was to evaluate influence of fungal laccase on atrazine adsorption–desorption behavior in different soils.

Experimental Part

SOILS

Three soil surface samples (0–5 cm) used in the adsorption–desorption experiments were collected from forest or agricultural sites of different soil geographical zones and included sod-podzolic soil (Moscow region, forest site; related to Spodosols), gray forest soil (Tula region, agricultural site; related to Alfisols), and chernozem (Kursk region, agricultural site; related to Mollisols). Soil samples were air-dried, passed through a 1 mm sieve, and stored at room temperature prior to testing. Their physical–chemical properties are summarized in Table 1.

Based on the soil texture analysis, sod-podzolic and gray forest soil were classified as silt loam, and chernozem was classified as silt clay loam.

ASSAY OF LACCASE ACTIVITY IN SOILS

To perform assay of laccase activity, soils under study samples were extracted with 140 mM sodium pyrophosphate (pH 7.1) at soil to solution ratio 1:10 (w/w) at 25°C for 24 h in accordance with Bonmati et al. (1998) followed by enzyme activity assay using PerkinElmer spectrophotometer (USA) in temperature-controlled 1-cm cuvette at 25°C.

Laccase activity was measured using syringaldazine (4-hydroxy-3, 5-dimethoxybenzaldehyde azine, Sigma, USA) as a substrate. The reaction mixture contained 3 mL of 0.1 M syringaldazine in 0.1 M acetate buffer (pH 4.5). The reaction was initiated by the addition of 0.05 mL of

TABLE 1. Physical and chemical properties of soils

| Soil | pH | OC, % | Exchangeable bases, mM eq./100 g | | | |
			Ca	Mg	K	Na
Sod-podzolic	4.7±0.1	3.77±0.02	3.1±0.5	1.2±0.5	0.27±0.03	nd[a]
Gray forest	6.7±0.1	2.01±0.02	11±2	9±2	0.49±0.03	nd
Chernozem	6.6±0.1	5.79±0.02	31±4	7±5	0.47±0.06	0.4±0.2

[a] Below detection limit

sodium pyrophosphate extract from soil and the increase in absorbance was monitored at 530 nm. One unit of enzymatic activity was defined as the amount of enzyme required to cause a change in absorbance of 0.1 per minute at 25°C.

CHEMICALS

Atrazine

Atrazine (99.97%) was purchased from Dr. Ehrenstorfer GmbH (Germany). Stock solution of 1 g/L was prepared in methanol (Sigma, USA) and stored in the dark at 4°C.

Laccase

For this study, laccase (EC 1.10.3.2) form the strain of basidiomycetes *Coriolus hirsutus* 075 (Wulf. Ex. Fr.) Quel. of the *Polyporaceae* family producing high-activity extracellular laccase was used. The strain from the Collection of the Komarov Botanical Institute, Russian Academy of Sciences (St. Petersburg) was kindly provided by Dr. V. Gavrilova. Extracellular laccase was isolated from the culture medium and purified in accordance with (Koroleva (Skorobogat'ko) et al., 1998). The control of laccase homogeneity was carried out with poly-acrylamide gel (PAAG) electrophoresis under nondenaturing conditions as described in Westermeier (2001). The pH optima of the major isoenzyme determined using pyrocatechol as substrate was 4.5 (Koroleva (Skorobogat'ko) et al., 1998). Stock solution of 17.5 g/L was prepared in 50 mM potassium phosphate buffer (pH 5.0) immediately before use.

STUDY OF ATRAZINE DEGRADATION IN THE PRESENCE OF LACCASE

To assess changes in adsorption–desorption behavior of atrazine due to the possibility of atrazine degradation in the presence of enzyme, coincubation of atrazine and laccase was conducted in 50 mM potassium phosphate buffer (pH 5.0) at $27\pm1°C$. Atrazine concentration was 5 mg/L, laccase concentration and activity were 3.5 g/L and 10^{-6} M, respectively. After 24, 48, 75, 96, 120, 144, and 168 h of incubation solutions were sampled, filtered through cellulose filter with cutoff 5 kD, and subjected for atrazine analysis using high-performance liquid chromatography (HPLC).

ADSORPTION–DESORPTION EXPERIMENTS

Sorption–desorption experiments were conducted at six sorbate concentrations. The initial atrazine concentrations were 1, 2, 3, 5, 8, and 10 mg/L. Equilibrium time was determined as 24 h by preliminary experiments.

Adsorption experiments were conducted in plastic centrifuge tubes closed with caps. Tubes were first filled with 2 g of soil, and then 10 mL 50 mM potassium phosphate buffer (pH 5.0) was added. The tubes were shaken vigorously, placed on rotator Intelli-Mixer RM-2 (Elmi, Latvia), and left at $27\pm1°C$ for equilibration between soil and phosphate buffer. After 24 h, atrazine solution was added correspondingly with the scheme of experiments. If required, laccase solution was added to create final concentration of 3.5 g/L. Tree replicates were used for each initial atrazine concentration. Then tubes were shaken vigorously and placed on rotator for 24 h at $27\pm1°C$. At the end of the equilibration period, the tubes were centrifuged at 1,200 g for 30 min to separate the solid and liquid phases. A 0.05-mL aliquot was removed from supernatant, filtered through cellulose filter with cutoff 5 kD, and subjected for atrazine analysis using HPLC.

Remaining atrazine solution was then removed with a pipette and replaced with clean potassium phosphate buffer for desorption. The tubes containing clean buffer were recapped and placed on the rotator at $27\pm1°C$ for atrazine desorption. After 24 h, the tubes were centrifuged and the supernatant was analyzed for desorbed atrazine using HPLC. The liquid was then removed, and the procedure was repeated until atrazine concentrations fell below the detection limit. In general, seven desorption steps were performed.

ATRAZINE ASSAY USING HPLC ANALYSIS

The determination of atrazine and its main metabolites desethylatrazine, desethyldesisopropylatrazine, 2-hydroxyatrazine, desethyl-2-hydroxyatrazine, and desisopropyl-2-hydroxyatrazine was conducted using HPLC technique as described elsewhere (Carabias-Martinez et al., 2002) with little modifications. HPLC was performed on a chromatograph Beckman Coulter System Gold (USA), equipped with two pumps, a membrane degasser, and a diode-array detector. The reverse phase C18 Ultrasphere ODS Beckman column (USA) 4.6 mm × 25 cm was used for separation. The diode array detector was set at 210, 225, 230, and 245 nm. The mobile phase consisted of acetonitrile (solvent A) – 1 mM phosphate buffer at pH 7.0 (solvent B) linear gradient from 2% to 98% of solvent A in 35 min. Flow rate was 1 mL/min and the volume injected was 20 µL. The analytical column was kept at constant temperature 30°C.

Results and Discussion

LACCASE ACTIVITY IN SOILS USED

No laccase activity was detected under selected conditions. That was probably due to air-drying and long-term storage of soil samples. The obtained results allowed us supposing that only introduced laccase influenced on adsorption–desorption behavior of atrazine rather than inherent soil laccase.

DEGRADATION OF ATRAZINE IN THE PRESENCE OF LACCASE

After coincubation of laccase and atrazine, no changes in atrazine concentration were detected. Data on atrazine degradation in the presence of fungal laccase are given in Fig. 1.

As it can be seen from Fig. 1, atrazine concentration in the solution did not change all over the time of the experiment. At that, atrazine metabolites were not detected in the solution. That finding was evident for the fact that laccase did not induce atrazine degradation in the solution under selected conditions.

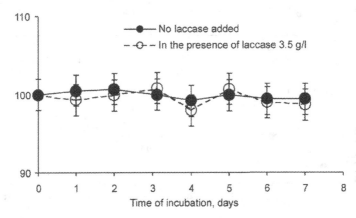

Fig. 1. Scavenging of atrazine in solution in the presence (empty markers) and absence (filled markers) of fungal laccase. Bars represent standard deviations

NONENZYMATIC ADSORPTION–DESORPTION OF ATRAZINE
IN DIFFERENT SOILS

Adsorption–desorption isotherms for atrazine in different soils are presented in Fig. 2. A summary of results for atrazine adsorption–desorption data in different soils studied is presented in Table 2. Adsorption isotherms were not linear as indicated by the Freundlich $n<1$ in all the cases. The latter was indicative of adsorption by heterogeneous media where high energy sites were occupied first, followed by adsorption at lower energy sites (Weber et al., 1996).

Atrazine Freundlich adsorption constants K_F for studied soils varied in the range 0.81–5.54, which is in accordance with previously published data

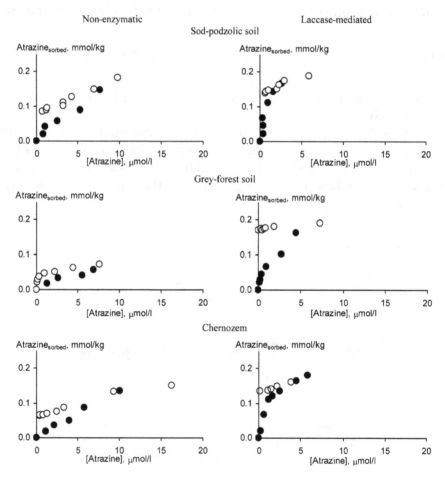

Fig. 2. Nonenzymatic and laccase-mediated adsorption (filled markers) and desorption (empty markers) of atrazine in three different soils

TABLE 2. Freundlich isotherm parameters (n and K_F) and hysteresis index H values for atrazine adsorption–desorption in different soils without fungal laccase addition

Soil	Adsorption			Desorption			
	K_F	n	R^2	K_F	n	R^2	H
Sod-podzolic	4.51	0.73	0.93	0.57	0.27	0.89	2.7
Gray forest	0.81	0.56	0.96	1.69	0.53	0.94	1.1
Chernozem	5.54	0.83	0.98	0.36	0.23	0.89	3.6

(Lesan and Bhandari, 2000). As a consequence of marginally higher organic carbon content, chernozem soil exhibited a larger sorption capacity for atrazine than other soils.

Desorption of atrazine from the soils was rather similar for sod-podzolic soil and chernozem while that was different significantly from gray forest soil where only negligible desorption hysteresis was observed. Desorption hysteresis was revealed in all soils with the chernozem soil manifesting a significantly greater hysteresis ($H = 3.6$) and gray forest soil exhibiting the smallest value of hysteresis ($H = 1.1$). The value of H close to 1 in case with gray forest soil reflected the fact that almost all adsorbed atrazine was readily desorbed. On the other hand, $H > 1$ was observed likewise for sod-podzolic and chernozem that indicated for partially irreversible adsorbed atrazine. The hypothesis on partial irreversible atrazine sorption could also be confirmed by the finding that no products of atrazine degradation were found over duration of the experiment. The latter meant that atrazine scavenging in solution resulted only from adsorption but not degradation process.

The gray forest soil was collected from the field site and contained freshly decaying biomass and consequently humic substances lacking in aromatic structures as compared with the other soils. This may be the reason why only negligible desorption hysteresis was observed in this case as hydrophobic binding between humic substances and atrazine was supposed to be a leading mechanism of their interaction.

LACCASE-MEDIATED ADSORPTION–DESORPTION OF ATRAZINE IN DIFFERENT SOILS

Adsorption–desorption data for atrazine in different soils in the presence of fungal laccase are presented in Fig. 2. A summary of results for atrazine adsorption–desorption behavior in different soils in the presence of fungal laccase is presented in Table 3. Similar to atrazine adsorption by soils without enzyme addition, isotherms in the presence of fungal laccase were not linear

TABLE 3. Freundlich isotherm parameters (n and K_F) and hysteresis index H values for atrazine adsorption–desorption in different soils in the presence of fungal laccase

Soil	Adsorption			Desorption			
	K_F	n	R^2	K_F	n	R^2	H
Sod-podzolic	5.80	0.61	0.74	0.36	0.13	0.93	4.7
Gray forest	3.13	0.56	0.99	0.21	0.02	0.74	28.5
Chernozem	6.80	0.66	0.86	0.22	0.06	0.68	10.9

either. The values of Freundlich n value varied in the range 0.56–0.66 being less than those obtained from adsorption experiments without laccase. The latter was indicative for complex adsorption process of atrazine by soils in this case rather than simple partitioning.

As it can be seen from Fig. 2 and Tables 2 and 3, laccase introduction resulted in a dramatic increase in atrazine adsorption. Atrazine Freundlich constants K_F for studied soils in the presence of fungal laccase varied in the range 3.13–6.80 exceeding those for adsorption experiments without enzyme.

Desorption of atrazine was also extremely reduced in the presence of fungal laccase. In the presence of enzyme atrazine was observed to desorb to a significantly greater extent than in case when laccase was not added, especially in soil where the lowest organic matter content was found. H values for atrazine adsorption mediated by laccase varied from 4.7 to 28.5, which exceeded those values 2–27 times for nonenzymatic adsorption. It appeared that adsorption of atrazine in all soils studied was enhanced when laccase was present. Like in case of nonenzymatic adsorption, no products of atrazine degradation were found over duration of the experiment. Therefore, enhanced laccase-mediated adsorption of atrazine in soils and increased hysteresis of that process could be attributed to the covalent bonding of atrazine to soil organic matter by oxidative coupling mechanism. This is an important observation as it points to the potential of laccase enzymes to effectively retard the mobility of atrazine in soils and groundwater.

Conclusions

The study was aimed to present and discuss results from ongoing studies evaluating the effects of fungal laccase in the fate of atrazine adsorption and desorption in soils. Adsorption–desorption of atrazine was studied in three soils with varying properties including organic matter types and contents. Effect of the addition of laccase enzyme on the adsorption and desorption behavior of atrazine was evaluated. It was observed that adsorption isotherms for all cases were nonlinear, indicative of adsorption by heterogeneous media where high energy sites were occupied first, followed by adsorption at lower energy sites. Atrazine desorption in sod-podzolic soil and chernozem exhibited some hysteresis in the absence of laccase; gray forest showed little hysteresis ($H = 1.1$) for atrazine. Addition of fungal laccase resulted in dramatic increases in adsorption for all soils studied. Freundlich K_F values increased with laccase addition possibly as a result of enzyme-catalyzed oxidative coupling of atrazine to soil organic matter. Desorption was reduced to little or negligible: hysteresis, as reflected by the H values, was significantly enhanced upon enzyme addition. Hysteresis indexes derived from laccase-mediated adsorption experiments 2–27 times exceeded those values for nonenzymatic adsorption. Results of this study illustrate the capability of engineered humification processes, such as those catalyzed by laccase enzymes, to significantly alter the fate and transport of atrazine in soil, sediment, and groundwater systems.

Acknowledgment

This study was financially supported by the Russian Foundation for Basic Researched (RFBR) under Grant No. 04-04-49679.

References

Barriuso, E., and Koskinen, W.C., 1996, Incorporating nonextractable atrazine residues into soil size fractions as a function of time, *Soil Sci. Soc. Am. J.* 60:150–157.

Barriuso, E., Laird, D.A., Koskinen, W.C., and Dowdy, R.H., 1994, Atrazine desorption from smectites, *Soil Sci. Soc. Am. J.* 1632–1638.

Ben-Hur, M., Letey, J., Farmer, W.J., Williams, C.F., and Nelson, S.D., 2003, Soluble and solid organic matter effects on atrazine adsorption in cultivated soils, *Soil Sci. Soc. Am. J.* 67:1140–1146.

Bollag, J.-M., 1992, Decontaminating soils with enzymes, *Environ. Sci. Technol.* 26:1876–1881.

Bonmati, M., Ceccanti, B., and Nannipieri, P., 1998, Protease extraction from soil by sodium pyrophosphate and chemical characterization of the extracts, *Soil Biol. Biochem.* 30(14):2113–2125.

Calderbank, A., 1989, The occurrence and significance of bound residues in soil, *Rev. Environ. Contam. Toxicol.* 108:71–98.

Carabias-Martinez, R., Rodriguez-Gonzalo, E., Herrero-Hernandez, E., Roman, F.J.S.-S., and Flores, M.G.P., 2002, Determination of herbicides and metabolites by solid-phase extraction and liquid chromatography. Evaluation of pollution due to herbicides in surface and groundwaters, *J. Chromatogr. A* 950:157–166.

Celis, R., Cornejo, J., Hermosin, M.C., and Koskinen, W.C., 1997, Sorption of two polar herbicides in soils and soil clays suspensions, *Soil Sci. Soc. Am. J.* 61:436–443.

Criquet, S., Farnet, A.M., Tagger, S., and Le Petit, J., 2000, Annual variations of phenoloxidase activities in an evergreen oak litter: Influence of certain biotic and abiotic factors, *Soil Biol. Biochem.* 32:1505–1513.

Croll, B.T., 1991, Pesticides in surface and underground waters, *J. Inst. Water Env. Manag.* 5:389–395.

Gao, J.P., Maguhn, J., Spitzauer, P., and Kettrup, A., 1998, Sorption of pesticides in the sediment of the Teufelsweiher poond (southern Germany). I: Equilibrium assessments, effect of organic carbon content and pH. *Wat. Res.* 32:1662–1672.

Gianfreda, L., and Bollag, J.-M., 1994, Effect of soils on the behavior of immobilized enzymes, *Soil Sci. Soc. Am. J.* 58:1672–1681.

Graymore, M., Stagnitti, F., and Allinson, G., 2001, Impacts of atrazine in aquatic ecosystems, *Environ. Int.* 26:483–495.

Huber, W., 1993, Ecotoxicological relevance of atrazine in aquatic systems, *Environ. Toxicol. Chem.* 12:1865–1881.

Karickhoff, S.W., 1984, Organic pollutant sorption in aquatic systems, *J. Hydraul. Eng. Div. Am. Soc. Civ. Eng.* 110:707–735.

Koroleva (Skorobogat'ko), O., Stepanova, E., Gavrilova, V., Morozova, O., Lubimova, N., Dzchafarova, A., Yaropolov, A., and Makower, A., 1998, Purification and characterization of the constitutive form of laccase from the basidiomycete *Coriolus hirsutus* and effect of inducers on laccase synthesis, *J. Biotechnol. Appl. Biochem.* 28:47–54.

Lee, L.S., Rao, P.S.C., Nkedi-Kizza, P., and Delfino, J.J., 1990, Influence of solvent and sorbent characteristics on distribution of pentachlorophenol in octanol-water and soil-water systems, *Environ. Sci. Technol.* 24:654–661.

Legrand, M.F., Costentin, E., and Bruchet, A., 1991, Occurrence of 38 pesticides in various french surface and ground waters, *Environ. Technol.* 12:985–996.

Lesan, H.M., and Bhandari, A., 2000, Evaluation of atrazine binding to surface soils, in: *Proceedings of the 2000 Conference on Hazardous Waste Research*, pp. 76–89.

Loiseau, L., and Barriuso, I., 2002, Characterization of the atrazine's bound (nonextractable) residues using fractionation techniques for soil organic matter, *Environ. Sci. Technol.* 36:683–689.

Magezan, G.N., Bolan, N.S., and Lee, R., 2003, Adsorption of atrazine and phosphate as affected by soil depth in allophanic and non-allophanic soils, *N Z J Agric. Res.* 46:155–163.

Mayer, A.M., and Staples, R.C., 2002, Laccase: New functions for an old enzyme, *Phytochemistry* 60:551–565.

Mersie, W., and Seybold, C., 1996, Adsorption and desorption of atrazine, deethylatrazine, deisopropilatrazine, and hydroxyatrazine on levy wetland soil, *J. Agric. Food Chem.* 44:1925–1929.

Mersie, W., Seybold, C., Tierney, D., and McName, C., 1998, Effect of temperature, disturbance and incubation time on release and degradation of atrazine in water columns over two types of sediments, *Chemosphere* 36:1867–1881.

Moorman, T.B., Jayachandran, K., and Reungsang, A., 2001, Adsorption and desorption of atrazine in soils and subsurface sediments, *Soil Sci.* 166:921–929.

Moreau, C., and Mouvet, C., 1997, Sorption and desorption of atrazine, deethylatrazine, and hydroxyatrazine by soil and aquifer solids, *J. Environ. Qual.* 26:416–424.

Munier-Lamy, C., Feuvrier, M.P., and Choné, T., 2002, Degradation of ^{14}C-atrazine bound residues in brown soil and rendzina fractions. *J. Environ. Qual.* 31:241–247.

Nannipieri, P., and Bollag, J.-M., 1991, Use of enzymes to detoxify pesticide-contaminated soils and waters, *J. Environ. Qual.* 20:510–517.

Sarkar, J.M., Malcolm, R.L., and Bollag, J.-M., 1988, Enzymatic coupling of 2,4-dichlorophenol to stream fulvic acid in the presence of oxidoreductases, *Soil Sci. Soc. Am. J.* 52:688–694.

Seta, A.K., and Karathanasis, A.D., 1997, Atrazine adsorption by soil colloids and co-transport through subsurface environment, *Soil Sci. Soc. Am. J.* 61:612–617.

Sjobald, R.D., Minard, R.D., and Bollag, J.-M., 1976, Polymerization of 1-naphthol and related phenolic compounds by an extracellular fungal enzyme, *Pestic. Biochem. Physiol.* 6:457–463.

Solomon, K.R., Baker, D.B., Richards, R.P., Dixon, K.R., Klaine, S.J., La Point, T.W., Kendall, R.J., Weisskopf, C.P., Giddings, J.M., Giesy, J.P., Hall, Jr., L.W., and Williams,

W.M., 1996, Ecological risk assessment of atrazine in North American surface waters, *Environ. Toxicol. Chem.* 15:31–76.

Steinheimer, T.R., 1993, HPLC determination of atrazine and principal degradates in agricultural soils and associated surface and ground water, *J. Agric. Food Chem.* 41:588–595.

Suflita, J.M., and Bollag, J.-M., 1980, Polymerization of phenolic compounds by a soil-enzyme complex, *Soil. Sci. Soc. Am. J.* 45:297–302.

Voudrais, E.V., and Reinhard, M., 1986, Abiotic organic reactions at mineral surfaces, *Geochem. Process. Mineral Surf.* 323:462–486.

Wang, T.S.C., Huang, P.M., Chou, C.-H., and Chen, J.-H., 1986, The role of soil minerals in the abiotic polymerization of phenolic compounds and formation of humic substances, *Interact. Soil Miner. Nat. Org. Microb.* 17:251–281.

Wauchope, R.D., Yeh, S, Linders, H.J., Kloskowski, R., Tanaka, K., Rubin, B., Katayama, A., Kördel, W., Gerstl, Z., Lane, M., and Unsworth, J.B., 2002, Review. Pesticide soil sorption parameters: Theory, measurement, uses, limitations and reliability, *Pest Manage. Sci.* 58:419–445.

Weber, Jr., DiGian, W.J., and DiGiano, F.A., 1996, *Process Dynamics in Environmental Systems*, John Wiley & Sons, Inc., New York.

Westermeier, R., 2001, *Electrophoresis in Practice.* VCH Verlagsgesellschaft, Weinheim and VCH Publishers Inc., New York, p. 368.

CHAPTER 30

PROBLEM OF MICROELEMENTOZE AND TECHNOLOGY ALLOWING ITS ELIMINATION WITH THE HELP OF GEOTHERMAL MINERALIZED SOURCES: NEW TECHNOLOGY OF MICROELENTOZE ELIMINATION

K.T. NORKULOVA*

Tashkent State Technical University, University Street, 2, Tashkent, Uzbekistan

Abstract: The purpose of this leaflet is to give information on ways of manufacturing of mineralized specifically concentrated waters from natural waters containing valuable mineral substances and possessing, in particular, high physiological value, in order to eliminate iodine deficiency.

Keywords: biosphere endemism; anthropogenic intervention; synergy; iodine deficiency; fractions; enrich; brome

Introduction

Research of microelementoze distribution and its elimination are major problems for the prognosis and treatment of flora and fauna condition of the chosen site of the Earth (Norkulova et al., 2003; Norkulova, 2002). In this respect Central Asia is an original area for study of many features, as it is remote enough from sea coast and is by itself an example of biosphere endemism of the Earth. Many processes on a background of influence of anthropogenic intervention have the special parameters in such ecosystems. It is necessary to study these processes on the basis of applied scientific and practical works with further distribution of certain recommendations.

In 2004 in the Tashkent state technical university the Republican scientific and practical conference on a theme "Technosphere, man and microelements " was held. The results of the conference have shown that only combining of efforts and integrated recommendations can target the main directions of practical measures on ecosystems improvement (Norkulova, 2005). These issues can be solved with the help of modern technologies. The

---

*To whom correspondence should be addressed. e-mail: narkulova_tstu@yahoo.com, b_orif_sh@mail.ru

C. Mothersill et al. (eds.), Multiple Stressors: A Challenge for the Future, 405–408.
© 2007 *Springer.*

reasonable strategy constructed on the basis of various directions of
synergy is necessary. As an example, we shall consider a problem of iodine
deficiency. During many millions of years water carries away with itself iodine
and other microelements to the bottom levels of the Earth, from mountains to
lowlands and reservoirs, that is, continuously the level of iodine concentration
at ocean raises, and on the huge territories of land reduces.

On Iodine Deficiency in Central Asia

The mankind began to notice iodine deficiency of the population and
domestic animals, therefore the methods of elimination of iodine deficiency
have appeared. For the necessary sites of the Earth to give maximal flour-
ishing of life the normalization in distribution of microelements (ME), in
particular, iodine, is necessary.

Despite of the problem of iodine deficiency in the region, there are huge
stocks of underground sources of iodine and other substances, which joining
ME circulation, can improve the condition of Central Asia ecosystem. Search
and strategy of optimum methods of scientific development, supervision, and
conclusions of research works allow finding practically implemented variants
for ecosystem treatment. The merge of ecology, technology, hydrogeology,
medicine, botany, and other disciplines on the basis of scientific development
opens new opportunities of ecology development.

Models of Microelements Migrations

The models of microelements migrations in closed ecosystems are analog of
the nonlinear cooperative phenomena in self-developing systems, where "slow"
and "fast" dynamic variables participate. Here we notice anthropogenic intervention, measured during tens of years, and also parallel natural processes
lasting hundred millions years. The partial management of ME condition in
eco-chain allows to organize the relatively balanced levels of concentration at
the expense of dynamic accumulation in the peak of a pyramid.

The microelements consumed by the people and fauna as a result of
circulation of substances, get into the structure of ground, water, and
through them into flora, partially coming back. If their leaving part comes
back, in eco-chain the necessary level of ME concentration is supported at
the expense of controlled anthropogenic ME balancing.

The system of self-development in the distribution of ME ecosystem at
anthropogenic influence starts deviating from that picture of independent
ecosystem generated by conditions of the wild nature.

The anthropogenic intervention slowly changes the picture of distribution
and acts as a small dynamic variable of the external background. At certain

approximations, it is possible to present evolution of concentration in the chosen object j of that ME on the background of other microelement communities which are taking place in the given ecosystem, that is

$$\frac{\partial n}{\partial t} j = \hat{L}\left(n_j, n_k, f_i, f_k, t, \vec{x}\right),$$ (1)

where
$\hat{L}$ is a certain nonlinear differential operator
t is the time measured by years
$\vec{x} = \vec{x}\,(x, y, z)$ are coordinates
f_i is the function of the influence of an anthropogenic origin.

   If we shall set a kind of function or $f_j - (t, \vec{x})$ equation of return dependence, that is

$$\frac{\partial f_j}{\partial t} = \hat{G}\left(n_j, n_k, t, \vec{x}\right),$$ (2)

where $\hat{G}$ is a nonlinear operator, then in view of boundary and entry conditions we shall receive a system for researching the nonlinear cooperative phenomena, that is, one of the most interesting analytical tasks of synergy.

   In many regions, under ground, on the depth of 2–3 km there are geothermal waters, in which spectrum and concentration by dissolved microelements sometimes exceed the usual sea water.

New Method for Enriching Specifically the Chosen Objects

We have developed and patented a method allowing to enrich specifically the chosen objects and to receive ME concentration as pastes, and mineral water with fine solid salt fractions. The offered way concerns the manufacture of mineralized specifically concentrated waters from natural waters containing valuable mineral substances and possessing, in particular, high physiological value. So, in the course of concentration the offered minitechnology allows to receive some fractions of various salts. The application of special methods of cooling and dehydration of the oversaturated complex solutions of salts allowed dividing them into fractions, in particular, at the expense of differentiation on groups the concentrates enriched by brome, and also by iodine from iodine–brome sources have been obtained.

   According to numerous scientific sources, Uzbekistan has mineral waters of various chemical compositions from more than 1,600 deep chinks. This circumstance offered by authors for using to develop the mineral waters resources. The put task is solved so, that obtained by the authors mineralized highly concentrated water at general common mineralization 420,000–460,000 mg/dm^3 contains Na-cations and Ca – 40,000–50,000 mg/dm^3, magnesium cations – 26,000–30,500 mg/dm^3, calcium cations – 68,000–80,000 mg/dm^3, hydrocarbonates – 60–75 mg/dm^3, sulfates – 90–110 mg/dm^3, anions of chlorine – 260,000–300,000 mg/dm^3, anions of iodine – 450–550 mg/dm^3, anions of brome – 250–350 mg/dm^3.

The concentrate of mineral water obtained by the authors contains micro-elements in the tied condition. On organoleptic parameters the concentrate represents a colorless transparent liquid without a smell, with some natural sediment of mineral substances and taste characteristic for a complex of the dissolved substances. When mixed with water the concentrate is quickly dissolved and does not form the solid sediment that allows expanding the assortment of iodine containing products on its basis.

References

Norkulova K.T., Mamatkulov M.M., 2003. Dynamics of crystallization in vacuum–evaporation systems. *Newsletter of TSTU*, No. 2, pp. 11–13.

Norkulova K.T., 2002. Scientific and technical engineering with the purpose of prevention of iodine-deficiency in Central Asia, 2002. *«NATO Conference» Integration of s&t system of the Central Asian republics to the western world*, May 16–18, Ankara, Turkey.

Norkulova K.T., 2005. On some mechanical ways improving heat exchange in barometric chambers. *Urgent problems of the mechanics and mechanical engineering: proceedings of the international conference*, Volume II, Almaty, pp 120–123.

CHAPTER 31

ECOLOGY-RELATED MICROBIOLOGICAL AND BIOCHEMICAL PARAMETERS IN ASSESSING SOILS EXPOSED TO ANTHROPOGENIC POLLUTION

ZDENEK FILIP AND KATERINA DEMNEROVA*

Institute of Chemical Technology, Department of Biochemistry and Microbiology, Technicka 3-5, 166 28 Prague 6, Czech Republic

> Motto:
> *I always avoid prophesying beforehand,*
> *because it is a much better policy to prophesy*
> *after the event has already taken place.*
>
> Sir Winston Churchill

Abstract: Soil is the foundation of the entire biosphere, and the most important complex interface for the global transformation and interchange of matter and energy. Also, only a healthy, that is, biologically active soil warrants the sufficient production of food for the growing human population. Yet, for decades soils have increasingly been subject to pollution and other adverse side effects of different human activities which negatively affect soil organisms, and especially microorganisms, which play a key role in ecologically important biogeochemical processes. Under stress conditions caused, for example, by a long-term dissemination of chemical pollutants, the development and biochemical activities of soil microorganisms undergo several alterations. To prevent negative ecological consequences, microbiologically related parameters should be involved in the indication of soil quality. In order to determine some reliable indicators, an international project was established and performed for 3 years in which scientists from the Czech Republic, Hungary, Russia, Slovak Republic, and Germany were participating. All participating teams applied a number of standardized methods for selected microbiological and biochemical soil parameters in their laboratories. Soil samples from both unaffected, and differently polluted soils were repeatedly analyzed. After evaluation of more than

---

*To whom correspondence should be addressed. e-mail: katerina.demnerova@vscht.cz

C. Mothersill et al. (eds.), Multiple Stressors: A Challenge for the Future, 409–428.

20 individual parameters it was concluded that nitrogen-fixing bacteria, total microbial biomass, soil respiration (CO_2 release), enzymatic activity (dehydrogenase), and in part also the humification activity of soil microorganisms could be used as quite sensitive indicators of soil quality. However, seasonal oscillations in the values of the individual parameters occurred, and should be appropriately respected in a comparative evaluation of results achieved. In further research, critical limits of the most prospective parameters should be recognized with respect to individual soil types and different soil uses.

Keywords: soil quality; indicators; biological methods

Introduction

In the last years, soil quality became a matter of awareness throughout the world. The roots of this issue go back several decades ago, and are connected with the growth in anthropogenic exploitation of soil resources in agriculture, forestry, industry, and engineering. Indeed, there was an individual "prophecy" made in the 1970s of the last century, for example, by Filip (1973) and Kovda (1975), who pointed to many risky developments in soil quality, and the possible consequences for environment and mankind, but this remained unheard. The main reason was the widely spread believe in an almost endless resilience and self-cleaning capacity of soil. In between, however, "the event" (see in motto) has taken place on a broad scale, and in many regions we have to compete with problems associated with a seriously deteriorated soil. In the 1980/1990s, the first moves toward the development of a well-aimed soil protection were made in Germany and the Netherlands, and also later in the European Community (Thormann, 1984; Barth and L'Heremite, 1987; Howard, 1993). In 1999, an important step was done in Germany by putting in force of a Federal Soil Protection Act. Now, a deficiency in reliable indicators of soil quality still exists which creates problems both for environmentalists and law makers. There is a need to define the present degree of soil deterioration in order to apply effective remediation measures. Also, a need exists for early warning indicators which should help to predict development in soil quality, and possibly prevent soil degradation in years to come. Here, we present some results of our estimations, and discuss similar attempts of other authors, whether some microbiological–biochemical soil characteristics are capable of indicating anthropogenically caused alteration in soil quality. But first, some global change syndromes should be mentioned for their close relation to soil quality.

The Syndrome Approach of Environmental Stress

In 1999, the NATO Committee on the Challenges of Modern Society adopted a Syndrome Approach originally developed by the German Government's Advisory Council on Global Change. The syndrome-based concept starts from the assumption that environmental stress is a part of dynamic human–nature interactions. Sixteen syndromes have been identified and divided into three subgroups "resource use," "development," and "sinks," the majority of which demonstrate relationship with soil quality (Table 1).

TABLE 1. An overview of global change syndromes

Syndromes	Causes and/or effects phenomena
Utilization syndromes	
Sahel syndrome	Overcultivation of marginal land
Overexploitation syndrome	Overexploitation of natural ecosystems
Rural exodus syndrome	Environmental degradation through abandonment of traditional agricultural practices
Dust bowl syndrome	Nonsustainable agroindustrial use of soils and water bodies
Katanga syndrome	Environmental degradation through the extraction of nonrenewable resources
Mass tourism syndrome	Destruction of nature for recreational ends
Scorched earth syndrome	Environmental destruction through war and military actions
Development syndromes	
Aral Sea syndrome	Environmental damage of natural landscapes as result of large-scale projects
Green revolution syndrome	Environmental degradation through the introduction of inappropriate farming methods
Asian tigers syndrome	Disregard of environmental standards in the course of rapid economic growth
Favela syndrome	Environmental degradation through uncontrolled urban growth
Urban sprawl syndrome	Destruction of landscapes through planned expansion of urban infrastructures
Major accident syndrome	Singular anthropogenic environmental disasters with long-term impacts
Sink syndromes	
Smokestack syndrome	Environmental degradation through large-scale diffusion of long-lived substances
Waste dumping syndrome	Environmental degradation through controlled and uncontrolled disposal of waste
Contaminated land syndrome	Contamination of environmental assets at industrial locations

Source: From Lietzmann and Vest (1999)

Typical symptoms of the *Sahel syndrome* include soil degradation, deser-tification, destabilization of ecosystems, loss of biodiversity, and threats to food security. The *overexploitation syndrome* involves both terrestrial and aquatic ecosystems which are exploited beyond their regenerative capac-ity, resulting in soil erosion, water scarcity, and increasing incidence of natural disasters. The *rural exodus syndrome* refers to environmental stress caused by abandonment of previously sustainable land-use practices, and its socioeconomic consequences. The *dust bowl syndrome* is a specific causal complex in which environmental destruction is caused by nonsustainable use of soils, bodies of water as biomass production factors, and involving intensive deployment of energy, capital, and technology. The *Katanga syn-drome* includes the environmental stress (sometimes irreversible) caused by intensive mining of nonrenewable resources above and below the ground, with no consideration given to preservation of the natural environment. The resulting symptoms include soil degradation and creation of contaminated sites. The *mass tourism syndrome* describes the network causes and effects generated by the steady growth of global tourism, and leading, for example, to high consumption of resources, fragmentation of landscapes, soil erosion, and inadequate disposal of sewage and waste. The *scorched earth syndrome* is attributed to the environmental degradation resulting from direct and indirect impacts of military activities, and includes soil degradation and con-tamination caused by fuels and explosives. The *Aral Sea syndrome* describes the impact of large scale or extensive reshaping of seminatural areas result-ing in symptoms such as loss of biodiversity and soil degradation. The *green revolution syndrome* is the result of a centrally planed intensification of agri-culture, however, with negative impacts on natural basis for production, due to soil erosion and degradation. The *Asian tiger syndrome* is a consequence of a rapid economical growth in the "newly industrializing states," and which has negative effects on the local environmental quality in total. The *Favela syndrome* generally refers to a process of unplanned and environmentally harmful urbanization, mainly in peripheral areas of "megacities," which leads, for example, to soil pollution and degradation, accumulation of waste, and many other negative effects. The *urban sprawl syndrome* refers to urban expansion with far-reaching environmental impacts such as fragmentation of ecosystems, soil sealing, contamination, and compaction. The central feature of the *major accident syndrome* is a localized disasters caused by humans, which has transboundary effects on environment resulting in contamina-tion of soil, water, and air, and in health hazards. The *smokestack syndrome* describes the remote effects of substance emissions following disposal on the environmental media, for example, acid enrichment in soils resulting from emission of NH_3, SO_2, and NO_x, and accumulation of persistent pesticides in soil and food chain. This syndrome can be seen in Central Europe with rapid forest decline. The *waste dumping syndrome* involves the localization, compaction, and accumulation of waste in landfills which can lead to contamination of soil and groundwater with harmful effects on drinking water quality and human health. This syndrome poses a serious hazard in many developing regions. The symptoms of the *contaminated land syndrome* generally include a loss of biodiversity, deposits of pollutants in soil, water, and air, and finally loss of ecologically important soil functions. Contaminated sites are usually found at agglomerated regions where heavy industrial, commercial, or military activities occur.

The importance of the syndrome-based approach for scientists and policy makers is that it may serve as a starting point for the development of the respective indicators, and for early intervention in order to avoid escalation of the environmental problems, and possibly also of related conflicts.

Environmental Importance of Soil Organisms and Their Impairment by Anthropogenic Pollutants

Soil is the foundation of the entire biosphere, and in a global scale, it contains by far the highest numbers and the greatest diversity of organisms. In a trivial comparison, there are at least twice as much or even more microorganisms in a teaspoon of soil then the total population of humans on the Earth. For the main groups of soil organisms, the following numbers of species have been estimated by Pankhurst (1997): bacteria 30,000; fungi 1 500,000: algae 60,000; protozoa 10,000; nematodes 5,000,000; earthworms 3,000. From the results of a restriction analysis of DNA extracted from soil samples, Torsvik et al. (1997) postulated that some 6,000 and up to 18,000 microbial species may exist in a single 1 g soil. Different organisms not only inhabit soil environments but also actively and rather quickly contribute to the transformation of their natural habitats. Approximate time scales for biological processes involved in the development of terrestrial ecosystems have been estimated by 1–100 years, while for physically processes the time required may take more then 10,000 years (Dobson et al., 1997). In Fig. 1, the linkages between different communities of soil organisms, their biochemical activities, and ecologically important processes such as mineralization and transformation of plant residues and other organic materials are shown.

There is no doubt that soil organisms maintain the predominant part of the matter and energy transfer in the cycles of carbon, sulfur, phosphorus, and other bioelements. If it were not for microorganisms, substances such as cellulose and lignin would not be recycled but accumulate in the environment with disastrous consequences for entire life on earth. The microbial turnover of carbon amounts some 43 Gt (gigatonnes) per year which is approximately equal to a total net biomass production (Smith and Paul, 1990).

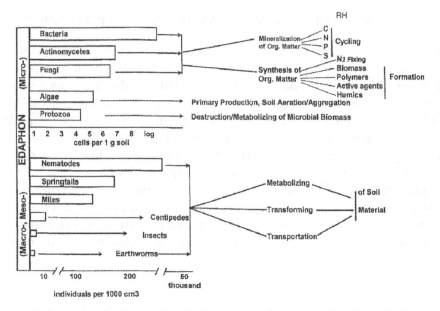

Fig. 1. Main groups of soil organisms, their approximate counts, and ecologically important activities. (From Filip, 2002.)

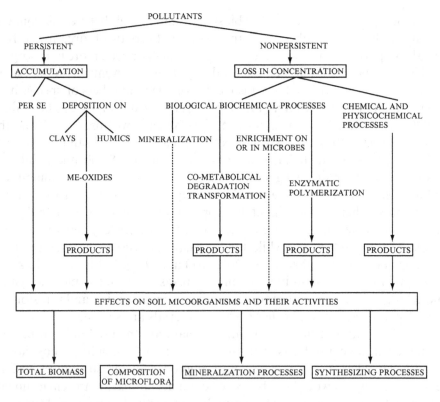

Fig. 2. Behavior of pollutants in soil and their possible effects on microorganisms and ecologically important microbial processes. (From Filip, 1998.)

In a review based on numerous individual studies, Filip (1995) reported evident effects of chemical contaminants on (i) size and composition of soil microbial populations as well as populations of micro- and mesofauna, (ii) turnover of carbonaceous substrata, (iii) turnover of nitrogenous substrata, (iv) influences on enzyme activities, (v) degradation of plant residues, and (vi) on the resistance of soil microorganisms against a heavy metal soil contamination. The author also emphasized the importance of site-specific nonbiotic factors that strongly influence the effectiveness of chemical contaminants on soil organisms and their metabolic activities. In a brief overview, the behavior of chemicals in soil, and their interactions with microorganisms are shown in Fig. 2.

Parameters of Soil Quality

Since the life on Earth is based upon carbon, CO_2 is the main final product of microbial mineralization of organic compounds. Soils represent a major source in liberating CO_2 to the atmosphere. Conclusively, the mineralization activity of soil microorganisms expressed in CO_2 release should be considered as an important parameter of soil quality.

From the energetic point of view, mineralization of organic matter represents a catabolic process releasing energy for anabolic (synthetizing) activities of soil organisms. One can assume that anabolic activities of soil microorganisms may also affect the global cycling of carbon. Norby (1997) calculated that only about 45% of the globally increased CO_2 production remains in the atmosphere while the main part is apparently missing. That missing carbon is proposed to be assigned to soil environments either into a labile, short-lived pool such as microbial biomass, or into a refractory soil organic matter such as humic substances. Thus, the assessment of anabolic activity of soil microorganisms, that is, the estimation of a balance between C bound in soil microbial biomass and humic substances on the one hand, and the amount of CO_2 released from soil on the other hand, can effectively contribute to an objective evaluation of the global carbon budget.

Among the major elements required for all forms of life, nitrogen is unique in that it exists in vast amounts in the atmosphere. From that stock, N_2 must be bound into a living matter by the activity of free-living and/or symbiotic microorganisms mainly in soils. For the enormous importance of the N_2 fixing in global cycle of nitrogen, its assessment should not fail in the evaluation of soil quality.

Beside the N_2 fixation, also other key processes of the N cycle such as ammonification of nitrogen bound in organic compounds, oxidation of

NH_4^+-N to NO_3^--N (nitrification), and reduction of NO_3/NO_2 to N_2 (deni-trification) represent microbially mediated soil processes which can be used as indicators in ecologically based system of assessing soil quality. This is because an anthropogenic alteration of the previously mentioned primary processes may have multiple consequences such as (i) an increased concentration of the greenhouse gas N_2O and/or other NO_x gases globally, (ii) losses in soil nutrient, (iii) acidification of soils, and (iv) increased transfer of N from soil to aquatic environments (Vitousek et al., 1997).

Not only microbially mediated cycles of C and N may become strongly affected by human activities. Similar is true also for the S, P, and other bioelements. However, the key roles of C and N in the biosphere, and also an availability of well-proved analytical methods may account for the priority involvement of C and N transformations in the biological assessment of soil quality.

Anthropogenic impacts on soil organisms and their activities may cause a risk to biodiversity of soil organisms. Different experimental methodical approaches such as estimation of phospholipid fatty acid profiles (PLFA), DNA hybridization, or DNA/RNA fingerprinting techniques may reveal an insight into microbial diversity in soil. However, until yet evidences are rather scarce on the dependence of the functioning of soil biogeochemical processes on an undisturbed biodiversity in the soil environment. Also, no standardized and easy-to-use methods are available that could make a part of a routine soil monitoring. Nevertheless, the preservation of biodiversity should be attempted with respect to a sustainable conservation of soil quality. In general, a higher biodiversity stands for longer food chains, more cases of symbiosis, and greater possibilities for negative feedback control, which could reduce oscillations and hence increase stability of a soil ecosystem (Bianchi and Bianchi, 1995). In the present time, however, a large-scaled soil monitoring of responses of the microbial community to anthropogenic stresses should be rather based on well-proved common-in-use microbiological techniques.

In a soil plough layer up to $9\,t\,ha^{-1}$ (d.w.) of bacterial biomass and about the same amount of fungi exists even in biologically less active soils such as podzol (Nikitin and Kunc, 1988). From an average size of a bacterial cell, a total surface of about $500\,ha$ bacteria per hectare of arable soil has been postulated (Kas, 1966). These numbers indicate that there is a high amount of living biomass in soil, and this can respond to effects caused by pollutants and other anthropogenic factors. Short- or long-term changes may occur to the total microbial biomass, individual groups of microorganisms, and also to different microbially mediated processes of basic ecological importance (Filip, 1995). The respective biological parameters should be considered as indicators in assessing soil quality.

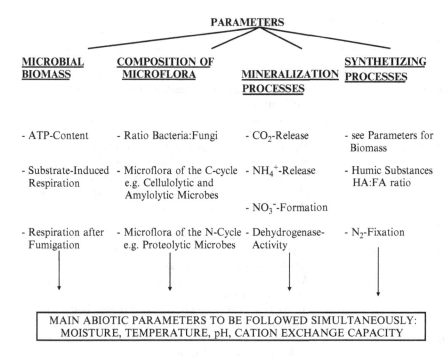

Fig. 3. Selected parameters to evaluate effects of chemical pollutants on ecologically important soil characteristics. (From Filip, 2002.)

In Fig. 3, the most relevant biological and also physical–chemical parameters which should be respected in the monitoring of soil quality are presented.

Experience from International Soil Testing Using Different Methods

In an international approach of 3 years in duration, the effectiveness in the indication of soil quality of different microbiological and biochemical methods was tested by five groups of scientists from the Czech Republic, Hungary, Russia, Slovakia, and Germany (Filip, 2002). In Table 2, basic characteristics of some of the total 49 repeatedly examined soil sites can be seen. Soil samples were collected three or four times a year from a respective plough layer, and a fraction <2 mm in size was analyzed in the laboratory. Individual analyses were performed uniformly using a laboratory manual which was compiled from internationally recognized collections of methods (Filip, 1994). Laboratory testing only was used in the investigations in order to equalize regularly occurring seasonal variations in soil moisture and

TABLE 2. Examples of soil sites included in testing

Country	Soils
Czech Republic	
	Brown Soil/Luvic Phaeozem; Dystric Cambisol
"Clean" sites	Natural and renovated grassland with or without NPK fertilizers
	Arable land with or without organic or mineral NPK fertilizers
Polluted sites	Natural grassland near to a motorway (N_{ox})
	Natural grassland in an urban area (organic pollutants)
	Arable land (SO_2 and other airborne pollutants)
	Mine spoil/anthropic Soil (containing ash from a power plant running with a brown coal)
Hungary	
	Chernozem/Calcic Chernozem
"Clean" sites	Arable land on loess (Univ. Exp. Station)
	Arable land on tuff and loess
Polluted sites	Arable land near to a power plant running with lignite (N_{ox}, heavy metals)
	Arable land near to an oil refinery (hydrocarbons, heavy metals)
Russia	
	Soddy-Podzolic Soil/Podzol
"Clean" sites	Arable land weakly cultivated (Univ. Moscow Exp. Station)
Polluted sites	Same land contaminated with 500 ppm Pb
	Same land contaminated with 2,000 ppm Pb
	Same land contaminated with 500 ppm Pb, Cd, Zn
Slovakia	
	Pseudogley-Chernozem/Stagnic Phaeozem
"Clean" sites	Arable land
Polluted sites	Arable land near to a metallurgy plant (Pb, Cd, As, Cu)
	Arable land near to a metallurgy plant (MgO)
	Arable land polluted by mineral oil

Source: From Filip (2002)

temperature which could eventually mask differences caused by different anthropogenic soil stressors.

In Table 3, the relative sensitivity of the most reliable parameters is shown. Nitrogen-fixing bacteria, the enzyme activity (dehydrogenase), and respiration activity (CO_2 release) appeared among the most sensitive methods to indicate anthropogenically caused soil stress. Sometimes, nitrification activity also appeared a useful indicator. In a soil artificially contaminated with Pb, a significant decrease in values of the individual parameters could be observed (Fig. 4). On the other hand, in soil samples

TABLE 3. Relative sensitivity of microbiological and biochemical parameters for the assessment of soil quality based on long-term analyses from 49 differently anthropogenically affected soils

Parameter	Relative sensitivity[a]
Microbial biomass	+/++
Composition of microbial communities	
Copiotrophic bacteria (colony forming units)	+/++
Oligotrophic bacteria	++
Actinomycetes	++
Microscopic fungi	++
Proteolytic spore-forming bacteria	−/+
Cellulose decomposer	+/++
N_2-fixing bacteria	++++
Pseudomonads	−/+
Biochemical process-linked activities	
Respiration (CO_2 release)	+++
Ammonification (NH_4^+ release)	++
Nitrification/denitrification	++/+++
Dehydrogenase activity	+++/++++
Humification activity	++

Source: From Filip (2002)
[a]Sensitivity (relative to control soil): − = Not detected; + = Low; ++++ = Maximum

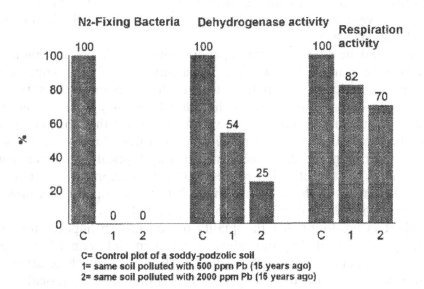

Fig. 4. Relative values of N_2-fixing bacteria numbers, and dehydrogenase and respiration activities in a Pb-contaminated soil. (From Filip, 2002.)

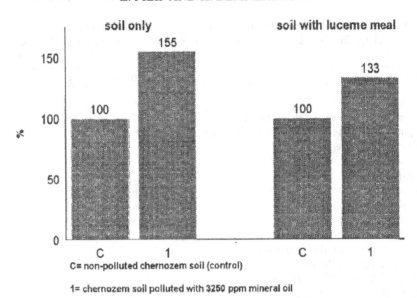

Fig. 5. Relative values of a respiration activity (CO_2 release) in soil samples polluted with mineral oil. (From Filip, 1998.)

contaminated by mineral oil an evident increase of respiration activity was measured (Fig. 5). A strong seasonal variability was observed even for the most sensitive parameter such as N_2-fixing bacteria. Their highest indicative values were obtained in summer (June), and the lowest ones usually in March.

It should be admitted that depending on the soil use, contradictory results were sometimes obtained. Mean amounts of soil biomass, for example, were lower in contaminated meadow and forest soils as compared with control sites, and on the other hand they were enhanced in arable and urban soils. Differences were found also in the capability of soil microbial communities to mineralize added substrate (lucerne meal) and in short-term changes in the contents and quality (optical density) of humic substances. The humification process was mainly suppressed in polluted soils where most of the added organic substrate was either mineralized or remained nonutilized during an incubation period.

There is no doubt that the quality of soil has an impact on (i) soil fertility, (ii) food quality and safety, (iii) human and animal health, and (iv) environmental quality (Parr et al., 1992). Each of these attributes deserves to be evaluated by specific indicators. In a large-scale view, such indicators may include (i) vegetative cover close to the soil; (ii) soil organic matter; (iii) biological activity and soil biodiversity; (iv) soil structure and porosity; (v) available water capacity; (vi) plant available nutrients; (vii) cation-exchange capacity; (viii) soil acidity; (ix) soil salinity, and (x) depth of rooting and crop vigor (Shaxson, 1998). The physical and physicochemical parameters mainly represent the end-point values, and this only in a retrospective view.

In order to obtain data on biological aspects of soil quality quickly, several authors adopted different aquatic tests for soil quality testing (Kördel, 2000; Pfeifer et al., 2000). Waterborne organisms such as algae, *Daphnia magna*, and also different freshwater fish have been recommended for use (Hund, 1994; Debus and Hund, 1997). Graefe (1997) and some other authors (see Vortragstagung, 1996) attempted to establish biological soil diagnostics on the classification, survival, and/or behavior of some soil invertebrates. Yakovlev (1997) recommended using the abundance in soil samples and the behavior in a soil suspension of selected protozoa (amoebae, Colpodida) and algae (*Heterotrix* sp.) in his characterization of virgin and anthropogenically affected soils. Römbke and Moltmann (1996) referred even to tests with higher vertebrates (birds, mammals), soil saprophagous invertebrates, plants, and (for pesticide side effects) pollinators, such as honeybees as utilizable in the evaluation of soils. Perhaps, some of such test strategies might have an advantage of simplicity in use and easy standardization. However, they seem to heavily oversimplify both biological and abiotic structural complexity and heterogeneity of soil, and mainly they disregard soil ecological functions. Thus, they could be rather useful in estimating toxicity of individual chemical pollutants than evaluating soil quality.

Indeed, an ecological approach to assessing soil quality can include investigations of individual populations of organisms if they have been clearly recognized as an important constituent of the sustainable biodiversity in a soil ecosystem, or an important agent in soil ecological processes such as nitrogen fixation or carbon mineralization. Certainly, with several million species of invertebrates living in soil and thousands species of bacteria in every gram of soil (Pankhurst, 1997; Torsvik et al., 1997) it is hardly possible to estimate any changes in biodiversity with reliable and easy applicable methods. Also, it remains questionable whether there is a true correlation between the degree of biodiversity and soil ecological functions (Wardle and Giller, 1996). Contrary to this, there is no doubt that firm linkages exist between different microbial communities, their biochemical activities, and ecologically important soil processes, such as mineralization and transformation of plant residues and other organic materials (Fig. 1). In a similar respect, Wilke et al. (2000) reported the usefulness of soil microbial biomass, respiration activity, and aerobic as well as anaerobic processes involved in the cycling of essential nutrients such as carbon, nitrogen, phosphor, and sulfur as parameters in assessing of soil quality. More recently, and following an extensive international discussion, a comprehensive approach into characterizing of soil ecological values by a microbiological approach has been proposed by Andren et al. (2004), and this is summarized in Table 4. Also similarly to our own methodical approach as shown in Fig. 3, Filzek et al. (2004) underline the necessity of estimating nonbiotic pedological factors that undoubtedly affect soil response to chemical pollution. By sampling soil along a transect from a primary Cd–Pb–Zn smelting works, disturbances even in fractions of soil humus, and up to a 1.5 km distance from smelter, have been found. Similarly, Filip and Kubat (2004) reported significant differences in humic acid and fulvic acid contents in a long-term affected, that is, chemically fertilized arable soils.

The actual question to be further elucidated is that of critical limits for the individual biological and biochemical parameters of soil quality. For the evaluation of results obtained in the individual tests, the respective effects can be rated according their importance, and this in relation to values obtained in control variants that remained untreated. Römbke and Moltmann (1996) suggested respective percentages in a range as shown in Table 5. The individual data, however, seem not applicable for the evaluation of soil ecologic functions. Dahlin et al. (1997) demonstrated some difficulties in solving this problem on soils with a low level of heavy metals contamination. There is also a remarkable degree of spatial variability of analytical data which could be obtained even from pedogenically homogeneous soil sites (Robertson et al., 1997). Specific solutions based on statistical methods have been suggested (Smith et al., 1993; Halvorson et al., 1996) but as an alternative, perhaps, methodological improvements in the soil quality testing should further be attempted, and including novel physicochemical and molecular-biological methods, if practicable enough for a broad-scale use in the monitoring of soil quality.

A Few Actual Data from the Current Literature

Actually, in spite of ecologically based approaches as summarized in Table 5, some authors still use to employ tests with single organisms to assess the effects of different chemicals in soil. Robidoux et al. (2004) used soil mesocosms with earthworms *Lumbricus terrestris* and *Eisenia andrei* as test organisms to assess toxicity of explosives in soil from an antitank firing range. Results indicated that the survival of earthworms was reduced up to 100% following a 28-day soil exposure. Bundy et al. (2004) sampled earthworms across an environmental gradient of metal contamination for an ecotoxicological assessment. Both indigenous (*Lumbricus rubellus, L. terrestris*) and introduced species (*E. andrei*) were analyzed for small molecule metabolites

TABLE 4. List of methodical approaches and tests to characterize soil ecological capital using natural microbial characteristics

Methodical approach	Tests: normalized, calculated, upon research
Microbial biomass	*Normalized tests:* ISO 14240-1 (substrate-induced respiration) ISO 14240-2 (fumigation-extraction) *Calculated test:* Microbial quotient *Tests upon research:* Fumigation-incubation Specific group biomass (e.g., ergosterol determination for fungi) Physiological method (initial mineralization rate of glucose added) ATP method (adenosine triphosphate concentration in soil)
Microbial activity	*Normalized tests:* ISO 15685 (nitrification – NH_4 oxidation) ISO 14238 (N mineralization and nitrification – incubation method) ISO 14239 (C mineralization = respiration – incubation method) ISO 16072 (basal respiration of heterotrophs) ISO 17155 (respiration curves) *Calculated tests:* Respiratory quotient Fungal/bacterial respiratory ratio *Tests upon research:* Denitrification (e.g., N_2O production – anaerobic conditions) N fixation (e.g., C_2H_4 production – incubation – nitrogenase activity) Mycorrhizae (percentage of root colonized in a test plant)
Enzymatic activity	*Normalized tests:* ISO 23753-1 (dehydrogenase activity – TTC method) ISO 23753-2 (dehydrogenase activity – ITC method) *Tests upon research:* Enzymatic activity tests (other than dehydrogenase: phosphatase, sulfatase, peptidase, urease, esterase, cellulase, amylase, xylanase, laccase, peroxydase, maltase, saccharase, cellobiase) Enzyme index
Root pathogens	*Tests upon research:* Detection of root pathogens
Microbial diversity	*Tests upon research:* Community level physiological profiles (e.g., BIOLOG system) Nucleic acid analysis (e.g., soil DNA extraction and PCR, RPFT, DGGE) Dilution plating and culturing methods Ester-linked fatty acids estimation (e.g., PFLA, FAME, MIDI, SLB – chemical microbial signatures)

Source: Accommodated from Andren et al. (2004)

TABLE 5. Rating of effects obtained in ecotoxicological tests as related to unaffected controls

	Mortality or lowered beneficence in %	
Rating	Laboratory tests	Semifield and field tests
Not harmful	<50	<25
Slightly harmful	50–79	25–49
Moderately harmful	80–99	50–75
Seriously harmful	>99	>75

Source: From Römbke and Moltmann (1996; modified)

as possible biomarkers by ^{1}H NMR spectroscopy. The spectral data revealed that biochemical changes were induced across the metal contamination gradient. Native worms of *L. rubellus* from the most polluted site were associated with an increase in the relative concentration of maltose and histidine but the concentrations were significantly reduced by metal contamination in *L. terrestris*. These results show that controversy in results may be obtained in tests based on the use of single test species. Fountain and Hopkin (2004) examined the effects of metal contamination (Cd, Cu, Pb, Zn) on natural population of collembola (springtails) in five soil sites. The highest number of species was found at the most contaminated site, although the collembola population also had a comparatively low evenness value with just two species clearly dominating.

Coja et al. (2006) found a collembola *Folsomia candida* and a carabid beetle *Poecilus cupreus* quite useful in testing lethal and/or sublethal effects of benzoxazolinone, some degradation products, and related pesticides. Nevertheless, it is to be stated repeatedly that results obtained in single organisms tests cannot mirror ecological importance of possible effects of chemicals on soil in a desirable way. In these respects microbiological and enzymatic biomarkers as used for soil monitoring, for example Garcia-Gil et al. (2004), represent much more prospective approach. However, also by employing strong ecologically related tests, an extreme thorough evaluation of results obtained in a test battery should be made. Sometimes, the results may differ from each other as shown recently, for example by Tobor-Kaplon et al. (2006). The most polluted soil (Zn, Cd) had the lowest stability to additional pollution with Pb and a heat stress with regard to respiration. However, bacterial growth rates were affected in a different way than respiration. Some controversial results as well as unsolved research question as connected with methods for evaluating human impacts on soil microorganisms and their activity were also highlighted in a recent review by Joergensen and Emmerling (2006). In general, however, the authors confirm the opinion, we expressed already earlier (Filip, 1973, 1998, 2002), that is, that a total microbial biomass, structure of microbial community, and different activity rates such as respiration and N mineralization represent the most useful parameters for assessing of soil ecological functions, and thus, they should be primarily considered in the soil examination and in a systematic long-term soil monitoring.

Conclusions

Because the entire biosphere strongly depends on the biologically based transforming processes of matter and energy in soil, there is a need for biological methods to testing and evaluating soil quality. Both ours and other authors' results underline the importance of process-related microbiological and biochemical parameters in this respect. The data obtained should be put in relation with physicochemical soil characteristics which naturally oscillate during a year. A concentrated research should be performed in order to recognize objective limits which will respect but not overcharge the natural soil resiliency. In all these difficult steps which directly and/or indirectly influence not only our daily life but also the existence of the total biosphere, an extensive international cooperation is a must. An important step in this very direction was made by the Commission of the European Communities very recently: On September 22, 2006, the Commission issued proposal for a "Directive of the European Parliament and the Council Establishing a Framework for the Protection of Soil" (see Commission, 2006) which formally opens the way for an effective soil protection throughout the European Union.

> *Prolog:*
> *SOIL FOR EVER*
> *People ask question different in level:*
> *Tell, what is soil? Is it just dirt or rather a value?*
> *What ever!*
> *For you and me who know the answer, however,*
> *And even for those who did not care yet (and will maybe never),*
> *Soil is and remains "The Fundament of Life" (like it was ever).*
> *Thus, let us call: "Long lives the soil!" And even more:*
> *KEEP SOIL ALIVE FOR EVER!*
>
> Z. Filip

Acknowledgment

One of us (Z.F.) wishes to express his gratitude to the European Commission, Brussels, for the granting of a Marie Curie Chair in Environmental Microbiology and Biotechnology, tenable at the Institute of Chemical Technology Prague, Czech Republic.

References

Andren, O., Baritz, R., Brandao, C., Breure, T., Feix, I., Franko, U., Gronlund, A., Leifeld, J., and Maly, S. 2004, Soil biodiversity, in: Reports of the Technical Working Groups established under the Thematic Strategy for Soil Protection, Vol. III, *Organic Matter and Biodiversity*, L. Van-Camp, B. Bujarrabal, A.-R. Gentile, R.J.A. Jones, L. Montanarella, C. Olazabal, and S.-K. Selvaradjou, eds., EUR 21319 EN/2, Office for Official Publications of the European Communities, Luxembourg, 872 pp.

Barth, H., and L'Heremite, P., 1987, *Scientific Basis for Soil Protection in the European Community*, Elsevier, London.

Bianchi, A., and Bianchi, M., 1995, Bacterial diversity and ecosystem maintenance, in: *Microbial Diversity and Ecosystem Function*, D. Allsop, R.R. Colwell, and D.L. Hawksworth, eds., CAB International, Wallingford, pp. 185–198.

Bundy, J.G., Spurgeon, D.J., Svendsen, C., Hankard, P.K., Weeks, J.M., Osborn, D., Lindon, J.C., and Nicholsson, J.K., 2004, Environmental metabonomics: Applying combination biomarker analysis in earthworms at metal contaminated site, *Ecotoxicology* 13: 797–806.

Coja, T., Idinger, J., and Blümel, S., 2006, Effects of the benzoxazoline BOA, selected degradation products and structure related pesticides on soil organisms, *Ecotoxicology* 15: 61–72.

Commission of the European Communities, 2006, *Proposal for a Directive of the European Parliament and the Council Establishing a Framework for the Protection of Soil and Amending Directive 2004/35/EC,* COM (2006) 232 Final, 30 pp.

Dahlin, S., Witter, E., Mortensson, A., Turner, A., and Baath, E., 1997, Where is the limit? Changes in the microbiological properties of agricultural soils at low levels of metal contamination. *Soil Biol. Biochem.* 29: 1405–1415.

Debus, R., and Hund, K., 1997, Ecotoxicological tests for effect assessment of bioavailable portion of soil, in: *Bioavailability as a Key Property in Terrestrial Ecotoxicity Assessment and Evaluation*, M. Herrchen, R. Debus, and R. Pramanik-Strehlow, eds., Fraunhofer IRB Verlag, Stuttgart, pp. 97–107.

Dobson, A.P., Bradshaw, A.D., and Baker, A.J.M., 1997, Hopes for the future: Restoration ecology and conservation biology, *Science* 277: 515–521.

Filip, Z., 1973, Healthy soil – Foundation of healthy environment, *Vesmir* 52: 291–293 (in Czech).

Filip, Z., 1994, *Methods of Soil Analyses for an International Research Project Development and Evaluation of Biological Methods for the Characterization of Undisturbed and Anthropogenically Polluted Soils*, Umweltbundesamt, Langen, 48 pp.

Filip, Z., 1995, Einfluss chemischer Kontaminanten (insbesondere Schwermetalle) auf die Bodenorganismen und ihre ökologisch bedeutenden Aktivitäten. *UWSF-Z. Umweltchem. Ökotox.* 7: 92–102.

Filip, Z., 1998, Soil quality assessment: An ecological attempt using microbiological and biochemical procedures, *Adv. GeoEcol.* 31: 21–27.

Filip, Z., 2002, International approach to assessing soil quality by ecologically-related biological parameters, *Agric. Ecosyst. Environ.* 88: 169–174.

Filip, Z., and Kubat, J., 2004, Mineralisation and humification of plant matter in soil samples as a tool in the testing of soil quality, *Arch. Agron. Soil Sci.* 50: 91–97.

Filzek, P.D.B., Spurgeon, D.J., Broll, G., Svendsen, C., Hankard, P.K., Kammenga, J.E., Donker, M.H., and Weeks, J.M., 2004, Pedological characterization of sites along a transect from primary cadmium/lead/zinc smelting works, *Ecotoxicology* 13: 725–737.

Fountain, M.T., and Hopkin, S.P., 2004, Biodiversity of collembola in urban soils and the use of *Folsomia candida* to assess soil quality, *Ecotoxicology* 13: 555–572.

Garcia-Gil, J.C., Polo, A., and Kobza, J., 2004, Microbiological and enzymatic biomarkers for soil monitoring, in: *Proc. 3rd Soil Sci. Days in Slovakia Int. Conf.*, J. Sobocka and P. Jambor, eds., Res. Inst. Soil Sci. & Soil Protection, Bratislava (ISBN 80-89128-11-4), pp. 77–84.

Graefe, U., 1997, Von der Spezies zum Ökosystem: Der Bewertungsschritt bei der bodenbiologischen Diagnose, *Abh. Ber. Naturkundemus.-Görlitz* 69: 45–53.

Halvorson, J.J., Smith, J.L., and Papendick R.I., 1996, Integration of multiple soil parameters to evaluate soil quality: A field example, *Biol. Fertil. Soils* 21: 207–214.

Howard, P.J.A., 1993, Soil protection and soil quality assessment in the EC, *Sci. Tot. Environ.* 129: 219–239.

Hund, K., 1994, *Entwicklung von biologischen Testsystemen zur Kennzeichnung der Bodenqualität*, Texte 45/94, Umweltbundesamt, Berlin.

Joergensen, R.G., and Emmerling, Ch., 2006, Methods for evaluating human impact on soil microorganisms based on their activity, biomass, and diversity in agricultural soils, *J. Plant Nutr. Soil Sci.* 169: 295–309.

Kas, V., 1966, *Mikroorganismen im Boden*, A. Ziemsen Verlag, Wittenberg Lutherstadt, 208 pp.

Kovda, V.A., 1975, *Biogeochemical Cycles in Nature and their Disturbance caused by Humans*, Nauka, Moscow (in Russian).

Kördel, W., 2000, Validation of ecotoxicological tests for the assessment of soil quality, in: *Contaminated Soils 2000*, Proc. 7th FZK/TNO Conf. on Contaminated Soils, Sept. 18–20 2000 Leipzig. Thomas Telford, London, pp. 878–881.

Lietzmann, K.M., and Vest, G.D., eds., 1999, *Environmental Security in an International Context*, NATO Committee on the Challenges of Modern Society, Report No. 232, 174 pp.

Nikitin, D.I., and Kunc, F., 1988, Structure of microbial associations and some mechanisms of their autoregulation, in: *Soil Microbial Associations*, V. Vancura and F. Kunc, eds., Elsevier, Amsterdam, pp. 157–190.

Norby, Z., 1997, Inside the black box, *Nature* 388: 522–523.

Pankhurst, C.E., 1997, Biodiversity of soil organisms as an indicator of soil health, in: *Biological Indicators of Soil Health*, C. Pankhurst, ed., CAB International, Wallingford, pp. 297–324.

Parr, J.F., Papendick, R.I., Hornick, S.B., and Meyer, R.E., 1992, Soil quality: Attributes and relationship to alternative and sustainable agriculture, *Am. J. Alter. Agric.* 7: 5–11.

Pfeifer, F., Haake, F., Kördel, W., and Eisenträger, A., 2000, Untersuchungen zur Rückhaltefunktion von Böden mit aquatischen Testsystemen, in: *Toxikologische Beurteilung von Böden*, St. Heiden, R. Erb, W. Dott, and A. Eisenträger, Hrsg., Spektrum, Heidelberg, pp. 1–18.

Robertson, G.P., Klingensmith, K.M., Klug, M.J., Paul, E.A., Crum, J.R., and Ellis, B.G., 1997, Soil resources, microbial activity and primary production across an agricultural ecosystem, *Ecol. Appl.* 7: 158–170.

Robidoux, P.Y., Dubois, Ch., Hawari, J., and Sunahara, G., 2004, Assessment of soil toxicity from an antitank firing range using *Lumbricus terrestris* and *Eisenia andrei* in mesocosm and laboratory studies, *Ecotoxicology* 13: 603–614.

Römbke, J., and Moltmann, J.F., 1996, *Applied Ecotoxicology*, CRC Lewis Publ., Boca Raton, p. 282.

Shaxson, T.F., 1998, Concepts and indicators for assessment of sustainable land use, *Adv. GeoEcol.* 31: 11–19.

Smith, J.L., Halvorson, J.J., and Papendick, R.I., 1993, Using multiple-variable indicator kriging for evaluating soil quality, *Soil Sci. Soc. Am. J.* 57: 743–749.

Smith, J.L., and Paul, E.A., 1990, The significance of soil microbial biomass, in: *Soil Biochemistry*, J.M. Bollag and G. Stotzky, eds., Vol. 6, Dekker, New York, pp. 357–396.

Thormann, A., 1984, Bodenschutz – Teil einer vorsorgenden Umweltpolitik, *Z. Kulturtechnik Flurbereinigung* 25: 195–202.

Tobor-Kaplon, M.A., Bloem, J., Römkens, P.F.A.M., and De Ruiter, P.C., 2006, Functional stability of microbial communities in contaminated soils near a zinc smelter (Budel, The Netherlands), *Ecotoxicology* 15: 187–197.

Torsvik, V.L., Daae, F.L., Goksoyr, J., Sorheim, R., and Overas, L., 1997, Diversity of bacteria in soil and marine environments, in: *Progress in Microbial Ecology*, M.T. Martins, ed., SBM-Brazil. Soc. Microbiol./ICOME, Sao Paulo, pp. 115–120.

Vitousek, P.M., Aber, D.J., Howard, R.W., Linkens, G.E., Matson, P.A., Schindler, D.W., Schlesinger, W.H., and Tilman, D.G., 1997, Human alteration of the global nitrogen cycle: Sources and consequences, *Ecol. Appl.* 7: 737–750.

Vortragstagung, 1996, Neue Konzepte in der Bodenbiologie. *Mitt. Dtsch. Bodenkd. Ges.* 81, DBG Oldenburg, 384 pp.

Wardle, D.A., and Giller, K.E., 1996, The quest for a contemporary ecological dimension to soil biology, *Soil Biol. Biochem.* 18: 1549–1554.

Wilke, B.-M., Winkel, B., and Pauli, W., 2000, Mikrobiologische verfahren zur Beurteilung der Lebensraumfunktionmvon Böden, in: *Toxikologische Beurteilung von Böden,* St. Heiden, R. Erb, W. Dott, and A. Eisenträger, Hrsg., Spektrum, Heidelberg, pp. 43–57.

Yakovlev, A.S., 1997, *Biological Diagnostics of Natural and Anthropogenically Affected Soils,* Sc.D. Thesis, Moscow State Univ., Moscow (in Russian).

CHAPTER 32

MOLECULAR AND CELLULAR EFFECTS OF CHRONIC LOW DOSE-RATE IONIZING RADIATION EXPOSURE IN MICE

ANDREYAN N. OSIPOV

N.I. Vavilov Institute of General Genetics RAS, Gubkin str. 3, Moscow 119991, Russia
e-mail: andreyan.osipov@rambler.ru

Abstract: The results of the three independent experiments on study of molecular and cellular effects in CBA/lac male mice chronically (up to 1 year) exposed to γ-radiation at a dose rate of 61 cGy/year are reviewed in this chapter. It was shown that the DNA–protein cross-links level was statistically increased only at early terms of irradiation (up to 40 days (~6.8 cGy)). The results of the comet assay study on spleen cells showed that very low dose-rate irradiation resulted in statistically significant increase in DNA strand breaks level, starting from a dose of ~20 cGy (120 days). Further prolongation of exposure time and, hence, increase of a total dose did not, however, lead to further increase in the extent of DNA strand breaks level. A dose–response curve for DNA single-strand breaks is good fitted by a polynomial regression $y = 0.6209 + 0.0313^*x - 0.0004^*x^2$, where y is the average comet index and x is a dose in cGy. At the days 120, 270, and 365 of the chronic irradiation (20, 45, and 61 cGy, respectively), approximately twofold increase over a control level in the apoptotic cell fraction was observed. It was found that chronic action of low dose-rate γ-radiation led to a change in the sensitivity of spleen cells to H_2O_2 exposure. A weakening of cellular antioxidant potential and/or repair capacity has been observed at early terms of irradiation (up to 80–120 days). In contrast, prolongation of irradiation resulted in activation of defense system in spleen cells. This effect could be attributed to a development of adaptation processes triggered upon accumulation of a certain dose. A «bystander effect», response of unirradiated directly cells due to signaling originating from irradiated cells, can be possibly involved in the effects observed in this study.

Keywords: DNA–protein cross-links; single-strand DNA breaks; apoptosis; micronuclei; mice; ionizing radiation; low dose-rate irradiation; low doses

C. Mothersill et al. (eds.), Multiple Stressors: A Challenge for the Future, 429–438.

Introduction

All living organisms, including human, are continuously exposed to ionizing radiation (IR) from natural sources. However, development of nuclear technologies and associated intentional (e.g., Hiroshima and Nagasaki) and accidental (e.g., Chernobyl) releases of radioisotopes have led to increase in a background level of IR. This technogenic part of IR exposure has risen significantly over last few decades.

While there are no doubts about negative biological effects of high-dose IR, debates about whether low-dose IR exposure is harmful or beneficial (hormetic) are still continuing among scientific community. Analysis of available literature indicates that low-dose IR exposure induces a complex of biochemical and biophysical reactions in animals (Calabrese and Baldwin, 2000; Mothersill and Seymour, 2003). It is not clear, however, whether those changes are consequences of organism adaptation to increase in IR background, and whether low doses cause any significant genetic alterations.

In this chapter, the results of the three independent experiments on study of molecular and cellular effects in CBA/lac male mice chronically (up to 1 year) exposed to γ-radiation at a dose rate of 61 cGy/year are reviewed.

Experimental Design

In chronic IR exposure experiments, 4–5-week-old CBA/lac male mice weighting 12–14 g (purchased from "pitomnik-Stolbovaya") were used. Mice were placed in plastic cages 14 days prior to IR exposure. Distribution of animals into control and experimental groups was random. Mice were given standard dry feed and water *ad libitum*. Experiments with chronic low dose-rate irradiation were carried out from 2000 to 2004. Three independent experiments with identical conditions were performed, utilizing totally 420 mice.

Experimental animals were chronically exposed to IR from a γ-ray unit "UOG-1" (VNIIFTRI, Russia) equipped with a ^{137}Cs source (activity 7.2×10^8 Bq) mounted in a steel container and specifically designed for long-term irradiation of biological objects. The IR source was placed above exposed targets. Chronic irradiation of animals was performed at a dose rate of 0.07 mGy/h (distance from mouse bedding to the γ-radiation source was 64 cm, filter lens #5). Variability of a dose rate within area to be irradiated (1 m^2) did not increase 10%. Irradiation was continuous with 10–15 min daily break for hygiene procedures. Control dosimetry was performed using thermoluminescent detectors TLD-100 (Sweden) and DTG-4 (Russia).

To deliver to animals, total cumulative doses of 6.7, 13.4, 20.2, 35.3, 45.4, and 61.3 cGy, low dose-rate irradiation was performed for 40, 80, 120, 210, 270, and 365 days, respectively.

Upon completion of chronic low dose-rate irradiation, mice were sacrificed and spleens were removed and processed for subsequent analysis. Suspension of spleen cells in phosphate buffered saline (pH 7.4) containing 0.14 M NaCl, 2.7 mM KCl, 3 mM NaN_3, was filtered through nylon mesh at 4°C.

The DNA–protein cross-links level (DPC) was determined by the K^+/SDS assay (Zhitkovich and Costa, 1992) with minor modifications (Osipov et al., 2000). The amount of DNA was measured with the Hoechst 33258 reagent using a FL-2110 fluorimetric analyzer (Solar, Belarus) with excitation 365 nm and emission 460 nm. The level of DPC was determined as a ratio of the amount of DNA in the supernatant to the total DNA in the sample.

Alkali single cell gel electrophoresis was carried out as described by Singh et al. (1988). According to the assay, the number of alkali labile sites and single-strand breaks (SSBs) are proportional to the number of DNA fragments and to distance DNA migrated from the nucleus after alkali electrophoresis of agarose-immobilized single cells. Fluorescent dye Hoechst 33258 (Sigma Chemical Co, St. Louis, MO, USA) was used to visualize DNA. Analysis was performed using the "Lumam I-2" fluorescent microscope (LOMO, Russia). Hundred comets were counted from each slide. Comets were divided into classes 0–4 (0 corresponded to no visible tail, 4 to total migration of DNA from the nucleus into the tail) depending on the shape (diameter, tail length, etc.). This method of visual damage is considered as a valid way for DNA damage analysis (Kobayashi et al., 1995). Results of the visual classification were subsequently confirmed using the analytic package image analysis software (Kinetic Imaging, Liverpool, UK).

A number of comets in each class were recorded and the average comet index (ACI) was calculated as: $ACI = (1 \cdot n1 + 2 \cdot n2 + 3 \cdot n3 + 4 \cdot n4)/\Sigma$, where $n1$–$n4$ are number of comets in classes 1–4 and Σ is the sum of counted comets, including comets in class 0.

Percentage of apoptotic cells was determined by the DNA diffusion assay described elsewhere (Singh, 2000).

For study of the induction of DNA damage in spleen cells by hydrogen peroxide, the cell suspension (1×10^6 cells/mL in PBS) were incubated with H_2O_2 (0.5 and 5 mM) for 30 min at 37°C. The level of DNA strand breaks was estimated by the DNA precipitation assay developed by Olive (1988) and adapted for fluorometric DNA measurement.

Statistical analysis of experimental results was performed using Student t-test. The results are presented as mean ± standard error.

Experimental Data

DNA–PROTEIN CROSS-LINKS

DPC are important type of damage in cells, induced by some chemical and physical agents (Oleinick et al., 1987). DPC represent bulky molecular lesions, which are hardly repaired, thereby interfering with normal functioning of the nuclear chromatin and causing serious genetic consequences whenever these lesions fall in noted essential DNA regions.

Figure 1 shows exposure time-dependent changes of DPC level in spleen cells of mice irradiated at a dose rate of 61 cGy/year.

The increased percentage of DNA tightly bound to proteins in irradiated relative to unirradiated control animals are considered to reflect induced by γ-irradiation a formation of DPC. As seen in Fig. 1 temporal (doses) dependencies of DPC levels are nonlinear. The maximum levels of DPC were recorded at day 10 and 30–40 (total doses 1.7, 5.1–6.8 cGy, respectively). The DPC level ratio of experimental to control animals at maximum points was ~1.5–1.7.

It is possible that the mechanism of DNA–protein cross-links formations at low doses of external γ-radiation is nonspecific and reflects the structural rearrangements of chromatin. In this case, topoisomerases may be the proteins involved in the DNA–protein cross-links formations under low dose rate irradiation.

SINGLE-STRAND DNA BREAKS

The comet assay study on spleen cells showed that very low dose-rate irradiation resulted in statistically significant increase in single-strand DNA

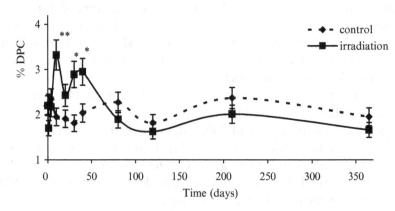

Fig. 1. Exposure time-dependent changes of percentage of DNA tightly bound to proteins (% DPC) in spleen cells of mice exposed to γ-radiation at a dose rate of 61 cGy/ year.* – Statistically significant with $P < 0.05$. ** – Statistically significant with $P < 0.01$

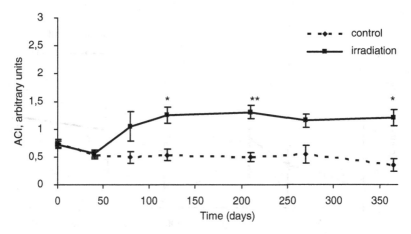

Fig. 2. Chronic irradiation induced time-dependent changes in the single-strand DNA breaks level in mouse spleen cells estimated by alkaline comet assay. ACI – Average comet index. *$P < 0.05$, **$P < 0.01$

breaks level, starting from a dose of 20 cGy (Fig. 2). Further prolongation of exposure time and, hence, increase of a total dose did not, however, lead to further increase in the extent of DNA strand breaks level. A dose–response curve for DNA SSBs is good fitted by a polynomial regression: $y = 0.6209 + 0.0313^*x - 0.0004^*x^2$, where y is the ACI, x is a dose in cGy.

PERCENT OF APOPTOTIC CELLS

A minor part of cells that has an extremely high level of DNA damage (e.g., apoptotic cells) would supposedly contribute substantially to an overall DNA damage level within an entire cellular population. To take into account the contribution of an apoptotic cell subpopulation to a final readout of DNA breaks in our experiments, we measured the percentage of apoptotic spleen lymphocytes from mice exposed to very low dose-rate IR or untreated animals using the "DNA diffusion" assay. At the days 120, 270, and 365 of the chronic irradiation (20, 45, and 61 cGy, respectively), approximately twofold increase over a control level in the apoptotic cell fraction was observed (Fig. 3). As expected, a correlation ($r = 0.86$; $P < 0.05$) between an overall level of DNA damage and percentage of apoptotic cells was noticed. These observations prompted us to recalculate overall DNA damage levels (the ACI coefficient) in irradiated versus untreated groups. When performed without counting highly damaged cells (comets within classes 3 and 4), the comet assay yielded in less, but still statistically significant, difference in DNA damage levels between irradiated (20–61 cGy) and untreated mice (data not shown).

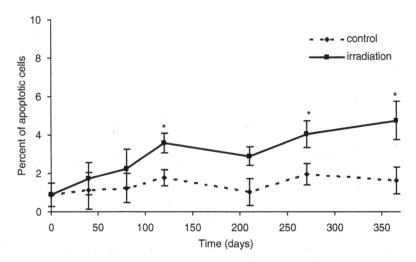

Fig. 3. Percent of apoptotic spleen lymphocytes in control mice and mice exposed to chronic very low dose-rate ionizing radiation for the different times

CELLULAR SENSITIVITY TO ADDITIONAL EXPOSURE

To study the exposure-time changes of the spleen cells sensitivity to additional exposure in the irradiated mice, we have investigated the induction of DNA damage in these cells by hydrogen peroxide. Hydrogen peroxide is a normal cell metabolite which in the presence of redox-active metals (e.g., Fe^{2+}) produced the formation of the highly toxic hydroxyl radical. An increase in sensitivity of the spleen cells from irradiated mice to hydrogen peroxide was revealed at 40 days of experiment (Fig. 4). This fact can probably be explained by the weakening of antioxidant potential/repair capacity in these cells. By 80 days, difference in responses of experimental and control mice splenocytes was decreased. The level of H_2O_2-induced DNA breaks in spleen cells of irradiated animals was slightly higher than that in control cells. Prolongation of irradiation of animals probably activated the defense systems of spleen lymphocytes that were expressed in the decrease of their sensitivity to H_2O_2 exposure at 210 days of experiment (Fig. 4). Thus, our results indicate that chronic action of low dose-rate γ-radiation led to a change in the sensitivity of mice spleen cells to H_2O_2 exposure. A weakening of antioxidant/repair cell potential has been observed at early terms of irradiation. In contrast, prolongation of irradiation resulted in activation of defense system in spleen cells. This effect could be attributed to a development of adaptation processes triggered upon accumulation of a certain dose.

Discussion

An IR dose of 1 Gy induces ~1,000 DNA SSBs per cell (Billen, 1990). Simple calculations based on these data show that irradiation with a dose rate of

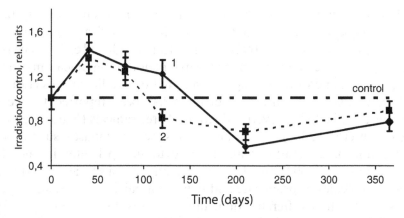

Fig. 4. Dynamics of the sensitivity of the spleen cells from mice continuously exposed to low dose-rate γ-radiation to hydrogen peroxide exposure. Data plotted as the ratio of effects in cells of the experimental animals and that in the control animals

0.07 mGy/h (~0.17 cGy/day) will induce about two DNA SSBs per cell per day, whereas a number of endogenous SSBs resulted from normal oxygen metabolism is ~1.2×10^5 per cell per day (Billen, 1990). It is, therefore, obvious that low dose-rate IR-induced DNA SSBs can hardly influence a total number of DNA breaks. Nature of spontaneous and induced DNA breaks is rather similar. IR-induced primary lesions could lead to genotoxic effects only in case if spatial distribution of the lesions along the chromatin and effectiveness of their repair is different from those spontaneously induced. Studies of the last decade suggest that the particular areas within chromatin possessing hypersensitivity to IR do exist (Oleinick et al., 1994). Double-stranded DNA clusters composed of multiple lesions on opposing DNA strands within a few helical turns are of particular danger to cells, since it is hard for DNA repair machinery to repair them (Goodhead, 1994). These clusters are thought to be crucial IR-induced DNA lesions leading to double-strand breaks (DSBs), and eventually to mutagenesis and cell death (Ahnstrom and Bryant, 1982). Sutherland et al. (2002) demonstrated that as low as 10 cGy IR caused an increase in clustered DNA damage level in human monocytes. Consistent with our results, nondividing primary human fibroblasts exposed to 1 mGy of IR were not able to repair DNA DSBs for several days, while effectiveness of DSB repair after higher doses was much better (Rothkamm and Lobrich, 2003).

An indirect mechanism, also known as a «bystander effect», response of unirradiated directly cells due to signaling originating from irradiated cells (Hall, 2003), can be possibly involved in the effects observed in this study. In this scenario, lesions within supersensitive chromatin regions in a minor, apoptotic cell population, appear to trigger a cascade of metabolic processes in different cell populations on both organ and organism levels. Bystander effects have been demonstrated after both low-LET and high-LET IR exposures (Mothersill et al., 2002). A signal from irradiated cells can be transmitted by direct intercellular contacts (gap junction communications), as well as by cytokines and/or oxygen species secreted by irradiated cells (Lorimore and Wright, 2003). A variety of changes has been reported to occur in bystander cells, including overproduction of free radicals (Narayanan et al., 1997; Leach et al., 2001), induction of stress-related kinases, such as JNK, ERK1/2, and others (Little et al., 2002), cytokines β-1-integrin and IL-1α (Osterreicher et al., 2003). In addition, reactive oxygen species (ROS) can act as signal molecules to propagate and regulate a particular cellular response, such as proliferation, differentiation, and apoptosis (Lehnert and Iyer, 2002). It is well known that actively transcribed DNA sequences are much more susceptible to DNA damage than those in compact chromatin regions due to unlimited accessibility of them for ROS (Chiu et al., 1982; Warters et al., 1987). Increase in actively transcribed genes, together with an increase in ROS production can, therefore, lead to elevated DNA damage. On the other hand, DNA damage in active genes is repaired faster and more efficiently compared with that in silent genes (Oleinick et al., 1984; Bohr, 1987).

Our speculation is supported by results of monitoring reparative and replicative DNA syntheses in mouse bone marrow cells reported by Mazurik et al. (2002), within a collaborative effort with our group, performed on the same mice used in our present study. The authors demonstrated that chronic low dose-rate irradiation of mice substantially induced reparative and replicative DNA syntheses in bone marrow cells (60% and 67% increase; $P<0.01$, $P<0.01$, respectively). As mentioned earlier, activation of DNA replication and repair is associated with increase in DNA strand breaks level. Besides, significant positive correlation ($r=0.87$; $P<0.01$) between DNA strand breaks and superoxide anion-radical content in bone marrow cells of the irradiated mice was shown, indicating additional production of DNA damage by ROS due to the loss of a part of structural proteins and conformational changes in expression sites of the chromatin during gene expression (Mazurik et al., 2002).

It was suggested that the bystander effect has an alternative, protective, feature due to elimination of highly damaged, potentially dangerous cells from a cell population (Belyakov et al., 2002; Prise et al., 2002). In accordance with this line of evidence, an adaptive response, an effect of increased radioresistance to high IR dose acquired after exposure to low doses, was associated with overproduction of ROS (Lehnert and Iyer, 2002). We demonstrated that increase in DNA breaks level by the days 120–365 of low-level IR exposure is accompanied by elevated resistance to hydrogen peroxide treatment. It is possible that the elevated cell resistance could be explained by either activation of DNA repair or elimination of a supersensitive cell population.

Thus, overall increase in the level of DNA breaks in mouse spleen lymphocytes as a result of chronic low dose-rate IR exposure can be possibly explained by structural rearrangement of the chromatin during gene expression activation, free-radical overproduction, and DNA repair activation. Although insignificant, a contribution of apoptotic cells to an overall level of DNA damage was also recorded, providing further support for the proposed mechanisms of low dose-rate radiation-induced effects observed in this study.

The experimental data analysis allows the stage mechanism of the cellular response to chronic low dose-rate ionizing radiation exposure developing with an increase in the exposure time (dose) of irradiation. First stage (doses up to 10–20 cGy) – accumulation of DNA damages (in particularly DNA–protein cross-links) in nonactive (bulk) chromatin, an increase in the cellular sensitivity to additional exposure; second stage (doses of 0.2–0.5 (0.6) Gy) – active response of cells to the damages and as consequence an increase in the quantity of DNA strand breaks, caused by activation of transcription and DNA repair, overproduction of the ROS, and apoptosis induction; acquisitive of the cellular resistance to additional exposure; balance between the DNA damages formation and their repair. And last (hypothetic) third stage (doses upper of 0.5 (0.6) Gy) – additional formation of DNA damages by free radicals due to chromatin conformation changes and exhausting free-radicals defense systems can lead to an increase in the cytogenetic disturbances frequency.

References

Ahnstrom G., and Bryant P.E., 1982, DNA double-strand breaks generated by the repair of X-ray damage in Chinese hamster cells, *Int. J. Radiat. Biol.* 41(6):671–676.

Belyakov O.V., Folkard M., Mothersill C., Prise K.M., and Michael B.D., 2002, Bystander-induced apoptosis and premature differentiation in primary urothelial explants after charged particle microbeam irradiation, *Radiat. Prot. Dosimetry* 99(1–4):249–251.

Billen D., 1990, Spontaneous DNA damage and its significance for the "negligible dose" controversy in radiation protection, *Radiat. Res.* 124:242–245.

Bohr V.A., 1987, Preferential DNA repair in active genes, *Dan. Med. Bull.* 34(6):309–320.

Calabrese E.J., and Baldwin L.A., 2000, Radiation hormesis: its historical foundations as a biological hypothesis, *Hum. Exp. Toxicol.* 19:41–75.

Chiu S.M., Oleinick N.L., Friedman L.R., and Stambrook P.J., 1982, Hypersensitivity of DNA in transcriptionally active chromatin to ionizing radiation, *Biochim. Biophys. Acta* 699(1):15–21.

Goodhead D.T., 1994, Initial events in the cellular effects of ionizing radiation: clustered damage in DNA, *Int. J. Rad. Biol.* 65:7–17.

Hall E.J., 2003, The bystander effect, *Health Phys.* 85(1):31–35.

Kobayashi H., Sugiyama C., Morikawa Y., Hayashi M., and Sofuni T., 1995, A comparison between manual microscopic analysis and computerized image analysis in the single cell gel electrophoresis assay, *MMS Commun.* 3:103–115.

Leach, J.K., Van Tuyle, G., Lin P.S., Schmidt-Uurich, R., and Mikkelsen, R.B., 2001, *Cancer Res.* 61:3894–3901.

Lehnert B.E., and Iyer R., 2002, Exposure to low-level chemicals and ionizing radiation: reactive oxygen species and cellular pathways, *Hum. Exp. Toxicol.* 21(2):65–69.

Little J.B., Azzam E.I., de Toledo S.M., and Nagasawa H., 2002, Bystander effects: intercellular transmission of radiation damage signals, *Radiat. Prot. Dosimetry* 99(1–4):159–162.

Lorimore S.A., and Wright E.G., 2003, Radiation-induced genomic instability and bystander effects: related inflammatory-type responses to radiation-induced stress and injury? A review. *Int. J. Radiat. Biol.* 79(1):15–25.

Mazurik V.K., Mikhailov V.F., and Ushenkova L.N., 2002, Dynamic component of maintaining genomic stability in murine bone marrow cells after chronic low-intensity irradiation lasting one year, *Radiat. Biol. Radioecol.* 42(6):669–674.

Mothersill C., O'Malley K., and Seymour C.B., 2002, Characterisation of a bystander effect induced in human tissue explant cultures by low let radiation, *Radiat. Prot. Dosimetry* 99(1–4):163–167.

Mothersill C., and Seymour C., 2003, Low-dose radiation effects: experimental hematology and the changing paradigm. *Exp. Hematol.* 31(6):437–445.

Narayanan, P.K., Goodwin, E.Hr., and Lehnert, B.E., 1997, Alpha particles initiate biological production of superoxide anions & hydrogen peroxide in human cells. *Cancer Res.* 57:3963–3971.

Oleinick N.L., Balasubramaniam U., Xue L., and Chiu S., 1994, Nuclear structure and the microdistribution of radiation damage in DNA, *Int. J. Radiat. Biol.* 66:523–529.

Oleinick N.L., Chiu S.M., and Friedman L.R., 1984, Gamma radiation as a probe of chromatin structure: damage to and repair of active chromatin in the metaphase chromosome, *Radiat. Res.* 98(3):629–641.

Oleinick N.L., Chiu S.-M., Ramakrishnan N., and Xue L.-Y., 1987, The formation, identification, and significance of DNA–protein cross-links in mammalian cells, *Brit. J. Cancer* 55:135–140.

Olive, P.L., 1988, DNA precipitation assay: a Rapid & Simple method for detecting DNA damage in mammalian cells, *Environ. Mol. Mutagen* 11:487–495.

Osipov A.N., Sypin V.D., Puchkov P.V., Razumova A.S., and Kuznetsova E.M., 2000, Changes in the level of DNA–protein cross-links in spleen lymphocytes of mice exposed to low-intensity γ-radiation at low doses, *Radiat. Biol. Radioecol.* 40(5):516–519.

Osterreicher J., Skopek J., Jahns J., Hildebrandt G., Psutka J., Vilasova Z., Tanner J.M., Vogt J., and Butz T., 2003, Beta1-integrin and IL-1alpha expression as bystander effect of medium from irradiated cells: the pilot study, *Acta Histochem.* 105(3):223–230.

Prise K.M., Belyakov O.V., Newman H.C., Patel S., Schettino G., Folkard M., and Michael B.D., 2002, Non-targeted effects of radiation: bystander responses in cell and tissue models, *Radiat. Prot. Dosimetry* 99(1–4):223–226.

Rothkamm K., and Lobrich M., 2003, Evidence for a lack of DNA double-strand break repair in human cells exposed to very low x-ray doses, *Proc. Natl. Acad. Sci. USA* 100(9):5057–5062.

Singh N.P., 2000, A simple method for accurate estimation of apoptotic cells, *Exp. Cell Res.* 256(1):328–337.

Singh N.P., McCoy M.T., Tice R.R., and Schneider E.L., 1988, A simple technique for quantification of low levels of DNA damage in individual cells, *Exp. Cell Res.* 175:184–191.

Sutherland B.M., Bennett P.V., Cintron-Torres N., Hada M., Trunk J., Monteleone D., Sutherland J.C., Laval J., Stanislaus M., and Gewirtz A., 2002, Clustered DNA damages induced in human hematopoietic cells by low doses of ionizing radiation, *J. Radiat. Res. (Tokyo)* 43(Suppl):149–152.

Warters R.L., Lyons B.W., Chiu S.M., and Oleinick N.L., 1987, Induction of DNA strand breaks in transcriptionally active DNA sequences of mouse cells by low doses of ionizing radiation, *Mutat. Res.* 180(1):21–29.

Zhitkovich A., and Costa M., 1992, A simple, sensitive assay to detect DNA–protein cross-links in intact cells and in vivo, *Carcinogenesis* 13:1485–1489.

SECTION 6

MULTIPLE STRESSORS – RISK ASSESSMENT AND LEGAL/ETHICAL ASPECTS

CHAPTER 33

MODELING THE BEST USE OF INVESTMENTS FOR MINIMIZING RISKS OF MULTIPLE STRESSORS ON THE ENVIRONMENT

GANNA KHARLAMOVA

Kiev National Taras Shevchenko University, Kiev, Ukraine
e-mail: akharlam@svitonline.com

Abstract: This chapter contains the assumption about a novel perspective on the relationship between foreign direct investment (FDI) and the stringency of environmental policies. We have created, by means of powerful econometric apparatus, possible models for equality of positive and negative influences of FDI on environmental security.

Keywords: foreign direct investments; environmental policies; ecological-economic model.

Introduction

Problems of preservation of the environment are mainly the topic for investigation for physicists, chemists, and biologists. But as one of the main parts of state safety and as a component of a nation's level of development, a country's environmental strategy should be developed also from the position of economic theory.

There are next to no publications concerning economic tools for the evaluation of the impact of environmental regulations in the world. This problem is nearly ignored in transition economies like Ukraine and other Commonwealth of Independent States (CIS) states.

Foreign Direct Investment and Environmental Policy: Theoretical Aspects

Attraction of FDI is becoming increasingly important for developing and transition economies. However, this is often based on the implicit assumption that greater inflows of FDI will bring certain benefits to the country's economy.

The character of environmental security problems today has changed. The emphasis today has shifted toward how to minimize the burden on environment caused by business activity. Moreover, many environment problems are global in scale, complexly interlaced.

441

C. Mothersill et al. (eds.), Multiple Stressors: A Challenge for the Future, 441–448.

There is a necessity in a well-grounded system of regulation of environment safety. Ecological standards nowadays are not only technical, but also economic, possibilities. At the same time there is a need in circumstances for reaching of economic optimum between productive efficiency, external costs, and ecological damages.

Developing effective environment policy for a sustainable future should imply the analysis of the environment and investing, especially foreign investing (eco-business trends, environment related industries, etc.). Gets comprehensible that further operating of an economy at condition of the absence goal-directed actions for account of the ecological factor in its structure, threatens arising the ecological blast in Ukraine, and other transition economy.

But to realize "pros and cons" of FDI as a source for minimizing risks of multiple stressors on the environment, one should realize global tendencies in investment and macroecological-economic aspects.

The circumstances of shortage of internal capital resources in transition countries force these countries mostly to rely on foreign investments to address environment issues. Spillovers due to foreign corporate presence include technology diffusion and development of less "pollute" productions.

Investments (by means of their impact on economic growth) in a recipient-state and transfer of new ecologically safe technologies must stimulate improvements in environmental security. However "the reverse of the medal" is often in fact the "pollution haven." This means that investing companies move operations to transition countries to take advantage of less stringent environmental regulations than in other developed countries. In addition, all countries may purposely undervalue their environment in order to attract new investment. Either way this can lead to excessive levels of pollution and environmental degradation.

The topic of this research is to determine the equilibrium of "benefits" and "costs" of foreign capital interference in issues of environmental security regulation in recipient-states (transition economies). The answer to the problem shares most risk-related methodologies, IT tools, and data sources, so they can be dealt with a synergistically coordinated way.

Ecological-Economic Models

In this research, modeling of foreign direct investment (FDI) effects on the ecology effective economy divided on two stages:

1. Creating of dynamic and optimization models
2. Creating of multifactor cross-section regression models.

As to dynamic and optimization models, there are two main directions in building of ecological-economic models:

1. With account of ecological factor in economic-mathematical models

2. With account of anthropogenic impact in models of ecosystems.

Models of the first type have a traditional structure of economic-mathematical models; include additional variables and connections that characterize ecological subsystem.

At the basis of second type models is a model of mathematical ecology, and anthropogenic activity is considered as exogenous impact on ecosystem.

The classic representatives of both types of ecological-economic models are Leontiev-Ford model and Mono-Irusalimskiy model correspondently.

The character of ecological-economical models is *controllability* – the presence of vacant exogenic variables that one can define oneself. As usual, combinations of values of defined variables are combined in scenarios of regulations of ecological-economic systems.

For receiving the adequate and reliable results of ecological-economic modeling, we attempt to describe system "environment-economy" – to make a model of ecological-economic production function of maximization of output.

So the general model of one-sector ecological-economic production

function we propose like this:
$$f(x) \to \max$$
$$\phi(x) \le a$$
$$\Psi(x) \le b$$
$$x \text{ from } T_x$$

where $f(x)$ is the income from sale of products' vector x; $\varphi(x)$, $\psi(x)$ are the vectors of costs of economic a and ecological b resources; ecological resource b is the actual damage (in money equivalent) to society as a result of pollution of firms or additional costs for compensation for such damages.

According the same scheme – two-sector ecological-economic production function in the case of liner dependence of income, costs of economic and ecological resources from intensity of production is like

this:
$$p_0 mx \to \max$$
$$A_1 x + A_2 y \le a$$
$$B_1 x - B_2 y \le b$$
$$0 \le x \le 1$$
$$0 \le y \le 1$$

So we have the task of parametric liner programming (under terms of perfect competition), where:

p_0 – a price of output

m – a vector-row of main production capacity

x – a vector-column of an intensity of main production technology

y – a vector-column of an intensity of technology in auxiliary process (destruction of pollutants)

a – a vector-column of available economic resources

$A_1 \geq 0$ – a matrix of standards of resource expenses in main production

$A_2 \geq 0$ – a matrix of standards of resource expenses in auxiliary process

b – a vector-column of limits on emissions (ecological resource)

$B_1 \geq 0$ – a matrix of standards of emissions of pollutants in main production

$B_2 \geq 0$ – a matrix of standards of destruction of pollutants in auxiliary process.

Empirical Analysis

Generally, statistical studies (Keller, 2002; Kharlamova, 2005; Mabey and McNally 1992), show that the effect of FDI on environment cannot be clearly identified and there is a need in a mathematically based modeling of situations. So this research makes an attempt to show the possibility of the creation of new ecological-economic models that give a possibility to optimize production at the presence of ecological restrictions; for further creation of ecological and economic regulation policies.

Dependence of ecological pollution level from FDI inward in Ukraine is described on Fig. 1, where variable FDI – FDI in Ukraine ($mln. USA)

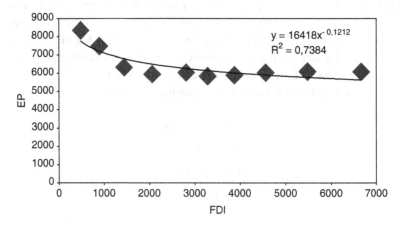

Fig. 1. Dependence of ecological pollution level from FDI inward in Ukraine in 1995–2004.

(1995–2004), variable EP – harmful emissions in environment of Ukraine (thous. t.) (1995–2004).

This research used panel data from 20 countries (taking in account CIS transition states): USA, Canada, Japan, Ukraine, Russia, Belarus, Moldova Rep., France, Germany, UK, Bulgaria, Czech Rep., Estonia, Hungary, Latvia, Lithuania, Poland, Romania, Slovak Rep., and Slovenia. The cross-country data support the estimation of multifactor regression models with fixed effects for evaluating FDI-environment mutual influence in the time period 1995–2004:

$$REGS_{it} = \alpha_i + \beta X + \varepsilon_{it}$$

where

$REGS_{it}$ – variable estimating the volume of ecological regulation in a state-recipient i in t year

α_i – constant time fixed effect

X – a vector of independent variables

ε_{it} – error of estimation.

Like in the work of Cole et al. (2004), a dependent variable – $REGS_{it}$ – we received by multiplying the lead content in gasoline variable by −1. Thus, an increase in REGS represents an increase in the stringency of regulations (i.e., a decrease in lead content) in all models.

As factor variables were choused:

FDI – FDI inflows (thous. US$)

CORRUPT – the level of government corruption (data of annual International Country Risk Guide)

FDICORRUPT – variable determining effect of interaction FDI and government corruption. Assumption: estimated coefficient β^*FDI expected to be positive, while coefficient of FDICORRUPT expected to be negative

GDP – GDP of a state-recipient (thous. US$)

URBANPOPsh – a share of urban population of a state-recipient (%). Foreword: urban population of a state mostly suffer from industrial pollution. The most negative ecological effect from "pollute" FDI inward is in the very urbanized states

MANUFsh – a share of industrial workers in the whole amount of workers in a state-recipient (%)

GDPgr – GDP growth rate (as a variable of an economy growth rate) (%)

EAP – present economically active population of a state-recipient, as a variable that descript a scale of a state-recipient.

The source of statistical data: the National Committee of Statistics of Ukraine, the Global Corruption Report (2003, 2004), the Little Green Data

Book (2003, 2004), the World Bank Reports (2003, 2004), Lovei (1998), World Investment Reports (2003, 2004).
Results indicate:

- FDI can have a positive impact on the strength of environmental regulations for all analyzed states
- The variable of the connection effect of FDI inward with the parameter of corruption level in the recipient-country is significant and negative in models for all states. It has been observed that if the degree of corruptibility is sufficiently high (low), FDI leads to less (more) stringent environmental policy, and FDI thus contributes (mitigates) to the creation of a pollution haven
- Variable of interrelation effect FDICORRUPT is significant and negative in all models, that confirms assumption about dependence of a FDI effect on the environment from corruption in a government of a state-recipient
- Variables GDPgr and GDP make positive impact on dependent variable and are significant in all models
- Variables EAP, URBANPOPsh, and MANUFsh detected as negative impacted on dependent variable and significant in all models
- Variable CORRUPT is statistically nonsignificant in model for CEE states
- Models are statistically appropriate for further forecasts
- Sensitivity analysis shows that variables URBANPOPsh and MANUFsh ARE NOT important parameters in the models
- As to Ukraine: correlation analysis FDI and $REGS_{it}$ shows inverse negative relationship of FDI inward and environment pollution.

The building of the "FDI-environment improvement" model for macrodata of transition economies was based on the classic "predator-sacrifice" model. The model showed that investments with innovation contributing to more than half of the production of firms of main productive economic industries of a recipient-state leads to decreasing of harmful emissions of these firms at not less than 20%.

Conclusions

Undoubtedly, economic tools can greater and more effectively affect environmental security in the state at macro- as at microlevels. However, FDI is unlikely unless investors have a reasonable understanding of the environment in which they will be operating.

Overall, every aspect of host countries' economic and governance practices affects the environment climate. On the answer "What may host countries do about it?" the answer should be:

The following policy action toward macroeconomic stability and ecological predictability should be priority:

- Pursue sound macroeconomic policies geared to sustained high economic growth and environment standards, attraction of "not pollute" and economically effective FDI inward.

- Strengthen domestic business climate in order to make domestic financial resources available to supplement and complement foreign investment for improving environment security.

Moreover, the combination of governmental policies and the national system of economic transfer toward the integration of the environment and sustainability could be considered as a conceptual State strategy. Authorities need to consider the following challenge: enshrine the principle of discrimination for "pollute" FDI in national legislation and implement procedures to enforce it through all levels of government and public administration.

There is a necessity in conduction of such incentives:

- To strengthen encouraging laws for investors who improve environment conditions of their business activity in any sector of economy of state-recipient FDI (especially, in tourism and forestry)

- To reform current and planned investment contracts in such a way to avoid "race to bottom" in environment legislation and in usage of natural resources

- To form the structure of international regulation and coordination for guarantees of positive impact of FDI on a stable economic and ecological development.

Any of these regulation advices does not need the creation of great and grave organization structure. But all advices can strengthen the connection between FDI inward and the level of environment safety in the state-recipient with further economic growth possibilities.

References

Cole, M.A., Elliott, R.J.R., and Fredriksson, P.G., Endogenous Pollution Havens: Does FDI Influence Environmental Regulations?//Research Paper 2004/20, Research paper series Internationalization of Economic Policy, 26 p.

Keller, W. and Levinson, A., Environmental regulations and FDI inflows to the U.S. States, *Review of Economics and Statistics* 84(4): 691–703, 2002.

Kharlamova, G., The Modeling of Mutual Influences of Ecological and Investing Processes// Abstr. of Scientific-Practical Conference "Models and Informational Technologies in Management of Social, Economic, Technical and Environmental Systems", 20–21 April 2005, Lugansk, Ukraine, pp. 140–144.

Kharlamova, G., To the Problem of Modeling Environmental Process in Interdependence with Investments//The Bulletin of Vladimir Dale East-Ukrainian National University, No. 5 (87), 2005, pp. 214–219.

Lovei, M., Phasing out Lead from Gasoline Worldwide Experience and Policy Implications// The World Bank, Washington, D.C., USA, 1998, 59 p.

Mabey, N. and McNally, R., Foreign Direct Investment and the Environment: From Pollution Havens to Sustainable Development//WWF-UK Report, Panda House, Weyside Park Godalming, 1998, 100 p.

The National Committee of Statistics of Ukraine. Ukraine in figures – Various issues. http://www.ukrstat.gov.ua

The Global Corruption Report 2003, Transparency International, 2003, Berlin: Transparency International http://www.globalcorruptionreport.org

The Global Corruption Report 2004, Transparency International, 2004, Berlin: Transparency International http://www.globalcorruptionreport.org

The Little Green Data Book 2003, World Development Indicators, 2003, The World Bank Group www.worldbank.org

The Little Green Data Book 2004, World Development Indicators, 2004, The World Bank Group www.worldbank.org

The World Bank. World Development Indicators – Various issues. http://www.worldbank.org/data

World Investment Report 2003, FDI policies for development: National and International perspectives. United Nations – NY and Geneva, 2003, 322 p.

World Investment Report 2004, The Shift towards Services/United Nations. NY, Geneva, 2004.

CHAPTER 34

LEARNING FROM CHERNOBYL: PAST AND PRESENT RESPONSES

OLEG UDOVYK

National Institute for Strategic Studies, Kyiv, Ukraine
e-mail: oleg_udovyk@hotmail.com

Abstract: This chapter reviews the past and present national and international mitigation programs, their respective major areas of activities, financing, and implementation, as well as adequacy and efficiency of the strategies used in addressing the problems that followed the Chernobyl catastrophe 26 April 1986.

Keywords: Chernobyl catastrophe; response; programs of mitigation

Early Handling the Chernobyl Catastrophe

The first priorities in early handling the Chernobyl catastrophe related to the sealing off the area of greatest contamination; resettling of people from the severely contaminated zones; protecting population against the use of contaminated food, water, and air and against a spillover of such contamination; implementing a thorough system of medical surveillance, screening analysis, and treatment; and establishing a complex system of financial, medical, and social support for ~3 million people defined as its "victims."(Brumfiel, 2006)

Given the magnitude of the disaster and the long-term nature of radiation contamination and illness, it should come as no surprise that all these operations required extraordinary measures and massive humanitarian aid, and that their delivery in its early stages was in many ways a quasi-military operation.

However, what must have been the only way of delivery the necessary assistance and of containment a nuclear explosion in the first period after the catastrophe became a serious impediment to social and economic recovery and a source of social apathy and psychological and interpersonal traumas for many Chernobyl sufferers, Chernobyl resettles, and their new neighbors (Mousseau et al., 2005; Williams and Baverstock, 2006).

C. Mothersill et al. (eds.), Multiple Stressors: A Challenge for the Future, 449–453.

The Present Chernobyl Assistance System

The present "top-down" paradigm of Chernobyl assistance system generated among many positive things also aid dependency, widespread social apathy and a "Chernobyl victim" syndrome.

Moreover, following the worst tradition of the Soviet central planning, with no respect for realism and setting unachievable long- and medium-term targets wherever short run supply shortfalls are gravest, financially over ambitious targets have been set not only in each and every round of deciding priorities and targets of the Government of Ukraine's Chernobyl programs ever since 1991.

The full volumes of assistance and privileges granted by laws were reconfirmed by the Constitution of Ukraine. Its Articles 16 and 22 taken together make benefits and privileges once granted by any law irrevocable, which opens way for only too easy but very effective political demagoguery, without any respect for financial constrains.

Consequently, the vicious circle of inadequate means and irrevocable commitments offers no chances of effective overcoming not only of economic consequences of the Chernobyl catastrophe, but – no less important – its social consequences.

The New Chernobyl Strategic Paradigm

It was proposed to square this vicious circle through revising the present paradigm of strategy of assistance in such a way (and in line with the present UNDP *Chernobyl Recovery and Development Program, UKR/02/2005*) as to assist affected people and communities to initiate their own recovery through organizing themselves into self-governing organizations to take the lead in planning, managing, and implementing their own social, economic, and ecological rehabilitation and development.

More specifically, it was put forward the following measures that could help achieve shifting of the Chernobyl strategy paradigm:

First, considering that the Government Chernobyl programs presently under operation are based on completely outdated legislation which effectively deters even development of a concept paper of a new strategy for overcoming the consequences of Chernobyl catastrophe, a number of revisions of the present Chernobyl-related legislation are put forward. It must go beyond its harmonization with that in other areas, such as taxation, regional development, land use, environment protection, and so on, and go in step with revisions in NPMs, should the latter continue, lest mutual consistency between the former and the latter is lost (Giles, 2006; Howard, 2006).

Moreover, the new legal framework should be suitable for and compatible with the new strategy of social and economic rehabilitation and development of Chernobyl territories. Consequently, revisions in the Law on Legal Regime of Chernobyl affected territories should account for improvement in the radiation position in the past 20 years and should therefore lift the present restrictions on economic and other activities.

Revisions in the Law on the Status should help introduce a new and far better targeted system of social, medical, and resettlement-related assistance. Regarding economic environment that would encourage business operations and investments, legislation that presently applies to everywhere else in Ukraine but Chernobyl territories must apply to Chernobyl territories as well.

Moreover special powers should be attributed to local governments there, at the expense of powers presently vested in central and/or oblast governments, especially in safeguarding the participatory nature of development and implementation of plans of area and local development.

The same refers to more fiscal decentralization, the right of rayon's (and community) councils to impose local contributions, greater flexibility in the use of present legislation in force in Ukraine regarding special economic zones, territories of priority development, and so on.

Second, it was emphasized the need for sound information on levels of the present radiological contamination, as well as that of output produced there, whether marketed or produced for own use, and whether sold in the Chernobyl affected territories or outside them.

This information must be reliable, sufficiently disaggregated, open to the public (and the media) scrutiny, and supported by a closed monitoring system.

It should include increased involvement of Ukrainian and international experts and its undisputed reliability is a significant condition for success in shifting the Chernobyl strategy priorities, in improvement of targeting of the assistance system to Chernobyl sufferers and for all economic recovery and development projects.

Third, the system of social assistance (medical and resettlement assistance too) should undergo not only improved targeting (far not accomplished yet) but – more importantly – a conceptual redesigning.

Considering the political space for limitation of the Chernobyl-related social assistance, we distinguished between assistance linked to radiation overdose absorbed by individuals, and radiation intensity linked with territory.

The former, once legitimacy of claims is reconfirmed and its total value is added to that of medical and resettlement-related assistance, could become a subject of buy-up operations by the treasury, at a discount, in all cases where individual Chernobyl sufferers were prepared to use the lump-sum down payment of undisputable present and future claims for start-ups of their individual business.

A parallel solution is also proposed regarding claims that relate to territorial privileges, as are measures that could help mitigate opposition against such contracts.

It is also discussed some technical issues related to who should legally be entitled to put forward such an offer on the part of the treasury, what should be the discount between total volume of outstanding and future claims and the offer to buy them up (for a strictly defined purpose and under condition of surrendering any further claims in future), what should be the composition of the panel for interviewing individual sufferers, and what sort of institutions would be responsible for establishing an environment that would encourage individual sufferers to positively respond to any such initiative, and so on.

We examined economic recovery schemes that could be applied in the Chernobyl affected territories. Instrument-wise they need to follow a standard menu of options for area and local (regional) development planning which is as much relevant to Chernobyl as to Ukraine as a whole: improved business environment and public governance (more transparency and less corruption), simplified business registration procedures, improved commercial and other judiciary operations, improved access to commercial finance, and so on.

We found three factors on specific importance for rehabilitation and development of Chernobyl territories.

The first is road, gas, water, and other infrastructure. These are costs that no single business can sustain, certainly no small- or medium-sized that are to become a true engine of growth and employment. Lack of infrastructure is also impediment for new investments, whether Ukrainian or foreign.

The second is sales markets. There are hardly any instances of successful restructuring and development strategies in their absence. It is also for this reason that introduction of the system of international product certification (which is not, of course, Chernobyl-specific, but which may be of far greater significance in the Chernobyl-affected territories than elsewhere in Ukraine) is especially important, as are special measures for marketing output from territories that were in the past or continue to be radioactively contaminated.

The third is the need, possibly more in Chernobyl than elsewhere in Ukraine, for Private–Public-Partnerships. It is important for participatory nature of any development projects and initiatives. It is important because it can offer better management and business oversight of these projects. It is important because it may offer additional financial and other inputs in an environment of finance and other resource shortage contrasted with large and urgent needs.

Conclusions and Recommendations

Following the 2006 *National Workshop on Chernobyl*, we recommended to maintain policy development processes regarding Chernobyl recovery toward:

- Intensifying policy dialogue with concerned government agencies on macro and sector policy issues related to the elaboration of the new Chernobyl strategy concept paper

- Prioritizing the Chernobyl region within the UN Country Office in Ukraine when areas are being selected for pilot interventions, and developing and implementing special policy instruments that would be adjusted to special characteristics of the localities of intervention

- Enhancing the integration of UN activities and local and national government initiatives, particularly with those that support the implementation of the new strategy

- Strengthening local partnerships with donors and related resource mobilization

- Enhancing public relations activities to support policy dialogue and promote the new paradigm of UN strategy for Chernobyl recovery, and through a targeted public information campaign, launched especially at the regional and rayon levels, to help better understand and internalize the essence of the new strategy thereby reducing the potential political opposition against it.

References

G. Brumfiel, Nuclear waste: Forward planning, Nature 440, 987–989, 2006.

Editorial, Recycling the past, Nature 439.

J. Giles, Nuclear power: When the price is right, Nature 440, 984–986, 2006.

B. Howard, Nuclear reactions, Nature 437, 955–955, 2005.

T.A. Mousseau, N. Nelson, and V. Shestopalov, Don't underestimate the death rate from Chernobyl, Nature 437, 1089–1089, 2005 (Correspondence).

News, Special Report: Counting the dead, Nature 440, 982–983, 2006.

D. Williams, and K. Baverstock, Chernobyl and the future: Too soon for a final diagnosis, Nature 440, 993–994, 2006.

CHAPTER 35

UNCERTAINTIES FROM MULTIPLE STRESSORS: CHALLENGES IN ECOLOGICAL RISK ASSESSMENT

DEBORAH OUGHTON

Department of Plant and Environmental Sciences, P.O. Box 5003, Norwegian University of Life Sciences, 1432 Aas, Norway. e-mail: deborah.oughton@umb.no

Abstract: This paper gives an overview of some of the main sources of uncertainty in Ecological Risk Assessment (ERA). These include numerical or statistical uncertainties (equivalent to *inexactness*), model and scenario uncertainties (*unreliability*), epistemological uncertainties (*ignorance* or *indeterminacy*) and ethical-social uncertainties. The paper argues that scientists and policy makers need to appreciate the deeper dimensions of uncertainty, and to go beyond a simple, numerical evaluation of uncertainty. Many of the challenges that multiple stressors raise for risk assessors reflect the fact that the uncertainties associated with multiple stressors are difficult to quantify and irreducible. The focus should be on a better understanding of the basic biological processes and mechanisms by which stressors interact.

Keywords: Ecological Risk Assessment, uncertainty, variability.

Introduction

Ecological risk assessment (ERA) is concerned with the evaluation of the impacts of environmental stressors on non-human species and ecosystems. Historically, radiation protection has dealt almost exclusively with the risks to humans. However, in recent years there has been an general consensus on the need to address effects on non-human species, and a number of projects and frameworks for protection of the environment from ionising radiation have been proposed (Pentreath, ICRP, ERICA, FASSET). Most of these approaches follow the four stages of ERA: (i) hazard identification or problem formulation; (ii) exposure assessment; (iii) effects assessment and (iv) risk characterisation (Fig. 1).

Like any complex environmental problem, ERA is confounded by a variety of sources of uncertainty. At all stages, from problem formulation to risk characterisation, the assessment is dependent on models, scenarios,

C. Mothersill et al. (eds.), Multiple Stressors: A Challenge for the Future, 455–465.
© 2007 *Springer.*

assumptions and extrapolations. These include technical uncertainties related to the data used, conceptual uncertainties associated with models and scenarios, as well as social uncertainties such as economic impacts, the interpretation of legislation and the acceptability of the assessment results to stakeholders. For radioactive substances, the knowledge base underlying the assessment is characterised by uncertainties on parameters (e.g., large data gaps on transfer coefficients and lack of knowledge on ecologically relevant effects such as reproduction), multi-causality (i.e. observed and predicted effects are not exclusive to one stressor) and imperfect understanding (e.g. complexity in extrapolation from individual to population to ecosystem effects).

Environmental risk assessment needs to deal openly with the deeper dimensions of uncertainty and acknowledge that uncertainty is intrinsic to complex systems. Many of the uncertainties associated with environmental risk cannot be quantified, and unforeseen complexities often make it the case that more research will not result in a reduction in uncertainty. This chapter presents an overview of the various types of uncertainties influencing assessment of the ecological risks associated with radioactive substances, arguing that the uncertainties associated with multiple stressors fall into the category of conceptual and epistemological uncertainties.

Ecological Risk Assessment

ERA as a science has undergone considerable development during the last decades. There is currently a general agreement that risk assessment is best addressed in four stages (Environment Canada, 1997; Suter et al., 2000; EC, 2003), described as (Fig. 1):

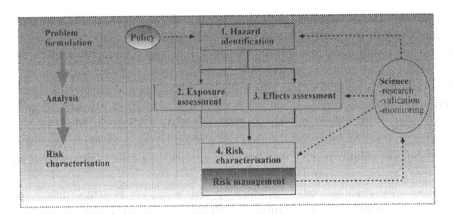

Fig. 1. Ecological risk assessment.

1. *Hazard Identification/Problem Formulation*: Identification of the inherent capacity of a substance to cause adverse effects including description of the source and affected environment, as well as identification of what is to be protected.

2. *Exposure Assessment*: Estimation of the concentration/dose to which environmental compartments (e.g. aquatic, terrestrial and air compartments) have been, are, or may be exposed. This estimation entails the determination of the sources, emission routes and degradation pathways of the substance, and distribution between the various compartments.

3. *Effects Assessment*: Estimation of the relationship between dose, or level of exposure to the substance, and the incidence and severity of an effect.

4. *Risk Characterisation*: Estimation of the incidence, severity and probability of effects likely to occur in the affected ecosystem, due to actual or predicted exposure to a substance.

Ideally, risk characterisation should produce a quantitative estimate of the risk in exposed population or estimates of the potential risk under different plausible exposure scenarios. In short, the risk characterisation stage attempts to make sense of the available information on exposure and effects and to describe what it means (ERICA, 2005a). The most straightforward and widespread approach is the use of a risk quotient (RQ), wherein predicted environmental concentrations (PEC) are compared with predicted no effect concentrations (PNEC) to give ratios for the different environmental compartments considered, that is

$$RQ = PEC/PNEC.$$

RQ values lower than one are generally deemed to be acceptable; values greater than one either require reconsideration, such as further information and/or testing for refinement of PECs and/or PNECs, or suggest the need for action, that is risk reduction (EC, 2003a).

In summary, ERA is concerned with three main questions:

1. What is the concentration of pollutants in the environment?

2. What organisms are exposed to pollutants and to what dose?

3. What are the effects of that exposure on population, ecosystem stability and/or biodiversity?

There are a number of differences in assessing the ecological risk of ionising radiation as compared with assessing risks to humans, for example, different driving forces (protection of ecosystems and biodiversity rather than protection of the individual) and a focus on different biological end points (reproduction rather than cancer). However, there are also a number of issues that are relevant for both cases: ecosystem transfer; dose–response relationships; biological interaction mechanisms; non-targeted effects and the impact of multiple stressors.

Uncertainties

TYPES OF UNCERTAINTIES

There is no universally accepted terminology for classifying the different types and sources of uncertainty. The most widely used distinction with respect to statistical uncertainties on scientific and technical information is that between uncertainty and variability:

- *Uncertainty* (*Type I uncertainty*) – lack of scientific knowledge about specific factors, parameters or models (that can partly be reduced through further study). Includes parameter, model and scenario uncertainties. For random variables, the uncertain belief about the likelihood of the variable having different values can be represented by probability distribution.

- *Variability* (*Type II uncertainty*) – natural variability arising from true heterogeneity, such as the observable variation resulting from randomness in nature (e.g. genetic variability or natural climate variability), as well as population, temporal and spatial variability. It is sometimes described as "noise" or "baseline variation". Statistical measures include mean, variance and standard variation, frequency and probability distribution. Natural variability is not usually reduced by further research. Although, for circumstances where a population has not been sufficiently sampled to provide satisfactory information on variability, this is an example of a Type 1 uncertainty about variability.

A number of workers have suggested that the concept of uncertainty needs to go beyond that of technical and data uncertainty to include factors such as conceptual uncertainties, indeterminancy and social uncertainties (e.g. Wynne, 1992; CERRIE, 2004). Whereas data uncertainties can be expressed using statistical methods, such as error and probability distribution functions, and are best described as inexactness or impression, conceptual uncertainties tend to be difficult to quantify, and introduce a more inherent unreliability into risk assessment. The inclusion of social uncertainties adds a further dimension to the evaluation of the assessment and includes both quantifiable (e.g. economic) and qualitative (stakeholder acceptability) aspects. Building on this, Walker et al. (2002) suggest that uncertainty in risk assessment can be classified along the following dimensions: its *location* (where it occurs), its *level* (whether it can best be classified as statistical uncertainty, scenario uncertainty or recognised ignorance) and its *nature* (knowledge related uncertainty or inherent variability). The nature dimension can be equated to the standard classification described earlier. The location and level dimensions are described later.

LOCATION

In ERA, the location of the uncertainty can be linked to the four stages of risk assessment: problem formulation (including the source characterisation, stakeholder consultation, design of conceptual model and selection of criteria); exposure analysis, effects analysis and risk characterisation. The problem formulation includes a variety of conceptual and societal uncertainties, whereas the exposure and effects analysis largely concerns uncertainties related to model parameters and inputs. The level and nature of uncertainty associated with a model parameter can change depending on location. For example, effects analysis might consist of a statistical analysis of the distribution of LD50 over a variety of end points and species, whereas in risk characterisation there are additional uncertainties concerning the applicability of the effects data to the ecosystem in question, selection of no-effect concentration or dose, extrapolation issues and use of safety factors. Alternatively, on site monitoring may change the range and precision of a number of parameters associated with exposure estimation.

LEVEL

According to Walker at al (2002), the "level" dimension gives an indication of the degree to which the uncertainty lends itself to characterisation using statistical methods namely the difference between statistical/technical uncertainty (inexactness or imprecision), model/scenario/conceptual uncertainty (unreliability) and epistemological uncertainty (ignorance or indeterminacy).

Statistical (or technical, data or numerical) uncertainty arises from uncertainties in the values of physical quantities used in calculations, most obviously in the data for input to models, but also in the parameters used within the models themselves and the model output. Examples in ERA include discharge rates, transfer coefficients and PNEC. Statistical uncertainty can be described using standard techniques, and includes factors such as range, standard deviation, precision and distribution type. These types of statistical error can be dealt with mathematically in model calculation, ranging from straightforward error propagation to probabilistic risk analysis and Monte Carlo analysis. The errors on data may be random or biased. Random errors may be estimated and treated statistically. Bias is often generated by use of subjective judgements, for example, deliberate "pessimism", "conservative estimation", "caution" and so

on. Examples of subjective bias in risk assessment include selection of the
95% percentile or the application of a safety or uncertainty factor.

Model and Scenario Uncertainties include mechanistic uncertainties that arise
from the use of numerical/mathematical models to represent physical systems,
both environmental and biological, and conceptual uncertainties from the overall
structure and concept of the model and scenario (transfer pathways, ecosystem
interactions). Uncertainties with models arise because models are imperfect rep-
resentations of real systems, and problems arise because the model structure may
be inadequate, there may be a mismatch between model and reality, or the model
may be over-complex in relation to knowledge. A model may produce uncertainty
in its predictions from flaws in its structure, including parameter values, or from
inadequacies in its concept. Scenario uncertainties can also include social and
economic uncertainties.

Ignorance is used to describe cases where there is no applicable data.
However the demarcation between this and model uncertainty may be
blurred as the models or analogues might be applied to give a "guesstimate"
or extrapolation. *Indeterminacy* reflects the state of affairs where it is simply
not possible to give any scientifically grounded estimate of probabilities or
states of affairs. The long-term ecosystem impact of the loss of a keystone
species, or social impacts of a technology are examples.

KNOWLEDGE AND UNCERTAINTY

For many types of uncertainties, increased research does not necessarily lead
to a reduction in uncertainty. While more data may reduce the standard error
of the mean, or site-specific measurements may help increase the reliability
of assessments, it is often the case that new observations result in an increase
in uncertainty. When there is a small dataset for a particular parameter, new
data may lie beyond existing limits case or context differences may make data
unreliable (e.g. when existing data for caesium ecosystem transfer was found
to be inapplicable to behaviour in upland ecosystems) or models can be
shown to be flawed (e.g. non-linear dose–response relationships at low dose).
The multiple stressor case is a good example of increased research increasing
uncertainty. Studies show that, when combined with other stressors, the
biological effects of radiation may be less than, greater than or equal to
those predicted by additive models. These types of uncertainty are not easily
described using standard statistical models, and reduce the precision with
which evaluate the impacts of pollutants. Nevertheless, it can be argued that
a better understanding of the interactions between stressors can elucidate
the mechanisms behind the resulting biological effects, and thus reduce
knowledge-related uncertainty within ecotoxicology. This can be exemplified
by considering the different types of uncertainty in ERA.

Uncertainties in Ecological Risk Assessment

NUMERICAL UNCERTAINTIES IN CALCULATION OF CONCENTRATIONS, EXPOSURE AND DOSE

Many of the parameters used in ERA have large statistical uncertainties associated with them. This includes the parameters used in transfer and assessment models, such as concentration ratios (CR), sediment/soil:water partitioning coefficients (Kd) and dose conversion coefficients (DCC) and input data such as estimates of expected concentrations in ecosystem compartments. Statistically, for radionuclides where there is a reasonable level of data such as radiocaesium and radiostrontium, the range in CR and Kd is known to be large, with ranges of two to three orders of magnitude not uncommon. However, for most radionuclides and species, there is little or no data available and the assessment must rely on analogue data or worst case estimates. Hence there are both variability- and knowledge-based uncertainties. However, it can be argued that where data is available, the reasons for variations are rather well understood, with factors like source speciation, soil type, vegetation type, water chemistry and trophic level all known to influence parameters. And in many cases site-specific analysis can reduce the error on parameters. Thus, although numerically rather large, these type of uncertainties can be described well by statistical methods, can be reduced by further research and are not deemed a source of scientific controversy.

Dose conversion coefficients are generally ascribed relatively low uncertainty, particularly when associated with external irradiation and internal gamma emitters. In theory, as long as the concentration of radionuclides is known, the calculation is simply a question of physics and the energy deposition (i.e. dose in Grays) can be estimated with relatively high precision. However, radiation biologists also acknowledge that there is a larger uncertainty with dose calculations from internal beta and alpha emitters as compared with gamma emitters, due to inhomogeneous distribution and, for humans, with the conversion from Grays to Sieverts. Whereas the Gray is a measure of deposited energy (J/kg), the Sievert relates that energy deposition to expected detriment and requires a number of weighting factors that introduce uncertainty, unreliability and controversy into the calculation. This is illustrated by the estimates of the uncertainties in dose conversion coefficients for humans by the CERRIE committee (Fig. 2). CERRIE pointed out that the ranges were much greater for alpha and beta emitters as compared with gamma emitters, but they also noted that the uncertainty associated with alpha and beta emitters included *model uncertainty* associated with internal distribution and the inhomogeneous deposition of energy (CERRIE, 2004). For humans,

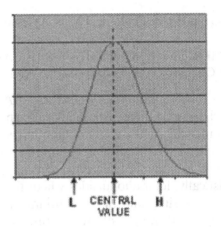

UNCERTAINTY in DOSE
COEFFICIENTS

* Central value correct within
  multiplying factor of 2-3

* In general, factor of ~10 between
  L and H (95% confidence interval)

* Specific factor estimates:
  * 3-6 for ¹³⁷Cs - all tissues
  * 7-8 for thyroid - ¹³¹I
  * 20-40 bone marrow - ⁹⁰Sr
  * >1000 bone marrow - ²³⁹Pu

L　CENTRAL　H
VALUE

Fig. 2. Uncertainties in dose estimates for human risk assessment. (From CERRIE, 2004.)

the dose estimation in Sieverts is essentially an estimate of cancer risk. It should not be confused with a physical, and scientifically robust calculation of dose, but recognised as being a rather woolly estimation of the detriment associated with an expected biological effect, and dependent on a number of assumptions. Of these, the radiation weighting factor (Wr) is arguably one of the most contentious parameters in radiation protection. However, it is better seen as a source of uncertainty in effect estimates rather than dose calculation.

UNCERTAINTIES IN ESTIMATION OF EFFECTS

The estimation of effects associated with exposure to ionising radiation has both statistical and model uncertainties. Most of the data on non-human species has been gathered with the intention of shedding light on the link between ionising radiation and cancer, and there are large data gaps for many species and effects (FASSET, 2003; Real et al., 2004). There are very few ecologically relevant studies – less than 1% of the available data on non-humans looked at ecologically important end points such as reproduction (ERICA, 2006a) – and few studies looked at chronic effects (ERICA, 2006b). However, where there is available data it is possible to construct statistical analysis of the data, such as the species sensitivity distribution (SSD) of responses to ionising radiation (Garnier Laplace et al., 2006) and the distribution of relative biological effects for alpha particles (Fig. 3).

Unfortunately, and in contrast to the uncertainties for CR and Kd, the reasons for the observed variability in the data is not well understood, many species and end points are missing from the dataset and there is considerable controversy on the models and assumptions involved. For example, there is no

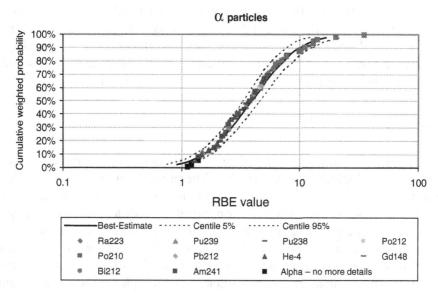

Fig. 3. The distribution of relative biological effectiveness (RBE) for alpha particles in non-human species. (From ERICA, 2006b).

data on marine mammals even though mammals are thought to be the most sensitive species (ERICA, 2006b). Although the range of values associated with the weighting factors for alpha and beta exposures are at least an order of magnitude lower than those reported for Kd and CR, one might argue that model uncertainties, the lack of understanding on mechanisms, the variable dose rate–response relationships, and the unreliability of extrapolating relative biological effectiveness for different end points, make weighting factors a far more contentious issue. However, it has also been proposed that the focus on weighting factors is out of proportion to the actual (i.e. numerical) uncertainty that they contribute to the overall risk assessments (Avila et al., 2004). Interestingly, for the human risk case, where one has at least only one biological end point to contend with, ICRP has stated that weighting factors should only be used for prospective risk assessments and that no uncertainty should be attached to the parameter (ICRP, 2004). This might be said to simply reflect the non-scientific nature of the parameters, putting them on a par with the somewhat arbitary "safety factors" used in other areas of environmental risk assessment.

Furthermore, the effects data from SSD is further used in risk characterisation to construct no observed effect levels that are in turn based on an assumption that an acceptable ecological risk is one where a chronic exposure will bring about a less than 10% change in more than 95% of the species (ERICA, 2006b). In other words, a 10% change in 5% of species is

unacceptable. Although this benchmark is in line with EU recommendations
(EU, 2003), there is still considerable controversy with the assumption, not
least because the 5% of species may include threatened or endangered species. Hence there are additional social and ethical uncertainties associated
with such use of the knowledge in decision making and policy.

Conclusion

The complexities attached to multiple stressors create uncertainties at every
level of an Ecological Risk Assessment. The presence of other stressors may
impact on the mobility and uptake of radionucides, thus creating variability
in transfer factors and biological concentration coefficients. For example,
acid rain can change the pH of soils, organic pollutants may change redox
potentials and heavy metals may interfere with biological membrane transport. These types of uncertainty can be quantified and potentially reduced
by further study. The more complex impacts of multiple stressors arise within
the effects assessment. The emerging knowledge about the biological mechanisms and interactions of stressors serves to challenge many of the models
and concepts used in ERA. Most striking is perhaps the observation that
combinations of stressors may be additive, synergistic or antagonistic: outcomes may be positive or negative. This undermines the basic assumption in
Environmental Impact Assessments that risks can be added, weakens the concept of dose linearity for radiation, and challenges the simplistic, reductionist
models of biological response to stressors at low doses. This is a classic case
of increased knowledge not reducing uncertainty. On the other hand, a better
understanding of the interactions of different stressors may help to increase
our understanding of biological mechanisms and responses to pollutants. But
it is important to note that these types of model, conceptual and epistemological uncertainty do not lend themselves to simple statistical quantification,
and are likely to be overlooked in numerical evaluations of uncertainties and
error. Indeed, for many organisms, ecosystems and endpoints, the uncertainties raised by multiple stressors are best described as unreliability or ignorance,
and cannot be dealt with by probabilistic risk assessment. "We know that we
don't know" needs to be appreciated in the risk evaluation.

References

Avila, R., Beresford, N.A., Aguero, A., Broed, R., Brown, J., Iosjpe, M., Robels, B., Suanez,
    A. (2004). Study of the uncertainty in estimation of the exposure of non-human biota to
    ionising radiation. *J. Radiol. Prot.* 24: A105–122.

CERRIE (2004). The UK Committee for Examining Radiation Risks from Internal Emitters. Report to UK Department of Environment. Available from www.cerrie.org.

Environment Canada (1997). Environmental assessments of the priority substances under the Canadian environmental protection act. Guidance manual, version 1.0. EPS 2/CC/3E., Chemicals Evaluation Division, Commercial Chemicals Evaluation Branch, Environment Canada.

ERICA (2005). Overview of Ecological Risk Characterization Methodology. Deliverable 4b. European Commission, 6th Framework, Contract N°FI6R-CT-2003-508847. Björk, M. & Gilek, M. (Eds).

ERICA (2006a). Scientific Uncertainties. ERICA Deliverable D7e. European Commission, 6th Framework, Contract N°FI6R-CT-2003-508847. Oughton, D. and Brevik, H. (Eds).

ERICA (2006b). Derivation of Predicted No Effect Dose Rate values for ecosystems (and their sub-organisational levels) exposed to radioactive substances. Deliverable D5. European Commission, 6th Framework, Contract N°FI6R-CT-2003-508847. Garnier-Laplace, J. & Gilbin, R. (Eds).

European Commission (2003). Technical guidance document in support of Commission Directive 93/67/EEC f risk assessment for new notified substances and Commission Regulation (EC) No 1488/94 on risk assessment for existing substances, Directive 98/8/EC of the European Parliament and of the Council concerning the placing of biocidal products on the market. Part II. Luxembourg: Office for Official Publication of the European Communities. Report nr EUR 20418 EN/2.

FASSET (2003). Radiation effects on plants and animals Deliverable 4. FASSET Project Contract FIGE-CT-2000-00102, Woodhead, D. & Zinger, I. (Eds).

Garnier-Laplace, J., Della-Vedova, C., Gilbin, R., Copplestone, D., Ciffroy, P. (2006). First Derivation of Predicted-No-Effect Values for Freshwater and Terrestrial Ecosystems Exposed to Radioactive Substances. *Environ. Sci. Technol.* 40: 6498–6505.

ICRP (2005). Basis for Dosimetric Qualities Used in Radiometric Protection (Committee 2 Web document – draft only).

Réal, A., Sundell-Bergman, S., Knowles, J.F., Woodhead, D.S., Zinger, I. (2004). Effects of ionising radiation exposure on plants, fish and mammals: Relevant data for environmental radiation protection. *J Radiol. Prot.* 24: 123–137.

Suter, G.W. (1993). *Ecological Risk Assessment.* Lewis Publishers, Chelsea.

Walker, W.E., Harremoes, P., Rotmans, J., Van der Sluijs, J.P., Van Asselt, M.B.A., Janssen, P., Krayer von Krauss, M.P. (2002). Defining Uncertainty. A conceptual basis for uncertainty management in model-based decision support. *Integrated Assessment*, 4: 5–17.

Wynne, B. (1992). Uncertainty and Environmental Learning. *Global Environmental Change*, 2: 111–127.

CHAPTER 36

NUCLEAR POLLUTION EXPOSURE AND RISK ASSESSMENT – THE CASE OF NUCLEAR REACTORS ACCIDENTS INVOLVING RADIOACTIVE EMISSION

BOGDAN CONSTANTINESCU* AND ROXANA BUGOI

"Horia Hulubei" National Institute of Nuclear Physics and Engineering, PO BOX MG-6, Bucharest 077125, Romania e-mail: bconst@ifin.nipne.ro

Abstract: Preventive policies for areas potentially exposed to radioactive contamination require management decisions which weigh the benefits of prevention against the risks and disruptions associated with their implementation. A framework is needed that integrates risk assessment and engineering options, compares options for risk reduction, communicates uncertainty, and effectively allows reiteration of the decision-making process. In the first part of the chapter, a brief description of 1986 Chernobyl accident and its Romania's related experience is done. A by-product of the environmental contamination was the contamination of foodstuffs produced in the affected areas. After Chernobyl accident, scientists who are not experts in radiation effects have attributed various biological and health effects to radiation exposure. These changes cannot be attributed to radiation exposure, especially when the normal incidence is unknown, and is much more likely to be due to psychological factors and stress. Attributing these effects to radiation not only increases the psychological pressure in the population and provokes additional stress-related health problems; it also undermines confidence in the competence of the radiation specialists. These observations are similar not only for the Former Soviet Union (FSU) regions, but also for Romania. In the second part, the case of a Bulgaria–Romania reactor is presented. It is widely recognized that environmental stress, especially environmental degradation could contribute, under certain political, economical, and social conditions, to the appearance of serious conflicts mainly in the developing countries. Romania and Bulgaria are a potential example in this regard, in relation to their existing Nuclear Power Plants (NPP). Response strategies to this potential threatening environmental change must be discussed. Four basic approaches that can be taken to enhance human security in both countries are considered. The first is fundamentally a preventive strategy oriented toward minimizing, if not entirely averting the potential nuclear environmental changes threaten human security. The other three approaches

C. Mothersill et al. (eds.), Multiple Stressors: A Challenge for the Future, 467–475.

presume that a (serious) incident will materialize, and thus are designed to reduce the vulnerability of human communities to it by avoiding the impacts of the changes, by creating defenses against the impacts or by simply adapting to the changes. All these options must be considered in relation to 1986 Chernobyl accident experience.

Keywords: nuclear pollution; Chernobyl; social adaptation; response strategies

Introduction

Preventive policies for areas potentially exposed to radioactive contamination by chemicals or physically disturbed by industrial development or military operations require management decisions which weigh the benefits of prevention against the risks and disruptions associated with their implementation. A framework is needed that integrates risk assessment and engineering options, compares options for risk reduction, communicates uncertainty, and effectively allows reiteration of the decision-making process.

In the first part of the chapter, a brief description of 1986 Chernobyl accident and its Romania's related experience is presented, while in the second part, the case of a Bulgaria–Romania reactor is described.

On 26 April 1986, the Chernobyl nuclear power station, in Ukraine, suffered a major accident which was followed by a prolonged release to the atmosphere of large quantities of radioactive substances. This caused acute radiation injuries and deaths among plant workers and firemen. It also led to radiation exposure to thousands of persons involved in rescue and clean-up operations. A by-product of the environmental contamination was the contamination of foodstuffs produced in the affected areas. After Chernobyl accident, scientists who are not well versatile in radiation effects have attributed various biological and health effects to radiation exposure. These changes cannot be attributed to radiation exposure, especially when the normal incidence is unknown, and is much more likely to be due to psychological factors and stress. Attributing these effects to radiation not only increases the psychological pressure in the population and provokes additional stress-related health problems, it also undermines confidence in the competence of the radiation specialists. These observations are similar not only for the Former Soviet Union (FSU) regions, but also for Romania.

It is widely recognized that environmental stress, especially environmental degradation could contribute, under certain political, economical, and social conditions, to the appearance of serious conflicts mainly in the developing countries. Romania and Bulgaria are a potential example in this regard, in relation to their existing Nuclear Power Plants (NPP): two nonenveloped WWER – 440/230, 400 MW and two enveloped WWER – 1,000, 1,000 MW pressurized water reactor (PWR) Soviet Units in Kozlodouy near Danube (100 km from Bucharest) in Bulgaria and one 660 MW enveloped PHWR Canadian CANDU Unit in Romania. If the Romanian existing plant is very new (1996) and its CANDU type is unanimously recognized as having a high level of security, the two nonenveloped Soviet–Bulgarian reactors are old and small (up to now) incidents are often reported for them. These facts cause anxiety in both countries, seriously affected by Chernobyl accident. Response strategies to this potential threatening environmental change must be discussed. Four basic approaches that can be taken to enhance human security in both countries are considered. The first is fundamentally a preventive strategy oriented toward minimizing, if not entirely averting the potential nuclear environmental changes threaten human security. The other three approaches presume that a (serious) incident will materialize, and thus are designed to reduce the vulnerability of human communities to it by avoiding the impacts of the changes, by creating defenses against the impacts, or by simply adapting to the changes. All these options must be considered in relation to 1986 Chernobyl accident experience.

Social Responses in Romania to the Chernobyl Accident

On 26 April 1986, the Chernobyl nuclear power station from Ukraine suffered a major accident that was followed by a prolonged release to the atmosphere of large quantities of radioactive substances. The specific features of the release favored a widespread distribution of radioactivity throughout the Northern hemisphere, mainly across Europe. A contributing factor was the variation of meteorological conditions and wind regimes during the period of release. Activity transported by the multiple plumes from Chernobyl was measured not only in Northern and Southern Europe, but also in Canada, Japan, and the USA. Only the southern hemisphere remained free of contamination. This caused acute radiation injuries and deaths among plant workers and firemen. It also led to radiation exposure to thousands of persons involved in rescue and clean-up operations. There was severe radioactive contamination in the area (10^{19} Bq total released radioactivity – approximately 300 times as for Hiroshima, from which 10^{18} Bq ^{131}I and 10^{17} Bq ^{134}Cs + ^{137}Cs), resulting in the evacuation of people from a 30-km zone around the power plant. It became clear over the months following the accident that radioactive contamination varying severity had also occurred in extensive areas of the Eastern Europe, including Romania.

From a biological point of view, the most significant radioactive substances in the emissions from the accident were iodine, strontium, and plutonium. Different problems arise with each radioactive substance. Radioactive iodine is short-lived and practically had disappeared some weeks after the accident. Its significance is due to the fact that, if inhaled or ingested, it accumulates in the thyroid gland, where it may deliver large radiation doses as it decays. The doses may result in impaired thyroid function and, many years after the exposure, in thyroid cancer. Cesium is the element that clearly dominates the long-term radiological situation after the Chernobyl accident. Due to its penetrating radiation, cesium deposited on the ground may give an external dose. It may also enter the food chain and give an internal dose. It is eliminated metabolically in a matter of months. Cesium is relatively easy to measure. Plutonium and strontium, on the other hand, present difficulties in measurement, but there is relatively little strontium in the fallout and it does not give a dose unless ingested or inhaled. Very little plutonium traveled far from the reactor site, and because of its chemical stability, it does not find its way easily into food chains.

In the first weeks following the accident, lethal doses were reached in local biota, notably for coniferous trees and voles (small mice) in the area within a few kilometers of the reactor. By autumn 1986, dose rates had fallen by a factor of 100, and by 1989, these local ecosystems had begun to recover. No sustained severe impacts on animal populations or ecosystems have been observed. Possible long-term genetic impacts and their significance remain to be studied (Gonzales, 1996).

A by-product of the environmental contamination was the contamination of foodstuffs produced in the affected areas. Although for some time after the accident key foodstuffs showed activity levels exceeding the maximum levels permitted by the Codex Alimentarius (which is established by FAO and WHO, sets the maximum permitted level of radioactivity for foodstuffs moving in international trading, e.g., 370 Bq/kg of radiocesium for milk products and 600 Bq/kg for any other food). Exceptionally, wild food products – such as mushrooms, berries and game – from forests in the more affected areas as well as fish from some European lakes also exceeded Codex levels.

In Romania, simple recommendations to minimize the contamination via foodstuffs included: greens' strong washing in the first 2 weeks after the accident, no milk and cheese in the first 6 weeks, limited eggs and meat consumption in the first 3 months, and destruction of fodder in the first 2 weeks. The use of underground water was recommended. Fortunately, the consumption of contaminated products in May–June 1986 was strongly limited by monitoring and warnings, so the contribution to the internal dose was quite low. It could be appreciated that the population's supplementary

Chernobyl internal irradiation in 1986 was compatible with usually applied medical irradiation, for example, less than 2 mSv (Constantinescu et al., 1988, 1993).

In the case of Chernobyl, as in many other radiological incidents, psychological effects have predominated. Information about severity and significance of this contamination was often sparse and uneven; public opinion was uncertain and even many doctors were not sure how to interpret information that did become available. As a result, there was a loss of confidence in the information and in the countermeasures recommended.

In general, the most widespread countermeasures were those which were not expected to impose, in short time for which they were in effect, a significant burden on lifestyles or the economy. These included advice to wash fresh vegetables and fruit before consumption, advice not to use rainwater for drinking or cooking, and programs of monitoring citizens returning from potentially contaminated areas. In reality, experience has shown that even these types of measures had, in some cases, a negative impact which was not insignificant (The International Chernobyl Project, 1991).

Protective actions having a more significant impact on dietary habits and imposing a relatively important economic and regulatory burden included restrictions or prohibitions on the marketing and consumption of milk, dairy products, fresh leafy vegetables and some types of meat, as well as the control of the outdoor grazing of dairy cattle. There was a minor disruption to normal life and economic activity in the affected areas. In particular, agricultural and forestry production was partially disturbed and some production losses were incurred.

After Chernobyl accident, scientists who are not well versatile in radiation effects have attributed various biological and health effects to radiation exposure. These changes cannot be attributed to radiation exposure, especially when the normal incidence is unknown, and is much more likely to be due to psychological factors and stress. Attributing these effects to radiation not only increases the psychological pressure in the population and provokes additional stress-related health problems, but also undermines confidence in the competence of the radiation specialists. These observations are similar not only for the FSU regions, but also for Romania.

The nature of these effects is complicated and it is wrong to dismiss them as irrational or to label them as "radiophobia." Many factors contribute to the development of this widespread association with nuclear bombs, or a lack of openness in the past on the part of governments, or the absence of intelligible explanations by scientists. Such effects are real and understandable, particularly in a mainly rural population whose work and recreation are closely interwoven with the land where restrictions may have had to be imposed by the authorities. Physicians and other people who might be looking for guidance have often been confused. The result is that rumors multiply, fears increase, and any health problem is quickly attributed to a nuclear cause. Uncorroborated narratives may become commonly held wisdom and unverifiable statistical data may be accepted with insufficient scrutiny.

For our country, as psychological effects, we can mention a small rise (10–15%) in spontaneous abortions in some regions and a slight decline in pregnancy rates following the disaster. Similar effects were reported for Sweden, Norway, and some Russian regions. There are no data about induced abortions, strictly prohibited in 1986 in Romania.

We must underline the necessity of a prompt, correct, and sincere information by the governmental authorities, essential to establish a solid, confidential relation with the population and to minimize psychological effects. As concerning the long-term effects (cancer and genetic anomalies), a serious international scientific effort to study, to cure and to avoid these phenomena is strongly recommended.

Considerations on Risk Assessment in the Case of Nuclear Reactors Accidents Involving Radioactive Emission

It is widely recognized that environmental stress, especially environmental degradation could contribute, under certain political, economical, and social conditions, to the appearance of serious conflicts mainly in the developing countries, for example, in Central and Eastern Europe. Romania and Bulgaria are a potential example in this regard, in relation to their existing NPP: two nonenveloped WWER – 440/230, 400 MW (very old model developed during 1960s) and two enveloped WWER – 1,000, 1,000 MW (model developed in the early 1980s) PWR Soviet Units in Kozlodouy near Danube (100 km from Bucharest) in Bulgaria and one 660 MW enveloped pressurized heavy water reactor (PHWR) Canadian CANDU Unit in Romania (The International Chernobyl Project, 1991). If the Romanian existing plant is very new (1996) and its CANDU type is unanimously recognized as having a high level of security, the four nonenveloped Soviet–Bulgarian reactors are old and small (up to now) incidents are often reported for them. These facts cause anxiety in both countries, strongly affected by the Chernobyl accident. Living in the shadow of an unsafe nuclear reactor is not much fun. But living in the dark and cold may be worse. The newest pair of reactors (5 and 6) at Kozlodouy are safer – both are of the VVER 1,000 MW design. However, they often have to be taken out of use for running repairs, so they tend to operate at only 50% of capacity.

Response strategies to this potential threatening environmental change must be discussed. Four basic approaches that can be taken to enhance human security in both countries are considered (Soroos, 1997). The first is fundamentally a preventive strategy oriented toward minimizing, if not entirely averting the potential nuclear environmental changes threaten human security. The other three approaches presume that a (serious) incident will materialize, and thus are designed to reduce the vulnerability of human communities to it by avoiding the impacts of the changes, by creating defenses against the impacts, or by simply adapting to the changes. All these options must be considered in relation to 1986 Chernobyl accident experience (large area regional radiological impact, with serious consequences for health, agriculture, and environment).

The first strategy – Preventing or Limiting Environmental Changes – involves special high costs technical and economical efforts (improving of security systems, even replacing of older reactors with modern nuclear ones or with classical power plants, or importing the necessary energy in the frame of a regional (international) arrangement).

The second strategy – Avoidance of Impacts – seeks to avoid being in a position to be impacted by environmental threats should they materialize, for example by not locating a home in an exclusion zone around the reactors – or by not using foodstuffs produced in such a zone.

The third strategy – Defense against Impacts – seeks to reduce vulnerability to environmental threats, not by avoiding them, but by taking measures that protect populations against adverse impacts. One example is the preventive distribution of potassium iodine (KI) tablets for thyroid protection against ^{131}I radioisotope in the most potentially exposed area around the reactors. We could also consider the realization of a permanent surveillance network for the radioactivity in the most exposed areas.

The final strategy – Adaptation to Impacts – is adaptation or reaction to environmental changes once they take place. The experience achieved during the Chernobyl accident is very useful in this case. A simple method to minimize the contamination via foodstuffs (the most prominent effect) was deduced: greens' strong washing in the first 2 weeks after the accident, no milk and cheese in the first 6 weeks, limited eggs and meat consumption in the first 3 months, and destruction of fodder in the first 2 weeks. The use of underground water is recommended. In the case of Chernobyl, as in many other radiological incidents, psychological effects have predominated. Information about severity and significance of this contamination was often sparse and uneven; public opinion was uncertain and even many doctors were not sure how to interpret information that did become available. We must underline the necessity of a prompt, correct, and sincere information by the governmental authorities, essential to establish a solid, confidential relation with the population and to minimize psychological effects.

However, for an efficient application of these strategies, we must find the answer to the following questions:

- What are the relative economic costs of these alternative strategies, especially in connection to our countries integration in EU
- Which proportion of the society resources should be invested, as opposed to other priorities (food security, relative high rate of morbidity and mortality), etc.)
- To which extent the response strategies adapted by one state add to the environmental or economical securities of the other state (approximately 40% of Bulgarian electrical energy is nuclear energy)
- Under what circumstances are states likely to opt for international cooperation as opposed to self-reliance in the pursuit of environmental security (see the potential role of International Atomic Energy Agency – Vienna, European Union, or, why not, NATO)
- What roles can nonstate actors, such as nongovernmental organizations and corporations are expected to play in advancing environmental security.

All these aspects must be related to promoting bilateral (Bulgaria–Romania) and international collaboration among scientists, politicians, and academics, contributing to the integration of our both countries into the International and Development Community.

It is possible that societies will implement a combination of these response strategies, either by design or more likely in an uncoordinated way. In some cases, a combination strategy may be a rational course of action. For our case, it may be necessary, with priority, to organize adaptive responses to the changes that cannot be prevented (accidental nuclear contamination).

What proportion of a society's resources should be invested in pursuing environmental forms security, as opposed to other priorities? Most human aspirations can be placed in one or two broad categories of human values – development or security. What responsibilities do states or societies have to assist others to cope with environmental threats? Developed countries have been reluctant to acknowledge a broader range of human rights, such as a right to development proposed by representatives of developing countries, partly out of concern that they may stimulate expectations for additional international assistance. Similar reservations will inevitably arise and environmental security should also be aggressively promoted as a human right. However, Chernobyl accident has clearly shown that the transboundary radioactive pollution is a real danger for all European countries, so in our particular case, the assistance of European Union is also a self-protection.

What roles can nonstate actors, such as nongovernmental organizations, are expected to play in advancing environmental security? Despite numerous pronouncements of the demise of the territorial state, national governments continue to play a predominant role in advancing the human security of their citizens. It is governments that are empowered to make and enforce policies and regulations that reign in those who would undermine environmental stability in the pursuit of their individual self-interest. Likewise, it is national governments that enter into negotiations with other countries on binding forms of international environmental cooperation. The objective is to avert, as far as practicable, exposure to people in the area from future radiation doses. This limitation must be achieved through various methods, all of which may have some negative effects or costs of a health, socioeconomic, psychological, or political nature.

Many possible actions may be considered. Some are quite simple. If there is a minor radioactive release, people may be advised to stay inside their homes and keep windows and doors shut. At the other extreme, whole areas of land or entire towns may have to be evacuated and large quantities of food excluded from human consumption.

It is oversimplistic to assume that all measures, which would reduce future dose, are beneficial and therefore should always be fully implements. For example, one of the measures to be reconsidered is that of resettling people elsewhere. Moving people to an area of lower radioactive contamination will probably reduce dose. Since future dose is considered to have a proportional effect on future level of risk, relocation should reduce the risk of long-term radiation effects. However, it is known that the stress of extensive changes in lifestyle can have very serious psychosocial and even physical effects on people. A balance must be struck between potential reduction in dose and possible harm that might be avoided on the one hand and the possible detrimental and disruptive effects of the resettlement on the other.

References

Constantinescu, B., Dumitru, C., Puscalau, M., 1993, Some aspects concerning the relationship environmental radioactivity – population in Romania after 1986, Rom. Rep. Phys. 45 (7–8): 487–495.

Constantinescu, B., Galeriu, D., Ivanov, E., Pascovici, G., Plostinaru, D., 1988, Determination of ^{131}I, ^{134}Cs, ^{137}Cs in plants and cheese after Chernobyl accident in Romania, J. Radioanal. Nucl. Chem. Lett. 128 (1): 15–21.

Gonzales, A. J., 1996, Chernobyl-Ten years after, IAEA Bulletin 3.

Soroos, M., 1997, Strategies for Enhancing Human Security in the Face of Global Change, NATO ARW "Environmental Change, Adaptation and Security", Budapest, Hungary, October 9–12.

The International Chernobyl Project Assessment of Radiological Consequences and Evaluation of Protective Measures, 1991 – Technical Reports, IAEA, Vienna.

CHAPTER 37

MULTIPLE STRESSORS AND THE LEGAL CHALLENGE

COLIN SEYMOUR AND CARMEL MOTHERSILL

*Medical Physics and Applied Radiation Sciences Department,
McMaster University, Hamilton, Ontario, Canada.
e-mail: seymouc@mcmaster.ca*

Abstract: The purpose of this chapter is to examine the difficulties associated with legislating in areas where there is incomplete or no knowledge.

Keywords: multiple stressors; legal implications; causation

The function of law is to regulate human commerce and activities. Although the law applies to populations, it operates at the individual level. It remains the responsibility of every citizen to uphold the law. This presupposes that the individual knows the law, and understands the application of the law. The application must be constant and certain, as the law must be applied equally to all individuals. Lawyers strive for legal certainty, but in environmental law this comes into conflict with scientific uncertainty. In this chapter, the conflict between the requirements of legal certainty and environmental legislation will be examined.

There is a Greek myth about a box, which held all the woes of the world. The box was owned by Pandora. Ultimately curiosity overcame prudence, and the box was opened, releasing misery into the world. Once opened, the box could not be shut again. This is a useful analogy for environmental legislation, as once a chemical or organism has been released into the environment it may cause unforeseen effects that cannot be corrected.

The problem for the lawyer or legislator is how should we regulate the opening of the box when it is not known what will come out or what the effects will be. Historically law is based on certainty, and the concept of contingency law based on altering situations has not been utilized in any legal philosophical system. The certainty of the law contributes to the stability of society. For example, murder has been regarded as a wrongful act for thousands of years, throughout many societies. Murder is regarded as a criminal act, and as such is punished by the state. Affairs between individuals

477

C. Mothersill et al. (eds.), Multiple Stressors: A Challenge for the Future, 477–481.
© 2007 *Springer.*

are largely based on contractual agreements, enforceable through the civil courts, without government intervention.

Companies are generally regarded as legal entities, and can initiate or settle legal actions in their own right. Although companies have to obey the law, the range of sanctions against a company is more limited. If the company offence is particularly offensive to the legal system, individual members of the company may be held liable. Both companies and individuals are liable in law for any harm they inflict on others, within certain limitations.

An individual who has been harmed must prove the cause of his harm, that he has suffered loss and damage, and that the harm was reasonably foreseeable. An example is an individual who falls down stairs, because of liquid floor cleaner on the stairs. If the individual is injured, then both the cause and the damage are proven, and the issue is one of reasonable foreseeability. If the liquid floor cleaner was left on the stairs and there was no indication of its presence, then the accident was foreseeable. If however there was a warning cone, a sign that cleaning was in progress, and the cleaner was actually working on the stairs, then the accident was not reasonably foreseeable. In the first case the individual could recover damages, whereas in the second case he could not.

The cause of damage in the above example was obvious, but complications can arise. The doctrine of Res Ipsa Loquitur, or the events speak for themselves, was developed after a gentleman was discovered lying unconscious outside a beer factory, soaked in beer and covered with broken pieces of barrel. Even though the man had no memory of the accident, and there were no witnesses, it was held that it was so obvious what had occurred that it was as if the events spoke for themselves. The common denominator in both examples given is the temporal link between the time of the accident and the observed result. The cause and effect happen in within a short time frame. In some environmental situations, this short time frame also occurs, as for example the Exxon Valdez oil spill, or the Bhopal Union Carbide plant. Another strong link is the visibility of the accident – oil on an arctic beach is noticeable, as are people dropping dead.

With environmental issues the links may not be as clear. The chemical or toxin may not be visible, and its effect may not be apparent until considerable time has elapsed. Generally, the longer the time interval, the harder it is to prove cause and effect.

If there is one pollutant, and ultimately a measurable effect, then it will be possible to demonstrate cause and effect. But what if one pollutant causes multiple effects on different species or the effect on one species causes the irreversible decline of a second species? In such cases the law will attempt to establish the causa causans – the cause of causes. The majority of blame would thus fall on this primary cause, and contributory blame on the secondary causes. This approach assumes a linearity of effect that causes a and b; c and d can be added together to produce the total effect. This approach will fail though if substance a was not thought to have harmful effects at the time it was discharged, or if the discharge of substance a had ceased years ago, or it could not be foreseen that b, c, or d would later be discharged in the same area, or that b, c, or d would produce cumulative effects.

To follow this example further, let us suppose that a factory owner has obtained the requisite licenses to discharge substance d. This he does, and animals on a surrounding farm begin to die. Assays demonstrate the presence of substance d in the dead animals, and so cause and effect would appear to be proven. But no assays are undertaken for substances a, b, and c, and even if there were the possibility of complex synergistic effects over time between the substances has not been considered.

It is becoming a legal principle that the polluter pays, but in this example only the last polluter would pay. Suggesting that the previous polluters had been dumping illegally could further muddy the example, but although this complicates it does not obfuscate the main point that the person perceived as causing the pollution is not strictly legally liable.

To summarize this line of argument, although the law can be adapted to apportion blame between multiple contaminants, it can only work with certainty if the contaminants have additive effects on the same species (the farmer's cow), and the contamination and effects occur over a reasonable time period. If in the environment the contaminants cause effects on different species over different time frames, altering the flora and fauna of an ecosystem in a subtle way, is this harmful? If these changes sensitize a population, so as to become exquisitely sensitive to emissions legally allowable, should those responsible for the emissions be liable? In a multistressor situation, should the primary cause be the first stressor or the largest stressor? Should this be measured by total amount of contaminant or by biological effect?

In the previous example it is not immediately apparent whether the true cause is the first pollutant or the last, and both positions are arguable. The arguments are further complicated if the assumptions of additivity and linearity previously made are challenged. It is appropriate here to introduce as an example the model developed by Engels, who together with Marx introduced change into the world. Linearity of thought is evident in Western philosophy, where the general assumption is that the desired end point will be reached through logically sequential steps, which build on each other. In Eastern philosophy, often symbolized by the Taoist ying/yang circle with interlocked black and white halves, each containing a small portion of the other colour, relationships are more dynamic and less linear. For example, in ancient Greek the word for high is the same as that for deep – the difference is a matter of individual perspective.

In the model proposed by Engels in his book *The Dialectics of Nature*, there were three fundamental rules. The first, known as the interpenetration of opposites, proposed that every object and process was composed of opposing forces. The second, known as the law of transformation of quantity into quality, proposed that small changes of quantity (gradual change) leads to a turning point where one opposite overcomes the other, causing a qualitative (sudden) change. The third, the law of negation postulates that change moves in spirals.

This model can be used to demonstrate that linearity of effect may not always occur even with one toxin. As an example, the LD50 is a commonly used toxicological assessment. At this particular dose, it is estimated that 50% of the test animals will die. Using Engel's model, each additional dose of toxin would cause a quantitative change until there was a qualitative change in the state of the animal. If the animal recovered though, it would respond to the toxin differently, an effect not predicted by a linear model but which has been demonstrated practically through adaptive experiments.

In environmental terms, the model predicts that once the dynamic of an ecosystem has been altered it can never be restored to its former state. It also predicts that small, seemingly innocuous changes, can cause a sudden change in state. This could also be regarded as being analogous to a chaotic system, where gradual changes cause a bifurcation point to be reached.

Without getting into too much detail on the model, its significance is that simple additive effects may not occur even using the same chemical. Using different chemicals, the effects may be synergistic or agonistic. Effectively this means effects cannot be predicted – the outcome is scientifically uncertain.

If the science is uncertain, how can law be framed to protect the environment whilst allowing legitimate commercial activities?

There are some current options. The first is the use of zoning, where areas are designated agricultural, industrial, housing, areas of outstanding natural beauty, or other but similar designations. Polluting industry is then kept in certain areas. Although this is an option used in many countries, there is no way of stopping waterborne, or airborne, pollution spreading into other areas, or even into other countries.

The second is to establish an environmental Doomesday book, so that current type, amount of species, and distributions are noted, and hence future change can be monitored.

The precautionary principle can also be used. Although not a scientific concept, it can be used within a legal framework, and to a certain extent already is. An environmental impact assessment must often be carried out before permissions for discharge are granted. The difficulties remain as to whether any effects are truly predictable using the existing methodologies, or whether the actual effects will be unpredictable and temporarily removed.

Another possible confusing effect is that of hormesis. Hormesis may be defined as a situation where low dose effects are opposite to high dose effects, and may be summarized by paraphrasing Paracelsus, that all substances are poisons, the effect is in the dose. This has obvious implications in the dispersal of substances, as at dilute concentrations they may be beneficial but at higher concentrations they may be toxic.

Different chemicals may not follow the same biochemical pathways, and may result in different end points. It would be useful to have a common end point, which would have to be fairly universal. An example of this would be stress, a word that is commonly used but because of the multiplicity of components is rarely completely defined.

Population stress as an end point would be difficult to define, but could perhaps loosely be described as a failure to thrive. Once that point has been reached in the test population, the pollution burden would have to be reduced to allow recovery (assuming no adaptive responses).

This could be achieved through a percentage reduction of all legal pollutants, or on the last polluter to the system, first removed basis. Population recovery could then occur, assuming a linear model of growth. Legally the position would be interesting as only the last polluter would benefit from a percentage reduction.

Complex mixtures of pollutants could also form new and novel compounds with unexpected properties. It might be difficult to identify a new compound on routine assays to monitor established and authorized pollutants. It would also be difficult to establish degrees of culpability.

To conclude, it would be prudent to have the onus of proof fall upon the party wishing to either add to or subtract from the existing environment (add chemicals or extract minerals) demonstrating the environmental impact in the existing environmental conditions. In order to demonstrate any change, it is also necessary to know the existing health of the local environment. This is also a problematic area.

SECTION 7

ROUND TABLE DISCUSSION SUMMARY

Round Table Discussion

There were four topics proposed for discussion

1. What is the evidence for genetic mutations produced by radiation?
2. What biomarkers of "risk" can be used for the environment?
3. Can we propose a hypothesis-driven research program in the science area which could point the way for regulators?
4. How can we approach the problem of monostressor regulation in a multistressor reality?

What is the Evidence for Genetic Mutations Produced by Radiation?

This surprising question was not easily answered. Advocates of the use of minisatellite and microsatellite mutations argued that these genomic changes were evidence of radiation-induced mutations but the consensus was that the epidemiology suggests that at most there is a 1% per Sv rate of hereditary effects due to radiation. This figure is actually an estimate based on models not real data. This controversial position led others to suggest that multiple stressors were really important here. Many pointed to the seminal work of Lord and colleagues showing that in mice, paternal irradiation made off-spring more sensitive to a chemical carcinogen challenge. The importance of epigenetic effects and intrinsic genetic background (e.g., ability to send aberrant cells into apoptosis) was also highlighted.

What Biomarkers of "Risk" can be Used for the Environment?

The issue here is that for human risk assessment, the yield of dicentric chromosomes usually assessed in stimulated lymphocytes following exposure to a toxic or radiation insult, is accepted as a "gold standard." This end point is strongly correlated with carcinogenic potential of the insult. The problem is that for environmental exposures to stressors, cancer-inducing potential is not really an issue. Relevant end points need to be linked to reproductive success (fertility and fecundity) and habitat stress is a major issue. There was much discussion on how these parameters could be assessed. The use of the micronucleus assay was discussed at length. An EU project sought to determine baseline

483

C. Mothersill et al. (eds.), Multiple Stressors: A Challenge for the Future, 483–484.

frequencies of micronuclei for a radiation project and found considerable baseline variation in unirradiated individuals. They postulated a multiple stressor cause due to other agents in the environment. The group interpreted this to mean that measurement of micronucleus frequency in populations might be a very good measure of environmental stress from multipollutant exposure.

Further discussion favored developing the validation of the micronucleus assay as an end point for multiple stressor effects in the environment as a research question in the research plan being developed (RECOMMENDATION 1).

Another end point discussed was the bystander signal. This is known to be produced by radiation and heavy metals and is suspected in other stressor scenarios. There was concern that the signal has not been identified and that it leads to a dynamic process which might be "hard to catch" and which is dependent on genetic background. The latter property was viewed by some as a positive attribute since it means genetically susceptible organisms could be selected. The validation of the end point in humans, mice, and fish was also seen as positive. Screening of exposed Chernobyl workers and residents 20 years later showed evidence of bystander factors in blood capable of producing micronuclei in the reporter cells.

Further discussion favored coupling the micronucleus assay with the bystander reporter assay as a suitable end point to explore in a research program (RECOMMENDATION 2).

There was considerable discussion on the usefulness of these universal stress markers, which are very sensitive early warning systems. They pose legal problems as they obviously cannot be linked to specific causal agents, but many saw this as the whole point. They identify stress burden but not the specific cause. Ultimately, it may be that we need universal markers to take remedial action in time but specific markers of exposure (biodosimeters) to prosecute polluters. Dissociating the protection/remediation area from the prosecution area will be challenging but is essential if we are to protect environmental resources adequately. "Who pays?" will be an issue.

The round table recommended that a task group be set up comprising legal and regulatory representatives as well as NGOs and scientists, to tease out the issues (RECOMMENDATION 3).